RECOLLECTIONS FROM
THE EARLY DAYS OF CHAOS THEORY

RECOLLECTIONS FROM **THE EARLY DAYS OF CHAOS THEORY**

Editor

Christophe Letellier
The University of Rouen Normandy, France

World Scientific

NEW JERSEY • LONDON • SINGAPORE • BEIJING • SHANGHAI • HONG KONG • TAIPEI • CHENNAI • TOKYO

Published by

World Scientific Publishing Co. Pte. Ltd.
5 Toh Tuck Link, Singapore 596224
USA office: 27 Warren Street, Suite 401-402, Hackensack, NJ 07601
UK office: 57 Shelton Street, Covent Garden, London WC2H 9HE

British Library Cataloguing-in-Publication Data
A catalogue record for this book is available from the British Library.

RECOLLECTIONS FROM THE EARLY DAYS OF CHAOS THEORY

ISBN 978-981-98-2312-3 (hardcover)
ISBN 978-981-98-2313-0 (ebook for institutions)
ISBN 978-981-98-2314-7 (ebook for individuals)

For any available supplementary material, please visit
https://www.worldscientific.com/worldscibooks/10.1142/14581#t=suppl

Desk Editor: Rhaimie Wahap

Typeset by Stallion Press
Email: enquiries@stallionpress.com

https://doi.org/10.1142/9789819823130_fmatter

PREFACE

At the start of my career, I had the good fortune to begin by teaching the history of science (mainly the Greek astronomy and its big transformation at the Renaissance) and epistemology: this made me aware of the importance of the context in which a theory emerges, and the importance of digging up those who have left an important mark on their field. Very often, reading the seminal works sheds light on the deeper meaning of certain concepts and why they were introduced this way and that. Naturally, I immersed myself in the works of Poincaré very early on... And going to at least consult the articles cited became a reflex. Later, I had the good fortune to cross paths with Otto E. Rössler (in 2006) and it was a real encounter. We soon began to exchange views not only on the foundations of the theory, but also on its history... Then I began to ask Otto about his own career, and that was the start of some very long sessions going through his archives (Otto and his wife Reimara keep absolutely everything). Diving back into his articles, which I had enjoyed during my Ph.D. thesis, was a real surprise: there was a lot more there than I had remembered. As I went along, I revisited his articles, then those of Lorenz, and many others...

Once you have skimmed through the articles of a Rössler or a Lorenz, it quickly becomes tempting to reconstruct the history from them. I will never forget the example of Lorenz's 1963 paper quoting Nemytskii and Stepanov's book entitled *Qualitative theory of differential equations* (1960): digging into this book explains absolutely nothing about Lorenz's path, since in his book *The essence of chaos* (1994), he explained that it was a reviewer who asked him to quote this work, a practice commonly encountered during the process of peer review of papers before their publication. Working on Otto Rössler's archives has shown me how rich it is to work directly with the researcher in question, who admittedly also has his own prism for breaking down the light of his story, but who, at least, provides the major influences that left enough of a mark on him for the memory to still be imprinted decades later.

For these reasons, on the occasion of Otto's 80th birthday, I wanted to seize the opportunity to hear from those who had contributed to the emergence of this chaos theory in the 1970s and 1980s... Many responded, happy to thank Otto for his contribution to the field. Some were not. The list of contributors of this book is therefore the results of my own prism on the history of chaos, the natural selection among the contributors I contacted according to their positive or negative answer... Obviously, some major contributors are missing. And I do not pretend to be exhaustive. This is a given list of contributors. During the (too) long process of editing this book, I had the opportunity to discover many personalities, all different from each other, but always passionate, often reporting similar approaches leading them to this subject of chaos. Their memoirs are delivered in this book, unfiltered. We worked without any particular framework, often starting with a free first draft that grew version after version under the impetus of my questions. If only for this last point, it is a particular version of the history — it would be more correct to say, stories — of chaos that is offered here. I have encouraged contributors to quote as many sources as possible. The aim was not to provide an introduction to the various concepts that come together in this story, but rather to report the little stories behind the development of this theory. I would like to take this opportunity to thank all the people I spoke to in the course of editing this book, for their kindness, their pleasure in recollecting forgotten moments, and the encounters that have shaped their various careers.

As Yves Pomeau told me, the word *history* is borrowed from Herodotus. In his writings, it means stories in the sense of *what is told*. This aspect is indeed present in many contributions. A priori, "stories" does not look too much for an impossible account of what happened, but appeals rather to what everyone remembers. The contributors should be considered as authors in in the tradition of Herodotus! Moreover, as Tom Carroll stated: "*recollections* sounds good. Some of us have been around long enough that our memories can't be trusted."

As you may have gathered, this is not about the truth. Everyone has their own, according to their own memory. The aim of this book is to give the curious a better idea of what this period was like, when the paradigm was being put in place. The idea was to make an original contribution to the corpus of this discipline... I think we have succeeded... I hope the reader will take as much pleasure as we did in shaping it.

Here is a non-exhaustive list (by alphabetical order) of peoples that I would like to thank for their help for producing this book. Ralph Abraham, Hugues Aroux, Tassos Bountis, Tom Carroll, Giulio Casati, Pierre Collet, Michel Dieuzaide, James Crutchfield, Kathryn Gentilello (Georgia Tech Archives), Robert Gilmore, Leon Glass, Celso Grebogi, Pierre Hénon, Hanspeter Herzel, Philip Holmes, Kensuke Ikeda, Felix Izrailev, Jürgen Kurths, René Lozi, Sylvain Mangiarotti, Nikla Manz, Christian Mira, Lars F. Olsen, Norman Packard, Ulrich Parlitz, Louis Pecora, Arkady Pikovsky, Yves Pomeau, Ludmilla Pulkin, Otto E. Rössler, Reimara Rössler, Bill Seaman, Robert Shaw, Dima Shepelyansky, Haris Skokos, Nataliya Stankevich, Ruedi Stoop, Ichiro Tsuda, Yoshisuke Ueda, and Franco Vivaldi.

Rouen, March 9, 2025
Christophe Letellier

Psychedelic view of chaos by Hugues Aroux, Rouen Normandy University, 2020.

https://doi.org/10.1142/9789819823130_fmatter

CONTENTS

Preface		v
Chapter 1.	Stepan Pavlovich Pulkin *Ludmila Pulkina*	1
Chapter 2.	Edward Lorenz's Contribution to Chaos Theory *Christophe Letellier*	11
Chapter 3.	Joseph Ford: The Evangelist of (Hamiltonian) Chaos *Christophe Letellier*	59
Chapter 4.	The Smale Program *Ralph Abraham*	107
Chapter 5.	Boris Chirikov — Sputnik of Chaos *Dima L. Shepelyansky*	123
Chapter 6.	In the Shadow of Chaos *Felix Izrailev*	137
Chapter 7.	My Dance with Chaos *Otto E. Rössler*	159

Chapter 8. Contributions to the Theory of Spontaneously Evolving Systems I: A Simple Model Class 205
Otto E. Rössler

Chapter 9. Chaos People 235
Franco Vivaldi

Chapter 10. Chaos and Me 249
Yves Pomeau

Chapter 11. How I Became Involved in Chaos 267
Philip Holmes

Chapter 12. How Chaos Shaped My Academic Life 273
René Lozi

Chapter 13. My Road to Chaos 333
Leon Glass

Chapter 14. Encountering Chaos through Light, and Beyond 343
Kensuke Ikeda

Chapter 15. First Chaotic Steps 377
Arkady Pikovsky

Chapter 16. Tübingen Blues 385
Lars F. Olsen

Chapter 17. From Chemical Chaos to Chaotic Brain 397
Ichiro Tsuda

Chapter 18. In Resonance with Chaos 405
Ulrich Parlitz

Chapter 19. Reminiscience of Chaos 415
Celso Grebogi

Chapter 20. From Noise to Chaos to Biology 429
Hanspeter Herzel

Chapter 21. A Knotted Road to Chaos 439
Robert Gilmore

Chapter 22. Chaos Research at the Naval Research Laboratory 453
Louis M. Pecora and Tom Carroll

https://doi.org/10.1142/9789819823130_0001

Chapter 1

STEPAN PAVLOVICH PULKIN

Ludmila Pulkina

Samara National Research University
louise@samdiff.ru

This chapter is a very brief account about Stepan Pulkin who published a paper in 1950 which triggered by the end of the 1960s early works in chaos by the Toulouse group. This paper is here translated in English.

1. From Partial Differential Equations to Discrete Maps

My father, Stepan Pavlovich Pulkin was graduated from Irkutsk (Siberia) in 1931 and was invited by Prof. Sergey Borisovich Sverzhenskii for teaching at Samara Pedagogical and Teachers' Institute. Samara became Kuibyshev in 1935 and returned Samara again in 1991.

In the 1930s, due to the development of the theory of nonlinear oscillations as exemplified, for instance, by the work of Kolmogorov,[1] or Aleksandr Vitt,[2,3] interest in iterative methods has increased. At the same time, my father began his scientific activity. No wonder that studing of properties of broken lines became his first object of investigations.[4–6] In 1936 he came in Moskow and continued his work at Moskow State University under the supervision of Ivan Georgievich Petrovskii who, from 1940 to 1944, was the Dean of Mechanics and Mathematics Faculty of Moscow University. Petrovskii suggested Pulkin to study iterated simple functions. In 1938, Pulkin submited and defended a thesis (kandidatskaya dissertatiya in Russian which is equivalent to the Ph.D. thesis) entitled *General study of piecewise iteration for an arbitrary curve with finite number of singularities.* The works in this direction continued successfully until the very end of 40th.

January 11, 1907

May 21, 1980

Fig. 1. Stepan Pavlovich Pulkin.

Ivan Georgievich Petrovskii (1901–1973) played a significant role in the scientific life of my father. Petrovskii was working mainly in the field of partial differential equations. Among others, he contributed to the solution of the 19th problem posed by Hilbert: "*Are the solutions of regular problems in the calculus of variations always necessarily analytic?*".[7] He also contributed to the topology of algebraic curves and surfaces. In January 1950, Pulkin received a letter from Ivan Georgievich (Fig. 2).

This letter was sent just before Frankl — one of the editors of the *Matematicheskii sbornik* — published his paper on the Tricomi problem.[9] This letter exhibits the Tricomi problem for the mixed elliptic-hyperbolic type equation[10]

$$y\frac{\partial^2 u}{\partial x^2} + \frac{\partial^2 u}{\partial y^2} = 0\,. \tag{1}$$

Solving this equation was considered by Francesco Tricomi (1897–1978). Further scientific way of my father has been strongly influenced by this letter. Ivan Georgievich insistently recommended to study mixed-type partial differential equations and boundary problems for such equations. Note that during 1950s, there was a considerable interest in this kind of problems. So my father began to study boundary problems for mixed-type PDE's and got significant results.[11–14] In 1959, he defended a doctoral thesis entitled *Investigations on mixed-type PDE's.*

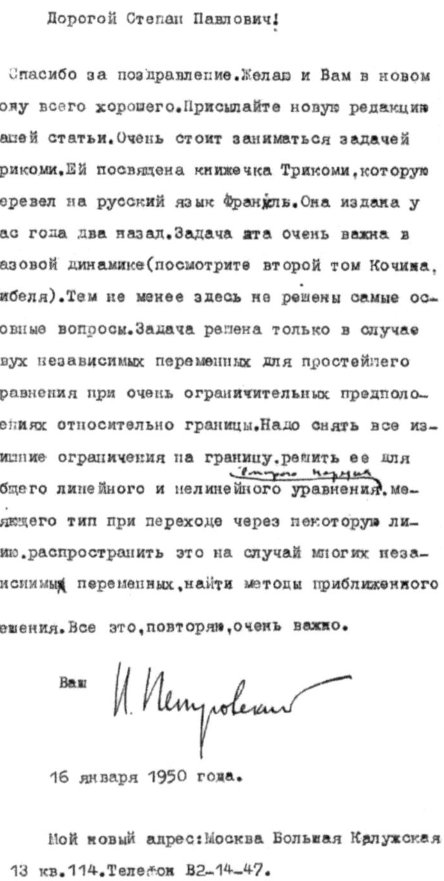

Дорогой Степан Павлович!

Спасибо за поздравление.Желаю и Вам в новом году всего хорошего.Присылайте новую редакцию Вашей статьи.Очень стоит заниматься задачей Трикоми.Ей посвящена книжечка Трикоми,которую перевел на русский язык Франкль.Она издана у нас года два назад.Задача эта очень важна в газовой динамике(посмотрите второй том Кочина, Кибеля).Тем не менее здесь не решены самые основные вопросы.Задача решена только в случае двух независимых переменных для простейшего уравнения при очень ограничительных предположениях относительно границы.Надо снять все излишние ограничения на границу.решить ее для общего линейного и нелинейного второго порядка уравнения.меняющего тип при переходе через некоторую линию.распространить это на случай многих независимых переменных,найти методы приближенного решения.Все это,повторяю,очень важно.

Ваш И. Петровский

16 января 1950 года.

Мой новый адрес:Москва Большая Калужская 13 кв.114.Телефон В2-14-47.

Dear Stepan Pavlovich,

Thank you for congratulations. I wish you all the best in the new year. Send me your new version of the paper. It is worth to deal with Tricomi problem. The book translated by F. [I.] Frankl is devoted to it. It is published in USSR two years ago. This problem is very important for gas dynamics (see vol. 2 of the book by [Nikolai Evgrafovich] Kochin and [Ilia Afanasevich] Kibel[8]). Nevertheless, the main points are not solved. The problem is solved for the simpliest equation in the case of two variables under very strong claims on the boundary. It ought to remove all unnecessary restrictions on the boundary, to solve it for general linear and nonlinear 2-order equations changing a type after passing a certain line, to extend the results to the case of many variables, to develop methods to find approximate solutions.
Again, this is very important.

Yours Petrovskii

January 16, 1950.

Fig. 2. Letter sent by Ivan Georgievich Petrovskii to Stepan Pavlovich Pulkin.

2. On Oscillating Iterated Sequences

Doklady Akademii Nauk SSSR, **73** (6), 1129–1132, 1950.

Translated by Christian Mira, Christophe Letellier and Ludmila Stepanovna Pulkina

Let $f(x)$ be a function of the real variable defined on the whole real axis. This work is devoted to the iteration of $f(x)$. As usual, we write it as $x_n = f(x_{n-1})$, x_0 being the initial point. We will say that x_0 generates the iterated sequence $\{x_n\}$. If $\{x_n\}$ converges towards a point ξ, $\lim x_n = \xi$, then $f(\xi) = \xi$, that is, ξ is a root of the equation $f(x) = x$, or fixed point

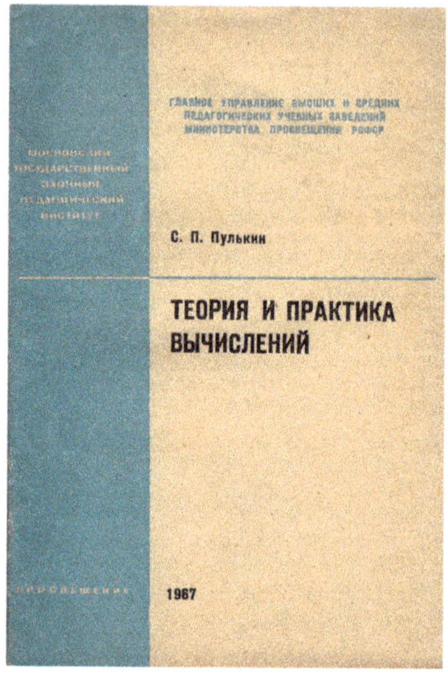

Fig. 3. Cover of the book published by Stepan Pavlovich Pulkin in 1973.

of the map $f(x)$. If $x_{n+p} = x_n$, $x_{n+k} \neq x_n$, for all $k < p$, then the points $x_n, x_{n+1}, \ldots, x_{n+p-1}$, correspond to a period-p cycle. For $p = 1$, the cycle is a fixed point. If the sequence presents a cycle, then every point of the sequence starting with one of them coincides with one point of this cycle.

Numerous works have been devoted to the study of the conditions for the convergence of iterated sequences. Julia and Fatou studied iteration of rational and integer functions in the complex plane in its whole, that is, not only a study of the condition of convergence, but also a study of every type of iterated sequences. When we perform an iterative process, starting from a point in a set E, we obtain a sequence of functions $\{f_n(x)\}$, with $f_n(x) = f(f_{n-1}(x))$ and $f_0(x) = f(x)$, defined in E.

The reference[5] is my attempt for a "global" study of iteration of the function of real variable.

Here, we will investigate functions of real variables satisfying the conditions as follows:

(1) The function $f(x)$ is piecewise continuous, and has first and second piecewise continuous derivatives.
(2) The function $f(x)$ has a finite number of maxima and minima.
(3) The curve $f(x)$ has a finite number of inflection points.

It is well known that fixed points and cycles can be of three types, namely attractive, repulsive and marginal. We will mean by neighborhood of a cycle the set of the neighborhoods of the points of this cycle which they do not intersect each other. An attractive cycle has a neighborhood such as every point of this neighborhood induces an iterated sequence with a monotone convergence toward the cycle. A repulsive cycle has a neighborhood such as every point of this neighborhood induces an iterated sequence which goes out from the boundaries of this neighborhood after a finite number of iterations.

We will designate by "neighborhood" of the point ξ of a cycle, a single side of its neighborhood [which is] a semi-open set B_ξ whose end is $\xi \in B_\xi$. The closure of some neighborhood of point ξ will be designated as the closed neighborhood of ξ.

Every point ξ of the cycle will be simultaneously considered with its arbitrary unilateral neighborhood, whatever its part (left or right), and we will designate as an oriented point (according to the left or right side) of the given point ξ. Thus, equivalently, the same point may define two different oriented points. By oriented neighborhood of point ξ', one mean every unilateral neighborhood of point ξ corresponding to the orientation (at the right or at the left of point ξ'). A point ξ' will be said to be a limit point of the oriented sequence $\{x_n\}$ if every neighborhood of point ξ' contains an infinite set of members of this sequence.

We will designate by a *limit point of class 1* a limit point isolated from the set of members of the iterated sequence. We will designate by a *limit point of class p* one of the limit points of class $p' < p$. We will designate by an *iterated sequence of class p*, an iterated sequence whose number of limit points of class p is finite.

We will say that the iterated sequence is oscillating with respect to a set of repulsive cycles $C_1, \ldots, C_l$, if all points of these cycles, and only those, are limit point of the highest class of the sequence. One may say that the iterated sequence is oscillating if there exists a set of repulsive cycles with respect to which it is oscillating.

In this paper, I establish few sufficient criteria for oscillation, and the existence of oscillating sequences of high class.

Let us consider the set of two second-order cycles, $C_1(\xi_1, \xi_2)$ and $C_2(\eta_1, \eta_2)$ where

$$\xi_2 = f(\xi_1),\ \eta_2 = f(\eta_1),\ \xi_1 = f(\xi_2),\ \text{and}\ \eta_1 = f(\eta_2)\,.$$

We assume that the function $f(x)$ is continuous at the right of the point ξ_2. Let us consider the right part of the neighborhood $\Delta_1^1 = [\xi_1, \xi_1 + h_1]$ du point ξ_1, where h_1 is sufficiently small to ensure that the function $f(x)$ is monotone and continuous in this neighborhood. Then the iterate of this neighborhood $f(\Delta_1^1)$ will be the unilateral neighborhood of ξ_2. Let us designate it as $\Delta_2^1 = [\xi_2, \xi_2 - h_2]$, where $h_2 > 0$. We suppose that the function $f(x)$ is continuous at point ξ_2 in the unilateral neighborhood Δ_2^1. One may choose h_1 such as in the neighborhood Δ_2^1, the function $f(x)$ is also monotone and continuous. $f(\Delta_2^1)$ will be the neighborhood of ξ_1. By definition, we set that $f(\Delta_2^1) = \Delta_3^1$ is the right neighborhood $[\xi_1, \xi_1 + h_1]$ of ξ_1 as Δ_1^1 is. Since C_1 is repulsive cycle, then h_1 can be chosen such as $h_1' > h_1$, and if $x_0 \in \Delta_1^1$, then the second iterate of x_0 will be out of $[\xi_1, x_0]$. The half-interval $\Delta_0^1 = \Delta_3^1 - \Delta_1^1$ will be designated as the the improper neighborhood of Δ_3^1.

For a given value h_1, it thus appears that, in the neighborhood of η_1, $f(x)$ is monotone and continuous in Δ_0^1 and $f(\Delta_0^1)$ is a unilateral neighborhood of η_1 that we will designate as Δ_1^2. For the repulsive cycle (η_1, η_2) and its neighborhood Δ_1^2, we repeat the reasoning developed for the cycle (ξ_1, ξ_2) and its neighborhood Δ_1^1. Let us consider that $f(\Delta_1^2) = \Delta_2^2$ with $f(\Delta_2^2) = \Delta_3^2 \supset \Delta_1^2$ and that $f(x)$ is monotone and continuous in each neighborhood Δ_1^2 and Δ_2^2, and in $\Delta_0^{(2)} = \Delta_3^2 - \Delta_1^2$. Finally, we assume that $f(\Delta_{0*}^2) = \Delta_1^1$.

We obtained a system of neighborhoods Δ_i^k $(i = 1,\ k = 1, 2)$ of points from the cycles, as well as a system of intervals Δ_0^1 and Δ_0^2, which are constituting a set F_Δ having the properties as follows.

(1) whatever is or will be $x_0 \in F_\Delta$, all its iterates belong to F_Δ; due to this, the elements of the iterated sequence cross all the neighborhoods Δ_i^k in a defined order of alternation.
(2) for every point $x_0 \in F_\Delta$, it is possible to find a sequence of antecedents belonging to F_Δ. One may choose this sequence such that it Is periodic and converging towards one of the cycles. The set F_Δ will be said fully invariant with respect to the system of given cycles and to the given set of their neighborhoods.

Theorem 1. *Let us consider a system of two repulsive cycles* (ξ_1, ξ_2) *and* (η_1, η_2)*, and a given system of neighorhood for which there exists a fully invariant set* F_Δ*. The set* F_Δ *contains a continuum of points which generates an iterated sequence which stays away from the considered system* *of cycles. Each neighborhood* Δ *contributes to the sum*

$$\Delta_i^k = E_\infty + E_0 + \sum_2^\infty E_k$$

where E_0 *is the set of point inducing the period-*∞ *cycles, that is, the increasing order antecedents of the points of these cycles.* E_k *is the set of points which induces the iterated sequences of class* k*.* E_∞ *is the set of points which induces a sequence of class* ∞*.*

The set E_0 *is countable.* E_∞ *and* E_k *for all* $k > 1$ *have the power of the continuum. Every interval belonging to* Δ_i^k *contains each of the sets* E_∞*,* E_0*, and* E_k $(k > 1)$*.*

Sketch for a demonstration of this theorem. Let us consider, for instance, the neighborhood Δ_1^1. Let us map this interval in itself by the mean of the inverse map $f^{-1}(x)$ such as the iterated sequence remains within the boundaries of the set F_Δ. All the possible images (consequents) from such a family form the "canonical" intervals $\delta^1_{k_1,k_2,\ldots,k_n}$, and with this

$$\Delta_1^1 = \sum_{k_1=1}^\infty \delta_{k_1},$$

$$\delta^1_{k_1,k_2,\ldots,k_{n-1}} = \sum_{k_n=1}^\infty \delta^1_{k_1,k_2,\ldots,k_{n-1},k_n}.$$

Each point x from the segment Δ_1^1 with the exeption of endpoints from the canonical intervals, one may characterize as an infinite sequence of natural numbers $[k_1, k_2, \ldots, k_n]$, this means that $x \in \delta^1_{k_1,k_2,\ldots,k_{n-1}}$, $(n = 1, 2, \ldots)$; we may call it countably-dimensional fraction. Let be $x \in \Delta_1^1$, $x = [k_1, k_2, \ldots, k_n]$; it appears that the elements of the iterated sequence inducing this point which is contained in Δ_1^1 are determined by uncountable fractions: $[k_{2\nu-1} - i, k_{2\nu}, k_{2\nu+1}, \ldots]$, $(i = 1, 2, \ldots, k_{2\nu-1} - 1$, and $\nu = 1, 2, \ldots)$.

The investigated struture of the set of elements from the sequence is in all the neighborhoods of the points of cycles generating the point $x \in \Delta_1^1$; with this representation by the mean of numerical fractions, we were convinced ourselves by the validity of the theorem.

Let $\{C_k\}$ $(k = 1, 2, \ldots, m)$ be a system of repulsive cycles. The set of neighborhoods of all points of C_k will be called a regular neighborhood Δ_k

of the cycle C_k if the following conditions hold:

(1) The neighborhoods Δ_i^k of points of a given cycle do not intersect each other.
(2) Δ_k contains no point of other cycles from the given system and has no points shared with the neighborhoods of other cycles from this system. de ce système.
(3) The function $f(x)$ is monotone and continuous in each of the neighborhoods $\Delta_{\xi_i}^k$.
(4) Among the neighborhoods $\Delta_{\xi_i}^k$, there exists one — and that we will designate as Δ_1^k (that is, it coincides with $\Delta_{\xi_i}^k$), such that $f(\Delta_{\xi_i}^k) = \Delta_{\xi_{i+1}}^k$, $(i = 1, 2, \ldots, a_k - 1)$; we will call it the main neighborhood of the point of C_k.
(5) For every point $x_0 \in \Delta_0^k$, n_k-iteration of it is located outside of the interval $[\xi_i, x_0]$.

Let us consider the system of repulsive cycles Considérons le système des cycles répulsifs C $= \{C_k\}$ and every system of right neighborhoods $\{\Delta_k\}$ $(k = 1, 2, \ldots, m)$. The set F_Δ of the real numbers, containing $\Delta = \sum_1^m \Delta_k$ will be said fully invariant with respect to the given system of cycles, and to the system of its neighborhoods, if

(1) every neighborhood Δ_k is regular;
(2) whatever the point $x \in F_\Delta$, the iterated sequence issued from this point belongs to F_Δ and each of its neighborhood Δ_k belong at least one point of this sequence;
(3) whatever the point $x \in F_\Delta$, the iterated sequence of the inverse function (sequence of antecedents) can be chosen such that each element of this sequence belongs to F_Δ and at least one of the elements belongs to Δ_k.

The set $F_\Delta - \sum_1^m \Delta_k$ will be called intermediary set.

Theorem 2. *Let C be a system of repulsive cycles and the system of their neighborhoods $\{\Delta_k\}$ $(k = 1, 2, \ldots, m)$ for which there exists a full invariant set F_Δ such as the closure of the intermediary set contains no cycle. The set F_Δ contains a continuum of points which generates an oscillating iterated sequence with respect to a given system of cycles. Each neighborhood $\Delta_{\xi_i}^k$ has the form*

$$\Delta_{\xi_i}^k = E_\infty + E_0 + \sum_2^\infty E_k$$

where E_0 is the set of points which generates the cycles of finite period (that is, the antecedents of increasing rank of the points of those cycles). E_k is the set of points which generates iterated sequences of class k. E_∞ is the set of points which generates sequences of ∞. The set E_0 is countable. For all $k > 1$ E_∞ and E_k have the power of a continuum. Each interval belonging to F_Δ contains points from each of the sets E_∞, E_0, and E_k $(k > 1)$.

This theorem obviously contains the validation of the existence of iterated sequences of finite and infinite class.

One may show some simple examples of functions for which this theorem is One may show some simple examples of functions for which this theorem is justified. For instance, $y = 4x - x^2$ has two repulsive fixed points, namely $x = 0$ and $x = 3$. The interval $[0; 4]$ is fully with respect to this system of fixed points.

References

1. A. N. Kolmogorov, I. G. Petrovskii, and N. S. Piskunov, A study of the diffusion equation with increase in the amount of substance, and its application to a biological problem, *Moscow University Mathematics Bulletin.* **1**, 1–26 (1937).
2. A. A. Vitt, Raspredelennye avtokolebatel'nye sistemy (Distributed self-oscillatory systems), *Zjurnal Tekhnicheskoi Fiziki.* **4**, 144–159 (1934).
3. A. A. Vitt, Kolebania skripichnoi struny (Oscillations of a violin string), *Zjurnal Tekhnicheskoi Fiziki.* **6**, 1459–1470 (1936).
4. S. P. Pulkin, On existence of broken lines of iteration with infinite set of limit points, *Scientific Notes of Pedagogical and Teachers' Institute (Kuibyshev).* **3**, 10–28 (1940).
5. S. P. Pulkin, On iterations of a function of one variable, *Izvestiya Akademii Nauk SSSR Seriya Matematicheskaya.* **71**, 106–108 (1942).
6. S. P. Pulkin, On a stability of iterative sequences, *Scientific Notes of Pedagogical and Teachers' Institute (Kuibyshev).* **7**, 149–159 (1943).
7. I. G. Petrowsky, Sur l'analyticité des solutions des systèmes d'équations différentielles, *Matematicheskii Sbornik.* **47**, 3–70 (1939).
8. N. E. Kochin, I. A. Kibel, and N. V. Rose, *Theoretical Hydromechanics*, 4th edn. Fizmatgiz, Moscow (1963) (in Russian).
9. F. I. Frankl, On a new boundary problem for the equation $y(\frac{\partial^2 z}{\partial x^2} + (\frac{\partial^2 z}{\partial y^2}) = 0$, *Prikladnaya Matematika i Mekhanika.* **152**, 99–116 (1951).
10. F. Tricomi, Sulle equazioni lineari alle derivate parziali di 2° ordine di tipo misto, *Atti della Accademia nazionale dei Lincei — Matematica e Applicazioni V.* **14**, 133–247 (1923).
11. S. P. Pulkin, On formulating the Tricomi problem in 3-dimensional space, *Notes of Pedagogical Institute.* **14**, 63–78 (1956).

12. S. P. Pulkin, The Tricomi problem for general Lavrentiev-Bitsadze equation, *Reports of the Academy of Sciences of USSR*. **118**, 38–41 (1958).
13. S. P. Pulkin, On the solvability of the Tricomi problem for the Chapligin equation, *Izvestiya vuzov. Mathematics*. **2**, 219–226 (1958).
14. S. P. Pulkin, On a uniqueness of the singular Gellerstedt problem, *Izvestiya vuzov. Mathematics*. **6**, 214–225 (1960).

https://doi.org/10.1142/9789819823130_0002

Chapter 2

EDWARD LORENZ'S CONTRIBUTION TO CHAOS THEORY

Christophe Letellier

Rouen Normandie University, Rouen, France
christophe.letellier@univ-rouen.fr

This chapter is devoted to the contribution of Edward Lorenz to chaos theory. It is based on some of his papers published in the 1960s and some other works that had some influence on his works.

1. Introduction

Chaos theory is a branch of the nonlinear dynamical systems theory. It would be difficult to determine the birth of a dynamical systems theory, but many great results were provided by Henri Poincaré who developed an original approach in a series of papers published between 1881 and 1886.[1–4] In this first work, he already left the quantitative approach focused on finding an approximated analytical solution, as general as possible, for a qualitative approach exhibiting some properties of a whole class of solutions. The qualitative approach is developed in the state space spanned by the variables occuring in the governing differential equations. Poincaré investigated how the solutions are organized in the state space by the singular points.[1–3] He distinguished four type of singular points in two-dimensional state space — node, saddle, focus and centre — with the help of a stability analysis. The concept of *limit cycle* was also introduced in this first series of papers. Later he generalized such study to periodic orbits with the help of a surface of section which allows to reduce the stability analysis of orbit to a stability analysis of one of its points.

When he investigated the restricted three-body problem, Poincaré introduced many other concepts. Following George William Hill,[5] he understood that the key concept to investigate aperiodic solutions was to use the unstable periodic orbits. Indeed, to obtain an approximated solution in his lunar theory, Hill did not start by considering a two-body problem, then adding some perturbation induced by the third body, but rather from a periodic orbit solution to the three-body problem (sun, earth and moon). Using this technique, Poincaré discovered the existence of particularly complicated solutions, the *homoclinic* orbits leaving the neighborhood of a singular point by its unstable manifold and returning to it by its stable manifold.[6,7] Investigating the properties of these particular solutions, he showed that the trajectory near a homoclinic orbit was sensitive to initial conditions, and organized in a quite complex way. Proving the existence of a homoclinic orbit is today considered as a proof for the existence of a chaotic trajectory.[8,9] In this memoir, Poincaré also stated his recurrence theorem:[6] Any d-dimensional dynamical system determines a flow $\phi_t : \mathbb{R}^d \mapsto \mathbb{R}^d$. If the flow has only bounded orbits, then, for each open set, any orbit that intersects this open set intersects it infinity often. This theorem also applies for the set of orbits of dissipative system which are within an invariant set. Poincaré also introduced the *analysis situs*.[10] His studies of the restricted three-body problem, he showed that a deterministic system may have aperiodic solutions which are sensitive to initial conditions. The main ingredients to define chaos were nearly provided; only the word was to maturate...

In his book *Science and Method*, Poincaré applied the consequence of these results to the problem of weather forecasting and try to address the question "*Why have meteorologists such difficulty in predicting the weather with any certainty?*"[11]

> A very small cause, which escapes us, determines a considerable effect which we cannot not see, and then we say that this effect is due to chance. If we knew exactly the laws of nature and the situation of the universe at the initial moment, we could predict exactly the situation of this same universe at a later moment. But, even when natural laws are no longer a secret to us, we will only be able to know the initial situation approximately. If this allows us to predict the future situation with the same approximation, this is all what we need, we say that the phenomenon has been foreseen, that it is governed by laws; but this is not always the case, it can happen that small differences in the initial conditions generate very large ones

> in the final phenomena; a small error on the first ones would produce a huge error on the latter. The prediction becomes impossible and we have the fortuitous phenomenon.

2. Early Developments of Weather Forecasting

In the 1920s, the problem of predicting the weather conditions became a problem of a first interest. Weather forecasting is a combination between science and technology to predict the conditions of the atmosphere for a given location and a given time. Attempts to predict the weather were completed for millennia. It became formally a science during the 19th century. Typically, weather forecasts require the collection of quantitative data about the current state of the atmosphere over the "neighborhood" of a given place and using meteorology to predict how the atmosphere will change. The development of weather forecasting is directly linked to the development of the telegraph communications. Indeed, the data must "travel" faster than the phenomena. This is thefefore on 1861 that *The Times* started to publish its daily weather forecasts. Modern numerical weather prediction requires an advanced understanding of the atmospheric physics, that is, not only of the ability to write not too complicated differential equations governing atmospheric flow[12] but also of developing numerical techniques (finite differentiation scheme in time as well as in space) to allow practical computations.[13]

Wilhelm Bjernkes stated the problem as follows.[14]

> If it is true, as any scientist believes, that subsequent states of the atmosphere develop from preceding ones according to physical laws, one will agree that the necessary and sufficient conditions for a rational solution of the problem of meteorological prediction are the following:
>
> (1) One has to know with sufficient accuracy the state of the atmosphere at a given time.
> (2) One has to know with sufficient accuracy the laws according to which one state of the atmosphere develops from another.
>
> Atmospheric processes are of a mixed mechanical and physical nature. For every single process, we can propose one or several mathematical equations derived from mechanical or physical principles. We will possess sufficient knowledge of the laws according to which atmospheric processes develop if we can write down as many equations independent from one another as there are unknown quantities. The state of the atmosphere at any point in time will be determined meteorologically when we can

calculate velocity, density, pressure, temperature and humidity of the air for any point at that particular time. [Since there are] three velocity components, there are seven unknown parameters to be calculated.

For calculating these parameters, we can propose the following equations:

(1) The three hydrodynamic equations of motion. These are differential relations among the three velocity components, density and air pressure.
(2) The continuity equation, which expresses the principle of the conservation of mass during motion. This equation is also a differential relation, namely between the velocity components and the density.
(3) The equation of state for the atmosphere, which is a finite relation among density, air pressure, temperature, and humidity of a given air mass.
(4) The two fundamental laws of thermodynamics, which allow us to write two differential relations that specify how energy and entropy of any air mass change in a change of state. These equations do not introduce any new unknowns into the original problem because energy and entropy are expressed by the same variables that appear in the equation of state and relate the changes of these quantities to changes of other known parameters. These other quantities are, firstly, the work done by an air mass, which is determined by the same variables that appear in the dynamical equations; secondly, the heat quantities received from or given off to the outside. These heat quantities will be constrained by physical data on incoming and outgoing radiation, and on the warming of the air in contact with the Earth's surface [conduction].

It should be emphasized that the problem is considerably simplified if there is no condensation or evaporation of water and thus water vapour contained in the air can be considered a constant constituent. [...] We can therefore set up seven equations independent from each other with the seven normally occurring variables. [...]

The main task of observational meteorology will be to provide simultaneous observations of all parts of the atmosphere, at the Earth's surface and aloft, over land and over sea. Based on the observations made, the first task of theoretical meteorology will then be to derive the clearest possible picture of the physical and dynamical state of the atmosphere at the time of the observations. This picture must be in a form that is appropriate to serve as a starting point for a weather prediction according to rational dynamical-physical methods.

For Bjernkes,[15]

We must apply the equations of theoretical physics not to ideal cases only, but to the actual existing atmospheric conditions as they are revealed by modern observations. These equations contain the laws

> according to which subsequent atmospheric conditions develop from those that precede them. It is for us to discover a method of practically utilising the knowledge contained in the equations. From the conditions revealed by the observations we must learn to compute those that will follow.

Lewis Fry Richardson established the numerical procedure to solve the required differential equations.[12] He quoted the work by Carl Runge.[16] This opened the possibility to solve the equations governing the atmosphere. A key step related to this is certainly the book published by Richardson in 1922 in which he developed a systematic mathematical procedure for predicting the weather and, after six weeks of work, he computed a change in the atmospheric pressur of 145 hPa in 6 h, a value which is too large to be realistic.[17] Richardson estimated that 64,000 of peoples would be required to compute in real time the evolution of the meteorology. Such a failure eclipsed for a few decades.

3. Edward Norton Lorenz: A Short Biography

Edward Norton Lorenz was born in West Harford (Connecticut) on May 23, 1917.[18–20] His father, Edward Henry Lorenz, was born in Hartford in 1882; he majored in mechanical engineering at the Massachusetts Institute of Technology, then located in downtown Boston. Ed became fascinated with numbers at an early age. Later, after he had learned multiplication, he took an interest in numbers that were perfect squares and could recite all such numbers between 1 and 10,000. He also enjoyed taking square roots using a longhand method, and even learned a method for extracting cube roots.

When he was seven, he had already developed an interest in maps and would even draw maps of places that he had invented. He was also found of astronomy. He was also interested in the weather, though it did not then occur to him that he would one day make a profession of it. He entered Dartmouth College (Hanover, New Hampshire) in 1934, having already decided that he wanted to major in mathematics. Among the 700 students entered at Dartmouth that year, only seven went on to major in mathematics. Lorenz enjoyed the logical clarity of mathematics. In 1938, he entered the graduate school of the mathematics department at Harvard. There, he attended to lectures about topics such as group theory, set theory, and combinatorial topology, taught by such luminaries as Saunders Mac Lane (1909–2005, founder of category theory), Marshall Stone (1909–1989, great contributor in functional analysis), and James Van Vleck (1899–1980,

Nobel prize for his work in the quantum theory of magnetism). Lorenz related that although he had originally been attracted to math by his love of numbers, during his graduate years he seldom saw any number other than 0, 1, and 2. He chose to work on a problem in mathematical physics under the guidance of George Birkhoff, the eminent mathematician well-known for his proof of Poincaré's Last Geometric Theorem,[21] a special case of the orbital three-body problem, which exhibits one of the key properties of chaos: sensitive dependence on initial conditions.[22] Lorenz's master's thesis was not about dynamical systems, but about Riemannian geometry.

In 1942, just a few months before he expected to receive his doctoral degree, the war intervened and he had to choose between being drafted and training to become part of a team of weather forecasters for the army. In March 1942, he was enrolled as a cadet in the Army Air Corps (now the Air Force) in a special eight-month master's program at Massachusetts Institute of Technolog (MIT), to be trained in weather forecasting. He realized that meteorology concerned the scientific understanding of the atmosphere, while weather forecasting was simply a particular practical application; weather forecasting was possible without a too precise understanding of the atmosphere, and investigating deeply the latter do not necessarily mean to be a forecaster. Nevertheless, the MIT program was devoted to both. Morning lessons were devoted to theory apprently relevant to forecasting although without any proof of their efficiency for improving forecasts. Afternoon lessons were focused to forecast from past weather maps as case studies. He met MIT faculty members as Hurd Willett (1902–1992) who contributed to the five-days forecast program with Carl-Gustav Rossby,[23] Henry Houghton known for his contribution on the interplay between the solar radiation and the atmosphere,[24] and Bernhard Haurwitz known for his contribution to dynamic meteorology,[25] among others. Lorenz was fascinated by the dichotomy between the physics of atmosphere and the practice of weather forecasting. At the end of the master's program, in November 1942, he remained at MIT as an instructor for the next year's program. This gave him the opportunity to attend advanced classes for becoming an experienced meteorologist. Nevertheless, he became convinced that something important was missing from the field.

After completing the final training course at MIT, Lorenz was sent to overseas. He first reported to Hawaii, where he attended another two months of training in tropical meteorology and then flew to Saipan (Pacific Ocean) in October 1944. At that time, Lorenz was quite close to the battle front.

With several Air Corps colleagues, he helped set up a weather forecasting operation in support of airborne bombing raids against Japan. He was working for forecasting upper-level winds, but his work was complicated by a severe lack of observations because the Air Corps were not very concerned by weather data. The only observations between Siberia and Saipan were taken by the crews of US aircraft themselves; Lorenz complained that the pilots, by lack of time, too often simply repeated the forecast as the observation. This was not very efficient for the next prediction. In the spring of 1945, the weather operation was transferred to the Japanese-held island of Guam where an important battle occured from July 21 to August 10, 1944: he was appointed head of the upper-air section where he worked until a few months after the end of the war.

Saipan battle: June 15-July 9, 1944

Okinawa battle: April 1st-June 22, 1945

At the end of the war, Lorenz took the decision to switch to meteorology rather than completing his doctorate in mathematics at Harvard. He reached this decision after many questioning and exchanges with

Henry Houghton who was the head of the meteorology department at MIT. Lorenz justified his decision as follows.[a]

> Mathematicians seem to have no difficulty in creating new concepts faster than the old ones become well understood, and there will undoubtedly always be many challenging problems to solve. Nevertheless, I believed that some of the unsolved meteorological problems were more fundamental, and I felt confident that I could contribute to some of their solutions.

Lorenz thus started more or less when a first numerical solution to the barotropic vorticity equation, an equation for the global atmospheric circulation, was provided by Jules Charney, Ragnar Fjørtoft and John von Neumann from the Institute for Advanced Study of Princeton.[26] This followed the introduction of some equations governing the baroclinic westerly current by Charney:[27] his important contribution was to show that "long waves" from the upper atmosphere affects the dynamics of the entire atmosphere. Charney was under the direct stimulation of Carl Rossby, and his pupil Victor Starr.[28–30] The computation was performed on an Electronic Numerical Integrator And Computer (ENIAC) and required 12 h of computations for a prediction of 24 h. This important result can be viewed as a first concretization of Richardson's dream. Victor Starr and Edward Lorenz (both from the MIT) were questioning the possibility of weather forecast using numerical simulations. They visited Charney's group to evaluate the acutal possibilities of such an approach.

In the early 1940s, Rossby did regret that meteorologists have to contend with observational data and cannot, as physicists, design an experimental setup to deeply understand relationships between cause and effect. In his studies,[31] starting from the fact that the atmosphere was under the influence of thermal convection and moving over a rough rotating surface, he performed a balance of the energy received from the sun, combining it with the common properties of a viscous fluid, the air, around a rotating sphere. He thus came to the conclusion that the atmosphere should be governed by three cells of convection, spread along the latitude (Fig. 1).

In the 1950s, some works were developed to extract the meteorological equations in relation with some simplified experimental devices. In this spirit, using a disk of air, F. Vettin reproduced thermal cyclones and a general trade-wind cell circulation more or less in agreement with the

[a] A Scientist by Choice. Speech delivered on the occasion of Lorenz's acceptance of the Kyoto Prize in 1991.

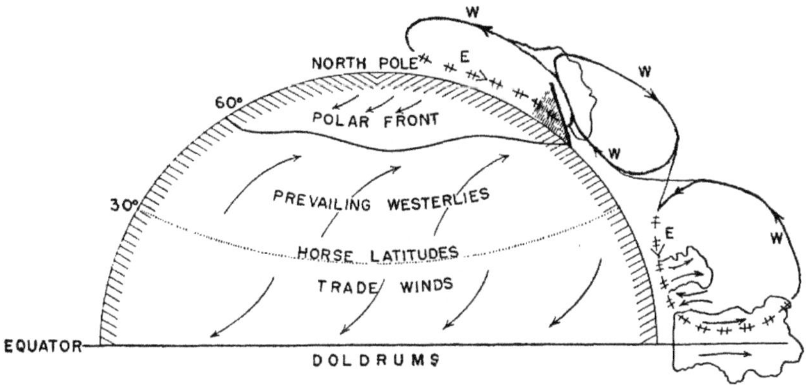

Fig. 1. "Doldrums" is a popular nautical term which refers to the belt around the earth near the equator where sailing ships sometimes get stuck on windless waters.

Source: After Rossby[31].

common understanding from the 1950s.[32] The key point for this was to develop some model, or experiments, in such a way that all terms in the one case are numerically identical to those in the other: this was the requirement to design experimental setup for better understanding the atmosphere. This type of discussions were already proposed by Helmholtz[33] and Rossby.[28] Such an approach led Dave Fultz to propose adimensional equations governing the atmosphere.[34] Dave Fultz's group (University of Chicago) designed an experimental setup to simulate the rotation of the heat point due to the rotation of the earth (Fig. 2).[35] The streamlines reveal the existence of two saddle points in the flow. The existence of such points is crucial since cause of sensitivity to initial conditions. Thermal convection in a rotating fluid was thus investigated by Raymond Hide during his Ph.D. thesis as an experimental device for the hydrodynamics of the general circulation of the Earth's atmosphere. Initially devoted to the study of the earth's magnetic field,[36] Hide changed his perspective when he understood that his studies had some bearing on meteorology. Hide designed a co-axial cylinders device with a source of heat at the outside cylinder and showed that some oscillations of the flow appear and can lead to some turbulent behavior.[37] Hide was then a professor of physics and geophysics at MIT from 1961 to 1967,[38] where he most likely met Lorenz. These two experiments were important for Lorenz when he derived his eponym system as discussed below.

Fig. 2. Streamline in a fluid within a layer of fluid ($h = 6.0$ cm) in a circular iron tank ($r = 15.7$ cm) around which a flame is counter-clockwise moved around ($\tau = 18.17$ s per revolution). The heating was initiated 7 min prior to the photograph. Mean temperature $\overline{T} = 29.8$°C with $\dot{T} = 0.325$°C.min^{-1}. The flame is located at the position it was when the picture used to draw the streamlines was shot.

Source: After Fultz *et al.*[35]

At this time Lorenz was already working on the global atmospheric circulation. He obtained one of his first important results by showing that a small fraction of the solar energy received by the atmosphere was converted into kinetic energy to balance friction.[39] To do so, the theoretical equations developed for governing the atmospheric circulation required significant simplification to be treated in a reasonable amount of time for real time weather forecast. Most often, the technique was based on a decomposition in Fourier modes, but Lorenz was more interested by understanding the phenomena than by getting accurate predictions:[40]

> It is only when we use systematically imperfect equations or initial conditions that we can begin to gain further understanding of the phenomena which we obseve. For if we omit the terms representing specified physical processes, such as friction, from the equations, or if we fail to include certain observable features, such as cloudiness, in the initial conditions, we may, by comparing the mathematical solutions with reality, gain some insight concerning the relative importance of the retained and omitted features. Of course, in so doing, we forgot the opportunity of simultaneously making the best attainable forecast.

Most likely, stimulated by discussions with Hide, Lorenz quoted two oversimplified experiments for reproducing the dynamics of a flow under conditions similar to those observed in the atmosphere: Fultz[35] and Hide[37] ones. This was a kind of justification for using simplified equations. Indeed, Lorenz oversimplified the problem by removing some terms from the governing equations or from the initial conditions. Although the problem was reduced to a few hundreds of equations, a quite reduced number with respect to the problem which has an infinite number of degrees of freedom, this remains quite untractable with the computers from the 1950s. Lorenz had to reach the point at which the simplified equations and initial conditions do no longer provide good predictions. Lorenz was therefore concerned by simplifiying as much as possible while keeping the desired phenomena. He finally oversimplify the problem to a two-dimensional horizontal flow of a non-viscous fluid and got the vorticity equation

$$\frac{\partial}{\partial t}\nabla^2\Psi = -\boldsymbol{k}\cdot\nabla\Psi\wedge\nabla(\nabla^2\Psi) \tag{1}$$

where Ψ is a stream function and $\boldsymbol{k}$ a unit vertical vector. This equation is equivalent to the barotropic vorticity equation which approximately governs the vertically-averaged horizontal flow in the atmosphere. This equation was expanded in a double Fourier series and, then limited to three modes. He plotted the evolution of the streamlines for 36 h (Fig. 3). He then observed that the patterns observed for $t > 30$ h were periodically repeated.

For Lorenz, the key point was to drastically simplify the equations at a point they can be integrated numerically, but retaining their nonlinear character. Thus, the simplified equations are realistic enough to save qualitatively some important phenomena in the atmosphere. Two years later,

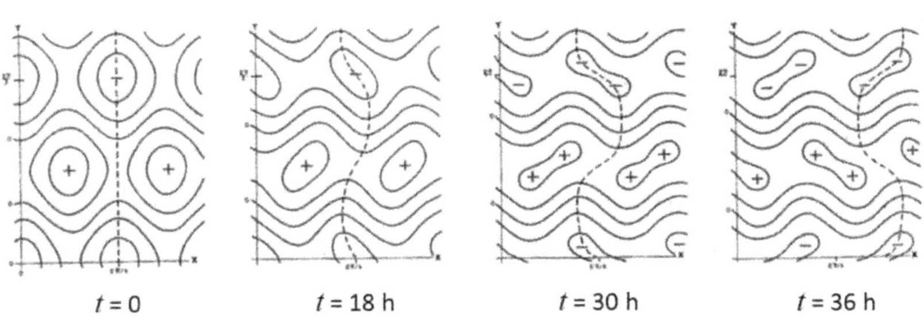

Fig. 3. Evolution of the streamlines of the bidimensional flow for a nonviscous fluid as simulated by Lorenz in 1960.

Source: Redrawn from Lorenz.[40]

to investigate a case closer to the general circulation of the atmosphere, Lorenz started from the same equations but for a two-layer model. He retained the twelve-dimensional deterministic nonlinear system[41]

$$\dot{X}_i = c_i + \sum_m b_{im} X_m + \sum_{m,n} a_{imn} X_m X_n \quad (i = \{1, \ldots, 12\}) \tag{2}$$

where the nonlinear terms represent the advective effects, and the linear ones represent the effects of friction and some of the effects of heating. The constant terms depends on the temperature field. Under the suggestions of Lorenz and Starr, similar equations were already investigated in 1959 by the oceanographer Kirk Bryan (today considered as the founder of numerical ocean modeling[42]).[43]

Using a Royal-McBee LGP-30, Lorenz was able to produce data for six year of atmospheric circulation in 24 h of computations. Lorenz thus observed large amplitude fluctuations with irregular periods of a month or two.[41] Lorenz noticed the randomness and the aperiodicity of the stream function, as clearly exhibited by the corresponding time series (Fig. 4). He also found a continuous Fourier spectrum. He qualified the predictions one day in advance as being nearly perfect, at two days in advance as very good and at four days as falling far short of perfection. Lorenz thus concluded as follows.[40]

> This work demonstrates the existence of deterministic systems, governed by equations whose nonlinearity resembles the nonlinearity of the atmosphere, which are not perfectly nor almost perfectly predictable by simple and simply determined linear formulas, if the period between successive observations is greater than half of the shortest significant period of oscillation.

In other words, although the sampling rate is satisfying the Nyquist criterion,[44] it is still not possible to correctly predict the simulated behaviors with linear techniques.

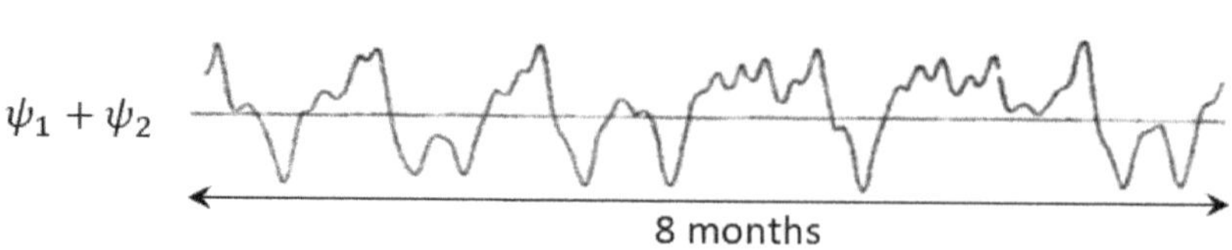

Fig. 4. Time series of the sum of the two stream functions Ψ_1 and Ψ_2 of each layer of the model for the global atmospheric circulation.

Source: Adapted from Lorenz.[123]

4. Predictability of Hydrodynamical Flow

Facing the impredictability of this 12-dimensional deterministic system, Lorenz investigated deeper such property with the question "*How predictable is hydrodynamic flow?*" His aim was to investigate meteorological predictions with the help of the techniques developed in the dynamical systems theory. One should not forget that he was the direct pupil of George Birkhoff who wrote the book entitled *dynamical systems.*[45] One may also remark that Viktor Nemytskii and Vyacheslav Stepanov's book[46] was also quoted but, as Lorenz clarified himself:[18]

> In preparing the paper I had been guided by Birkhoff's treatment of dynamical systems. When I submitted the paper, the reviewer, who proved to be quite conscientious and competent, noted that some of my results overlapped some of the works that had postdated Birkhoff's monograph, and he specifically mentioned the recently translated textbook *The qualitative theory of differential equations* by the mathematicians V. V. Nemytskii and V. V. Stepanov, originally published in 1946.

Consequently, Lorenz used only Birkhoff's book.

Remembering that the theory of dynamical systems comes from the astronomy as well exemplified with Poincaré's contribution,[7] Lorenz stated that *meteorologists are not in a class with astronomers*! He framed his study to the stationary flows which, at this time, were known to be periodic, quasi-periodic or aperiodic. For Lorenz, it is clear that a quasi-periodic behavior is a *recurrent* behavior (in the sense of Bikhoff) which is the finite sum of periodic functions: it is a *predictable* behavior. An aperiodic behavior can also repeat a state already visited, but the states following the two occurences do not remain close, except temporarily: it is an impredictable behavior.

At this time, limit cycle are well-known since their discovery by Poincaré. They were investigated by Andronov and his co-workers in various contexts,[47,48] or as the solutions of the van der Pol equations[49] as investigated by Karl Bonhöffer, for instance.[50] In the 1950s, quasi-periodic solutions were investigated with the help of analog computer by the Japanese Hayashi group.[51,52] All of these solutions are predictable, in the sense that "*predictability at a given range is essentially a measure of the errors in predicting that range.*"[53] Poincaré started to investigate some aperiodic solution in the neighborhood of a homoclinic orbit,[6] that he qualified as inextricably intricated and he gave up in drawing them.

To the best of our knowledge, the first picture of what Poincaré did not drew was proposed by Vladimir M. Mel'nikov in 1962.[54–56]

For Lorenz, the motivation for investigating aperiodic solution comes from the vertical motions of the ocean surface which are considered as having a non-periodic component driven by winds.[53] To give an idea of what he meant by aperiodic and non-predictable, Lorenz took the example of the dynamics underlying the sunspots, a dynamics which was *imperfectly understood*. For him, there are three types of predictability: (i) *intrinsic* predictability which depends only upon the flow itself, (ii) *attainable* predictability which is limited by the unavoidable inaccuracies in measurements, and (iii) *practical* predictability which is limited by our inability to identify which technique to use. Periodic motions do not present intrinsic errors for predicting them, quasi-periodic motions are practically predictable with little error. Aperiodic motions were left with a limited consideration in 1963, and their predictability was not determined.

To quantify the predictability of a nonlinear deterministic flow, Lorenz reused his 12-dimensional system investigated in 1962 but, then, in performing repeated numerical simulations, he observed a surprising feature:[53]

> During the course of this investigation we encountered a rather striking phenomenon. At various times we found it desirable to repeat portions of previous computations. For this purpose we took the values which the machine had printed at one particular time step, and entered these values into the machine as new initial conditions. Sometimes the machine did not repeat its previous performance. Fairly soon, small differences between the solutions would appear, and these would grow until eventually there was no resemblance between the two solutions.

The cause of these diverging results was the fact that his analog computer was computing with six digits but printed only three of them. Since he restarted his simulation from the printed numbers, three digits were omitted: this was the source of the perturbation between the two different simulations. For Lorenz, this meant clearly that the original solution was unstable, and that the long-term prediction would be a total failure.

To investigate such unstable solutions, he was reasoning in the state space (or phase space), that is, the evolution of the system is viewed as a trajectory in a space, *even when he was not visualizing state space*. He thus defined a limit cycle as a stability implying a periodicity, and an unstable periodic orbit as an aperiodicity implying an instability.

> A solution is unstable if two arbitrarily close states do not remain arbitrarily close. We have left open the possibility that the subsequent histories of two sufficiently close states may continue to bear some resemblance to each other, even though they do not remain arbitrarily close.

This is a clear hint to Poincaré's recurrence theorem:[6]

Theorem 1. *Let us assume a neighborhood $\mathcal{N}$ of an invariant set and that the volume is an integral invariant; if one considers a point $\boldsymbol{x}_0$ in this neighborhood $\mathcal{N}$, there exists some solutions which visit it an infinite number of times.*

Poincaré added two important corollaries.

Corollary 1. *Any trajectory issued from a point $\boldsymbol{x} \in \mathcal{N}$ crosses an infinite number of times this neighborhood $\mathcal{N}$.*

Corollary 2. *The theorem 1 still holds when the volume is not an integral invariant, until that there exists any positive invariant.*

This last corollary allows to apply the theorem 1 to dissipative dynamical systems, particularly when the solution is considered in the invariant set. This is in this sense that Lorenz made his remark. Consequently, Lorenz concluded that

> If a solution is purely nonperiodic, that is, if it has no periodic components, it is not only unstable, but two neighboring solutions must eventually become as far separated as two randomly chosen solutions. [...] Errors in observation will always prevent us from distinguishing between two sufficiently similar states. If these two states eventually evolve in different ways, no method of predicting the distant future from the present can give good predictions in both instances. [...] Hence no method of predicting the distant future from the present and past can give good predictions in both instances. [...] One meteorologist remarked that if the theory were correct, one flap of a sea gull's wings would be enough to alter the course of the weather forever.

When he wrote this 1963 paper, Lorenz had already started to work with his oversimplified model for the Rayleigh–Bénard convection. Indeed, as he himself detailed:[18]

> My search came to an abrupt end one afternoon in 1961 when I was visiting Barry Saltzmann at the Travelers Weather Center [...].

> Barry showed me a system of seven equations that he had been solving numerically. The equations were a bit like mine, but they modeled convective fluid motion driven by heating from below, such as might occur locally over warm terrain, instead of the global atmospheric circulation, which is driven mainly by horizontal differences in heating. He was interested in periodic solutions and had obtained a number of them, but he showed me one solution that refused to settle down.

So, let us remind Saltzman's contribution.[57]

Barry Saltzman did his graduate work at the MIT, where he obtained a master of science in meteorology in 1954. He then prepared a Ph.D. in meteorology in 1957[58] under the supervision of Victor Starr.[59] As a graduate student and a research scientist, Saltzman contributed to the *General Circulation Project* at MIT initiated by Starr in 1948 with foundings from the U.S. Air Force. This project triggered the development of the advanced computer models used for climate and weather prediction studies as those performed by Phillips.[60]

In his Ph.D. thesis, Saltzman introduced the use of spatial Fourier analysis to quantify nonlinear dynamical interactions between zonal scales, establishing a link to the theory of turbulence based Fourier analysis in the spirit of Andrei Kolmogorov,[61,62] George Batchelor,[63] and Fjørtoft,[64] among others. Saltzman generalized the Reynolds decomposition[65] of an arbitrary spatial field $\boldsymbol{u}(\boldsymbol{x})$ into its domain mean $\overline{\boldsymbol{u}}$ and deviation $\boldsymbol{u}*(\boldsymbol{x}) = \boldsymbol{u}(\boldsymbol{x})-\overline{\boldsymbol{u}}$, to derive a more detailed wavenumber decomposition $\overline{\boldsymbol{u}} = \tilde{\boldsymbol{u}}_0$, and

$$\boldsymbol{u}*(\boldsymbol{x}) = \sum_{m\neq 0} \tilde{\boldsymbol{u}}_m \, \mathrm{e}^{i2\pi m\, x} \tag{3}$$

where $m = \{0, \pm 1, \pm 2, \ldots\}$. Whereas others as Lorenz,[39] used the $(\boldsymbol{u}, \boldsymbol{u}*)$ decomposition to analyze interactions between the global scale $l_1 = 2\pi a \cos\varphi$ — where a is the radius of the earth and φ the latitude — and some smaller scales, Saltzman used a decomposition into interactions between each scale defined by their wavelengths $l_m = \frac{l_1}{|m|}$ $(m \neq 0)$. The originality of Saltzman was therefore the way he developed the decomposition.

In 1961, Saltzman moved for a position at the Travelers Research Center in Hartford (Connecticut), to continue fundamental research on the atmosphere. As Lorenz, Barry Saltzman was interested in simplified models of flows. He approximated the two-dimensional equations developed by Oberbeck,[66] and Boussinesq[67] governing roll convection between two isothermal free surfaces as experimentally investigated by Bénard.[68] These experiments were theoretically investigated by Rayleigh.[69] Saltzman's motivation came from the fact that most of atmospheric motion are caused by Rayleigh–Bénard convection. His motivation was also to get a system as simple as possible to understand atmospheric flows but, contrary to Lorenz, he was concerned by the reliability of the model with respect to experimental results.

Saltzman proposed the seven-dimensional system[57]

$$\begin{cases} \dot{x}_1 = \alpha_{11}x_1 + \alpha_{12}x_4 + \alpha_{13}x_2x_3 \\ \dot{x}_2 = \alpha_{21}x_2 + \alpha_{22}x_5 + \alpha_{23}x_1x_3 \\ \dot{x}_3 = \alpha_{31}x_3 + \alpha_{32}x_6 + \alpha_{33}x_1x_2 \\ \dot{x}_4 = \alpha_{41}\tilde{R}x_1 + \alpha_{42}x_4 + \alpha_{43}x_1x_7 + \alpha_{44}x_2x_6 + \alpha_{45}x_3x_5 \\ \dot{x}_5 = \alpha_{51}\tilde{R}x_2 + \alpha_{52}x_5 + \alpha_{53}x_1x_6 + \alpha_{54}x_3x_4 + \alpha_{55}x_2x_7 \\ \dot{x}_6 = \alpha_{61}\tilde{R}x_3 + \alpha_{62}x_6 + \alpha_{63}x_1x_5 + \alpha_{64}x_2x_4 \\ \dot{x}_7 = \alpha_{71}x_7 + \alpha_{72}x_1x_4 + \alpha_{73}x_2x_5 \end{cases} \tag{4}$$

where

$$\begin{cases} x_1 = X_1 \;\; = \Psi_1(3,1) \\ x_2 = X_7 \;\; = \Psi_1(4,1) \\ x_3 = X_{13} = \Psi_1(1,2) \\ x_4 = X_{30} = \theta_2(3,1) \\ x_5 = X_{32} = \theta_2(4,1) \\ x_6 = X_{38} = \theta_2(1,2) \\ x_7 = X_{50} = \theta_2(0,2). \end{cases} \tag{5}$$

are the components of a Fourier decomposition to approximate the fields of a vertical stream function $\Psi(t, \boldsymbol{x})$ (3 modes) and of the temperature deviation $\theta(t, \boldsymbol{x})$ (4 modes). This motion was two-dimensional: there is no motion along the roll axis due to the aspect ratio of the convection cells, $\frac{L}{H} = 6\sqrt{2}$. According to Saltzman, this model is *the simplest convection model capable of representing non-linear interaction between harmonic components of the stream field.*[57] The variables $x_1 = \text{Re } \hat{\Psi}_{1,1}$ and $x_4 = \text{Im } \hat{\theta}_{1,1}$ correspond to the convection intensity and the thermal field, respectively. The variable $x_7 = \text{Im } \hat{\theta}_{0,2}$ is associated with the difference from the vertical temperature gradient.

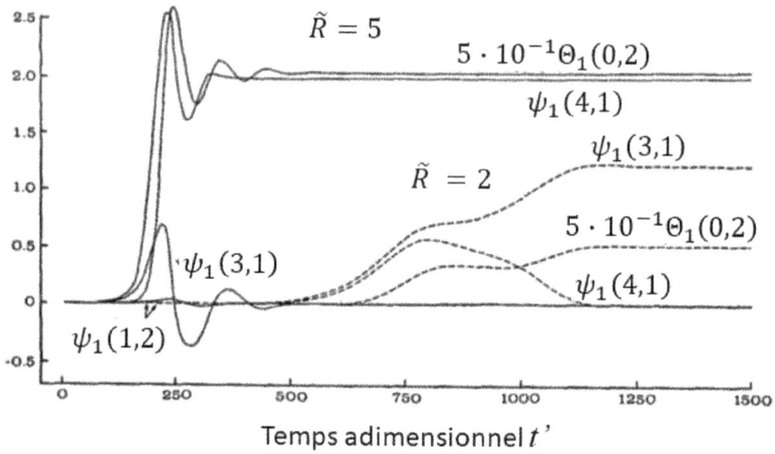

Fig. 5. Transient regime for two values of the Rayleigh number $\tilde{R} = 1$ and $\tilde{R} = 5$. *Source*: Processed from Saltzman.[57]

Saltzman limited his investigation to Rayleigh numbers slightly greater than the critical value, that is, to reduced Rayleigh number at about $\tilde{R} = R_c = 1$. He plotted the solutions to his model for $\tilde{R} = 1$ and $\tilde{R} = 5$ (Fig. 5). He thus focused his attention on the transient regimes between slight perturbed states (inducing some convection) to stable states (with no convection), that is, $1 \leqslant \tilde{R} \leqslant 10$. This was motivated by the fact that for this low reduced Rayleigh numbers, the solutions to the model (4) are still quite realistic.

From this plot, one may observe that, as Lorenz wrote,[70] *all except but three of the dependent variables eventually tended to zero, and these three variables underwent irregular, apparently nonperiodic fluctuations.* Lorenz visited Saltzman as previously mentioned and we do not know which curves he exactly saw. As discussed below, Saltzman performed some computations for larger values of the reduced Rayleigh number ($\tilde{R} > 20$) but he did not provide plots. If we numerically integrate the model (4) for $\tilde{R} = 15$ (Fig. 6), one may see what Lorenz concluded. Nevertheless, the three apparently irregular modes converge to zero too, but in a longer time than the four others.

For even larger reduced Rayleigh number $\tilde{R}$ (Fig. 7), Saltzman remarked that *oscillatory, overstable cellular motions are present and, consequently an alternating value of the heat transport about a time-mean value is*

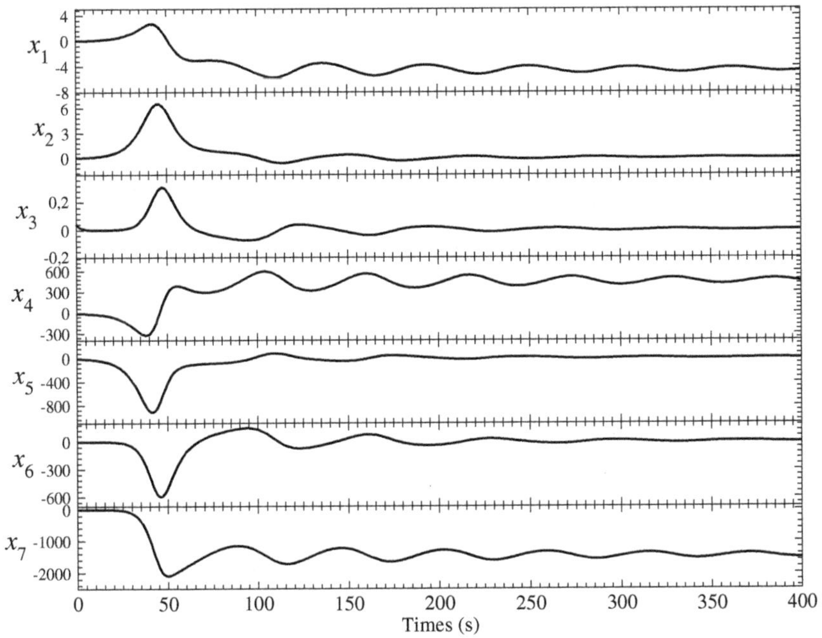

Fig. 6. Evolution of the seven Fourier modes involved in the model (4).

Note: Parameter value: $\tilde{R} = 15$.

found. The lack of sufficient degrees of freedom in the vertical undoubtedly contributes to this effect.[57] Saltzman considered that his Fourier decomposition was too much truncated to provide interesting (realistic) solutions, particularly for $\tilde{R} > 20$: he did not push further his study. One may remark that Saltzman's paper was anounced as the first of a couple (see the title). He never published the companion paper. Our assumption is that, once Lorenz's key paper was published in 1963, what he had in mind was no longer of interest.

For $\tilde{R} = 28$, the time series of the seven variables are actually aperiodic, in a sustained way. To have a better view of the nature of the solution, let us propose a representation of the dynamics in a projection of the state space (Fig. 8). The obtained structure clearly suggests a perturbed Lorenz attractor, where "perturbed" means that it has a higher-dimensional structure. To the best of our knowledge, it was never noticed that Saltzman saw aperiodic time series but did not understand their interest. One should not forget that, in the acknowledgements of his paper,

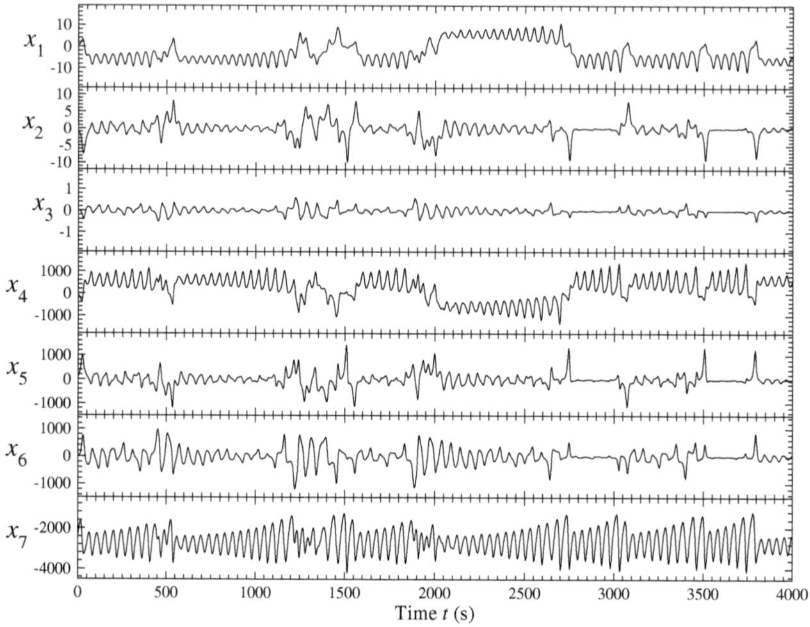

Fig. 7. Evolution of the seven Fourier modes involved in the model (4).
Note: Parameter value: $\tilde{R} = 28$.

Lorenz wrote that he was "*indedbted to Dr. Barry Saltzman for bringing his attention to the existence of nonperiodic solutions of the convection equations.*"

Lorenz restarted from the seven-dimensional model (4) and removed the four "slow" variables to get

$$\begin{cases} \dot{x}_1 = \alpha_{11}x_1 + \alpha_{12}x_4 \\ \dot{x}_4 = \alpha_{41}\tilde{R}x_1 + \alpha_{42}x_4 + \alpha_{43}x_1x_7 \\ \dot{x}_7 = \alpha_{71}x_7 + \alpha_{72}x_1x_4 \end{cases} \tag{6}$$

that is, as rewritten in the common way,

$$\begin{cases} \dot{x} = \alpha_{11}x + \alpha_{12}y \\ \dot{y} = \alpha_{41}\tilde{R}x + \alpha_{42}y + \alpha_{43}xz \\ \dot{z} = \alpha_{71}z + \alpha_{72}xy \end{cases} \tag{7}$$

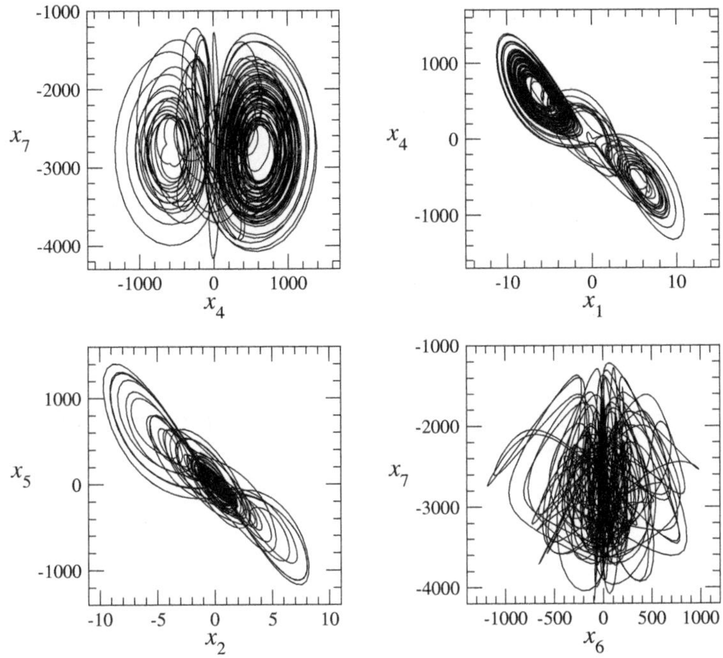

Fig. 8. Chaotic attractor produced by the seven-dimensional system (4) proposed by Saltzman.

Note: Parameter value: $\tilde{R} = 28$.

where $\alpha_{11} = -\alpha_{12} = \sigma = 10$, $\alpha_{41} = -\alpha_{42} = -\alpha_{43} = \alpha_{72} = 1$ and $\alpha_{71} = -b = -\frac{8}{3}$. In such a system, x corresponds to the intensity of convection, y is the difference in temperature between the left and the right sides of rolls, and z is the difference from the linear profile of temperature (observed without any convection). For Lorenz, the system (6) was the simplest system producing aperiodic solutions.

Lorenz was motivated by the irregular fluctuations of the instantaneous patterns observed in turbulent flows. One of his tasks was to provide an alternative to the statistical approach of turbulence. Since a quasi-periodic solution is a combination of periodic solutions with incommensurable periods, periodic solution is a particular case of quasi-periodic solutions. Consequently, an aperiodic solution does not belong to the class of quasi-periodic solutions; an unstable solution is sensitive to initial conditions. With all of these properties, we have more or less all what is required to define, today, a *chaotic* solution, a term that he did not used in 1963...

Lorenz started to remind the reader that his system is deterministic. It is dissipative since the divergence of the flow

$$\frac{\partial \dot{x}}{\partial x} + \frac{\partial \dot{y}}{\partial y} + \frac{\partial \dot{z}}{\partial z} = -\left(\sigma + b + 1\right) \tag{8}$$

is strictly negative for positive values of the parameters. He then showed that the aperiodic solutions are bounded by using the coordinate transformation

$$\left| \begin{array}{l} x' = x \\ y' = y \\ z' = z - R - \sigma. \end{array} \right.$$

He also investigated the stability of the singular points. For the point $S_0 = (0, 0, 0)$ and the parameter values as set above, he found three real eigenvalues, among which one is negative when $R > R_c = 1$, that is, when the threshold for instabilities is exceeded. For the two others singular points

$$\left| \begin{array}{l} x_\pm = \pm\sqrt{b(R-1)} \\ y_\pm = \pm\sqrt{b(R-1)} \\ z_\pm = R - 1, \end{array} \right. \tag{9}$$

the characteristic polynomial is

$$\lambda^3 + (\sigma + b + 1)\lambda^2 + (R + \sigma)b\lambda + 2\sigma b(R-1) = 0\,; \tag{10}$$

it has two complex conjugated eigenvalues [with Re $(\lambda) = 0$] when

$$R = \frac{\sigma(\sigma + b + 3)}{\sigma - b - 1}.$$

Lorenz showed that when $\sigma < b + 1$, the convection is always stable, and when $\sigma > b + 1$, the convection is unstable if R is sufficiently large. When $\sigma = 10$, and $b = \frac{8}{3}$, this conditions is $R > \frac{470}{19} = 24.74$. Typically, he applied Poincaré's program.

Lorenz also introduced the concept of *centrality*: a trajectory contained among its asymptotic orbits is said to be central. In a modern language, one would say that a central solution is a solution belonging to a set invariant under the action of the time. A stable and central trajectory is quasi-periodic or, in a reciprocal way, an aperiodic trajectory is unstable. If a solution is aperiodic, then two states differing by imperceptible quantities will evolve towards considerably different states. This property prevents from any long term prediction.

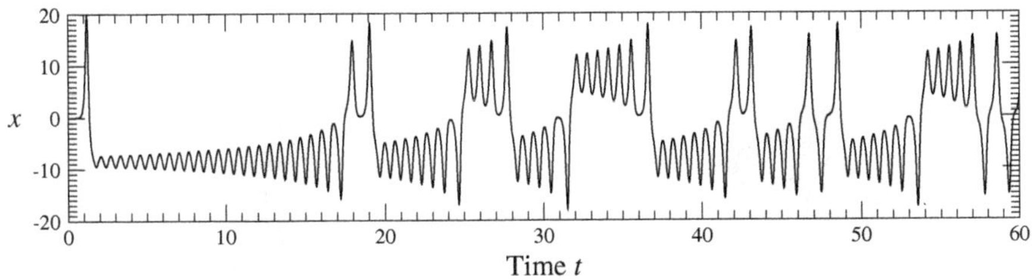

Fig. 9. Time series of the variable x produced by the three-dimensional system (7) for $R = 28$.

With his analog computer, Lorenz plotted a numerical integration of the system (7). He used initial conditions very close to the singular point S_0. Such a choice leads to a time series of variable x as shown in Fig. 9. Its aperiodic nature is clearly observable but Lorenz does not stop here his investigation. He developed also all what he learned from the nonlinear dynamical systems theory.

The first step for Lorenz was to plot the trajectory representing the time evolution of the system in the state space. He already mentioned that he was working in such a space but he did not plot them. This is not the first time that such a trajectory is plotted in the state space. In the 1950s, there are a few two-dimensional systems, more or less similar to the van der Pol system, investigated with trajectories plotted in the two-dimensional state space: among others, one could quote the 1926 paper by van der Pol,[49] the 1953 paper by Karl Bonhöffer,[71] or the 1961 paper by Richard FitzHugh, for instance.[72] Of course, since these solutions are produced by two-dimensional systems, they are periodic. In the 1953 paper by Paul Brock[73] or in the 1960 paper by Chihiro Hayashi's group,[74] the periodic solution is produced by a four-dimensional system since produced by a sinusoidally driven two-dimensional oscillator and plotted in the plane associated with the dissipative oscillator. Lorenz was thus the first to publish an aperiodic trajectory plotted in the state space. As reported in Abraham's chapter (see p. 111), Yoshisuke Ueda plotted an aperiodic solution in a plane projection of the state space associated with a sinusoidally driven van der Pol system but this plot was only excavated in 1993.[75]

Lorenz provided seven revolutions within the attractor, in two different plane projections: y-z and x-y as shown in Fig. 10. He did that using a non-conventional orientation of the x-axis (towards the bottom). He numbered

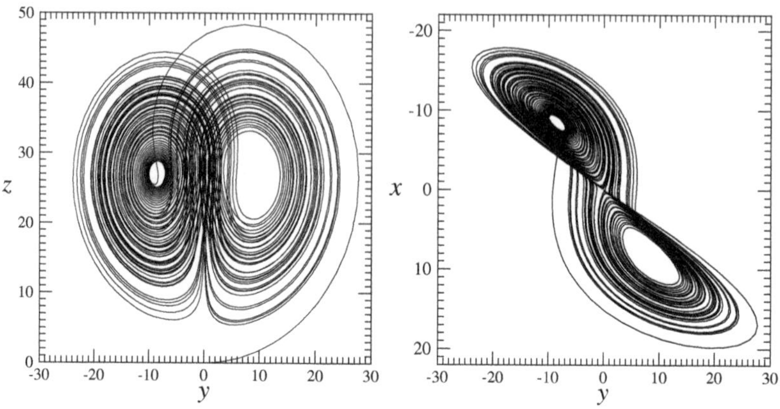

Fig. 10. Time series of the variable x produced by the three-dimensional system (7) for $R = 28$. Here there are more revolutions than originally plotted by Lorenz.

some states along the trajectory to indicate the direction of the flow: clockwise (anti-clockwise) for the left (right) wing in the y-z plane.

Lorenz, who showed that the volume visited by the trajectory is converging to zero, investigated what is called today the manifold in which they are structured by plotting isopleths, a technique very often used in meteorology. Isopleths are iso-curves, in the present case iso-x-values, plotted in the y-z plane (Fig. 11). Lorenz thus showed that the trajectory was developed within a pair of surfaces which merge in their bottom parts. As we will see, this is a topological approach and this *branched manifold* became a very useful tool to describe the architecture of chaotic attractors (see p. 446). In Lorenz's plot, the values along the axis are ten times larger what they should be: this is due to the fact that the Royal-McBee LGP-30 used by Lorenz was only printing three digits without any coma.

Then Lorenz used a technique developed by Norman Levinson — who was also at the MIT — for investigating a sinusoidally driven two-dimensional oscillator.[76] Starting from the equation investigated by Mary Lucy Cartwright (1900–1998) and John Edensor Littlewood (1885–1977),[77] Levinson introduced the piecewise linear driven oscillator

$$\epsilon\ddot{x} + \phi(x)\dot{x} + \epsilon x = b \sin t \tag{11}$$

where

$$\phi(x) = \left| \begin{array}{ll} +1 & \text{if } |x| > 1 \\ -1 & \text{otherwise.} \end{array} \right. \tag{12}$$

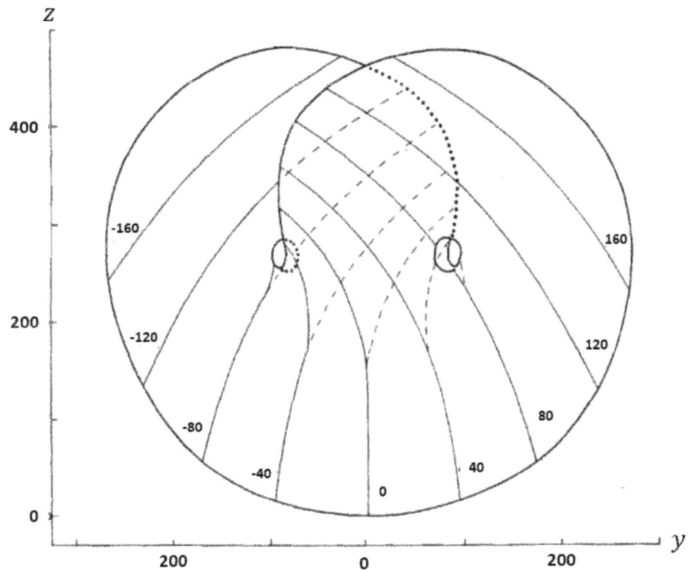

Fig. 11. Iso-x-values as drawn by Lorenz in the y-z plane for $R = 28$. Here the values reported along the axes are the actual values.

This system can be rewritten as the four-dimensional mixed system

$$\begin{cases} \dot{x} = y \\ \dot{y} = -x \dfrac{-\phi(x)\, y + u}{\epsilon} \\ \dot{u} = v \\ \dot{w} = -\omega^2 u, \end{cases} \tag{13}$$

thus appearing as a nonlinear dissipative oscillator driven by a linear conservative one. The system is therefore semi-dissipative (or semi-conservative).[78] Parameter b is hidden in the initial conditions $u_0 = 0$ and $v_0 = b$ (the two others are $x_0 = 0.5$ and $y_0 = 0$). There is no accurate values provided by Levinson for the parameters ϵ and b. He only states that $\epsilon > 0$ is a small constant and $0 < b < 1$.[79] Using a Poincaré section and the duration τ_n between two successive crossings with the Poincaré section, Levinson showed that the sequence $\{\tau_n\}$ is not periodic, leading to the conclusion that *the recurrent motions* [solution to the system (13)] *are either continuous periodic or else discontinuous recurrent.* Only two classes of solutions are thus here distinguished. As a background, Levinson used

Birkhoff's works,[80,81] an interesting contribution by Marie Charpentier to Birkhoff's curves,[82] and some results by Marston Morse on surface with negative curvature.[83]

With numerical simulations of the system (13), we only got periodic or quasi-periodic motions (Fig. 12). A first-return map to the Poincaré section

$$\mathcal{P}_{\rm L} = \left\{ (x_n, u_n, v_n) \in \mathbb{R}^3 \mid y_n = 1, \dot{y}_n > 0 \right\} \tag{14}$$

reveals the annular structure characteristic of quasi-periodic motions. Varying the initial conditions did not change the nature of the solution. We also varied the pulsation of the driving harmonic oscillator [Fig. 12(c)]: only periodic and quasi-periodic solutions were obtained. We observed the aperiodic nature of the duration between successive crossings with the Poincaré section [Fig. 12(d)].

Using an approach similar to those developed by Levinson, Lorenz used a first-return map to investigate how the trajectory stays in a given wing or makes a transition from one to the other. To do so, he used a first-return map to the maxima of the variable z. Before going further, it is important to note that such a choice does not correspond to a "classical" Poincaré section since the so-constructed Poincaré section has two components: one in each wing, see Fig. 10(a) where the two components of the Poincaré section are drawn. Two arguments could be provided to justify this: first, the order-2 rotation symmetry around the z-axis that the Lorenz system (7) has. This means that the vector field $\boldsymbol{f}(\boldsymbol{x})$ is equivariant, that is, $\boldsymbol{f}(\Gamma \cdot \boldsymbol{x}) = \Gamma \cdot \boldsymbol{f}(\boldsymbol{x})$ where Γ is a 3×3 matrix corresponding to the change of coordinates $(x, y, z) \mapsto (-x, -y, z)$.[84] The second argument is that the Lorenz attractor is bounded by a genus-3 torus:[85] Tsankov and Gilmore showed that an attractor bounded by a genus-g torus should be investigated with a $(g-1)$-component Poincaré section ($g > 1$). If constructing a first-return map can be well justified when the variable z of the Lorenz system is considered (and for some range of parameter values), this is far to be appropriate for most of the cases.[86,87]

Using the maxima of the variable z, Lorenz got what is sometimes called the "Lorenz map" [Fig. 13(a)]. It is a two-branch map with a non-smooth maximum (a cuspide). Typically, when the trajectory crosses the Poincaré section with a point located in the increasing branch, the trajectory is reinjected in the wing from which it is issued. Contrary to this, when the intersection with the Poincaré section is located in the decreasing branch, the trajectory switches from one wing to the other. Lorenz thus investigated the number of revolutions within a wing between two successive transitions.

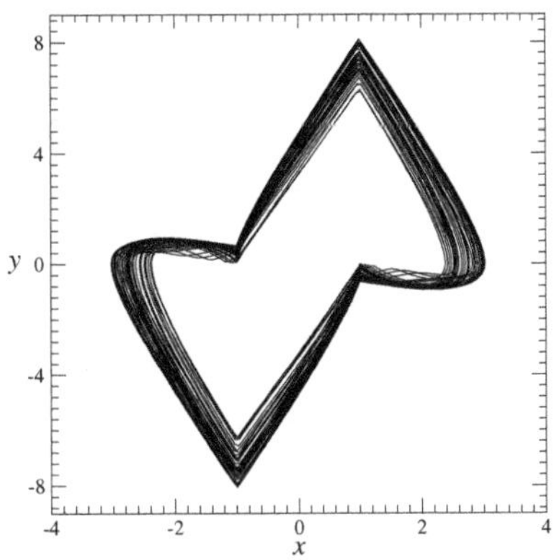

(a) State portrait

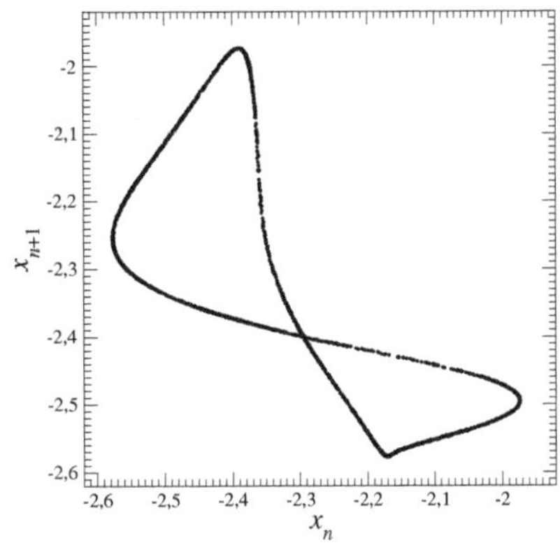

(b) First-return map

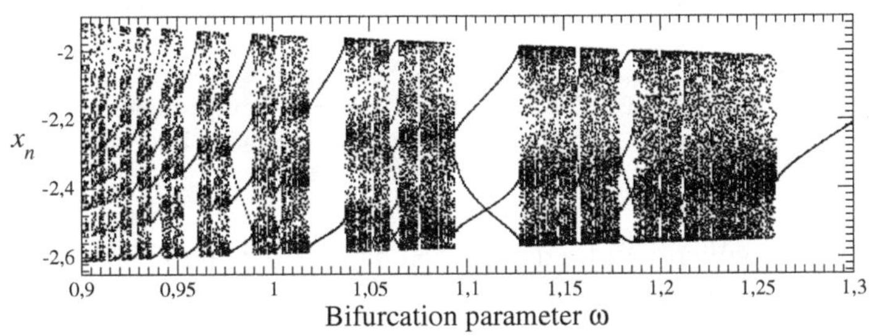

(c) Bifurcation diagram versus the pulsation ω

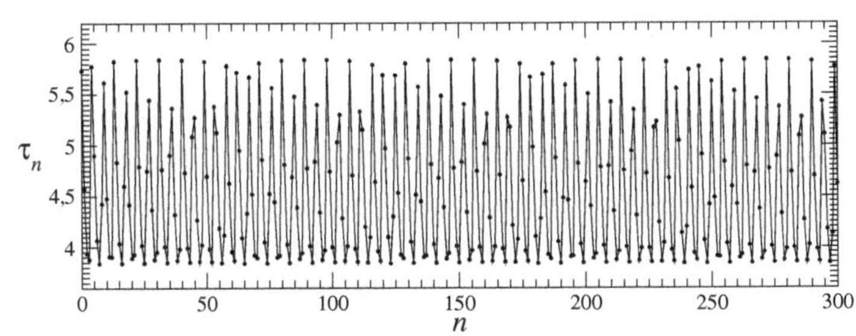

(d) Duration τ_n between two successive crossings

Fig. 12. Solution to the system (13) investigated by Levinson in his 1949 paper.
Notes: Parameter values: $\epsilon = 0.3$, $b = 0.3$ and $\omega = 1$ when not specified.

To do so, he idealized his map with the piecewise linear map

$$M_{n+1} = \left| \begin{array}{ll} \mu\, M_n & \text{if } M_n < \frac{1}{2} \\ \mu - \mu\, M_n & \text{otherwise,} \end{array} \right. \tag{15}$$

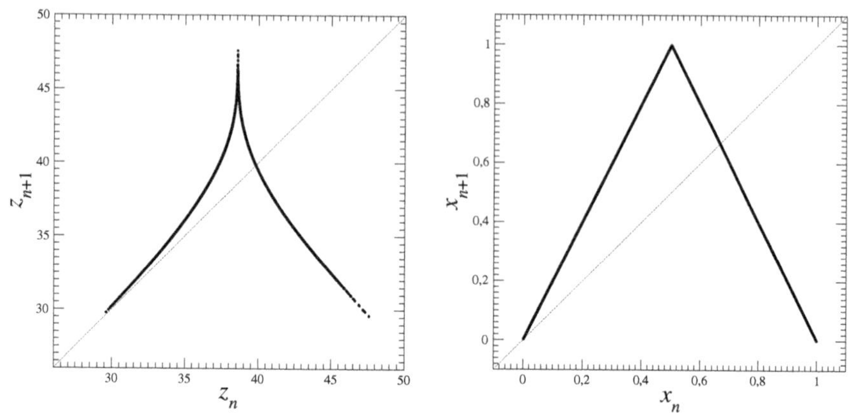

(a) Lorenz map (b) Tent map

Fig. 13. First-return map (a) computed from the Lorenz system (7) and (b) from the tent map (15) with $\mu = 1.999$. When $\mu = 2$, it is rather difficult to find initial conditions leading to an aperiodic orbit by iterating numerically the map (15).

with $\mu = 2$ and which will be designated as the tent map [Fig. 13(b)]. Such a map was first investigated by Stanislaw Ulam and Paul Stein.[88,89] It is important to note that Ulam was already aware about the ergodic nature of some solutions to simple deterministic systems in the 1950s.[90] He had a talk on this topic during the International Congress of Mathematicians (30 August–6 September 1950, Cambdrige) during a session entitled *Random processes in physics and communication*, just after Claude Shannon who introduced his entropy.[91] In particular, Ulam gave his views on some *specific problems connected with the relation between descriptions of physical phenomena by random processes and the theory of probabilities on the one hand, and the deterministic descriptions by methods of classical analysis in mathematical physics, on the other.* He stated, for instance, that by iterating the so-called Logistic map

$$x_{n+1} = \mu x_n (1 - x_n) \tag{16}$$

with $\mu = 4$, one obtains, for almost all x_0, the same ergodic distribution of iterates in the interval $[0, 1]$, a point he investigated in the 1940s with John von Neumann.[92]

Using the tent map, Lorenz investigated the recurrent map $M_n = m_n \pm 2^n M_0$ where m_n is even. He thus got the result that

(1) if $M_0 = \frac{u}{2^p}$ with u odd, then $M_{p-1} = \frac{1}{2}$ and $M_{p+1} = 0$: there is no convection for a denumerable set of trajectories;
(2) if $M_0 = \frac{u}{2^p v}$ where v is odd and $\frac{u}{v}$ irrational, then for $k > 0$, $M_{p+1+k} = \frac{u_k}{v}$ where u_k is even: this implies that there exists a $q > 0$ such that $\frac{u_{k+q}}{v} = \frac{u_k}{v}$: there is a denumerable set of periodic orbits.
(3) If M_0 is irrational, then for any $k > 0$, $M_{p+k} \neq M_p$: this means that there is non-denumerable set of aperiodic orbits.
(4) Let be two sequences issued from M_0 and $M_0' = M_0 + \epsilon$, then $M_k = M_k \pm 2^k$ and all the solutions are unstable; in other words, all solutions but a set of null measure are aperiodic.

By invetisgating the different numerical time series $\{M_n\}$, he found that the number N_p of possible period-p orbits. Thus, he got 1 period-1 orbit, 1 period-2 orbit, 2 period-3 orbits, and so on. These orbits are as follows.

	Transient	
Period-1 orbit	$\frac{1}{3}$	$\triangleright \frac{2}{3} \triangleright \frac{2}{3} \triangleright \frac{2}{3} \triangleright \frac{2}{3} \triangleright \frac{2}{3} \triangleright \frac{2}{3} \triangleright \cdots$
Period-2 orbit	$\frac{1}{5}$	$\triangleright \frac{2}{5} \triangleright \frac{4}{5} \triangleright \frac{2}{5} \triangleright \frac{4}{5} \triangleright \frac{2}{5} \triangleright \frac{4}{5} \triangleright \cdots$
Period-3 orbit	$\frac{1}{7}$	$\triangleright \frac{2}{7} \triangleright \frac{4}{7} \triangleright \frac{6}{7} \triangleright \frac{2}{7} \triangleright \frac{4}{7} \triangleright \frac{6}{7} \triangleright \cdots$
Period-3 orbit	$\frac{1}{9}$	$\triangleright \frac{2}{9} \triangleright \frac{4}{9} \triangleright \frac{8}{9} \triangleright \frac{2}{9} \triangleright \frac{4}{9} \triangleright \frac{8}{9} \triangleright \cdots$
Period-4 orbit	$\frac{1}{3 \times 5}$	$\triangleright \frac{2}{15} \triangleright \frac{4}{15} \triangleright \frac{8}{15} \triangleright \frac{14}{15} \triangleright \frac{2}{15} \triangleright \frac{4}{15} \triangleright \cdots$
Period-5 orbit	$\frac{1}{11}$	$\triangleright \frac{2}{11} \triangleright \frac{4}{11} \triangleright \frac{8}{11} \triangleright \frac{6}{11} \triangleright \frac{10}{11} \triangleright \frac{2}{11} \triangleright \cdots$

For instance, the two period-3 orbits differ in the fact that there are two $M_n > \frac{1}{2}$ in the first orbit, and only one in the second orbit: these two period-3 orbits are therefore clearly distinct. Then Lorenz located these periodic orbits in the first-return map built on the maxima of the variable z: he got

(1) period-1 orbit: $z_n \in \{39.8\}$;
(2) period-2 orbit: $z_n \in \{37.7, 41.0\}$;
(3) period-3 orbit: $z_n \in \{36.9, 39.1, 41.4\}$ and $z_n \in \{36.2, 38.0, 41.9\}$.

He checked numerically that these periodic orbits are *presumably unstable.* He thus performed a numerical extraction of the unstable periodic orbits embedded within the asymptotic behavior. This was then developed during the next three decades.[93]

Lorenz concluded his work on periodic orbits by stating that *all sequences except a set of measure zero are therefore nonperiodic, and correspond to nonperiodic trajectories.* In other words, trajectories which can be considered as representing a *deterministic nonperiodic flow* form a nondenumerable set. With such a conclusion, Lorenz revisited his description by the "*merging surface*" obtained with the isopleths. He showed in particular that the surface is in fact an infinite complex of surfaces (today, we would say that there is a foliation). He mentioned the description of those by using the decimal expansion of real numbers by using only 0 and 1. The sequence of 0 and 1 *corresponding to a particular surface contains a history of the trajectories lying in that surface*: this is exactly the approach which was used a few years later to describe the structure of periodic orbits in unimodal maps.[94] Lorenz used 0 (1) for the left (right) wing as used, for instance, by Colin Sparrow with the symbols L and R.[95] According to such a symbolic dynamics, aperiodic solutions are nondenumerable, while periodic corresponding to repeating decimal expansions, form a denumerable set.

Applying all these results to the atmosphere, which is *ostensibly nonperiodic*, it is not possible to make long-term prediction unless *the present conditions are know exactly.* Lorenz therefore concluded that *in view of the inevitable inaccuracy and incompleteness of weather observations, precise very-long-range forecasting would seem to be non-existent.*[70] Lorenz left open the question whether long-range could be a few days or a few centuries. Although Lorenz wrote[96]

> Thanks to the interest of Jule Charney, who by then had become a colleague of mine at M.I.T., my paper soon gained considerable attention among meteorologists.

Let us now report the first fifteen quotations of the seminal 1963 Lorenz's work. In what follows, if not specified, Lorenz's 1963 paper is meant when Lorenz is quoted.

(1) Lorenz's paper is quoted in a "hydrographische Bibliographie" published in the *Deutsche Hydrographische Zeitschrift* in the early 1964;[97]

(2) Quoting Landau[98] and Stuart[99] for the need of a nonlinear theory for flow presenting an exponential growth of disturbance, Aihara investigated the long term variations of large-scale atmospheric motions with the help of a simple two-layer model;[100] he quoted Lorenz for stating that his configuration does not provide nonperiodic motions.
(3) Starting from the "maximum simplification" of the equations by Lorenz, that is, a set of three ordinary differential equations, adding a fourth one component, Douglas Lilly investigated numerical instabilities that he tried to stabilize by improving the numerical scheme.[101]
(4) Questioning the simplification of the governing equations, Lee Segel investigated the degree at which nonlinear terms should be cut without strongly modifying the dynamics;[102] Lorenz is quoted as "the most interesting" study with a simplified model for meteorological flow.
(5) Li Peng, from the MIT, investigated a three-level quasi-geo- strophic model:[103,104] Lorenz is quoted for his procedure for establishing his model and for his numerical procedure. Peng present nonperiodic time series — once qualified as being irregular — but remained focused on the hydrodynamical problem and missed the dynamical one. Peng thanked Lorenz, Starr and Reginald E. Newell for helpful discussions.
(6) Under the supervision of Phillips, and then of Lorenz, John Young spent his Ph.D. thesis in studying a quasi-geostrophic two-level mode (19 variables):[105] he distinguished periodic and non-periodic behaviors, the latter being analyzed using power spectra. The fifth chapter is entitled "non-periodic flow" and contains some non-periodic time series.
(7) Richard Somerville investigated the Boussinesq equations for two-dimensional rectangular channel;[106] he quoted Lorenz for his simplifications of the governing equations and for his numerical procedure. There is no time series.
(8) In a simplified model, Louis Berkofksy and Ralph Shapiro (Air Force Cambridge research laboratories) investigated the solar-weather relationships.[107] Lorenz is quoted for having showed that the difference between two trajectories is "as large as the difference between arbitrary initial states," thus only focusing on a class of behaviors he described.
(9) Using the maximum simplification of the equations, Edward Epstein followed Lorenz (and Saltzmann) to produce a set of governing equations.[108] In particular, he revisited the minimum hydrodynamical

equations introduced by Lorenz.[40] His purpose was to push the idea that "since the probabilistic nature of the initialization is unavoidable, then so also is the probabilistic nature of the prediction." Among the six papers by Lorenz he quoted,[40, 70, 109–112] the one published in 1963 is rather anecdotal for Epstein.

(10) A. E. Cook and P. H. Roberts investigated the two-disc dynamo introduced by Rikitake.[113] They quoted Lorenz for similarities between the Rikitake system and the Lorenz one; they also used plane projection of the state space — the first non-periodic state portrait published since Lorenz's paper — and for constructing a "limit surface" similar to the surface obtained by using isopleths.[114] This is actually the first paper inheriting some uncommon techniques to investigate the dynamics from Lorenz.

(11) Thomas Gleeson (Florida State University) stimulated the work by Epstein previously discussed. His 1970 paper[115] is an extension of what he published earlier. Lorenz is quoted for having pushed the relation between statistical mechanics and meteorology.

(12) Reference[116]

(13) Douglas Dockery, another Ph.D. student supervised by Lorenz investigated a dish-pan model of the atmosphere,[117] following some Fultz and coworkers.[35] Lorenz is quoted for having demonstrated that "the atmosphere is an unstable system such that the initial errors will gro until they are the size of the errors for randomly chosen initial conditions." The period-doubling of errors is estimated to be 1.7 days, to compared to the 2.5 days obtained by Lorenz. He used a kind of symbolic dynamics.

(14) During a summer school devoted to *Dynamic meteorology*, Norman Phillips (MIT) gave a lecture entitled *Principles of large scale numerical weather prediction*,[118] during which he quoted Lorenz for his simplification of governing equation and his contribution to the predictability the atmospheric motions. Note that Lilly also gave a lecture at this summer school and also quoted Lorenz for the fact that "any introduced uncertainty continues to grow, so that two initial states differing by some infinitesimal amount will diverge until they differs as much as two randomly chosen states having the same statistical properties.[119]

(15) Richard Naistat and John Young (Lorenz's pupil) — both at the university of Wisconsin — quoted Lorenz for his double-approximation

forward difference.[120] They were using a linear one-dimensional model for studying the Saint-Patrick's day storm of 1965.

(16) J. S. Green (Imperial College, London) investigated a two-dimensional incompressible motion sinusoidally driven.[121] The purpose was to develop a theory for which turbulence is seen as small amplitude disturbances superimposed to a laminar flow in the spirit of Stuart.[99] It seems that Lorenz is here quoted under the suggestion of one of the referees.

(17) Lars Moen (Swedish Meteorological and Hydrological Institute) investigated a spectral quasi-geostrophic model solved by using orthogonal functions in the horizontal direction but finite differences in the vertical one:[122] the aim is to study the development of cyclone. Lorenz is quoted for his techniques to derive the governing equations and its numerical procedure to solve them.[40, 70, 123, 124]

(18) John McLaughlin and Paul Martin studied the development of turbulence for a low Prandtl number fluid in the Rayleigh–Bénard geometry.[125] Their purpose is to test the Ruelle–Takens scenario for the transition to turbulence.[126] Lorenz is quoted for his system which provides an example of subcritical Hopf bifurcation (investigated in details by McLaughlin and Martin). Their subsequent paper[127] is devoted to the study of the Lorenz system and to a larger-dimensional truncation: a significant part of the paper remind the reader what Lorenz did. The Lorenz surface, drawn using isopleths (Fig. 11), is proposed as "one of the simplest possible examples of the *strange attractor* discussed by Ruelle and Takens." They concluded that the 1963 Lorenz system present a subcritical Hopf bifurcation while larger-dimensional models — as the 12-dimensional Lorenz one[41] — present a supercritical Hopf bifurcation which supports the Ruelle–Takens scenario.

Lorenz had an impact in meteorology for pushing to a maximum simplification of the equations governing atmospheric motion to better understand their nature and for investigating the notion of predictability of deterministic systems with a great sensitivity to initial conditions in terms of statistics. He was often quoted for his numerical procedure for integrating the simplified equations. He is also recognized for having pointed out that weather forecasting is only reliable for short-term prediction (over a few days). This is only in the mid-70s that his innovative approach for characterizing the solutions of dynamical systems was recognized and developed: he provided a major link between Poincaré's and Birkhoff's heritage and chaos theory. As explained in Otto Rössler chapter (p. 159), his 1963 paper was crucial for the development of his characterization of chaotic attractors.

5. Deducing Climate from Simple Models

The earth atmosphere is presumably governed by a set of physical deterministic laws, which most often as are set of partial differential equations combined with appropriate boundary conditions. Lorenz thus assumed that the climate, that is, a set of long-term statistical properties of the atmosphere, *is determined by the same system of equations.*[128] He thus transformed the general problem in climatology of deducing the climate from their governing equations into the fundamental mathematical problem of deducing the statistics of solutions to a given system from the equations themselves. Due to the frequent occurrence of the expression "*change of climate*", he thus stated that *many investigators are concerned with averages over long but finite time intervals.* Of course, such a problem is limited to stationary systems for which solutions possess long-term statistical properties. Lorenz exhibited that there is no evidence that the atmospheric system is stationary and, consequently, *there is no a priori reason why a climate need exist.*[128] Indeed, the weather of the current century resemble to the weather of the past century, but this is certainly not the case of the weather over the 12,000 years, particularly when an ice age flourished. And even if the atmospheric dynamics would be stationary, the sensitivity to initial conditions would be sufficient to allow various possible climates, for which the corresponding solution would be unstable. According to Lorenz's approach, a climate would exist if it corresponds to a stable solution. He thus turned the problem in investigating the probability for a stable solution.

Starting from the assumption that a climate can be identified when it is associated with a periodic or a quasi-periodic solution, Lorenz thus investigated the prevalence of unstable solution in the simplest dynamical system under the form of a recurrence map

$$X_{n+1} = f(X_n). \tag{17}$$

To ensure that the solution is bounded, such a map must have at least a periodic point defined as

$$X_n = f(X_n). \tag{18}$$

The simplest function f is thus a quadratic map. Depending on the boundary conditions — a very typical approach for a hydrodynamicist — the map can be one of the maps as follows.

(1)	$X_{n+1} = aX_n - X_n^2$	where $X_n \in [0;4]$	if $0 \leqslant a \leqslant 4$;
(2)	$X_{n+1} = 4b\left(X_n - X_n^2\right)$	where $X_n \in [0;1]$	if $0 \leqslant b \leqslant 1$;
(3)	$X_{n+1} = c - \frac{1}{2}X_n^2$	where $X_n \in [-4;+4]$	if $0 \leqslant c \leqslant 4$.

Obviously, these three maps are smooth unimodal maps whose dynamics are orbitally equivalent for appropriate parameter values (they are topologically conjugate).

The stability of the solutions to one of these maps, say the first one, is investigated as follows. If $\{X_0, X_1, \ldots, X_k\}$ and $\{Y_0, Y_1, \ldots\}$ are two solutions and if $Y_n = X_n + \epsilon_n$ then $\epsilon_{n+1} = \lambda_n \epsilon_n - \epsilon_n^2$ where $\lambda_n = a - 2X_n = f'(X_n)$. When ϵ_0 is sufficiently small, $\Lambda = \prod_{n=0}^{k-1} \lambda_n$ but this does not apply for any type of solution because this quantity diverges when the solutions are aperiodic. This marker does not have the interesting property that the Lyapunov exponents, introduced a decade later,[129] have, since they remain bounded even for an aperiodic solution. In anyway, Lorenz used his exponent Λ for detecting stable solution when $|\Lambda| < 1$. He thus established that for $0 \leqslant a \leqslant 1$, the period-1 solution $X = 0$ is stable, for $1 \leqslant a \leqslant 3$, the period-1 solution $X = a - 1$ is stable. Thus two different climates are stable depending on the parameter value. Increasing further the parameter a, the period-1 solution $X = \frac{a+1}{2}$ is stable and corresponds to another stable climate. This is today viewed in one sight with a bifurcation diagram as plotted in Fig. 14. The first bifurcation diagram numerically computed was published by Robert May in 1976.[130]

With the exponential divergence

$$\Lambda = \lim_{n \to +\infty} \frac{1}{N} \sum_{n=0}^{N-1} \ln |\lambda_n| \tag{19}$$

for defining the characteristic exponent, Lorenz would have obtained a now classical Lyapunov exponent (Fig. 14).

Since his exponent Λ was not bounded for aperiodic solutions (unstable climates), Lorenz used the mean value

$$\overline{X}_n = \lim_{N \to +\infty} \frac{1}{N} \sum_{n=0}^{N-1} X_n \tag{20}$$

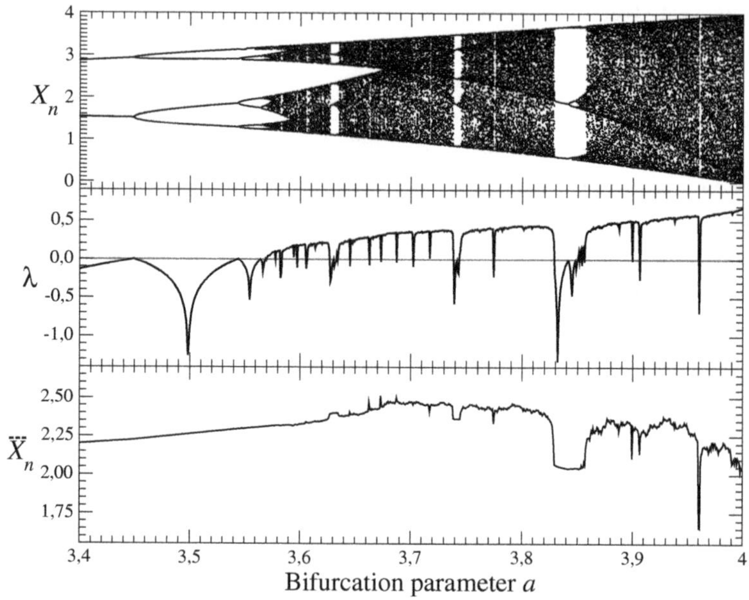

Fig. 14. Bifurcation diagram as a function of parameter a for the quadratic map $X_{n+1} = aX_n - X_n^2$ investigated by Lorenz in 1964. The Lyapunov exponent λ and the mean value $\overline{X}$ are also plotted as a function of parameter a.

Compared to the Lyapunov exponent, the mean value $\overline{X}_n$ presents the advantage to be be a continuous function of the parameter a for a given periodic solution: it is thus characteristic of the considered periodic solution. Nevertheless, Lorenz stated that the problem of finding how $\overline{X}_n$ depends on the parameter a was an open problem. It is still the case today.

Combining analytical computations with numerical simulations, Lorenz investigated the range for the periodic windows (see Table 1). He also establish that

(1) Every pair of a-values for which the solutions are aperiodic is separated by a continuum of periodic windows;
(2) the set of a-values for which the solution is periodic forms an almost everywhere dense set;
(3) the set of a-values for which the solution is aperiodic forms a nowhere dense set, that is, arbitrarily close to any a-value, there is a a-value for which the solution is periodic.
(4) there are no continum of a-value for which the solution is nonperiodic.

Table 1. Range of a-values for the first five periodic windows of the quadratic map investigated by Lorenz.

Period	$a_{\text{inf}} < a < a_{\text{sup}}$	Δa	$\overline{X}$
1	$1 < a < 3$	2	1.000
2	$3 < a < 3.449$	0.449	2.118
3	$3.8284 < a < 3.8401$	0.0107	2.059
4	$3.96010 < a < 3.96145$	0.00135	1.661
5	$3.99002583 < a < 3.990281$	0.0000227	1.391

Lorenz then concluded that almost a-values are associated with a periodic solution in the sense that there is a null probability that a randomly selected a-value corresponds to nonperiodic solution. He also found that the widths of the intervals for a period-p orbit decrease very rapidly as p increases. He also remarked that when $0 \leqslant a \leqslant 3.449$, all the solutions are periodic, but when $3.9 \leqslant a \leqslant 4$, a large majority of a-values correspond to aperiodic solutions.

To conclude Lorenz returned to the experiments conducted by Hide[37] and Fultz.[35] He established a link between the sequence of periodic and aperiodic orbits with the sequence of behaviors observed in the experimental flows. Thus, $X = 0$ would correspond to no wave (observed for $0 \leqslant a \leqslant 1$), the waves would exist for $1 \leqslant a \leqslant 3$ with a constant energy. Setting $a \in [3; 3.449]$ would correspond to oscillating waves, when $3.449 \leqslant a \leqslant 4$ would be associated with aperiodic waves or, equivalently, with waves vacillating irregularly. He remarked that these features resemble certain ones observed in hydrodynamical systems as in a Rayleigh–Bénard convection or in a flow with the transition from laminar to turbulent flow as the Reynolds number is increased.[128] Lorenz felt that

> this resemblance is no mere accident, but that the difference equation captures much of the mathematics, even if not the physics, of the transitions from one regime of flow to another, and, indeed, of the whole phenomenon of instability.

Nevertheless, for Lorenz, the sequence of bifurcations of the real atmosphere would be well captured by the quadratic map when $a \in [3.449, 4]$, that is, over the interval for which the aperiodic solutions are prevalent, confirming his 1963 result that the unstable solutions are prevalent, preventing any long-range-prediction. Since unstable solutions are better investigated with numerical integration, Lorenz is firmly convinced that numerical simulations must be used for deducing climate.

From a methodological point of view, Lorenz understood that, if a computing machine cannot prove a theorem, it can suggest a proposition to be proven. To support his view, he quoted Ulam who promoted the use of computing machine as a heuristic aid to mathematical reasoning.[88]

6. Conclusion

Among the first fifteen papers quoting the 1963 Lorenz paper, 11 are devoted to atmospheric motion (and/or meteorology), 2 about hydrodynamics, and 2 about geophysics. From them, and apart the two main contributions previously exhibited, Lorenz was quoted for (i) the link between nonperiodic solutions and the statistics for weather forecasting, and (ii) the dynamical treatment of these solutions, although that this contribution only came in the 1970s. Nonperiodic time series were plotted by Peng (1965) and Young (1966), both from MIT. This is only in 1971 that Cook and Roberts proposed plane projections of a nonperiodic state portrait, followed in 1975 by McLaughlin and Martin. From the latter work, the community working in fluid dynamics retained that the truncation used by Lorenz was too severe for providing realistic solutions. Nevertheless, these last authors were the first to make the link between the concept of of strange attractors introduced by Ruelle and Takens[126] and the now-called Lorenz attractor. It is interesting to note that Lorenz and Ruelle attended in 1971 (July 15–21) at the symposium held at the University of California at San Diego. Lorenz discussed his view about a deterministic turbulence which is nonpredictable due to "practical purposes".[131] He did present neither the solutions to his 1963 system, nor his strange attractor. Ruelle gave a talk related to his work with Takens on turbulence seen as a "strange attractor".[132] The link between nonperiodic solutions to the Lorenz system and strange attractor was missed.

McLaughlin and Martin had a strong influence, for instance, on Arkady Pikovsky (p. 377). Then, Art Winfree published a short paper in *Nature* quoting Lorenz for "impredictable aperiodic solution"[133] and, more importantly, in sending Lorenz's paper to Otto (see p. 180 and Ref. 134). Then Ruelle published his paper on the Lorenz attractor,[135] and popularized it among hydrodynamicists.

Lorenz thus produced the seeds of some major developments performed during the next three decades. For the topological approach, the surface obtained with the isopleths led to branched manifolds[136,137] and templates.[138–140] Using symbolic dynamics, this was already initiated by

Hadamard,[141] Marston Morse and Gustav A. Hedlund,[142] or Norman Levinson;[76] this was further developed by Nicholas Metropolis, Paul Stein and Myron Stein,[143] Pierre Collet and Jean-Pierre Eckmann,[94] or Bernard Derrida, Annie Gervois and Yves Pomeau,[144] among others. For the first time, Lorenz plotted an aperiodic trajectory in the state space. The computation of a first-return map also provides an important element in the architecture of chaotic attractors, particularly for strongly dissipative systems.[145] In all of that, Lorenz opened the door towards what became by the end of the 1970s, the chaos theory and was developed over at least two decades, 20 years later!

References

1. H. Poincaré, Sur les courbes définies par une équation différentielle (1), *Journal de Mathématiques Pures et Appliquées, III.* **7**, 375–422 (1881).
2. H. Poincaré, Sur les courbes définies par une équation différentielle (2), *Journal de Mathématiques Pures et Appliquées, III.* **8**, 251–296 (1882).
3. H. Poincaré, Sur les courbes définies par les équations différentielles (3), *Journal de Mathématiques Pures et Appliquées, IV.* **1**, 167–244 (1885).
4. H. Poincaré, Sur les courbes définies par les équations différentielles (4), *Journal de Mathématiques Pures et Appliquées, IV.* **2**, 151–217 (1886).
5. G. W. Hill, Researches on the lunar theory, *American Journal of Mathematics.* **1**, 5–26 (1878).
6. H. Poincaré, Sur le problème des trois corps et les équations de la dynamique, *Acta Mathematica.* **13**, 1–270 (1890).
7. H. Poincaré, *Les méthodes nouvelles de la mécanique céleste III.* Gauthier-Vilard, Paris (1899).
8. L. P. Shilnikov, A case of the existence of a denumerable set of periodic motions, *Soviet Mathematics Doklady.* **6**, 163–166 (1965).
9. S. Gonchenko, A. Kazakov, D. Turaev, and A. L. Shilnikov, Leonid shilnikov and mathematical theory of dynamical chaos, *Chaos.* **32**(1), 010402 (2022). doi: 10.1063/5.0080836.
10. H. Poincaré, Analysis situs, *Journal de l'École Polytechnique.* **1**, 1–121 (1895).
11. H. Poincaré, *Science and Method.* Thomas Nelson & Sons (1909).
12. L. F. Richardson, The approximate arithmetical solution by finite differences of physical problems involving differential equations, with an application to the stresses in a masonry dam, *Philosophical Transactions of the Royal Society A.* **210**, 459–470 (1911).
13. W. F. Sheppard, Central-difference formulae, *Proceedings of the London Mathematical Society.* **31**, 449–488 (1899).
14. V. Bjerknes, Das Problem der Wettervorhersage, betrachtet vom Standpunkte der Mechanik und der Physik, *Meteorologische Zeitschrift.* **21**, 1–7 (1904).

15. V. Bjerknes, Meteorology as an exact science, *Monthly Weather Review.* **42**, 11–14 (1914). doi: 10.1175/1520-0493(1914)42⟨11:MAAES⟩2.0.CO;2.
16. C. Runge, Über die numerische Auflösung von Differentialgleichungen, *Mathematische Annalen.* **46**, 167–178 (1895).
17. L. F. Richardson, *Weather Prediction by Numerical Process.* Cambridge University Press, Cambridge, England (1922).
18. E. N. Lorenz, The bulletin interviews professor Edward N. Lorenz, *World Meteorological Organization Bulletin.* **45**, 111–120 (1996).
19. T. N. Palmer, Edward Norton Lorenz, 23 May 1917 — 16 April 2008, *Biographical Memoirs of Fellows of the Royal Society.* **55**, 139–155 (2009). doi: 10.1098/rsbm.2009.0004.
20. K. Emmanuel, *Edward Norton Lorenz, 1917-2008.* National Academy of Sciences, Washington, D. C. (2011).
21. H. Poincaré, Sur un théorème de géométrie, *Rendiconti del Circolo Matematico di Palermo.* **33**, 375–407 (1912). doi: 10.1007/BF03015314.
22. G. D. Birkhoff, Proof of poincarés's geometric theorem, *Transactions of the American Mathematical Society.* **14**, 14–22 (1913).
23. C. H. Nielsen, Hurd Willett, forcaster extraordinaire, *Weatherwise.* **46**(4), 38–44 (1993). doi: 10.1080/00431672.1993.993025.
24. H. G. Houghton, *Physical Meteorology.* MIT Press, Cambridge, MA (1985).
25. B. Haurwitz, *Dynamic Meteorology.* McGraw-Hill Book Company, New York (1941).
26. J. G. Charney, R. Fjörtoft, and J. von Neumann, Numerical integration of the barotropic vorticity equation, *Tellus.* **2**(4), 237–254 (1950). doi: 10.1111/j.2153-3490.1950.tb00336.x.
27. J. G. Charney, The dynamics of long waves in a baroclinic westerly current, *Journal of Atmospheric Sciences.* **4**(5), 136–162 (1947). doi: 10.1175/1520-0469(1947)004⟨0136:TDOLWI⟩2.0.CO;2.
28. C.-G. Rossby, On the solution of problems of atmospheric motion by means of model experiments, *Monthly Weather Review.* **54**, 237–240 (1926).
29. V. P. Starr, An essay on the general circulation of the earths atmosphere, *Journal of Atmospheric Sciences.* **5**(2), 39–43 (1948). doi: 10.1175/1520-0469(1948)005⟨0039:AEOTGC⟩2.0.CO;2.
30. C.-G. Rossby and V. P. Starr, Interpretations of the angular-momentum principle as applied to the general circulation of the atmosphere, *Journal of Atmospheric Sciences.* **6**(4), 288–288 (1949). doi: 10.1175/1520-0469(1949)006⟨0288:IOTAMP⟩2.0.CO;2.
31. C.-G. Rossby, The scientific basis of modern meteorology. In *Yearbook of Agriculture – Climate and Men.* United States Department of Agriculture, Washington, D.C. (1941), pp. 599–655.
32. F. Vettin, Ueber den aufsteigenden Luftstrom, die Entstehung des Hagels und der Wirbel-Stürme, *Annalen der Physik.* **178** (10), 246–255 (1857). doi: 10.1002/andp.18571781005.
33. H. von Helmholtz, On a theorem relative to movements that are geometrically similar, together with an application to the problem of steering balloons. The Smithsonian Institution, Washington (1891), pp. 67–77.

Originally published in the Monatsberichte der Königlich Preussischen Akademie der Wissenschaften zu Berlin, 1873.
34. D. Fultz, Non-dimensional equations and modelling criteria for the atmosphere, *Journal of Atmospheric Sciences.* **8**(4), 262–267 (1951). doi: 10.1175/1520-0469(1951)008⟨0262:NDEAMC⟩2.0.CO;2.
35. D. Fultz, R. R. Long, G. V. Owens, W. Bohan, R. Kaylor, and J. Weil, *Studies of Thermal Convection in a Rotating Cylinder with Some Implications for Large-Scale Atmospheric Motions.* American Meteorological Society, Boston, MA (1959). doi: 10.1007/978-1-940033-37-2.
36. R. Hide, Some experiments on thermal convection in a rotating liquid, *Quarterly Journal of the Royal Meteorological Society.* **79**, 161– (1953). doi: 10.1002/qj.49707933916.
37. R. Hide and R. Stoneley, An experimental study of thermal convection in a rotating liquid, *Philosophical Transactions of the Royal Society of London A.* **250**(983), 441–478 (1958). doi: 10.1098/rsta.1958.0004.
38. P. L. Read, Raymond hide. 17 May 1929—6 September 2016, *Biographical Memoirs of Fellows of the Royal Society.* **67**, 191–215 (2019). doi: 10.1098/rsbm.2019.0016.
39. E. N. Lorenz, Available potential energy and the maintenance of the general circulation, *Tellus.* **7**(2), 157–167 (1955). doi: 10.1111/j.2153-3490.1955.tb01148.x.
40. E. N. Lorenz, Maximum simplification of the dynamic equations, *Tellus.* **12**(3), 243–254 (1960). doi: 10.1111/j.2153-3490.1960.tb01307.x.
41. E. N. Lorenz, The statistical prediction of solutions of dynamic equations. In *Proceedings of the International Symposium on Numerical Weather Prediction*, Tokyo, Japan (1962), pp. 629–635.
42. K. Bryan and M. D. Cox, A numerical investigation of the oceanic general circulation, *Tellus.* **19**(1), 54–80 (1967). doi: 10.1111/j.2153-3490.1967.tb01459.x.
43. K. Bryan, A numerical investigation of certain features of the general circulation, *Tellus.* **11**, 163–174 (1959). doi: 10.3402/tellusa.v11i2.9302.
44. H. Nyquist, Certain factors affecting telegraph speed, *Bell System Technical Journal.* **3**(2), 324–346 (1924). doi: 10.1002/j.1538-7305.1924.tb01361.x.
45. G. D. Birkhoff, *Dynamical Systems.* American Mathematical Society, New York (1927).
46. V. V. Nemytskii and V. V. Stepanov, *Qualitative Theory of Differential Equations.* Princeton University Press (1960). Translation of the 2nd Russian edition, 1949.
47. A. A. Andronov, Les cycles limites de poincaré et la théorie des oscillations auto-entretenues, *Comptes-Rendus de l'Académie des Sciences.* **189**, 559–561 (1929).
48. A. A. Andronov, A. A. Vitt, and S. E. Khaikin, *Teoriya kolebanii.* Fizmatgiz, Moscow (1959). *Theory of Oscillators.* Pergamon Press, 1966.
49. B. van der Pol, On "relaxation-oscillations", *Philosophical Magazine.* **2**(11), 978–992 (1926).

50. K. F. Bonhöffer, Activation of passive iron as a model for the excitation of nerve, *Journal of General Physiology.* **32**(1), 69–91 (1948).
51. C. Hayashi, Forced oscillations with nonlinear restoring force, *Journal of Applied Physics.* **24**(2), 198–207 (1953). doi: 10.1063/1.1721238.
52. C. Hayashi, *Nonlinear Oscillations in Physical Systems.* McGraw-Hill (1964).
53. E. N. Lorenz, The predictability of hydrodynamic flow, *Transactions of the New York Academy of Sciences.* **25**, 409–432 (1963).
54. V. K. Mel'nikov, On lines of force in a magnetic field, *Doklady Akademii Nauk SSSR.* **144**, 747–750 (1962).
55. V. K. Mel'nikov, On the stability of a center for time-periodic perturbations, *Trudy Moskovskogo Matematicheskogo Obshchestva.* **12**, 3–52 (1963) (in Russian).
56. V. K. Mel'nikov, Qualitative description of strong resonance in a nonlinear system, *Doklady Akademii Nauk SSSR.* **148**, 1257–1260 (1963).
57. B. Saltzman, Finite amplitude free convection as an initial value problem–I, *Journal of the Atmospheric Sciences.* **19**(4), 329–341 (1962). doi: 10.1175/1520-0469(1962)019⟨0329:FAFCAA⟩2.0.CO;2.
58. B. Saltzman, Equations governing the energetics of the larger scales of atmospheric turbulence in the domain of wave number, *Journal of the Atmospheric Sciences.* **14**, 513–523 (1957). doi: 10.1175/1520-0469(1957)014⟨0513:EGTEOT⟩2.0.CO;2.
59. K. A. Maasch, R. J. Oglesby, and A. Fournier, Barry Saltzman and the theory of climate, *Journal of Climate.* 2141–2150 (2005). doi: 10.1175/JCLI3383.1.
60. N. A. Phillips, The general circulation of the atmosphere: A numerical experiment, *Quarterly Journal of the Royal Meteorological Society.* **82**(352), 123–164 (1956). doi: 10.1002/qj.49708235202.
61. A. N. Kolmogorov, Dissipation of energy in locally isotropic turbulence, *Doklady Akademii Nauk SSSR.* **32**, 16–18 (1941).
62. A. N. Kolmogorov, The local structure of turbulence in incompressible viscous fluids for very large Reynolds number, *Doklady Akademii Nauk SSSR.* **30**, 9–13 (1941).
63. G. K. Batchelor, *The Theory of Homogeneous Turbulence.* Cambridge University Press (1953).
64. R. Fjørtoft, On the changes in the spectral distribution of kinetic energy for twodimensional, nondivergent flow, *Tellus.* **5**(3), 225–230 (1953). doi: 10.1111/j.2153-3490.1953.tb01051.x.
65. O. Reynolds, On the dynamical theory of incompressible viscous fluids and the determination of the criterion, *Philosophical Transactions of the Royal Society of London.* **186**, 123–164 (1895). doi: 10.1098/rsta.1895.0004.
66. A. Oberbeck, über die wärmeleitung der Flüssigkeiten bei Berücksichtigung der strömungen infolge von Temperaturdifferenzen, *Annalen der Physik.* **243**(6), 271–292 (1879). doi: 10.1002/andp.18792430606.
67. J. Boussinesq, *Théorie Analytique de la Chaleur*, Vol. 2. Gauthier-Villars, Paris, France (1903).

68. H. Bénard, Les tourbillons cellulaires dans une nappe liquide, *Revue Générale des Sciences Pures et Appliquées.* **11**, 1261–1271, 1309–1328 (1900).
69. L. Rayleigh, On the convective currents in a horizontal layer of fluid when the higher temperature is on the under side, *Philosophical Magazine VI.* **32**, 529–546 (1916).
70. E. N. Lorenz, Deterministic nonperiodic flow, *Journal of the Atmospheric Sciences.* **20**, 130–141 (1963). doi: 10.1175/1520-0469(1963)020⟨0130:DNF⟩2.0.CO;2.
71. K. F. Bonhöffer, Modelle der nervenerregung, *Die Naturewissenschaften.* **40**, 301–311 (1953).
72. R. FitzHugh, Impulses and physiological states in theoretical models of nerve membrane, *Biophysical Journal.* **1**, 445–466 (1961). doi: 10.1016/S0006-3495(61)86902-6.
73. P. Brock, The nature of solutions of a Rayleigh-type forced vibration equation with a large coefficient of damping, *Journal of Applied Physics.* **24**(8), 1004–1007 (1953). doi: 10.1063/1.1721425.
74. C. Hayashi, Y. Nishikawa, and M. Abe, Subharmonic oscillations of order one half, *IRE Transactions on Circuit Theory.* **7**(2), 102–111 (1960). doi: 10.1109/TCT.1960.1086644.
75. Y. Ueda, *The Road to Chaos.* In R. Abraham and H. B. Steward (eds.), *Science Frontier Express Series.* Aerial Press, Santa Cruz, CA (1993).
76. N. Levinson, A second-order differential equation with singular solutions, *Annals of Mathematics.* **50**(1), 127–153 (1949). doi: 10.2307/1969357.
77. M. L. Cartwright and J. E. Littlewood, On non-linear differential equations of the second order, I, *Journal of the London Mathematical Society.* **20**(3), 180 (1945).
78. O. Ménard, C. Letellier, J. Maquet, L. L. Sceller, and G. Gouesbet, Analysis of a non synchronized sinusoidally driven dynamical system, *International Journal of Bifurcation & Chaos.* **10** (7), 1759–1772 (2000).
79. N. Levinson, A simple second order differential equation with singular motions, *Proceedings of the National Academy of Sciences.* **34** (1), 13–15 (1948). doi: 10.1073/pnas.34.1.13.
80. G. D. Birkhoff, Sur le problème restreint des trois corps, *Annali della Scuola Normale Superiore di Pisa II.* **4**, 267–306 (1935).
81. G. D. Birkhoff, Nouvelles recherches sur les systèmes dynamiques, *Memorie della Pontificia Accademia dei Nuovi Lincei III.* **1**, 85–216 (1935).
82. M. Charpentier, Sur quelques propriétés des courbes de M. Birkhoff, *Bulletin de la Société Mathématique de France.* **62**, 193–224 (1934).
83. H. M. Morse, Recurrent geodesics on a surface of negative curvature, *Transactions of the American Mathematical Society.* **22**, 84–100 (1921).
84. C. Letellier, P. Dutertre, and G. Gouesbet, Characterization of the Lorenz system, taking into account the equivariance of the vector field, *Physical Review E.* **49**(4), 3492–3495 (1994). doi: 10.1103/PhysRevE.49.3492.
85. T. D. Tsankov and R. Gilmore, Topological aspects of the structure of chaotic attractors in $\mathbb{R}^3$, *Physical Review E.* **69**, 056206 (2004).

86. C. Letellier, Caractérisation topologique et reconstruction des attracteurs étranges. Ph.D. Thesis, University of Paris VII, Paris, France (May, 1994).
87. O. E. Rössler and C. Letellier, *Chaos: The World of Nonperiodic Oscillations.* Springer Nature, Switzerland AG (2020). doi: 10.1007/978-3-030-44305-4.
88. S. M. Ulam, *A Collection of Mathematical Problems.* Interscience, New York (1961).
89. P. R. Stein and S. M. Ulam, Non-linear transformation studies on electronic computers, *Rozprawy Matematyczne.* **39**, 1–65 (1964).
90. S. Ulam, Random processes and transformations. In *Proceedings of International Congress of Mathematicians*, Vol. 2. American Mathematical Society, Providence, R.I. (1952), pp. 264–275. 30 August–6 September, 1950.
91. C. Shannon, Some topics in information theory. In *Proceedings of International Congress of Mathematicians*, Vol. 2. American Mathematical Society, Providence, R.I. (1952), pp. 262–263. 30 August–6 September, 1950.
92. S. Ulam and J. von Neumann, On combination of stochastic and deterministic processes, *Bulletin of the American Mathematical Society.* **53**, 1120 (1947). Abstract 53-11-403.
93. P. Cvitanović and B. Eckhardt, Symmetry decomposition of chaotic dynamics, *Nonlinearity.* **6**, 277–311 (1993).
94. P. Collet and J.-P. Eckmann. In A. Jaffe and D. Ruelle (eds.), *Iterated Maps on the Interval as Dynamical Systems.* Progress in Physics, Vol. 1. Bikhäuser, Boston, Berlin (1981).
95. C. Sparrow, *The Lorenz Equations: Bifurcations, Chaos and Strange Attractors*, Vol. 41. Springer-Verlag, Berlin (1982). doi: 10.1007/978-1-4612-5767-7.
96. E. N. Lorenz, *The Essence of Chaos.* University of Washington Press (1993).
97. F. Model and H. Mädler, Hydrographische Bibliographie, *Deutsche Hydrographische Zeitschrift.* **17**, 89–104 (1964). doi: 10.1007/bf02311984.
98. L. D. Landau, On the problem of turbulence, *Doklady Akademii Nauk SSSR.* **44**, 339–342 (1944) (in Russian).
99. J. T. Stuart, On the non-linear mechanics of hydrodynamic stability, *Journal of Fluid Mechanics.* **4**(1), 1–21 (1958). doi: 10.1017/S0022112058000276.
100. M. Aihara, Growth of the baroclinic disturbance and asymptotic approach toward finite amplitude steady state, *Journal of the Meteorological Society of Japan II.* **42**(3), 139–162 (1964). doi: 10.2151/jmsj1923.42.3_139.
101. D. K. Lilly, On the computational stablility of numerical solutions of time-dependent non-linear geophysical fluid dynamics problems, *Monthly Weather Review.* **93**(1), 11–25 (1965). doi: 10.1175/1520-0493(1965)093⟨0011:OTCSON⟩2.3.CO;2.
102. L. A. Segel, The structure of non-linear cellular solutions to the Boussinesq equations, *Journal of Fluid Mechanics.* **21**(2), 345–358 (1965). doi: 10.1017/S0022112065000228.
103. L. Peng, Numerical experiments on planetary meridional temperature gradients contrary to radiational forcing, *Pure and Applied Geophysics.* **61**, 173–190 (1965). doi: 10.1007/bf00875299.

104. L. Peng, A simple numerical experiment concerning the general circulation in the lower stratosphere, *Pure and Applied Geophysics.* **61**, 191–218 (1965). doi: 10.1007/bf00875777.
105. J. A. Young, On the relation between zonal heating asymmetries and large-scale atmospheric fluctuations in space and time. Ph.D. Thesis, Massachusetts Institute of Technology, Cambridge, MA (February, 1966).
106. R. C. J. Somerville, A nonlinear spectral model of convection in a fluid unevenly heated from below, *Journal of Atmospheric Sciences.* **24**(6), 665–676 (1967). doi: 10.1175/1520-0469(1967)024⟨0665:ANSMOC⟩2.0.CO;2.
107. L. Berkofsky and R. Shapiro, Numerical experiments with a vertically parameterized model designed to study the effects of high-level heating on the large-scale circulation of the lower atmosphere. Technical Report 571, Office of Aerospace Research, United States Air Force (1968).
108. E. S. Epstein, Stochastic dynamic prediction, *Tellus.* **21**, 739–759 (1969). doi: 10.3402/tellusa.v21i6.10143.
109. E. N. Lorenz, A study of the predictability of a 28-variable atmospheric model, *Tellus.* **17**(3), 321–333 (1965). doi: 10.1111/j.2153-3490.1965.tb01424.x.
110. E. N. Lorenz, On the range of atmospheric predictability. In *Proceedings of the First National Conference on Statistical Meteorology* (1968), pp. 11–19.
111. E. N. Lorenz, Three approaches to atmospheric predictability, *Bulletin of the American Meteorological Society.* **50**, 345–351 (1969).
112. E. N. Lorenz, The predictability of a flow which possesses many scales of motion, *Tellus.* **21**(3), 289–307 (1969). doi: 10.1111/j.2153-3490.1969.tb00444.x.
113. T. Rikitake, Oscillations of a system of disk dynamos, *Mathematical Proceedings of the Cambridge Philosophical Society.* **54**(1), 89–105 (1958). doi: 10.1017/S0305004100033223.
114. A. E. Cook and P. H. Roberts, The Rikitake two-disc dynamo system, *Mathematical Proceedings of the Cambridge Philosophical Society.* **68**(2), 547–569 (1970). doi: 10.1017/S0305004100046338.
115. T. A. Gleeson, Statistical-dynamical predictions, *Journal of Applied Meteorology and Climatology.* **9**(3), 333–344 (1970). doi: 10.1175/1520-0450(1970)009⟨0333:SDP⟩2.0.CO;2.
116. W. V. R. Malkus, Non-periodic convection at high and low Prandtl number, *Mémoires de la Société Royale des Sciences de Liège.* **IV**, 125–128 (1972).
117. D. W. Dockery, An analytic study of the predictability of the flow in a dish-pan model of the atmosphere. Ph.D. Thesis, Massachusetts Institute of Technology, Cambridge, MA (August, 1972).
118. N. A. Phillips, Principles of large scale numerical weather prediction. In P. Morel (ed.), *Dynamic Meteorology.* D. Reidel Publishing Company, Dordrecht, Holland (1973), pp. 1–96. doi: 10.1007/978-94-010-2599-7. Lectures Delivered at the Summer School of Space Physics of the Centre National d'Etudes Spatiales, Held at Lannion, France, 7 August–12 September 1970.

119. D. K. Lilly, Lectures in sub-synoptic scales of motion and two-dimensional turbulence. In P. Morel (ed.), *Dynamic Meteorology*. D. Reidel Publishing Company, Dordrecht, Holland (1973), pp. 352–418. doi: 10.1007/978-94-010-2599-7. Lectures Delivered at the Summer School of Space Physics of the Centre National d'Etudes Spatiales, Held at Lannion, France, 7 August–12 September 1970.
120. R. J. Naistat and J. A. Young, A linear model of boundary layer flow applied to the St. Patrick's day storm of 1965, *Journal of Applied Meteorology and Climatology*. **12**(7), 1151–1162 (1973). doi: 10.1175/1520-0450(1973)012⟨1151:ALMOBL⟩2.0.CO;2.
121. J. S. A. Green, Two-dimensional turbulence near the viscous limit, *Journal of Fluid Mechanics*. **62**(2), 273–287 (1974). doi: 10.1017/S0022112074000681.
122. L. Moen, A spectral model for investigation of amplifying baroclinic waves, *Tellus*. **26**(4), 424–443 (1974). doi: 10.1111/j.2153-3490.1974.tb01621.x.
123. E. N. Lorenz, Energy and numerical weather prediction, *Tellus*. **12**(4), 364–373 (1960). doi: 10.3402/tellusa.v12i4.9420.
124. E. N. Lorenz, The mechanics of vacillation, *Journal of Atmospheric Sciences*. **20**(5), 448–465 (1963). doi: 10.1175/1520-0469(1963)020⟨0448:TMOV⟩2.0.CO;2.
125. J. B. McLaughlin and P. C. Martin, Transition to turbulence of a statically stressed fluid, *Physical Review Letters*. **33**(20), 1189–1192 (1974). doi: 10.1103/PhysRevLett.33.1189.
126. D. Ruelle and F. Takens, On the nature of turbulence, *Communications in Mathematical Physics*. **20**, 167–192 (1971).
127. J. B. McLaughlin and P. C. Martin, Transition to turbulence in a statically stressed fluid system, *Physical Review A*. **12**(1), 186–203 (1975). doi: 10.1103/PhysRevA.12.186.
128. E. N. Lorenz, The problem of deducing the climate from the governing equations, *Tellus*. **16**(1), 1–11 (1964). doi: 10.1111/j.2153-3490.1964.tb00136.x.
129. G. Benettin, L. Galgani, and J.-M. Strelcyn, Kolmogorov entropy and numerical experiments, *Physical Review A*. **14**, 2338–2345 (1976). doi: 10.1103/PhysRevA.14.2338.
130. R. M. May, Simple mathematical models with very complicated dynamics, *Nature*. **261**(5560), 459–467 (1976). doi: 10.1038/261459a0.
131. E. N. Lorenz, Investigating the predictability of turbulent motion. In *Statistical Models and Turbulence*. Springer-Verlag (1972), pp. 195–204.
132. D. Ruelle, Strange attractors as a mathematical explanation of turbulence. In *Statistical Models and Turbulence*. Springer-Verlag (1972), pp. 292–299. doi: 10.1007/3-540-05716-1_16.
133. A. T. Winfree, Unclocklike behaviour of biological clocks, *Nature*. **253**, 315–319 (1975). doi: 10.1038/253315a0.
134. C. Letellier and V. Messager, Influences on Otto E. Rössler's earliest paper on chaos, *International Journal of Bifurcation & Chaos*. **20**(11), 3585–3616 (2010).

135. D. Ruelle, The Lorenz attractor and the problem of turbulence. In *Proceedings of the Symposium "Quantum Dynamics: Models and Mathematics"*. Springer, Vienna, Austria (1976), pp. 221–239.
136. R. F. Williams, Expanding attractors, *Publications Mathématiques de l'Institut des Hautes Études Scientifiques.* **43**(1), 169–203 (1974). doi: 10.1007/BF02684369.
137. R. F. Williams, The Lorenz attractor, *Lecture Notes in Mathematics.* **615**, 94–112 (1977).
138. G. B. Mindlin and R. Gilmore, Topological analysis and synthesis of chaotic time series, *Physica D.* **58**(1), 229–242 (1992). doi: 10.1016/0167-2789(92)90111-Y.
139. R. W. Ghrist, Branched two-manifolds supporting all links, *Topology.* **36**(2), 423–448 (1997). doi: 10.1016/0040-9383(96)00006-7.
140. R. Gilmore, Topological analysis of chaotic dynamical systems, *Reviews of Modern Physics.* **70**(4), 1455–1529 (1998). doi: 10.1103/RevModPhys.70.1455.
141. J. Hadamard, Les surfaces à courbures opposées et leurs lignes géodésiques, *Journal de Mathématiques Pures et Appliquées V.* **4**, 27–74 (1898).
142. M. Morse and G. A. Hedlund, Symbolic dynamics, *American Journal of Mathematics.* **60**(4), 815–866 (1938). doi: 10.2307/2371264.
143. N. Metropolis, M. L. Stein, and P. R. Stein, On finite limit sets for transformations on the unit interval, *Journal of Combinatorial Theory A.* **15**, 25–44 (1973).
144. B. Derrida, A. Gervois, and Y. Pomeau, Iteration of endomorphisms on the real axis and representation of numbers, *Annales de l'Insitut Henri Poincaré A.* **29**, 305–356 (1978).
145. C. Letellier, Branched manifolds for the three types of unimodal maps, *Communications in Nonlinear Science and Numerical Simulation.* **101**, 105869 (2021). doi: 10.1016/j.cnsns.2021.105869.

https://doi.org/10.1142/9789819823130_0003

Chapter 3

JOSEPH FORD: THE EVANGELIST OF (HAMILTONIAN) CHAOS

Christophe Letellier

Rouen Normandie University, Rouen, France
christophe.letellier@univ-rouen.fr

This chapter is devoted to Joseph Ford's contribution to the development of chaos theory and, more specifically, to the branches we can call conservative (more largely, Hamiltonian chaos) chaos and quantum chaos, the latter being a subclass of the former. Here, Ford reveals himself to be a man of insatiable generosity and boundless altruism. His energy was spent for the benefit of the chaos community, leading to the dissemination of over 1000 abstracts between 1978 and 1980, up to the creation of *Physica D*. He was the privileged interlocutor of Boris Chirikov, and a close friend of Otto E. Rössler.

1. Introduction

Joseph Ford was born December 18, 1927 in Ashville, Buncombe County, North Carolina. He served in the Navy during the second World War. In 1956, Ford defended his Ph.D. thesis devoted to statistical mechanics at the Johns Hopkins University under the supervision of Theodore Berlin (1917–1962) who is known for his contribution to the Ising model with Mark Kac (1914–1984).[1] After two years as a research physicist at Union Carbide Corporation (a chemical and polymers company) in Niagara Falls, he switched for an academic career at the University of Miami in 1958. He then moved to Georgia Tech as associate professor in 1961 and rose through the ranks to become Regents' Professor of Physics in 1978. From September of

1973 to December of 1973, he was awarded a NATO Senior Fellowship at the University of Milano. Ford died April 26, 1995.

When he investigated the Fermi–Pasta–Ulam–Tsingou problem that lead him to describe some chaotic orbits, he viewed them from the point of view of statistical mechanics. He quickly understood that these particular trajectories were important and started to serve as a catalyst, encouraging many scientists (Boris Chirikov, Morikazu Toda) working in the field of statistical mechanics to focus their attention to this particular type of solutions and some related problems. Working with Hamiltonian systems, he was particularly influent on those working in quantum physics, celestial mechanics, and in mathematics.

In 1965, while he was an associate professor at Georgia Tech, Ford wrote a small paper in the *Tech Alumnus* about the way science is supported by the US government.[2] He inserted a brief description of the abilities a scientist may have (in the spirit of what was later developed by Bruno Latour[3]):

> A scientist [...] must first have an idea for solving some problem. This is the first and most difficult part. [...] A great sophistication is required to develop a feasible avenue of inquiry leading to a solution. [...] While a scientist finds this process as natural as breathing, the general public considers it very mysterious.
>
> Having developed his idea, the scientist must find an agency willing to consider it. [...] He then submits a formal proposal presenting his research idea in detail. The given agency (through a panel of scientists) then ranks in order of merit all the proposals being considered for support. [...] Frequently, the budget is amended by the agency.
>
> If all these hurdles are successfully passed, the proposal is awarded financial support in the form of a grant [...] and the scientist begins work. Regular progress reports are submitted as well as periodic fiscal reports.

He defended the idea that science was viewed, in US, as very useful by the end of World War II, stimulated by the contribution of meteorologists to the success of the D-day.[4] The main argument then used for supporting physical science was *national survival* which is, most likely according to Ford, "the only issue upon which both Republican and Democrat can agree."[2] Thus, in 1963, the budget for research was distributed as follows.

1.	Physical science	1,030 million
2.	Life Science	488 million
3.	Engineering	138 million
4.	Psycological science	49 million
5.	Mathematical science	43 million
6.	Social science	31 million

Since the social sciences are poor and the humanities receive almost nothing, the result of this allocation of funding is that "*we have created the technology for aleviating world hunger, but lack the knowledge of human behavior needed to give our technology to a backward people*".[2] Among other imbalance, he questioned the fact that high energy physics is preferred to a process of desalinization of sea water using nuclear reactors. He criticized the government which chooses the organizations which already have the skills — the spiral of the rich getting richer —, the center of the new world became the California with the Harvard-MIT. For him, Americans suffer "*from a myopia of competing with the Russians*". Among his personal acquaintances, he cannot find a scientist who supports the three-billion-dollar moon program. To sum up, Ford was promoting more supports for scientific research not directly related to national defense.

He cared deeply about writing well and speaking effectively. He had a remarkable ability to put his views and insights into lapidary form, and his speech owed much to the Southern Evangelists whom he must have heard when growing up in the rural South during the Depression. His generosity to his friends knew no bounds.

To thank his friend Boris Chirikov, Ford wanted to make him a present[a]:

> In 1993, in occasion of one of the conferences organized here in Como, we organized a dinner. Our plan was to make a surprise to Boris Chirikov since it was also his 65th birthday. In Siberia, he had a very poor computer and we planned to offer him a portable computer, much more powerful than the computer he had at his institute of nuclear physics in Novosibirsk. Indeed, at that times, the situation in Russia was a disaster. We were absolutely sure that Boris would be extremely happy and appreciate really much this present since he strongly needed for his normal research activity. We wanted to buy the computer sharing the expense among us. However,

[a]Message from Casati to Letellier, August 22, 2023.

Joe Ford decided to take care personally of this. He arrived from US not only with a big computer but also with the printer. It has been a big problem for Joe to bring to Como such object from US. Joe was already not in good health. I realized that it was a hard job form him. But he really wanted to make Boris happy so he managed to do this (Let aside the cost that at those times was quite relevant and was all his personal money).

This is to tell you how generous person Joe was.

So we went to the restaurant, put the computer and a printer behind the table covered by a blanket. Then at some point Joe stood up, made the speech in honour of Boris and at the end he uncovered the computer. We were all expecting a big exclamation of surprise and happiness from Boris. Instead, Boris barely said thanks, did not react at all and remained seated! That was all. Obviously, Joe was totally disappointed and could not understand this reaction.

The next day Boris came to me and said privately that he absolutely could not take the computer to Novosibirsk; he would meet problems and in any case the computer would be taken away from him from the Moscow border police. So he said that he would leave the computer to our Institute and, obviously, not to inform Joe about this because it was like rejecting his present.

The point is that when at the dinner Boris saw the computer, he immediately realized that he could not take this present to his home; he realized how much energy Joe has put on it, how valuable was the present for him and he became totally embarrassed and could not say any word.

Chirikov has always remained loyal to his country.[b]

Boris never discussed the political situation in Soviet union. When he was a child he was in Leningrad during the Leningrad blockade. Several of his colleagues were persecuted during Stalin period and after. So, he just did not mention politics. But he liked his country and always tried to do his best to help the institute of nuclear physics where he was working. In 1990, after the end of Soviet system, most physicists, including the members of the theoretical group of which he was the head, left the institute and emigrated abroad. He felt his duty to remain there, and he did so and remained there alone.

During the period of my collaboration with Boris in Siberia, the working conditions had some problematics aspects. Just to tell you one thing: when we finished our work on the kicked

[b]Casati to Letellier, March 04, 2024.

rotator in 1975, we set up to write a draft of the paper. So we seated down at the desk one near the other with a pen and some clear sheet of papers. I was waiting for him starting to write something but instead he gave the pen to me and said: you write the draft. I asked why? Then, he explained to me that I was going to take the draft to Italy. There was the possibility that, at the border, some not very clever guard, may become suspicious to see in my hand a scientific paper not written by myself and he could perhaps suspect that I was smuggling something and I could be in troubles. So, I was careful not to cross the border with papers not written with my hands.

By the way, this reminds me another thing. That paper with Boris (and also with Joe and Felix Izrailev) on the kicked rotor[5] is the most quoted paper of my career. Still Boris did not want to send the paper to any journal. He said that this was only a preliminary study and that we should understand better the problem. There was no pressure to publish papers, no h-index and the like. The only desire was to understand physics. It has been a very nice period. Finally, this paper was eventually published in 1979 in the proceedings of the 1977 Como conference. But it has been never sent to any journal.

Always ready to promote science, Ford freely and warmly exchanged with many great contributors to chaos theory (Chirikov, and Rössler among many others). As Giulio Casati remembers[c]:

> Joe has been a very good and generous person. He was enthusiastic about chaos. It all happened after a visit we had in Moscow and Novosibirsk after discussion with several mathematicians in Moscow and especially with Boris Chirikov. After these discussions, it became clear to us that exponential instability can make deterministic systems really unpredictable. For Joe this was a really striking thing and somehow changed his attitude. After that, he become, in his own words "*an evangelist of chaos.*" He started collecting all preprints in non-linear dynamics, spending days in making photcopies, putting in envelopes and send to colleagues all around the world. I advised him that he was spending to much energies in this. It has been so much work and somehow this affected his health.

The initiation of his work as an evangelist was initiated just after the organization of the 1977 Como conference, Ford wanted to promote in a more efficient way the results obtained by those investigating nonlinear

[c]Casati to Letellier, August 21, 2023.

dynamical systems: he thus conceived the delivery of *Nonlinear Science Abstracts* that he personally collected, organized, typed (see p. 237), Xeros copied and distributed. He thus contributed to the emergence of an interdisciplinary community sharing some techniques of analysis, independently from the origin of the system under study. The success of his tireless efforts resulted in the foundation of the first journal devoted to nonlinear dynamics, *Physica D*.

As the chaotic dynamics community gained in size and acceptance, Ford started to pursue philosophical ideas to explore the deeper consequences of deterministic chaos and yet again he found himself in the role of pioneer. Kurt Gödel (1906–1928) and Alan Turing (1912–1954) supplanted Henri Poincaré (1854–1912) in his daily thinking. He realized that the concept of chaos transcends the domains of mappings and differential equations, and repeatedly drew attention to "algorithmic complexity" — a concept also rooted in the notion of exponential instability — as a means of defining and assessing the fundamental limits of human ability to deal with chaotic systems. He started to investigate quantum chaos. Thus, in the final announcement of the 1983 Como conference,[d] with Giulio Casati, Ford considered the problem of quantum chaos as follows:

> Chaos (or randomness) in classically deterministic systems now appears in the title or as the subject area of many current pending conferences. To our knowledge, the upcoming Como conference will be the first gathering dedicated to the question of how and to what extent the notions of classical chaos can be extended into the quantum domain. Schrödinger's equation is a deterministic scheme for computing the time evolution of the wave function probability, just as Newton's equations provide the deterministic algorithm for computing the time evolution of the (q, p) system state. But the deterministic systems of Newtonian dynamics can exhibit strictly random time evolution: in consequence, quantum systems might be expected to exhibit a randomness over and above that already existing in the wave function.

Armed with his algorithmic complexity, Ford not only perceived but demonstrated by computer experiments that dynamical chaos as it is understood in classical mechanics (and implemented in ergodic theory in mathematics) does not exist in quantum mechanics.[6] As a genuine scientist,

[d] Letter from Casati and Ford, May 13, 1982.

Ford did not flinch: He courageously proceeded to challenge the very basis of quantum mechanics itself.[e]

> Another main point is that he became almost obsessed by the following: exponential unstable systems are unpredictable and therefore unsolvable. On the other hand, systems with zero Lyapounov exponents, that is with linear instability could therefore, in principle, be solved. An example is given by polygonal billiards, in particular by triangles. In triangles, there is no exponential instability and therefore could be analytically solved. As a consequence he started putting all his energies in finding eigenfunctions and eigenvalues of (generic) triangular billiard. It was a desperate (practically impossible) task. Even best mathematical scientists like Aleksandr I. Shnirel'man (then at Tel Aviv University) told him to forget this: it was practically an impossible task, too complicated. I tried to tell him many times, even in a sharp way, to give up this idea but he did not care. He wanted to analytically solve non integrable quantum systems like generic quantum triangles. Having such solutions it would be quite important since they could be much more useful than standard integrable systems as a basis to describe chaotic systems. Joe was very sorry that I did not like his idea and did not follow him along this line.

Sadly, this daring flight was interrupted by his death. Giulio Casati reports as follows.[f]

> one day I received a message from him (I do not remember if it was a letter or already a mail). The message said: "*Dear Giulio, I have two news, one good one bad. I start with the good one*" (and here he described a new idea he got that possibly could be useful to solve his problem, etc, etc.). "*Now the bad news: yesterday I went to the doctor and he told me that I have cancer in the lungs and I have only few days to live.*" After few days he died.

In issues of integrity and quality he was particularly uncompromising. He was a passionate man who devoted his life to his work: he was a compulsive worker.

> Chaos workers never die... they simply lose control of their variables!.[g]

---

[e]Casati to Letellier, February 27, 2024.
[f]Casati to Letellier, August 21, 2023.
[g]Email from Ford to Felix Izrailev, October 31, 1989.

2. Ford and the FPUT Problem

In the preface of the proceedings to the Como conference (June 10, 1977), Giulio Casati and Ford stated that "*completely deterministic, nonlinear systems can yield wildly chaotic solution behavior has, over the past two decades, been independently discovered and re-discovered by numerous scientists working in a host of distinct scientific disciplines.*" Indeed, according to Ford,[7] there is a branch from which emerged chaotic solutions and which is devoted to Hamiltonian systems, that is, most often to conservative systems. He wanted to provide a history, perhaps less known than the Poincaré–Birkhoff–Lorenz–Ruelle–Rössler branch:

> [...] my gratitude also extends to each new generation of students who view history through a telescope's far end and are therefore ever appreciative of knowledge that nonlinear dynamics did not begin with Feigenbaum or even Lorenz.

Both branches started with Poincaré's results showing that the three-body problem is not integrable[8] and it is associated with homoclinic or heteroclinic tangle leading to aperiodic (stochastic, ergodic or chaotic) behavior.[9] His continuator, George Birkhoff showed that many-body systems can often be solved analytically with a perturbation theory.[10]

Then, Enrico Fermi (1901–1954) proposed a theorem asserting that Hamiltonian systems are, in general, ergodic.[11–14] When he visited the Los Alamos Laboratory where the MANIAC I computer was working since 1952 (related to the division supervised by Nicholas Metropolis), Fermi was entertaining with John Pasta (1918–1981) and Stanislaw Ulam (1909–19084) in looking for a problem to investigate with the help of numerical simulations. They started by considering a small nonlinear array of mass points linked by light Hookean springs with a small nonlinear term: this problem was coded by Mary Tsingou (born in 1928).[15] Contrary to what was suggested by Fermi's theorem, the system behaves periodically with a time period smaller by many orders of magnitude than Poincaré's recurrence time. Unfortunately, Fermi died in November 1954 and the report was only later written by Pasta with some plots produced by Tsingou, and dated on November 2, 1955.[16]

In 1954, Andreï Kolmogorov (1903–1987) announced that, under small non-integrable perturbation of the Hamiltonian, nearly linear systems are, in general, quasi-periodic.[17] Under the supervision of Kolmogorov, Vladimir Arnol'd (1937–2010) investigated some stability problems in planet motions. Using a map from the circle onto itself,[18, 19] he considered

Fig. 1. Vladimir Arnol'd in Moscow during the 1970s.

the problem of small divisors, known by the astronomers to correspond to solutions associated with non-converging series.[20] He thus identified that there are *zones of instability* where *the separatrices of hyperbolic points intersecting each other create an intricate network.*[21] To give a picture of this, Arnol'd proposed a drawing of the heteroclinic tangle (Fig. 2), exactly the one Poincaré renounced to draw in his new methods for celestial mechanics.[9] Note that, for this type of drawing, Arnol'd was inspired by Vladimir Mel'nikov's study of plasma motion in a given magnetic field in which a drawing of an heteroclinic tangle is inserted.[22] In 1963, Mel'nikov published two other papers in which such drawings are inserted (see p. 24).[23,24] With them, the connection between the "zone of instability" (or stochastic motion) and heteroclinic tangle is proposed. Mel'nikov and Arnol'd quoted Poincaré's *méthodes nouvelles de la mécanique céleste.*[9] In his 1963 paper, Arnol'd thanks Boris Chirikov for his help, most likely for considering a simple map rather than the initial three-body problem.

In 1959, while investigating resonance between the Larmor rotation of charged particles, Boris Chirikov found a criterion for observing a "*stochastic*" variation of the current,[25] associated with apparent random changes in phase. At that time, Ford was puzzled by the surprising periodic orbit obtained by Fermi, Pasta, Ulam and Tsingou (FPUT) with a nonlinear system, performed an analytical study[26] of the problem by using a perturbation technique due to Nikolay Krylov (1879–1955) and Nikolay Bogolyubov (1909–1992).[27] He also performed new numerical simulations of this problem. Remembering that Poincaré showed that the three-body

Fig. 2. Heteroclinic tangle in the zone of instability of a Poincaré section of the circle map investigated by Arnol'd.[21]

Fig. 3. From the left to the right: Ludvig Dmitrievich Faddeev (1934–2017), Giulio Casati and Boris Chirikov in Moscow during the 1970s.

Source: Courtesy of Giulio Casati.

problem produces irregular solutions sensitive to initial conditions, and Carl Siegel (1896–1981) who showed that, in every neighborhood of an analytic Hamiltonian, integrable or not, there exists a non-integrable Hamiltonian,[28] Ford was convinced, as Fermi also believed, that a nonlinear system as the FPUT problem should produce stochastic behaviors. In 1963, Ford and John Waters concluded that "*weakly coupled oscillators share many properties of linear systems*".[29] They thus suggested that nonlinear systems may admit analytic solutions. In fact, in 1966, with Felix Izrailev,[30] Chirikov determined the domain of stochasticity in the system investigated by Fermi, Pasta, Ulam, and Tsingou.[16] Two years later, these physicists obtained numerical evidences for stochastic behaviors in the FPUT problem.[31]

3. Hénon and the KAM Theory

Ford and Water's results about weakly coupled oscillators led Morikazu Toda (1917–2010) to introduce a chain of particles interacting by nonlinear force and governed by

$$m\ddot{u}_n = -\phi'(u_n - u_{n-1}) + \phi'(u_{n+1} - u_n) \tag{1}$$

where

$$\phi(r) = \frac{a}{b}e^{-br} + ar + \text{ const.} \tag{2}$$

with two positive parameters a and b, the mutual displacement $r = u_n - u_{n-1}$.[32,33] $\phi'(r)$ is the derivative of ϕ with respect to r. Later, with Soptswood Stoddard and Jack Sturner, Ford proved the integrability of the so-called Toda lattice using a Poincaré section and the mean divergence between close trajectories as a function of time.[34]

In 1970, with Gary Lunsford, Ford showed that systems with small coupling strength can produce chaos.[36] In other words, a weak nonlinearity can destroy the quasi-periodic tori when there is a resonnance overlap, that is, when Chirikov's criterion is fulfilled. In this paper, Chirikov is thanked for the time he spent to teach about nonlinear oscillating systems. He also investigated this problem with Giulio Casati who remined this work as follows[h]:

> In 1972, after the usual military service period in Italy, I went for a trip in US. I knew some paper by Joe Ford so I went to visit him at Georgia Tech, Atlanta. In that occasion he invited me to spend a period as a postdoc with him. I started in the fall of 1972, and we began to look at the Toda lattice. Since the lattice has soliton solutions, we considered the possibility that the system was integrable. From numerical results, we got clear evidence that the system was likely to be integrable

[h] Email from Casati to Letellier, August 17, 2023.

(at that time, basically nothing was known, no Lyapunov exponents, etc.). In fact, I remember a call from Joel Lebowitz from a conference in Chicago; he said that he has heard some rumour about something like "KAM theorem" and he was inquiring what it was about. Still somehow, we got convinced that the system was integrable. So I started to take the system with $N = 2$ degrees of freedom to look for the second constant of motion, besides the energy. I tried for about two weeks and I failed. Then, we wrote a letter to Michel Hénon, the astronomer in Nice. After two weeks, we received his letter saying that I have put everything in place to investigate the problem, and in a couple of hours, he found the second constant of motion which he wrote down in the letter.[37] Obviously, I was very frustrated. The only consolation was the following: in my search for the second constant of motion, as it was usually done, I started in the limit of harmonic oscillations. It turned out that, if you follow this way, you will never get the constant of motion. Hénon was an astronomer and it was natural to him to start from the opposite limit, that of colliding bodies. If you do this, it is easy to find the second constant. So, this is how integrability of the Toda lattice was discovered.

In 1973, Morikazu Toda visited us in Atlanta (Fig. 4). I was curious to know how he did find the soliton solutions. Indeed,

(a) At a Japanese restaurant in Atlanta, 1973

(b) At the 1977 Como conference

Fig. 4. (a) Morikazu Toda and his wife Keiko, Joseph Ford, a colleague from Georgia Tech, and Giulio Casati. (b) Morikazu Toda and Nobuhiko Saito. Saito — also working with computer experiments — was exchanging with Ford since the late 1960s.[35]

Source: Courtesy of Giulio Casati.

> if you insert soliton expression in the Toda equations, you can verify that it is indeed a solution but this happens via a sequence of miraculous cancellations. So, I asked him how did he manage to discover such solution? His answer was (I am not sure if it was the truth or just a joke) that he started from the soliton expression, and he then wrote down the Toda lattice!

Michel Hénon was born in 1931 in Paris. In 1957, he started his Ph.D. thesis under the supervision of Evry Schatzman (1920–2010) at the Institut d'Astrophysique de Paris. In 1961, he defended his Ph.D. thesis, entitled *L'évolution dynamique des amas globulaires.*[38] In the early 1960s, Hénon came interested in this problem derived from the restricted three-body problem "*by the beautiful computations and drawings in Broucke's thesis.*"[39] Indeed, Roger Broucke (1932–2005) and his superivisor of André Deprit (1926–2006, Université Catholique de Louvain) were both stimulated to investigate the three-body problem[40] by their teacher George Lemaître (1894–1966).[41] In particular, Broucke exhibited different classes of periodic orbits in the restricted three-body problem from numerical studies performed with an Univac Solid State computer (USSC 90) at Brussels.[42] He investigated the particular case of the earth-moon with a mass ratio $\mu = 0.012155$.[43] This study was in the spirit of Poincaré's search for periodic orbits using the principle of least action,[44] that is, searching orbits with some constrain. He did that by referring to Birkhoff's transformation of coordinates.[43,45,46] As Poincaré in 1896,[47] Broucke paid attention to collision.

Before that, Hénon started his career in building analog computer for computing the internal structure of stars.[48] While his colleagues Evry Schatzman (1920–2010) and Jean-Claude Pecker (1923–2020) used calculation rules, with a precision of three significant digits, Hénon "*was more demanding than* [they] *were,* [and] *was greatly embarrassed by these rather rough ways of calculating* [Jean-Claude Pecker remembered that] *Michel threw himself into analog calculation, with derivations and integrations simulated in the operations by chokes and resistors. He set to work with a precise objective, a soldering iron and an entire IAP part. But techniques evolve quickly... After three years of hard work, Bull and IBM had invaded the digital computing market, and the Observatoire de Paris was equipped* [...] *with high-performance punched-card computers.*"[49] He described the principle of analog computing as follows.[48]

> It is [based on] a model. The magnitudes of the calculation are represented by proportional electrical voltages; the equations are represented by electrical circuits connecting these

Fig. 5. Michel Hénon at the Institut d'Astrophysique de Paris in 1956 or 1957, in the front of the analog computer he built during three years with 500,000 welds. Picture shot by Michel Dreux. Courtesy of Pierre Hénon.

> voltages to each other. In other words, we build an electrical system governed by a system of equations which is formally identical to the one we want to solve, and it is naturally easier to experiment on an electric circuit than on a star.

In 1963, with Carl Heiles, Michel Hénon investigated the system

$$\begin{cases} \dot{x} = V_x \\ \dot{y} = V_y \\ \dot{V}_x = -x - 2xy \\ \dot{V}_y = -y - x^2 + y^2 \end{cases} \tag{3}$$

for which the energy

$$E = \frac{1}{2}\left[\left(V_x^2 + V_y^2\right) + \left(x^2 + y^2 + 2x^2y - \frac{2}{3}y^3\right)\right] \tag{4}$$

is a constant of motion. This manuscript was received by the *Astronomical Journal* on August 7, 1963, that is, only a few months after Lorenz's paper. The system (3), now known Hénon–Heiles system, has two degrees of freedom, and is derived from the Hamiltonian function

$$\mathcal{H} = \frac{V_x^2 + V_y^2}{2} + U(x, y) \tag{5}$$

defining the energy E of the system and where the potential is

$$U(x, y) = \frac{x^2 + y^2}{2} + x^2 y - \frac{2}{3} y^3. \tag{6}$$

Hénon and Heiles provided pictures as the Poincaré section plotted in Fig. 6 of what they designated as the *chaotic sea* and that they described as follows.[50] The orbit is *quite complicated* and *the existence of a third integral is uncertain.* Such orbit appears when the energy is greater than a critical value, and corresponds to a quasi-periodic island for which the third integral *seems to exist*, and *the space left between these regions is the "ergodic" region where the third integral is non isolating.* The volume filled by the ergodic region increases with the energy. The plot from the original paper was drew, point by point, by Hénon's wife, Michèle Hénon [49, pp. 192–193]:

> Thousands of little dots drawn by me on a large sheet of graph paper spread out on our kitchen table, under the dictation of Michel reading huge listings [...] And for six months my evenings were punctuated by "abscissa" value, "ordinate" value = one little dot! And he had only look at the result once the last point had been plotted, and then he had exulted "*it's fantastic, it's totally unexpected, better than I expected*". But I can't guarantee that there weren't a few wrong points in the batch!

This Poincaré section is the first picture illustrating the KAM theory.

In 1965, Arnol'd proved a theorem which can be stated as follows. If two vector fields $\boldsymbol{F}$ and $\boldsymbol{G}$, with null divergence, are such as $[\boldsymbol{F}, \boldsymbol{G}] = \boldsymbol{0}$ and are not everywhere colinear, $\boldsymbol{F} \wedge \boldsymbol{G} \neq \boldsymbol{0}$, then, almost every trajectory of the vector field $\boldsymbol{F}$ are closed and everywhere dense on analytical tori T^2 embedded in M.[51] He then remarked that for flow such as $\boldsymbol{\nabla} \wedge \boldsymbol{V} = k\,\boldsymbol{V}$ with k constant, the streamlines can be topologically complicated, referring to Fig. 2. As an example of a system with an heteroclinic tangle, he proposed

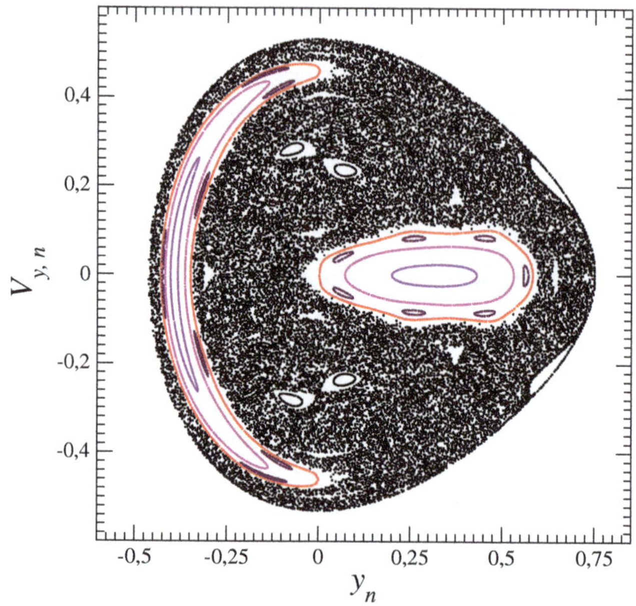

Fig. 6. Poincaré section of the solutions produced by the Hénon–Heiles system (3) from six different sets of initial conditions corresponding to the energy $E = 0.14085$.

the flow governed by the set of equations

$$\begin{cases} \dot{x} = A \sin z + C \cos y \\ \dot{y} = B \sin x + A \cos z \\ \dot{z} = C \sin y + B \cos x. \end{cases} \tag{7}$$

Arnol'd was then in France for a visit at the Institut Henri Poincaré (Paris). According to Michel Hénon, who had already published his numerical study of a restricted three-body problem (see below), Arnol'd wrote him a six-page letter with a list of questions related to the 1964 Hénon–Heiles' paper.[i] Arnol'd thus stimulated Hénon to investigate numerically the system (7), leading him to compute a Poincaré section with quasi-periodic island and chaotic sea (Fig. 7): Hénon was fast in providing a numerical confirmation of Arnol'd's result.[53]

[i]Letter from Arnol'd to Hénon, June 21, 1965.[52]

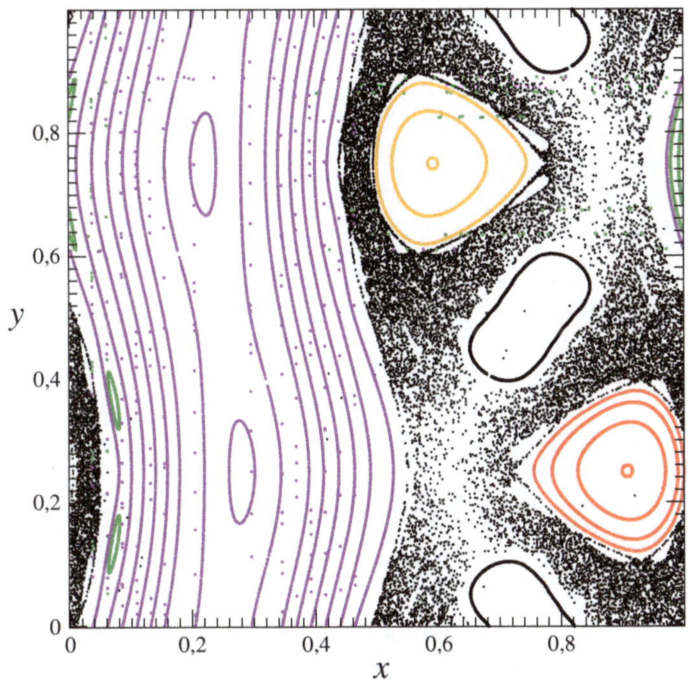

Fig. 7. Poincaré section ($z_{\rm P} = \pi$) of the simple Arnol'd system (7) producing solution with a complicated topology and similar to the one computed by Hénon in 1965.

Notes: Parameter values: $A = \sqrt{3}$, $B = \sqrt{2}$, and $C = 1$.

In the same spirit, in 1969, Hénon proposed the map[54]

$$\begin{cases} x_{n+1} = x_n \cos\alpha - \left(y_n - x_n^2\right) \sin\alpha \\ y_{n+1} = x_n \sin\alpha + \left(y_n - x_n^2\right) \cos\alpha. \end{cases} \tag{8}$$

For $\cos\alpha = 0.4$, he got a small chaotic sea as plotted in Fig. 8(a); another parameter value leads to a larger chaotic sea [Fig. 8(b)].

In 1969, Grayson Walker and Ford proposed an introduction to the KAM theory with the help of the results obtained by Hénon and Heiles.[55] They thus provided pictures of the chaotic sea in this latter system. Their conclusion was that the non-integrability of a system is not sufficient, contrary to what Poincaré's theorem may suggest, for irregular solutions. The so-called KAM theorem exhibits a threshold for the energy beyond which the stochastic trajectories dominate the state space.

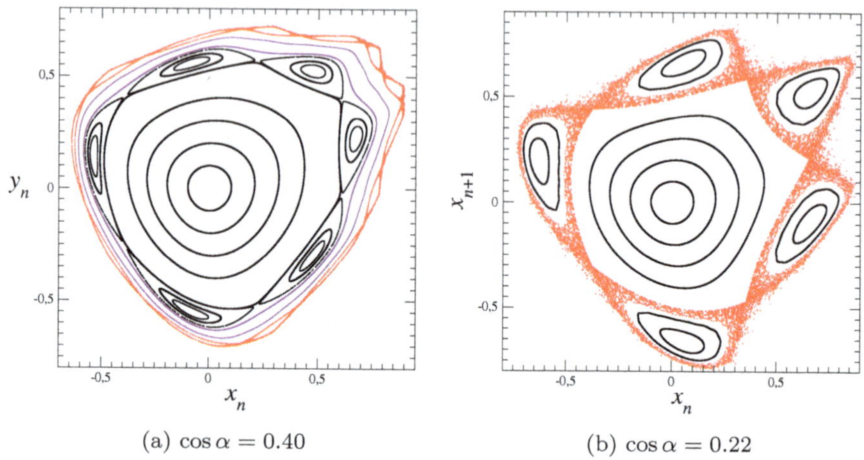

Fig. 8. Quasi-periodic islands (black and blue) and chaotic sea (red) produced by the 1969 Hénon map (8).

Among the various investigations of Hamiltonian systems, Hénon and Heiles' paper had a significant impact in astronomy and in other fields for two reasons: first, because they identified solutions associated with a non-isolating third integral of motion to the three-body problem and with ergodic properties; second, they computed Poincaré sections showing the quasi-periodic islands (corresponding to an isolating third integral) and the now so-called chaotic sea. Their influence can be seen, for instance, in the 1964 work by George Contopoulos (then at the University of Thessalniki, Greece) who, with Ludwijk Woltjer (Columbia University), quickly reacted to Hénon and Heiles' paper in confirming their results by investigating a two-dimensional harmonic oscillator potential corresponding to the motion of stars near the galactic plane which is periodic both in x and y (Fig. 9).[56]

Indeed, the paper published by Hénon and Heiles triggered many works. For instance, Basil Barnabis (Columbia University) checked their results with another potential and confirmed that increasing the energy increases the volume of the state space filled by the chaotic sea, in connection with a non-isolating third integral.[57] Note that Barnabis was in contact with Contopoulos and Woltjer.

In plasma physics, Marshall Rosenbluth (1927–2003, International Atomic Energy Agency, Trieste), who contributed to the development of

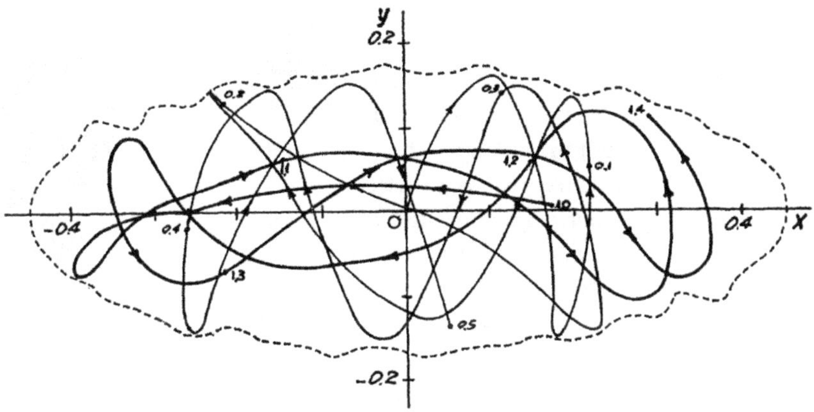

Fig. 9. Ergodic solution plotted by Contopoulos and Woltjer in the x-y plane projection of the four-dimensional state space.

an H-bomb, investigated with Roald Sagdeev, John Taylor and George Zaslavksy, flux lines in high temperature plasma and found motion of them that they qualifed as being Brownian-like.[58] They remarked that the Brownian motion destroys the invariant tori. Note that Rosenbluth worked in the 1950s with the dynamicist Nicholas Metropolis (Los Alamos).[59]

Another example of chaotic solution detected in an Hamiltonian system was provided by J. M. Anthony Danby (1929–2009, North Carolina State University) who met Contopoulos at Yale University in 1962 (Boris Garkinfel was also there) and worked with Delprit. In 1962, Danby published a well-quoted introductory book to celestial mechanics,[60] and a second one to foster the applications of computers to problems in astronomy.[61] In 1968, he pointed out the possible non-integrability of a system resulting from an axisymmetric potential for planet motion.[62] He actually focused on domain of the state space "from which mathematical angels are excluded," that is, for which analytical computations cannot be performed. He thus exhibited "wild" solutions which separate quasi-periodic islands. The wild solutions were associated with a *zone of instability* (Arnol'd's terminology) as predicted by Poincaré.[9] There is a sketch of the heteroclinic network in the spirit of those drew by Mel'nikov and Arnol'd (without any quotation to their works). In 1970, Danby further developed his investigations, consolidated by Walker and Ford's contribution,[55] to exhibit the relevance of chaotic sea. He used the words "chaotic," "ergodic," or "wild" to designate this type of motion.[63]

There are thus many works inspired by Hénon and Heiles, that are devoted to ergodic properties and establishing a link with a heteroclinic tangle.[64–67] With the four examples here selected, it is seen that aperiodic solutions — designated as stochastic or ergodic — were already investigated in the 1960s. The numerical techniques to distinguish quasi-periodic motion from stochastic ones were not yet common: Lorenz (see Chapter 2) and Hénon were the first to propose a first-return map and a Poincaré section, respectively, to exhibit the nature of the solutions. The former had a strong influence for the scientists working with dissipative systems, and the latter for those investigating conservative dynamics. The four examples (Contopoulos, Barnabis, Rosenbluth and Danby) previously discussed were related to the 1977 Como conference which played a decisive role as revealed in some chapters (see, for instance, Vivaldi's chapter, p. 246). In 1977, plotting Poincaré section was becoming common for the astronomers who were at the edges of chaos. As related by Casati, this led Hénon to accept the invitation to give a talk at the Como conference as follows[j]:

> By the way, Michel Hénon was an interesting clever person. When Joe Ford was in Como to visit me in 1976, we invited Hénon for private discussion and it has been very interesting. Then, in 1977 when, with Ford, we organized the first conference on quantum chaos, we invited Hénon, who came. Then, we asked him to present a talk and, at first, he refused since, as he said, he had nothing new to say. We insisted. Then, finally he accepted and said: "*ok, if you want I have some little thing to speak about.*" This little thing was the Hénon attractor![68]

4. Other Chaotic Maps

In 1969, Chirikov introduced the map

$$\begin{cases} \omega_{n+1} = \omega_n + \epsilon \cos 2\pi \, \Psi_n \\[2ex] \Psi_{n+1} = \Psi_n + \dfrac{T}{2\pi} \, \omega_{n+1} \pmod 1 \end{cases} \tag{9}$$

[j]Email from Casati to Letellier, August 17, 2023.

which quickly became the *standard map* in a slightly different form (see p. 125). The standard map first appeared in the form of a recursive relation for atoms in a periodic potential.[69,70] Chirikov plotted some "stochastic" time series. He applied his criterion for stochasticity to physical problems as the lines of force of the magnetic field in a stellarator, the solar system, or a turbulent flow. Chirikov also propagated the idea of using simple map like

$$r_{n+1} = k\, r_n \quad (\text{mod } 2^p) \tag{10}$$

as "pseudo-random generators".[71] The latter is inherited from an early work with M. Antipov and Felix Izrailev.[72] Indeed, the idea was not new and he quoted Derrick Lehmer for the pseudo-random generator

$$u_n = u_0\, 23^n \quad (\text{mod } 10^8 + 1) \tag{11}$$

that was tested on an ENIAC computer in 1951.[73] The interest for pseudo-random generator[74–76] was induced by the suggestion of the "Monte-Carlo" method by Metropolis and Ulam.[77]

Chirikov provided the first two-dimensional plot of the standard map (see p. 132) in the proceedings of the conference organized in 1973 by Christian Mira (Toulouse, France) and devoted to discrete map and their applications.[78] Before the emergence of chaos in electronic circuits during the late 70s and early 80s, there is an important contribution by Igor Gumowski and Mira (the "Toulouse Research Group") whose short history can be found in Ref. 79. Stimulated by the paper written by Stepan Pulkin (see the first chapter, p. 3) who showed that in one-dimensional noninvertible map, infinitely many unstable cycles may lead to bounded complex iterated sequences, Gumowski and Mira studied in 1968 the piecewise-linear map[80,81]

$$\left\{ \begin{aligned} x_{n+1} &= (1-\lambda)x_n + y_n \\ y_{n+1} &= y_n + \left| \begin{array}{ll} -2\lambda x_n - 0.9\lambda & x_n < -0.5 \\ -\dfrac{\lambda x_n}{5} & \text{if } |x_n| < 0.5 \\ -2\lambda x_n + 0.9\lambda & x_n > -0.5. \end{array} \right. \end{aligned} \right. \tag{12}$$

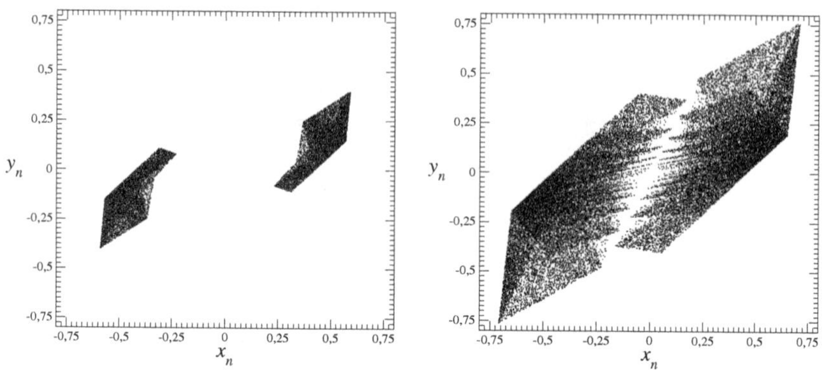

(a) $\lambda = 2.30$

(b) $\lambda = 2.39$

Fig. 10. Chaotic solution to the map (12) (a) as initially published, and (b) just before a boundary crisis.

For some λ-values, they obtained an attractive limit set made of bounded clouds of points as shown in Fig. 10(a) ($\lambda = 2.30$). In the original paper (1968),[80] Mira mentioned that

> The inverse map of [the map (12)] allows, in setting an initial condition in the neighborhood of 0 (stable double point), to remark that there exists a cycle with a very large period, perhaps infinite.

One year later, Gumowski and Mira designated these types of behaviors as "Pulkin phenomenon." This is only ten years later that the first chaotic solutions were widely investigated in electronic circuits.[82] Increasing slightly the λ-value to 2.39, the chaotic solution observed is just before a boundary crisis unifying the two sets of points [Fig. 10(b)].

At the 1973 Toulouse conference where Chirikov proposed his first plot of a chaotic map, Jean Bernussou, Liu Hsu and Christian Mira investigated the quasi-conservative map[83]

$$\begin{cases} x_{n+1} = y_n + \mu x_n + \dfrac{2(1-\mu)x_n^2}{1+x_n^2} + \alpha\left(1 - \beta y_n^2\right) y_n \\ y_n = -x_n + \mu x_{n+1} + \dfrac{2(1-\mu)x_{n+1}^2}{1+x_{n+1}^2} \end{cases} \tag{13}$$

and proposed beautiful pictures of *stochastic annular zone*, that is, of chaotic solutions (Fig. 11).

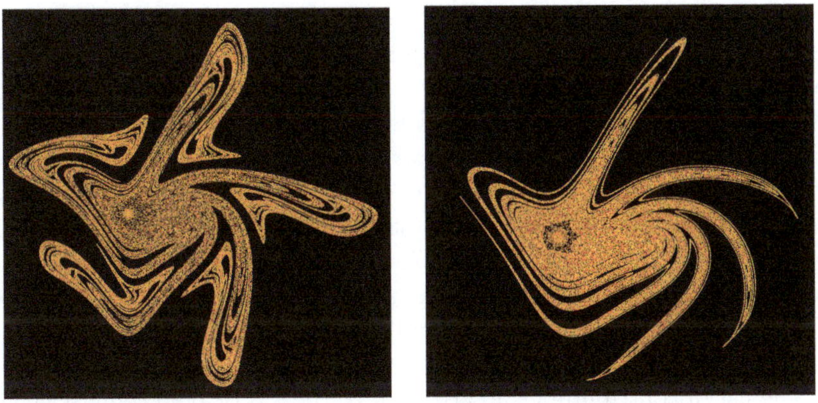

(a) $\beta = 1$ (b) $\beta = 10$

Fig. 11. The so-called "mythic bird" (a) solution to the map (13): it was the "logo" of the 1973 Toulouse conference. Another type of chaotic solution (b) to the same map.

Notes: Other parameter values: $\alpha = 0.005$, and $\mu = \frac{1}{8}$.

5. From the Como Conference to ALISE

The Como conference held on June 1977. On August 8, 1977, Chirikov wrote to Ford encouraging him to produce a collection of abstracts as follows: "*I believe that it is a brilliant idea about your ALISE (Abstract LIsting SErvice)*".[k] The first set of abstracts was sent on September 1977. It was made of 115 abstracts among which, only 10 cannot be retrieved today from the web, meaning that most of them were published. A third of the abstracts was sent by the attendees at the 1977 Como Conference organized by Giulio Casati and Joseph Ford, which held during the summer of 1977, that is, only a few weeks before the first delivery.

According to its title, *Stochastic behavior in classical and quantum Hamiltonian systems*, the Como conference was devoted to a rather theoretical approach. The term "stochastic" is here used in the sense of Chirikov. At least since 1934, this term designates some processes which are random by nature,[84] and not related to deterministic system. The term "chaotic", although already introduced in 1975,[85,86] did not spread yet in this community (see Arkady's chapter, p. 378).

[k]Letter from Chirikov to Ford, August 8, 1977.*

The aim of the organizers was to bring together astronomers, biologists, economists, physicists, and mathematicians; unfortunately, there was contribution neither in biology nor in economy. Most of the papers were devoted to conservative dynamics with the notable exception of Feigenbaum's work.[87] Among the abstracts, four were announced to appear in the proceedings of the 1977 Como conference but were actually published elsewhere: they were written by Christian Marchal from ONERA,[88] Louis Martinet from the Department of Astronomy of the Genova University showing Poincaré section "à la Hénon",[89,90] Eduard Prugovecki by then at the University of Toronto,[91] and Neville Temperley from Swansea University (whose contribution was not retrieved).

The list of contributors who attended to the 1977 Como conference and sent abstracts to Joe Ford for his first delivery are reported in Table 1 (the references are those retrieved from the web). Seven contributions to the proceedings of the Como conference do not have an abstract in the first delivery: thus, $\frac{2}{3}$ of the authors from these proceedings sent at least one abstract, as well as four other attendees (Marchal, Martinet, Prugovecki and Temperley).

Table 1. List of attendees to the 1977 Como conference who published a paper in the proceedings and sent an abstract to Joseph Ford for his delivery on September 1977.

Contributor (and Ref.)	Field
George Contopoulos[92]	Astronomy
A. J. Lichtenberg[93]	Quantum physics
Gary R. Smith[94]	Quantum physics
Martin Kummer[95]	Dynamical system
R. C. Churchill,[96] G. Pecelli,[97] and D. L. Rod[98]	Astronomy
M. Toda[99]	Optics
O. I. Bogoyavlensky[100]	Dynamical system
Mitchell J. Feigenbaum[101]	Dynamical system
Hermann Haken[102] and A. Wunderlin	Synergetics
J. L. van Hemmen[103]	Condensed matter
Ian C. Percival[104]	Quantum physics
R. A. Marcus,[105] D. W. Noid, and M. L. Koszykowski	Quantum physics
M. Tabor[106]	Quantum physics
R. T. Swimm, and J. B. Delos[107]	Quantum physics
Martin C. Gutzwiller[108]	Dynamical system
G. Casati,[5] R. V. Chirikov,[109] F. M. Izraelev, and Joe Ford	Quantum physics
R. H. G. Helleman,[110] and T. Bountis	Dynamical system

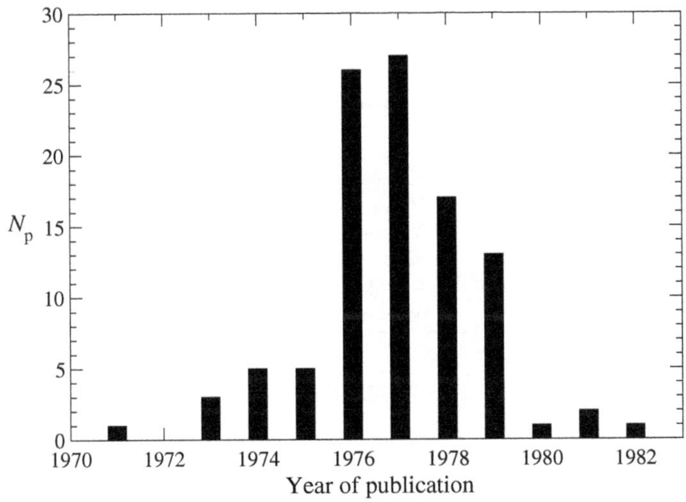

Fig. 12. Year of publication of the papers whose abstracts were inserted in the first delivery by Joseph Ford on September 1977.

In the first delivery, most of the papers were published between 1976 and 1979, only a very few before or after that period (see Fig. 12): the abstracts were therefore related to recent works (just published or already submitted). The mean number of quotations of these resulting papers is rather large ($\overline{N}_q = 480$) although with a large standard deviation ($\sigma_q = 1600$): 43 abstracts correspond to papers quoted more than 50 times. How Ford's delivery might have contributed to the success of these works would deserve further attention. It is interesting to note that, as known in scientific communities, conference papers are less quoted than journal papers. Only two contributions to the proceedings of the 1977 Como conference are often quoted: the paper by Ian Percival on spectra of molecules ($N_q = 562$) and the paper by Casati, Chirikov, Izrailev and Ford on quantum chaos ($N_q = 637$). Among the 110 abstracts, 7 were quoted more than 1,000 times as reported in Table 2. All these contributions are obviously well-known from dynamicists, at least from those who were active in the 1980s and 1990s.

The number of abstracts received by Ford increased quickly and led him to spend restless days (including week-ends, see p. 238) to type them. As detailed by Franco Vivaldi, this led to the creation of the journal *Physica D* which was launched in April 1980. The editors were

Table 2. The seven most quoted papers whose abstracts were inserted in the first set of abstracts sent by Joe Ford, shortly after the 1977 Como Conference.

1st Author	Year	Topics		# Quotations
Haken[102]	1977	Synergetics	—	11186
May[111]	1976	Chaos	Dissipative	8806
Chirikov[109]	1979	Plasma	Conservative	5249
Feigenbaum[101]	1978	Chaos	Dissipative	5041
Hénon[68]	1976	Chaos	Dissipative	3936
Berry[112]	1978	Astrophysics	Conservative	1534
May[113]	1976	Chaos	Dissipative	1484

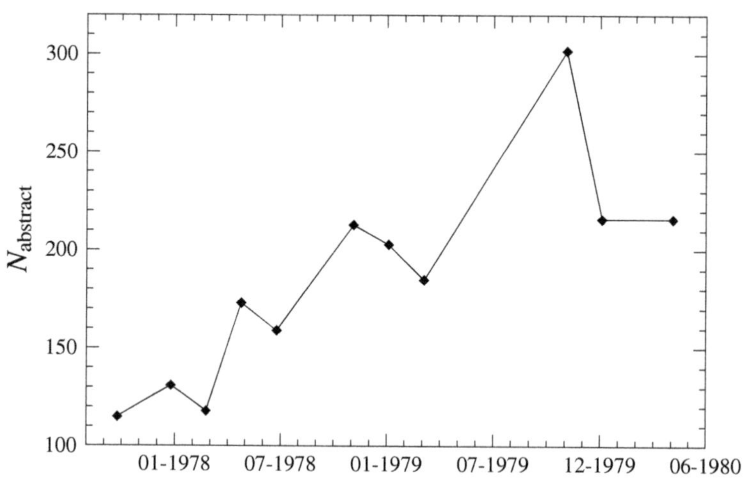

Fig. 13. The number of abstracts contained in each delivery by Ford between September 1977 and May 1980.

Hermann Flaschka (University of Arizona, Tucson, USA), Joseph Ford (Georgia Institute of Technology, Georgia, USA), Alan C. Newell (Clarkson College of Technology, New York, USA) and A. Scott (University of Wisconsin, USA). Among the advisory editors, were Paul C. Fife (Tucson, AZ), Hermann Haken (Stuttgart, Germany), Michel Hénon (Nice, France), Allan N. Kaufman (Berkeley, CA), Yakov Sinai (Moscow), Morikazu Toda (Yokahama), and Vladimir E. Zakharov (Moscow).

The table of contents of the first issue of *Physica D* was

(1) A. R. Bishop, J. A. Krumhans, and S. E. Trullinger, Solitons in condensed matter: A paradigm, pp. 1–44;

(2) D. J. Kaup, The inverse scattering solution for the full three dimensional three- wave resonant interaction, pp. 45–67;
(3) S. A. Orszag and J. B. McLaughlin, Evidence that random behavior is generic for nonlinear differential equations, pp. 68–79;
(4) M. Jimbo, T. Miwa, Y. Mõri, and M. Sato, Density matrix of an impenetrable Bose gas and the fifth Painlevé transcendent, pp. 80–158;
(5) A. Sym, The second soliton workshop (27 August–1 September 1979, Jadwisin, Poland): A short summary, pp. 159–163.

Ford was translating some manuscripts, for instance from Japanese-English into American-English and improving the presentation, to have a submittable work.[1]

Among these four papers published in the first issue of *Physica D*, let us focus our attention on the work devoted to a simple model for reproducing the "random" nature of turbulence. This is in fact a light development of one point investigated in a long lecture on the statistical approach of turbulence[114] given by Steven Orszag (1943–2011 from MIT) who sent an abstract of it for Ford's first delivery. His approach is based on Boussinesq's views of turbulence distinguishing the main flow — responsible for the volume crossing a section — which is deterministic and the small-scale which might be apparently random and unstable, and whose contribution to the main flow is null.[115] Fully developed turbulent flows are strongly sensitive to triggering perturbations but not their statistical properties. One of the interesting points developed by Orszag is the introduction of a set of differential equations

$$\dot{x}_i = x_{i+1}x_{i+2} + x_{i-1}x_{i-2} - 2\,x_{i+1}x_{i-1} \quad (i = 1, \ldots, 5) \tag{14}$$

where $x_i = x_{i+5}$ to mimic the "random" nature of turbulence. This is a chain of equations with quadratic monomials as in the FPUT problem. He designed this set of equations to be analogous to the Navier–Stokes equations in the sense that (i) only quadratic monomials are considered among the modes retained, (ii) the energy is a constant of motion since

$$\frac{\mathrm{d}}{\mathrm{d}t} \sum_{i=1}^{5} \frac{x_i^2}{2} = 0 \tag{15}$$

[1]Letter from Ford to Allan Kaufman and to Hermann Haken, September 18, 1981.

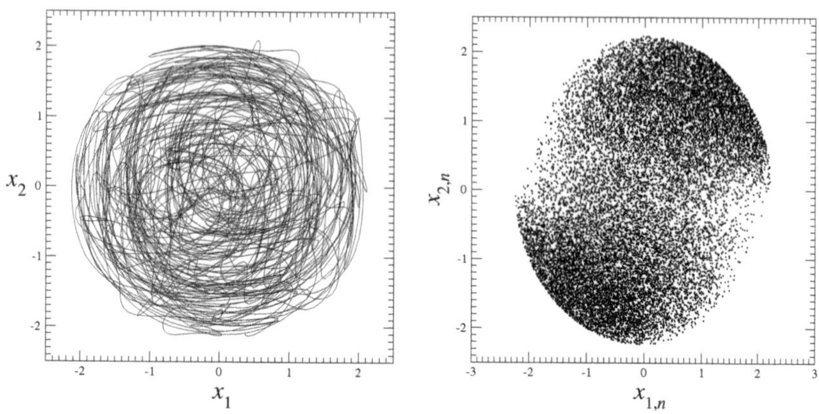

(a) Of the state space (b) Of the Poincaré section

Fig. 14. Plane projection (a) of a "random" trajectory produced by the quadratic model (14) used by Orszag for investigating the stastical properties of turbulence. A plane projection of the four-dimensional Poincaré section defined by $x_3 = 0$ (b) is also plotted. Initial conditions: $x_1(0) = 0.54$, $x_2(0) = -1.54$, $x_3(0) = -0.68$, $x_4(0) = -1.19$, and $x_5(0) = -0.68$, corresponding to the constant energy $E = 2.5$.

and (iii) Liouville's theorem

$$\sum_{i=1}^{5} \frac{\partial}{\partial x_i} \left(\frac{\mathrm{d}x_i}{\mathrm{d}t} \right) = 0 \tag{16}$$

applies, suggesting that an incompressible fluid is considered. He then used some results from the nonlinear dynamical systems theory to investigate the solutions of this system. He computed a trajectory [Fig. 14(a)] for which he showed that its statistical properties are in agreement with those of an ergodic solution. He thus concluded that the "deceivingly simple" model (14) demonstrates "the nature of random solutions to differential equations".

The paper which appeared in the first issue of *Physica D* was written with John McLaughlin who published before three papers about the transition to turbulence.[116–118] With Paul Martin, he tested the Ruelle–Takens scenario[119] in showing that, in a three Fourier modes decomposition of a Rayleigh–Bénard convection, the motion is periodic and not aperiodic as in higher-dimensional truncations.[116] These two authors mentioned the Lorenz system and reported some Lorenz's main results in a second paper[117] to conclude that the Lorenz system is too strongly truncated to correspond to experimental dynamics (see Pikovsky's chapter, p. 378, for its reception).

Their results seemed to support the Ruelle–Takens scenario. Orszag and McLaughlin were exchanging since 1975 (Orszag is acknowledged for helpful discussions in a 1976 paper[118]). What was added with respect to Orszag's initial lecture is a study in which the dimension of the model is varied to test the Ruelle–Takens scenario, the adding of small dissipative term to check the robustness of the results against small dissipation, and a Poincaré section (inherited from Lorenz's paper) to exhibit the ergodic nature of the solution [Fig. 14(b)]. Orszag and McLaughlin thus showed that they got similar results as those obtained by Lawrence Kells and Orszag with some Galerkin truncations of Navier–Stokes equations for the two-dimensional flow of an inviscid fluid. Chirikov and Ford are acknowledged for stimulating discussions and their works are quoted,[36,55,78,120] with those of Hénon,[37] and Arnol'd,[21] among others.

6. Ford as an Evangelist

Ford was spending a lot of his time for helping his colleagues and friends to publish their ideas. Let us take an arbitrary example based on an exchange with Giulio Casati.[m] This letter is in fact a ten pages report on a paper that was published in the proceedings of the Symposium on Synergetics which held at Schloss Elmau (Bavaria) on April 26–May 1, 1982.[121] Ford started to invite his friend to smooth his English, explaining that there is "*no necessity* [to] *go through the personal agony of improving* [his] *humorously Italian flavored English,* [but he needs] *only to have someone who does speak and write fluent English to correct* [his] *preprints.*" When Ford published his wide-audience paper entitled "How random is a coin toss?" in *Physics Today,* he expressed the regrets that Casati was considering it as being a *purely philosophical or conjectural tract.* It is interesting to note that, contrary to Casati, Ford was convinced by the fact that it was not possible to define accurately chaos and, indeed, his definition, based on predictability and computability, was quite critized.[122] If there are a few definitions for chaos, there is actually a possibility to propose some efficient definitions, although some counterexamples can always be provided.[123,124] In fact, Ford was so confident in the numerical test for randomness that, since chaotic dynamics "passes every computable test for randomness," chaos is random! It appears that Ford never thought otherwise than within the statistical mechanics, his initial field. Within a topological approach, chaos

[m]Letter from Ford to Casati, June 12, 1982.

is structured around a population of periodic orbits[9,125] while random motion is not. Ford was quite close to this when he asserted that "*most orbits of a random system are themselves also random.*" Indeed, the next step is to assert that an uncountable set of orbits of chaotic systems are periodic orbits, being stable or unstable. Ford's answer has the typical wording of a *predicator*!

Ford was also complaining that Casati did not mention the KAM theory in his review. He also remarked that quantum mechanics is always based on weakly-perturbed integrable systems, leaving it invalid for describing fully chaotic motion of quantum system. Breaking his initial argument in favor of quantum chaos that, according to the correspondence principle, quantum chaos should exist as the counterpart of classical chaos,[126] he asserted that chaotic quantum systems do not approach their classical counterparts when the Planck constant tends to zero. Then, Ford quoted the not yet solved problems in quantum mechanics as the duality between wave and particle. And so on for five more pages... If one has a look into Casati's paper, Ford's comments are clearly far beyond what is in it. He actually admitted that his comments were more a review of his recent thoughts than a response to Casati's review on quantum chaos.[n] For instance, Ford, self-proclamed "the evangelist of chaos," provided these maxims:

> Chaos, a state of things in which chance is supreme.[o]
>
> Chaos workers never die... they simply lose control of their variables![p]

7. Friendship Between Ford and Chirikov

The correspondence between Ford and Chirikov started as least on 1966.[q] During the summer of that year, Ford visited nine Soviet cities among which Leningrad, Moscow, Samarkand, Kiev and tried to meet Chirikov. In 1977, Ford was sending materials (most likely, reprints) to Chirikov. If these two scientists were stimulated by the same type of problems — integrability versus stochasticity, what is quantum chaos? — they had a very different ways of thinking that they opposed in a touching friendship. Perhaps this can be summarised by saying that Ford was more motivated by general ideas and their metaphysical consequences, while Chirikov favored

[n]Letter from Ford to Rössler, June 29, 1982.
[o]Webster's 1963 New Collegiate Dictionary.
[p]Letter from Ford to Otto, July 4th, 1994.
[q]Letter from Ford to Chirikov, April 12, 1966.*

the rigor of theory. Both cherished statistical mechanics. Ford lost himself in trying to solve paradoxes or unsolvable problems while Chirikov produced significant landmark results.

Ford knew that "*to accept the future, we must renounced of the past, a formidable challenge*".[127] Thus, quoting Lev Tolstoï (1828–1910)[128]:

> I know that most men, including those at ease with problems of the greatest complexity, can seldom accept even the simplest and most obvious truth if it be such as would oblige them to admit the falsity of conclusions which they have delighted in explaining to colleagues, which they have proudly taught to others, and which they have woven, thread by thread, into the fabric of their lives.

Where Ford provided pictures and metaphysics, Chirikov provided equations and concepts: compare Ford's "*what is chaos*"[127] to Chirikov's "*Dynamical mechanisms of noise.*"[129] The two scientists were opposed about quantum chaos. Ford viewed their differences as follows[r]:

> You are one of the conservatives who could prefer a gradual and confortable evolutionary chaos, while I clearly am a rebel seeking the future as rapidly as I can, perhaps more rapidly you might say.

In a letter sent to Chirikov, Ford wrote as being a referee of a recently published paper,[130] complaining about the time spent: "*How I can do my own research when I spend so much time doing work for friends? It is the price for fame I suppose*". He then wrote[s]

> The notion of transient chaos would represent little more than an historical footnote were it not for the fact that all known evidence indicates that quantum mechanics does not exhibit any true chaos, i.e., any randomness over and abobe that containted in the wave function itself or probabilities derived from it. One may now adopt either of two attitudes: quantum mechanics can adequately describe observation by careful use of pseudo-chaos and transient chaos. In this paper, Chirikov seeks to salvage quantum mechanics through transient chaos. It is a noble and worthwhile cause; but only time will reveal whether this view reflects the truth or only the final argument of quantal conservatism.

Ford claimed that his approach was inherited from the statistical mechanics[127]:

> in its strictest technical sense, chaos is merely a synonym for randomness as the algorithmic complexity theory of Andrei Kolmogorov,[131,132]

[r]Letter from Ford to Chirikov, July 1985.
[s]Letter from Ford to Chirikov, November 18, 1986.

> Gregory Chaitin,[133] and Ray Solomonov[134, 135] so clearly reveals" [that] a chaotic orbit generates a random cell-number sequence having maximal complexity. The chaotic orbit itself is therefore truly random in the sense that such an orbit cannot be computed by any finite algorithm and its information content is both infinite and incompressible.
>
> Our orbital computation is found to require just as much informational input as it provides in informational output. This means that our computers have now ceased computing or predicting anything because the output orbital data are so chaotic, so unpredictably random that our input information must, of necessity, be equivalent to a copy of the output. [...]
>
> For chaotic systems, determinism, existence-uniqueness remain, as they must, but predictability is lost.
>
> [...] analytically solvable systems having highly ordered motions like the harmonic oscillator or the Kepler problem have orbits which require only a relatively small amount of input information to reap a large quantity of output orbital information. [...] Thus, these analytically solvable systems are not only deterministic, they are also reasonably predictable and nonrandom. However, when the Newtonian system becomes sufficiently complex — a hard sphere gas or a roulette wheel, for example — our orbital computation is found to require just as much informational input as it provides in informational output.

Ford used the example of coin toss. Michael V. Berry notes that "it is amusing [...] that in the ideal roulette wheel and coin toss — two exemplars of unpredictable systems whose randomness arises largely from imperfect knowledge of initial conditions — the motion is in fact integrable ... so that the instability is of a linear type and therefore much weaker than in, say, a pinball machine [or a hard sphere gas]."[136] In fact, Ford understood that Chirikov considered his paper published in *Physics Today*[137] as "hard science" but rejected his resolution of the random-determinate paradox.[t] Chirikov was thanked for his feedback and encouragements with humor: *Praise should be shipped FOB, Novosibirsk; blame should be forwarded directly to Atlanta.*

Ford considered himself as a *revolutionary.*[u] Nevertheless he admitted that it was sometimes painful for him to reach a satisfactory level of rigor, as for instance, to have a rigorous 6th section in this *Physics Today* paper.

To end this section devoted to the special partnership between Ford and Chirikov, let us mention an event described by Ford as follows.[138]

[t]Letter from Ford to Casati, June 12, 1982.
[u]Letter from Ford to Casati, June 12, 1982.

[There is a] remarkable evening when Yascha Sinai and I stood stark naked and toe to toe in downton Moscow beating each other with birch boughs. "*Are there photographs of this scene*"? Surprisingly, yes. Yascha Sinai, Joe Ford and Giulio Casati all have copies available for inspection (maybe even distribution!, see Fig. 16).

The Georgia Institute of Technology
Atlanta, Georgia 30332–0430
School of Physics

(404) 894-5255

19 December 1988

Dear Boris,

CONGRATULATIONS!!!

You ought to be proud of yourself.

Hope that you and yours have the very absolute, most happiest... ever...

Merry Christmas to all and to all good night,

Joe

joseph ford
The Evangelist Of Chaos

P.S.1. Imagine that this letter came first.

P.S.2. Many have commented on the moving and tender prose style of your Preface. I, of course, tell them that it is but another manifestation of my influence on the chaos community.

Fig. 15. A typical letter written by Joseph Ford in a flourished style. Letter from Ford to Chirikov, December 19, 1988.

Source: Courtesy of Dima Shepelyanski.

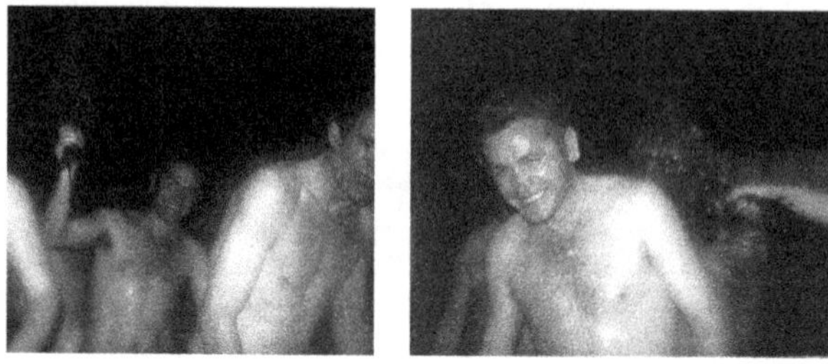

(a) Giulio beating Joe (b) Joe beating Giulio

Fig. 16. The memorable evening spend by Joe Ford, Giulio Casati and Yascha Sinai in a sauna at Moscow.
Sources: Pictures shot by Sinai. Courtesy of Giulio Casati.

> A goodly group was in attendance at the Russian sauna that evening, but I no longer remember precisely who was there. What I remember vividly is roasting over a well-stoked furnace for several eternities and then dashing out to die in a pool filled with ice cubes. As the evening progressed, iterative trips between furnace and refrigerator were interrupted only by brief stopovers for quick bites of mystery meat on dry brown bread and gulps of warm Russian beer.

8. Ford–Rössler Symbiosis

The first letter between Ford and Rössler which was found is dated on November 1978, Rössler was thanking Ford for his first delivery of the *Nonlinear Science abstracts.*[v] They meet each other at the International Symposium on Synergetics at Schloss Elmau organized by Hermann Haken. During that conference, Ford was enjoying discussions with Rössler[w]:

> I found it especially gratifying to learn that another shared some of my heretical thoughts. [...] There is another major heresy of mine that I did not mention at Schloss Elmau; I am quite convinced that the second law of thermodynamics is not of unlimited validity and that, therefore, a perpetual motion machine of the second kind is feasible. [...] Why then may we not permit an integrable system to absorb energy (into its controllable degrees of freedom) from a heat reservoir and then completely extract this energy in the form of work delivered to a work reservoir?

[v] Letter from Rössler to Ford, November 23, 1978.
[w] Letter from Ford to Rössler, June 8, 1981.

For Rössler too, this was "*one of the nicest scientific experiences* [he] *had in years*".[x] Ford and Rössler were quite close. In 1987, Ford wrote[y] "*I love you like a brother and respect you as my most esteemed colleague*!"

On September 7–11, 1981, Rössler gave in Bordeaux (France) a talk entitled *Chaos and chemistry*, which appeared in the proceedings of the conference about chemical dynamics.[139] Otto sent a preprint to Ford — acknowledged on October 5th, 1981. Ford was stimulated to publish it in his recently launched *Physica D*.[z] Alan C. Newell, one of the advisory editors, replied to Ford.[aa]

> *Physica D* should provide a forum for articles which stimulate scientific thinking along new directions. [...] However, one does not want to open the door to the crazies and therefore such papers must be written and argued very carefully. Some people are good at looking at things with fresh perspectives but at the same time cannot bring their thoughts to concrete levels without feedback from serious critics. [...] I have no argument with either [Rössler's manuscript] (which seems particularly appropriate for *Physica D*) or the concept of an idea paper. It should however reviewed by ideas people e.g. Kruskal, Feynman, Ruelle, Prigogine, Rice, Lebowitz, for the purpose of clarifying fuzzy thinking and fuzzy writing.

Regarding Otto's writing, in a private review of one of his manuscripts "an estimate of Planck's constant" written possibly in the early 1982, Ford wrote

> Your present manuscript contains a hunch supported by conjectures. Given the magnitude of the task you here undertake, this is the equivalent of attacking a large dragon with a penknife. Please understand that the hard facts in your possession are inadequate to survive the withering scrutiny of a hostile and cruel world.

Using the example of a numerical determination of the Planck constant by Galgani and Scott who felt into total disrepute and became considered as "crazy Italians",[66] Ford tried to discourage Otto to publish his manuscript. It seems that Ford was not successful since Otto published his manuscript in the Proceedings of a conference which held in Budapest.[140]

---

[x] Letter from Rössler to Ford, June 25, 1981.
[y] Letter from Ford to Rössler, November 29th, 1987.
[z] Letter from Rössler to Ford, January 10, 1981.
[aa] Letter from Alan Newell to Joseph Ford, September 1981.

Fig. 17. First Dynamics Days organized at La Jolla (California) (1982). From the left to the right: Tassos Bountis, Robert Helleman (Bountis' Ph.D. advisor and the initiator of the "Dynamics Days" series), Hans Groeneveld (mathematical physicist from Utrecht, Netherland), Joseph Ford, and Giulio Casati.

Source: By courtesy of Tassos Bountis.

In 1987, Ford was complaining about Rössler[bb]:

> Otto, you are as bright and shiny as a beautiful new button and as warmly cuddly as a blue-eyed teddy bear, but your hand-writing is an abomination in the eyes of God and is absolutely beyond my poor capacity to read. In consequence, I have to enlist the aid of three wise men and a camel (the camel does all the work) to gain a rough translation of your letters, and even then, the hardest parts remain untranslated. Moreover at this point, I have not even begun the struggle to understand your science. Is there not perhaps a typewriter available for your use somewhere there in Tübingen? If so, would you wondrously kind as to use it, thereby giving me the benefit of a running start toward gaining an understanding of the content of your letters?

[bb]Letter from Ford to Rössler, November 29th, 1987.

Fig. 18. Participants to the 1983 Como NATO Advanced Research Workshop on quantum chaos. June 20, 1983 in the front of the Villa Olmo. By courtesy of Otto E. Rössler. First row from the left: 2nd Michael Berry, 5th Tassos Bountis, 8th Joseph Ford, 12th Martin Kruskal, at the right end Giulio Casati. 2nd row: 5th B. Grammaticos, 6th B. Dorizzi, 2nd from the right Jean Belissard... Participants were Albeverio (Ruhr Universität), V. Aquilanti (Perugia, Italy), J. Belissard (Marseille, France), M. V. Berry (Bristol, UK), S. D. Bosanac (Zagreb, Croatia-Yugoslavia), C. M. Bowden (Huntsville, USA), H. A. Brown (Rolla, USA), R. J. Creswick (Columbia, USA), H. Dekken (Den Haag, Netherland), G. Dell'Antonio (Roma, Italy), J. B. Delos (Williamsburg, USA), K. Dysthe (Tromsc, Norway), B. Eckhardt (Atlanta, USA), G. S. Ezra (Ithaca, USA), M. Feigenbaum (Ithaca, USA), S. Fishman (Haifa, Israel), L. Garrido (Barcelona, Spain), M. Goldman (Colorado, USA), Graham (Essen, Germany), I. Guarneri (Pavia, Italy), M. Gutzwiller (IBM, USA), J. Hannay (Bristol, UK), H. Haken (Stuttgart, Germany), G. Hase (Amsterdam, Netherlands), R. Helleman (Twente, Netherlands), R. Jancel (Paris, France), R. V. Jensen (Yale, USA), A. Ladislau (Prague, Czechoslovakia), W. E. Lamb (Arizon, USA), A. N. Kaufman (California, USA), M. A. Keramat (Los Alamos, USA), J. R. Klauder (Bell Laboratories, USA), M. Kruskal (Princeton, USA), H. Lehnigk (London, UK), Z. Lewis (Kaserslautern, Germany), G. Lindblad (Sweden), B. Mandl (Twente, Netherlands), R. A. Marcus (CalTech, USA), I. Masayoshi (Kagoshima, Japan), B. Nagel (Stockholm, Sweden), J. W. Neuberger (Denton, USA), P. Pechukas (Columbia, USA), A. Peres (Technion, Israel), G. Propser (Milano, Italy), R. Ramaswamy (Tata, India), W. P. Reinhardt (USA), P. Richens (Bristol, UK), M. Robnik (Bonn, Germany), D. Rökaerts (Belgium), O. E. Rössler (Tübingen, Germany), J. A. Sanders (Amsterdam, Netherlands), K. E. Shuler (La Jolla, USA), Souillard (Paris, France), H. S. Taylor (Amsterdam, Netherlands), R. D. Taylor (Toronto, Italy), J. Tennyson (USA), F. Valz-Gris (Pavia, Italy), N. van Kampen (Amsterdam, Netherlands), van Zejts (Twente, Netherlands), J. J. M. Veraaschot (Germany), F. Verhulst (Amsterdam, Netherlands), W. M. Visscher (Los Alamos, USA), F. Vivaldi (Pavia, Italy), A. Voros (Saclay, France), J. A. Wheeler (USA), F. W. Wiegel (Twente, Netherlands), M. Wilkinson (Bristol, UK), S. S. M. Wong (Toronto, Canada), R. Wyatt (USA).

What Ford seemed to be lost in the front of Rössler's creativity, blaming himself for not being able to ask more than "*how, where, when, and especially why did you first define hyperchaos?*"

"*My good friend in Tübingen advised me never to trust, much less read, an article with no pictures.*"[cc] Consequently, Ford requested Rössler to send a picture "*to guide* [him] *through* [his] *article*", suggesting one of Rössler's choice appearing in *Alice in wonderland.*

9. Conclusion

Joseph Ford was one of the founders of the conservative chaos community. Unstinting in his efforts to disseminate not only his own ideas on the subject, but also the work of others, he was responsible for several critical actions leading to the emergence of the subject: revival of the FPUT problem, organization of the 1977 (and later 1983) Como conference and creation of the journal *Physica D.* He contributed to the development of the study of conservative chaos and to the emergence of quantum chaos. His first contribution on the FPUT problem triggered a revival of interest in the subject; his subsequent contributions helped to spread the ideas that emerged in the 1960s and 1970s. His definition of chaos in terms of its randomness was not adopted by chaoticians: some of his texts remain, but they seem a little confused today. Nevertheless, Ford remains indispensable for a good understanding of the emergence of conservative chaos and quantum chaos and their respective communities.

Acknowledgments

CL wishes to thank Giulio Casati for his enthousiast interactions and for providing so many nice comments and documents. He wishes to thank Pierre Hénon for providing a picture of his brother and some comments about it. He wishes to thank Haris Skokos and Tassos Bountis for stimulating discussions and some pictures. He thanks the Georgia Institute of Technology for some letters exchanged by Joseph Ford with Boris Chirikov and Otto E. Rössler; the latter is thanked for some other letters.

[cc]Letter from Ford to Rössler, June 8th, 1991.

References

1. T. H. Berlin and M. Kac, The spherical model of a ferromagnet, *Physical Review.* **86**, 821–835 (1952). doi: 10.1103/PhysRev.86.821.
2. J. Ford, Science and the taxpayer, *The Georgia Tech Alumnus.* **43**(5), 6–9 (1965).
3. B. Latour and S. Woolgar, *Laboratory Life: The Social Construction of Scientific Facts.* Sage Publications, Beverly Hills (1979).
4. A. Persson, Right for the wrong reason?: A new look at the 6 June 1944 D-Day forecast by a neutral swede, *Bulletin of the American Meteorological Society.* **101**(7), E993–E1006 (2020). doi: 10.1175/BAMS-D-18-0311.1.
5. G. Casati, B. V. Chirikov, F. M. Izraelev, and J. Ford, Stochastic behavior of a quantum pendulum under a periodic perturbation. In G. Casati and J. Ford (eds.), *Stochastic Behavior in Classical and Quantum Hamiltonian Systems.* Springer, Berlin (1979), pp. 334–352. doi: 10.1007/BFb0021757.
6. J. Ford, Quantum chaos, is there any? In *Directions in Chaos*, Vol. 2. World Scientific Publishing, Singapore (1988), pp. 128–147. doi: 10.1142/9789814415729_0006.
7. J. Ford, The Fermi-Pasta-Ulam problem! paradox turns discover, *Physics Reports.* **213**(5), 271–310 (1992).
8. H. Poincaré, Sur le problème des trois corps et les équations de la dynamique, *Acta Mathematica.* **13**, 1–270 (1890).
9. H. Poincaré, *Les méthodes nouvelles de la mécanique céleste III.* Gauthier-Vilard, Paris (1899).
10. G. D. Birkhoff, *Dynamical Systems.* American Mathematical Society, New York (1927).
11. E. Fermi, Beweis dass ein mechanisches Normalsystem im allgemeinen quasi-ergodisch ist, *Physikalische Zeitschrift.* **24**, 261–265 (1923).
12. E. Fermi, Dimostrazione che in generale un sistema meccanico normale è quasi-ergodico, *Nuovo Cimento.* **25**(1), 267–269 (1923). doi: 10.1007/BF02959600.
13. E. Fermi, Generalizzazione del teorema di Poincaré sopra la non esistenza di integrali uniformi di un sistema di equazioni canoniche normali, *Nuovo Cimento.* **26**, 105–115 (1923). doi: 10.1007/BF02959350.
14. E. Fermi, Über die Existenz quasi-ergodischer Systeme, *Physikalische Zeitschrift.* **25**, 166–167 (1924).
15. J. Tuck and M. Menzel, The superperiod of the nonlinear weighted string (FPU) problem, *Advances in Mathematics.* **9**(3), 399–407 (1972). doi: 10.1016/0001-8708(72)90024-2.
16. E. Fermi, J. Pasta, and S. Ulam, Studies of nonlinear problems. Technical Report Report LA-1940-1, Los Alamos Laboratory, Los Alamos (1955).
17. A. N. Kolmogorov, On conservation of conditionally periodic motions for a small change in Hamilton's function, *Doklady Akademii Nauk SSSR.* **98**, 527–530 (1954).

18. V. I. Arnol'd, Small denominators. I. Mapping the circle onto itself, *Izvestiya Akademii Nauk SSSR. Seriya Matematicheskaya.* **25**, 21–86 (1961).
19. V. I. Arnol'd, Proof of a theorem of A. N. Kolmogorov on the invariance of quasi-periodic motions under small perturbations of the Hamiltonian, *Russian Mathematical Surveys.* **18**, 9–36 (1963). doi: 10.1070/RM1963v018n05ABEH004130.
20. J. Barrow-Green, *Poincaré and the Three Body Problem.* History of Mathematics, Vol. 11. American Mathematical Society (1997).
21. V. I. Arnol'd, Small denominators and problems of stability of motion in classical and celestial mechanics, *Russian Mathematical Surveys.* **18**, 85–191 (1963). doi: 10.1070/RM1963v018n06ABEH001143.
22. V. K. Mel'nikov, On lines of force in a magnetic field, *Doklady Akademii Nauk SSSR.* **144**, 747–750 (1962).
23. V. K. Mel'nikov, On the stability of a center for time-periodic perturbations, *Trudy Moskovskogo Matematicheskogo Obshchestva.* **12**, 3–52 (1963) (in Russian).
24. V. K. Mel'nikov, Qualitative description of strong resonance in a nonlinear system, *Doklady Akademii Nauk SSSR.* **148**, 1257–1260 (1963).
25. B. V. Chirikov, Resonance processes in magnetic traps, *Atomnaya Energiya.* **6**, 630 (1959) (in Russian).
26. J. Ford, Equipartition of energy for nonlinear systems, *Journal of Mathematical Physics.* **2**(3), 387–393 (1961). doi: 10.1063/1.1703724.
27. N. Kryloff and N. Bogolyubov, *Introduction to Nonlinear Mechanics.* Princeton University Press, Princeton, New Jersey (1947).
28. C. L. Siegel, On the integrals of canonical systems, *Annals of Mathematics.* **42**(3), 806–822 (1941). doi: 10.2307/1969262.
29. J. Ford and J. Waters, Computer studies of energy sharing and ergodicity for nonlinear oscillator systems, *Journal of Mathematical Physics.* **4**(10), 1293–1306 (1963). doi: 10.1063/1.1703904.
30. B. V. Chirikov and F. M. Izrailev, Statistical properties of a non-linear string, *Soviet Physics–Doklady.* **11**, 30–32 (1966).
31. F. M. Izrailev, A. I. Khisamutdinov, and B. V. Chirikov, Numerical experiments with a chain of coupled anharmonic oscillators. Technical Report Report 252, Institute of Nuclear Physics, Novosibirsk, USSR (1968). English translation: LA-4440-TR, Los Alamos, 1970.
32. M. Toda, Vibration of a chain with nonlinear interaction, *Journal of the Physical Society of Japan.* **22**(2), 431–436 (1967). doi: 10.1143/JPSJ.22.431.
33. M. Toda, Waves in nonlinear lattice, *Progress of Theoretical Physics Supplement.* **45**, 174–200 (1970). doi: 10.1143/PTPS.45.174.
34. J. Ford, S. D. Stoddard, and J. S. Turner, On the integrability of the toda lattice, *Progress of Theoretical Physics.* **50**(5), 1547–1560 (1973). doi: 10.1143/PTP.50.1547.
35. N. Saitô, N. Ooyama, Y. Aizawa, and H. Hirooka, Computer experiments on ergodic problems in anharmonic lattice vibrations, *Progress of Theoretical Physics Supplement.* **45**, 209–230 (1970). doi: 10.1143/PTPS.45.209.

36. J. Ford and G. H. Lunsford, Stochastic behavior of resonant nearly linear oscillator systems in the limit of zero nonlinear coupling, *Physical Review A.* **1**(1), 59–70 (1970). doi: 10.1103/PhysRevA.1.59.
37. M. Hénon, Integrals of the Toda lattice, *Physical Review B.* **9**, 1921–1923 (1974). doi: 10.1103/PhysRevB.9.1921.
38. M. Hénon, Sur l'évolution dynamique des amas globulaires, *Annales d'Astrophysique.* **24**, 369–419 (1961).
39. M. Hénon, *Generating Families in the Restricted Three-Body Problem.* Springer, Berlin (1997).
40. A. Deprit and R. Broucke, Régularisation du problème restreint plan des trois corps par represéntations conformes, *Icarus.* **2**, 207–218 (1963). doi: 10.1016/0019-1035(63)90016-2.
41. G. Lemaître, Regularization of the three-body problem, *Vistas in Astronomy.* **1**, 207–215 (1955). doi: 10.1016/0083-6656(55)90028-3.
42. R. A. Broucke, Periodic orbits in the restricted three-body problem with earth-moon masses. Technical Report, Jet Propulsion Laboratory, California Institute of Technology (1968).
43. R. Broucke, Recherches d'orbites périodiques dans le problème restreint plan (système terre-lune). Ph.D. Thesis, Université Catholique de Louvain, Belgium (February, 1963).
44. H. Poincaré, Sur les solutions périodiques et le principe de moindre action, *Comptes-Rendus de l'Académie des Sciences.* **123**, 915–918 (1896).
45. G. D. Birkhoff, Regularization of the restricted three-body problem, *Rendiconti del Circolo Matematico di Palermo.* **39**, 265–334 (1915). doi: 10.1007/BF03015982.
46. R. Broucke, Regularizations of the plane restricted three-body problem, *Icarus.* **4**(1), 8–18 (1965). doi: 10.1016/0019-1035(65)90013-8.
47. A. Chenciner, A note by Poincaré, *Regular and Chaotic Dynamics.* **10**, 119–128 (2005). doi: 10.1070/RD2005v010n02ABEH000306.
48. M. Hénon, 44. Calcul de modèles avec une machine analogique. In *Liege International Astrophysical Colloquia.* Liege International Astrophysical Colloquia, Vol. 9 (1959), pp. 490–491.
49. J.-M. Alimi, R. Mohayaee, and J. Perez, *Une vie dédiée aux systèmes dynamiques : hommage à Michel Hénon.* Hermann, Paris, France (2016).
50. M. Hénon and C. Heiles, The applicability of the third integral of motion: Some numerical experiments, *The Astronomical Journal.* **69**(1), 73–79 (1964).
51. V. I. Arnol'd, Sur la topologie des écoulements stationnaires des fluides parfaits, *Comptes-Rendus de l'Académie des Sciences.* **261**, 17–20 (1965).
52. J. Laskar, Michel Hénon and the stability of the solar system (2014). Paper based on a talk during the workshop in honour of Michel Hénon, Institut Henri Poincaré, Paris, December 5, 2013.
53. M. Hénon, Sur la topologie des lignes de courant dans un cas particulier, *Comptes-Rendus de l'Académie des Sciences (Paris).* **262**, 312–314 (1966).

54. M. Hénon, Numerical study of quadratic area-preserving mappings, *Quarterly of Applied Mathematics.* **27**, 291–312 (1969). doi: 10.1090/qam/253513.
55. G. H. Walker and J. Ford, Amplitude instability and ergodic behavior for conservative nonlinear oscillator systems, *Physical Review.* **188**, 416–432 (1969). doi: 10.1103/PhysRev.188.416.
56. G. Contopoulos and L. Woltjer, The "third" Integral in Non-Smooth Potentials, *Astronomical Journal.* **140**, 1106 (1964). doi: 10.1086/148009.
57. B. Barbanis, On the isolating character of the "third" integral in a resonance case, *Astronomical Journal.* **71**, 415 (1966). doi: 10.1086/109946.
58. M. Rosenbluth, R. Sagdeev, J. Taylor, and G. Zaslavski, Destruction of magnetic surfaces by magnetic field irregularities, *Nuclear Fusion.* **6**(4), 297 (1966). doi: 10.1088/0029-5515/6/4/008.
59. N. Metropolis, M. L. Stein, and P. R. Stein, On finite limit sets for transformations on the unit interval, *Journal of Combinatorial Theory A.* **15**, 25–44 (1973).
60. J. M. A. Danby, *Fundamentals of Celestial Mechanics.* The Macmillan Company (1962).
61. J. M. A. Danby, *Astrophysics Simulations.* Wiley (1965).
62. J. M. A. Danby, Motion of a satellite of a very oblate planet, *Astronomical Journal.* **73**, 1031 (1968). doi: 10.1086/110765.
63. J. M. A. Danby, Wild dynamical systems, and the role of two or more small divisors. In G. E. O. Giacaglia (ed.), *Periodic Orbits, Stability and Resonances.* Springer Netherlands, Dordrecht (1970), pp. 272–285. doi: 10.1007/978-94-010-3323-7_23.
64. F. G. Gustavson, On constructing formal integrals of a Hamiltonian system near an equilibrium point, *Astronomical Journal.* **71**, 670 (1966). doi: 10.1086/110172.
65. J. H. Bartlett and C. A. Wagner, The stability of motion in a periodic cubic force field, *Celestial Mechanics.* **2**, 228–236 (1970). doi: 10.1007/BF01229497.
66. L. Galgani and A. Scotti, Planck-like distributions in classical nonlinear mechanics, *Physical Review Letters.* **28**, 1173–1176 (1972). doi: 10.1103/PhysRevLett.28.1173.
67. L. Galgani and A. Scotti, Recent progress in classical nonlinear dynamics, *La Rivista del Nuovo Cimento.* **2**, 189–209 (1972). doi: 10.1007/BF02747780.
68. M. Hénon, A two-dimensional mapping with a strange attractor, *Communications in Mathematical Physics.* **50**(1), 69–77 (1976).
69. T. A. Kontorova and Y. I. Frenkel, On the theory of plastic deformation and doubling, *Zhurnal Eksperimental'noi i Teoreticheskoi Fiziki.* **8**, 89 (1938).
70. S. Aubry, The twist map, the extended Frenkel-Kontorova model and the devil's staircase, *Physica D.* **7**(1), 240–258 (1983). doi: 10.1016/0167-2789(83)90129-X.

71. B. V. Chirikov, Research concerning the theory of non-linear resonance and stochasticity. Technical Report IYAF-267-TRANS-E, Institute of Nuclear Physics, Novosibirsk (1969). Translated at CERN from the Russian, CERN-Trans-71-40, 1971.
72. M. B. Antipov, F. M. Izrailev, and B. V. Chirikov, Statistical check of a generator of pseudo-random numbers, *Vychislitelnie Systemy.* **30**, 77–85 (1968).
73. D. H. Lehmer, Mathematical methods in large-scale computing units, *The Annals of the Computation Laboratory of Harvard University.* **26**, 141–146 (1951).
74. R. R. Coveyou, Serial correlation in the generation of pseudo-random numbers, *Journal of the Association for Computing Machinery.* **7** (1), 72–74 (1960). doi: 10.1145/321008.321018.
75. T. E. Hull and A. R. Dobell, Random number generators, *SIAM Review.* **4**(3), 230–254 (1962).
76. W. H. Payne, J. R. Rabung, and T. P. Bogyo, Coding the Lehmer pseudo-random number generator, *Communications of the Association for Computing Machinery.* **12**, 85–86 (1969). doi: 10.1145/362848.362860.
77. N. Metropolis and S. Ulam, The Monte Carlo method, *Journal of the Statistical Association.* **44**, 335–341 (1949).
78. B. V. Chirikov, Some numerical experiments with a nonlinear mapping: stochastic component. In J. Lagasse and C. Mira (eds.), *Transformations ponctuelles et leurs applications.* Editions du CNRS, Paris, France (1973), pp. 409–428.
79. C. Mira, Chapter 8: I. Gumowski and a Toulouse Research Group in the "prehistoric" times of chaotic dynamics. In R. Abaraham and Y. Ueda (eds.), *The Chaos Avant-Garde* (2000), pp. 95–197. doi: 10.1142/9789812386472_0008.
80. C. Mira, étude de la frontière de stabilité d'un point double stable d'une récurrence non linéaire autonome du deuxième ordre. In *Proceedings of the IFAC Symposium on Pulse-rate and Pulse-number Signals in Automatic Control* (1968), pp. D43–7II.
81. I. Gumowski and C. Mira, Sensitivity problems related to certain bifurcations in non-linear recurrence relations, *Automatica.* **5**(3), 303–317 (1969). doi: 10.1016/0005-1098(69)90073-9.
82. C. Letellier and J.-M. Ginoux, Development of the nonlinear dynamical systems theory from radio-engineering to electronics, *International Journal of Bifurcation & Chaos.* **19**(7), 2131–2163 (2009).
83. J. Bernussou, L. Hsu, and C. Mira, Quelques exemples de solutions stochastiques bornées dans les récurrences autonomes du 2nd ordre. In *Transformations ponctuelles et leurs applications*, Vol. 229. Colloques internationaux du CNRS, Toulouse (1973), pp. 195–220.
84. J. L. Doob, Stochastic processes and statistics, *Proceedings of the National Academy of Sciences.* **20**(6), 376–379 (1934). doi: 10.1073/pnas.20.6.376.

85. T.-Y. Li and J. A. Yorke, Period three implies chaos, *The American Mathematical Monthly.* **82**(10), 985–992 (1975).
86. O. E. Rössler, Chaotic behavior in simple reaction system, *Zeitschrift für Naturforschung A.* **31**, 259–264 (1976).
87. M. J. Feigenbaum, Universality in complex discrete dynamical systems. Technical Report LA-6816-PR, Los Alamos Theoretical Division (1977). Annual Report, pp. 98–102, July 1975–September 1976.
88. C. Marchal, Collisions of stars by oscillating orbits of the second kind, *Acta Astronautica.* **5**(10), 745–764 (1978). doi: 10.1016/0094-5765(78)90066-8.
89. L. Martinet, Heteroclinic stellar orbits and "wild" behaviour in our galaxy, *Astronomy & Astrophysics.* **32**, 329–333 (1974).
90. L. Martinet and F. Mayer, Galactic orbits and integrals of motion for stars of old galactic populations. III. Conclusions and applications, *Astronomy & Astrophysics.* **44**(1), 45–57 (1975).
91. E. Prugovečki, Stochastic phase spaces and master Liouville spaces in statistical mechanics, *Foundations of Physics.* **9**, 575–587 (1979). doi: 10.1007/BF00708370.
92. G. Contopoulos, Disappearance of integrals in systems of more than two degrees of freedom, *Celestial Mechanics.* **17**, 167–172 (1978). doi: 10.1007/BF01371328.
93. A. J. Lichtenberg, Adiabatic and stochastic motion of charged particles in the field of a single wave. In G. Casati and J. Ford (eds.), *Stochastic Behavior in Classical and Quantum Hamiltonian Systems.* Springer, Berlin (1979), pp. 18–34. doi: 10.1007/BFb0021734.
94. G. R. Smith, Overlap of bounce resonances and the motion of ions in a trapped-ion mode, *Physical Review Letters.* **38**, 970–973 (1977). doi: 10.1103/PhysRevLett.38.970.
95. M. Kummer, On resonant Hamiltonians with two degrees of freedom near an equilibrium point. In G. Casati and J. Ford (eds.), *Stochastic Behavior in Classical and Quantum Hamiltonian Systems.* Springer, Berlin (1979), pp. 57–75. doi: 10.1007/BFb0021738.
96. R. C. Churchill, G. Pecelli, S. Sacolick, and D. L. Rod, Coxistence of stable and random motion, *The Rocky Mountain Journal of Mathematics.* **7**(3), 445–456 (1977).
97. G. Pecelli and E. S. Thomas, An example of elliptic stability with large parameters: Lamé's equation and the Arnold-Moser-Rüssmann criterion, *Quarterly of Applied Mathematics.* **36**, 129–140 (1978). doi: 10.1090/qam/502471.
98. D. L. Rod, G. Pecelli, and R. C. Churchill, Hyperbolic periodic orbits, *Journal of Differential Equations.* **24**, 329–348 (1977).
99. M. Toda, A question about the localized mode due to a light impurity. In G. Casati and J. Ford (eds.), *Stochastic Behavior in Classical and Quantum Hamiltonian Systems.* Springer, Berlin (1979), pp. 145–150. doi: 10.1007/BFb0021741.

100. O. I. Bogoyavlensky, Nonlinear oscillation regimes in some physical problems. In G. Casati and J. Ford (eds.), *Stochastic Behavior in Classical and Quantum Hamiltonian Systems*. Springer, Berlin (1979), pp. 151–162.
101. M. J. Feigenbaum, Quantitative universality for a class of nonlinear transformations, *Journal of Statistical Physics*. **19**, 25–52 (1978). doi: 10.1007/BF01020332.
102. H. Haken, Synergetics, *Physics Bulletin*. **28**(9), 412–414 (1977). doi: 10.1088/0031-9112/28/9/027.
103. J. L. van Hemmen, Dynamics and ergodicity of the infinite harmonic crystal a review of some salient features. In G. Casati and J. Ford (eds.), *Stochastic Behavior in Classical and Quantum Hamiltonian Systems*. Springer, Berlin (1979), pp. 232–240. doi: 10.1007/BFb0021747.
104. I. C. Percival, Regular and irregular spectra of molecules. In G. Casati and J. Ford (eds.), *Stochastic Behavior in Classical and Quantum Hamiltonian Systems*. Springer, Berlin (1979), pp. 259–282. doi: 10.1007/BFb0021750.
105. R. A. Marcus, D. W. Noid, and M. L. Koszykowski, Semiclassical studies of bound states and molecular dynamics. In G. Casati and J. Ford (eds.), *Stochastic Behavior in Classical and Quantum Hamiltonian Systems*. Springer, Berlin (1979), pp. 283–292.
106. M. Tabor, The role of periodic orbits in semiclassical quantization. In G. Casati and J. Ford (eds.), *Stochastic Behavior in Classical and Quantum Hamiltonian Systems*. Springer, Berlin (1979), pp. 293–298. doi: 10.1007/BFb0021752.
107. J. Delos and R. Swimm, Semiclassical calculation of energy levels for non-separable systems, *Chemical Physics Letters*. **47** (1), 76–79 (1977). doi: 10.1016/0009-2614(77)85310-4.
108. M. C. Gutzwiller, Bernoulli sequences and trajectories in the anisotropic Kepler problem, *Journal of Mathematical Physics*. **18** (4), 806–823 (1977). doi: 10.1063/1.523310.
109. B. V. Chirikov, A universal instability of many-dimensional oscillator systems, *Physics Reports*. **52**(5), 263–379 (1979). doi: 10.1016/0370-1573(79)90023-1.
110. R. H. G. Helleman, On the iterative solution of a stochastic mapping. In U. Landman (ed.), *Statistical Mechanics and Statistical Methods in Theory and Application*. Springer US, Boston, MA (1977), pp. 343–370. doi: 10.1007/978-1-4613-4166-6_16.
111. R. M. May, Simple mathematical models with very complicated dynamics, *Nature*. **261**, 459–467 (1976).
112. M. V. Berry, Regular and irregular semiclassical wavefunctions, *Journal of Physics A*. **10**(12), 2083 (1977). doi: 10.1088/0305-4470/10/12/016.
113. R. M. May and G. F. Oster, Bifurcations and dynamic complexity in simple ecological models, *The American Naturalist*. **110**, 573–599 (1976).
114. S. A. Orszag, *Lectures on the Statistical Theory of Turbulence*. Number 31 in Flow Research Report, Flow Research Incorporated (1974).

115. J. Boussinesq, *Théorie de l'écoulement tourbillonnant et tumultueux des liquides dans les lits rectilignes à grande section.* Gauthier-Villars, Paris (1897).
116. J. B. McLaughlin and P. C. Martin, Transition to turbulence of a statically stressed fluid, *Physical Review Letters.* **33**(20), 1189–1192 (1974). doi: 10.1103/PhysRevLett.33.1189.
117. J. B. McLaughlin and P. C. Martin, Transition to turbulence in a statically stressed fluid system, *Physical Review A.* **12**(1), 186–203 (1975). doi: 10.1103/PhysRevA.12.186.
118. J. McLaughlin, Successive bifurcations leading to stochastic behavior, *Journal of Statistical Physics.* **15**, 307–326 (1976). doi: 10.1007/BF01023056.
119. D. Ruelle and F. Takens, On the nature of turbulence, *Communications in Mathematical Physics.* **20**, 167–192 (1971).
120. F. M. Izrailev and B. V. Chirikov, Some numerical experiments with a simple model of turbulence. In *Programming and Mathematical Methods for Solving the Physical Problems*, Dunba, October 30–November 4, 1973 (1974), pp. 266–277.
121. G. Casati, Chaos in quantum mechanics. In H. Haken (ed.), *Evolution of Order and Chaos.* Springer, Berlin (1982), pp. 255–261.
122. R. W. Batterman, Defining chaos, *Philosophy of Science.* **60**(1), 43–66 (1993).
123. R. L. Devaney, *An Introduction to Chaotic Dynamical Systems*, 3rd edn. Chapman and Hall/CRC (2021). doi: 10.1201/9780429280801.
124. R. Brown and L. O. Chua, Clarifying chaos: Examples and counterexamples, *International Journal of Bifurcation & Chaos.* **06** (02), 219–249 (1996). doi: 10.1142/S0218127496000023.
125. P. Cvitanović, Periodic orbits as the skeleton of classical and quantum chaos, *Physica D.* **51**(1), 138–151 (1991). doi: 10.1016/0167-2789(91)90227-Z.
126. J. Ford and G. Mantica, Does quantum mechanics obey the correspondence principle? Is it complete? *American Journal of Physics.* **60**(12), 1086–1098 (1992). doi: 10.1119/1.16954.
127. J. Ford, What is chaos, that we should be mindful of it?, In P. Davies (ed.), *The New Physics.* Cambridge University Press, New York (1989), pp. 348–372.
128. L. Tolstoy, *What Is Art and Essays on Art (1930).* Obscure Press (2006).
129. B. V. Chirikov, Dynamical mechanisms of noise. In *4th International Symposium on Selected Topics in Statistical Mechanics.* The Publisher, Dubna (1987).
130. B. V. Chirikov, Transient chaos in quantum and classical mechanics, *Foundations of Physics.* **16**, 39–49 (1986). doi: 10.1007/BF00735179.
131. A. N. Kolmogorov, Three approaches to the quantitative definition of information, *Problemy Peredachi Informatsii.* **1**, 3–11 (1965).
132. A. N. Kolmogorov, Three approaches to the quantitative definition of information, *International Journal of Computer Mathematics.* **2**(1–4), 157–168 (1968). doi: 10.1080/00207166808803030.

133. G. J. Chaitin, On the simplicity and speed of programs for computing infinite sets of natural numbers: Statistical considerations, *Journal of the Association for Computing Machinery.* **16**, 407–422 (1969).
134. R. J. Solomonov, A formal theory of inductive inference. I, *Information and Control.* **7**, 1–22 (1964).
135. R. J. Solomonov, A formal theory of inductive inference. I, *Information and Control.* **7**, 224–254 (1964).
136. M. V. Berry, Semi-classical mechanics of regular and irregular motion. In G. Iooss, R. G. H. Helleman, and R. Stora (eds.), *Chaotic Behavior of Deterministic Systems.* NATO, North Holland, Amsterdam (1983), pp. 171–271. Les Houches, Session 36.
137. J. Ford, How random is a coin toss? *Physics Today.* **36**(4), 40–47 (1983). doi: 10.1063/1.2915570.
138. J. Ford, G. Mantica, and G. H. Ristow, The arnol'd cat: Failure of the correspondence principle. In D. K. Campbell (ed.), *Chaos/Xaoc.* American Institute of Physics, New York, NY (1990), pp. 477–496.
139. O. E. Rössler, Chaos and chemistry. In C. Vidal and A. Pacault (eds.), *Nonlinear Phenomena in Chemical Dynamics*, Vol. 12. Springer, Berlin (1981). doi: 10.1007/978-3-642-81778-6_12.
140. O. E. Rössler, An estimate of Planck's constant. In P. Erdi (ed.), *Dynamical Phenomena in Neurochemistry.* Publications of the Institute of Theoretical Physics, Hungarian Academy of Sciences, Budapest (1985).

https://doi.org/10.1142/9789819823130_0004

Chapter 4

THE SMALE PROGRAM

Ralph Abraham

University of California, Santa Cruz, USA
abraham@vismath.org

Here is a journey between Stephen Smale's horseshoe map, one of the mathematical illustrations of chaos, that offers a topological framework to understand sensitive dependence on initial conditions, invariant sets, and strange attractors. Nearly at the same time, Japanese researchers emphasized experimental mathematics, nonlinear circuits, and applications to real-world systems. Thus, with the pioneering figure of Yoshisuke Ueda, they offered a distinctive approach which was focused on physical realizations of chaotic behavior through analog computing and electrical engineering. This was then followed by Otto E. Rössler and the Santa Cruz Chaos Cabal.

1. Introduction

We begin with an outline of the development of dynamical systems theory along the lines of differential topology, that is, the Smale program, 1958–1968. This was contemporary with the parallel development of the chaos program of Ueda,[1] Lorenz,[2] Gumowski and Mira,[3] and Rössler,[4–6] 1961–1975.

2. The Mexican Connection, 1956

Stephen Smale finished his Ph.D. thesis in differential topology in 1956, working with Raoul Bott at the University of Michigan. At that time I was there in Ann Arbor, finishing my undergraduate program in Engineering Mathematics. I was introduced to differential topology in a course by Bott, on the way to my Ph.D. on general relativity in 1960, working with Nathaniel Coburn.

Solomon Lefschetz, the legendary algebraic topologist, was very altruistic. Noting that mathematics in Latin America was not well developed, he began devoting half of every year in Mexico City to build up a graduate program in the math department of the National Autonomous University of Mexico (UNAM). He had become interested in the Russian literature on dynamical systems theory, written an influential book in 1957,[7] and begun a series of summer conferences in Mexico City on topology and dynamical systems.

After his Ph.D. in 1956, Smale attended Lefschetz' summer conference in Mexico City [8, p. 147]. There he met René Thom, Moe Hirsch, and Elon Lima, who played important roles in our story. Around 1958, Lima (Brazilian) finished his Ph.D. thesis on topology with Edwin Spanier in Chicago, and introduced Smale to Mauricio Peixoto. Peixoto was a Brazilian student of Lefschetz in Princeton, 1958–59. His theorem on the structural stability of flows in two dimensions,[9] based on work of G. F. De Baggis,[10] was an early breakthrough in dynamical systems theory.

I met Lefschetz at the UNAM in 1959, where I was writing my Ph.D. thesis and then in 1964, when I began at Princeton University, where he was chair of the Math department. During my time there, I saw the dynamical systems group attending a Summer meeting. Stephen Smale was among them, but we did not meet.

In 1960, for my first academic job, I arrived at Berkeley University which, in the Fall of 1960, suddenly had a brand new staff of important new math professors and visitors. Stephen Smale arrived from Princeton, along with the Ed Spanier (algebraic topology), Shing-Shen Chern (differential geometry), and Morris Hirsch (differential topology) from Chicago, René Thom (differential topology, Fields Medal, 1958) from Paris, Christopher Zeeman (topology, expositor of catastrophe theory) from Warwick, Mauricio Peixoto from Rio, Bob Williams (knot theory), Dick Palais (nonlinear functional analysis) and others comprising a research group on dynamical systems theory based on differential topology. Moe Hirsch (a student of Spanier) and I were among the newbies in this group. The Smale program was focused on the stable manifolds, structural stability, and conjugacy of diffeomorphisms. At his time we devoted much time reading and discussing the works of Poincaré and Birkhoff, especially concerning the stable curves of surface transformations and their transversal intersections.

Smale had proved the existence of stable and unstable manifolds, his first major result in this field.[11] He developed the horseshoe map, his second major result, also in this year. After the publication of the stable manifold theorem, Thom proved that transversal intersection of stable manifolds is a generic property of diffeomorphisms.[12] In the Summer of 1961, Smale presented his stable manifold theorem in Urbino, and later that summer, in Bonn, I presented my own proof of Thom's transversality theorem.[13]

The Smale program was boosted into orbit by an influential survey article by Smale.[14] The first 10 pages set out the foundations of the program: conjugacy of diffeomorphisms, fixed and periodic points, stable and generic properties, the nonwandering set, hyperbolic fixed points, stable manifolds, his stable manifold theorem, the Morse inequalities, and a structural labelled diagram later called the Smale diagram. Already on page 10 we find drawings of homoclinic intersections of stable and unstable manifolds for surface transformations, discovered by Poincaré and analyzed in detail by Birkhoff and Paul Smith.[15] Smale's ingenious simplification of the homoclinic tangle in the two-dimensional case, the horseshoe map, appears in page 25 (Fig. 1). The details of these ideas, and many other original concepts, may be found in the original publication, or its reprint in Smale's book.[8] In this epochal paper, Smale carefully credits his predecessors (in chronological order) — Poincaré,[16] Birkhoff,[17] Morse,[18] Andronov and Pontrjagin,[19] Thom,[20] Elsgolts,[21] Reeb,[22] and Peixoto.[23]

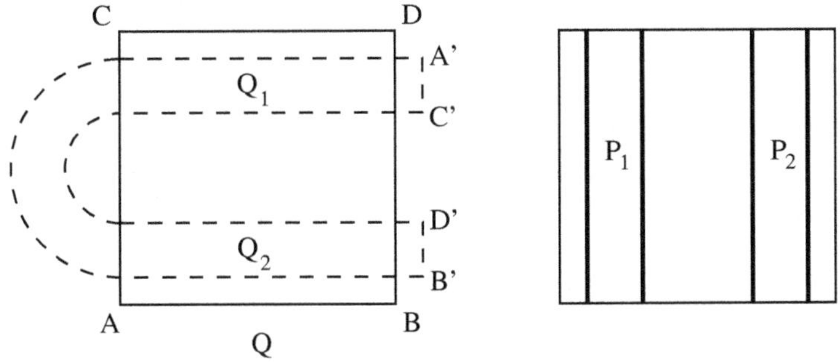

Fig. 1. Smale's horseshoe map. A global diffeomorphism maps the square Q into the region bounded by dotted lines with $G(\mathrm{A}) = \mathrm{A}'$, etc. Each component P_1 and P_2 of $g^{-1}(g(\mathrm{Q}) \cap \mathrm{Q})$ is such that g is a linear map with $g(\mathrm{P}_i) = \mathrm{Q}_i$ $(i = 1, 2)$.

After the first year of the Smale program in Berkeley (1961), Smale moved to Columbia University, where also there was a cadre of leading mathematicians, including Paul Smith, Sammy Eilenberg, and Serge Lang. In the September of 1961, he visited the Soviet Union and met the dynamics community there, especially Anosov. The next year, I also moved to Columbia. At this time, Smale began applying the methods of differential topology to the calculus of variations — the beginning of a new field which became known as global analysis. This comprised rewriting of all of classical analysis using differential topology and geometry. The calculus of variations, the theory of minimal surfaces, and mathematical physics (classical mechanics, Maxwell theory, and general relativity, in particular) were particularly extended using this new approach. At this time Mike Shub and Charlie Pugh joined our group.

In the Fall, Smale organized a seminar on global analysis. But early on, in October, the Cuban Missile Crisis intervened. For a few weeks, I took over the seminar, lecturing primarily on my proof of the Thom transversality theorem for global manifolds in the context of infinite-dimensional manifolds and mappings. I published the notes as a preprint.[24]

3. The Early French School

A year after Smale's return to Berkeley, I moved on to Princeton University, where Lefschetz still had huge influence, and other greats such as Jack Milnor were rising. Besides the University, there was the Institute of Advanced Studies, with its own math department, including permanent members like Armand Borel, and visitors including Dick Palais, Bernard Morin, and Al Kelley. I was able to teach graduate courses and seminars in dynamical systems and global analysis. In one of these courses, I rewrote much of celestial mechanics with the new language and technology of global analysis.[25] In another, I treated the transversality of stable manifolds in the global context of (infinite-dimensional) manifolds of mappings.[26] Jerrold Marsden and Joel Robbin were important collaborators in these courses and the related books. In 1966, Smale was in Moscow to accept his Fields Medal at the International Congress. In my fourth year, 1967–68, I decided to accept a position at UC Santa Cruz.

Around 1966, I began to receive letters from René Thom in which he reported regular progress in his creation of catastrophe theory. In this simple context of gradient (non-chaotic) dynamical systems, he made crucial use of the language of attractors, basins, and bifurcations, which

Fig. 2. Fall 1958. From the left, the third is Yoshisuke Ueda, the fifth is Prof. Chihiro Hayashi, and the sixth is Associate Prof. Michiyoshi Kuwahara. Behind, the four racks are the analog computer.

Fig. 3. Summer 1961. From the right, the first is Yoshisuke Ueda. Behind, the six racks are the analog computer.

became fundamental in the further evolution of dynamical systems theory, and later, chaos theory. He popularized a style of application of these notions, introduced earlier by Poincaré and his Russian followers. It has been observed by the chaos pioneer, Christian Mira, that the ideas had occurred before Thom in the Russian literature, from 1963 [27, p. 175].

Thom's difficult ideas appeared eventually in book form. The French original of 1972 was updated by Thom and translated to English by David Fowler in 1975.[28] The impact on the mathematical community was further facilitated by a series of exemplary articles by Chris Zeeman, beginning in 1976, and collected in book form in 1977.[29] The word *attractor* is indexed in Thom's book on 19 pages, and 37 pages in the index of Zeeman's book. The word *bifurcation* is indexed by Thom on 18 pages, and by Zeeman on 67 pages. Poincaré is mentioned on four pages by Thom. These words and ideas were burned into the working vocabulary of the entire scientific community from this time on, dominating all the applications of chaos theory which followed.

4. The Japanese School

In 1968, the *American Mathematical Society* had organized an four-week conference on global analysis, July 1–26. The proceedings were edited by Chern and Smale, in 1970. Fifty-three papers were included.[30] Dynamical systems theory was only one of several topics covered. Members of our community were nearly all there. This was the moment, I believe, at which our group finally became aware of the experimental work and simulations on chaotic attractors. Yoshisuke Ueda, discovered the first clearly chaotic attractor in analog simulation, the *Japanese attractor*,[31] for which he accurately drew the homoclinic tangle of inset and outset curves for the forced Duffing equation

$$\begin{cases} \dot{x} = y \\ \dot{y} = \mu\left(1 - \gamma x^2\right) y - x^3 + B\cos\nu t \end{cases} \tag{1}$$

obtained at Kyoto University, on November, 1961 (Fig. 4),[1] Edward Lorenz, discovered his chaotic attractor, at MIT, on 1963,[2] and Christian Mira, discovered his chaotic attractor in the iterated quadratic map of the plane,

$$\begin{cases} x_{n+1} = (1 - \lambda)x_n + y_n \\ y_{n+1} = y_n + f(x_n) \end{cases} \tag{2}$$

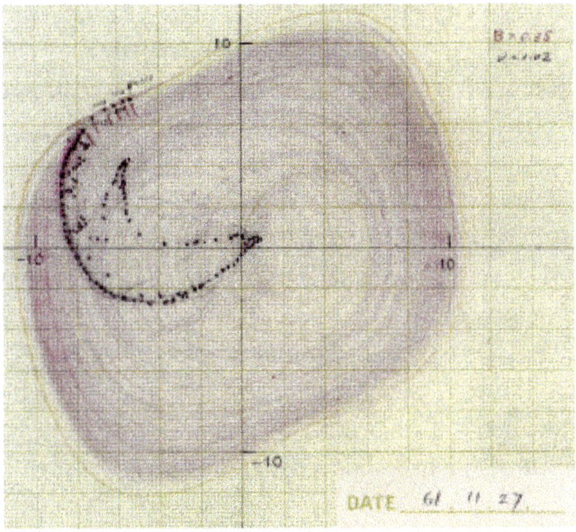

(a) Original plot obtained on November 27, 1961

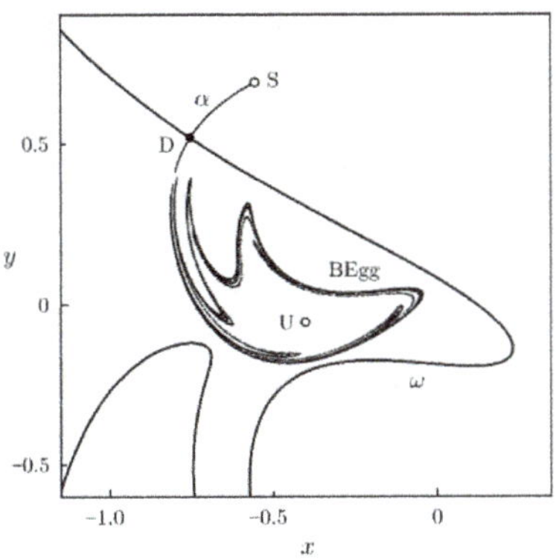

(b) Poincaré section of the Duffing attractor

Fig. 4. With amazing prescience, shortly after our study in Berkeley of homoclinic intersections for surface diffeomorphisms, Yoshisuke Ueda, then a graduate student at Kyoto University, discovered his chaotic attractor in a Poincaré section of the forced Duffing equation. He correctly obtained the homoclinic intersection.

Note: BEgg = Broken eggs.

where

$$f(x_n) = \left| \begin{array}{ll} -2\lambda x_n - 0.9\lambda & x_n < -0.5 \\ -\dfrac{\lambda x_n}{5} & \text{if } |x_n| < -0.5 \\ -2\lambda x_n + 0.9\lambda & x_n > 0.5 \end{array} \right. \qquad (3)$$

creating the theory of critical curves for iterated maps, Toulouse, 1968.[32,33] These discoveries sounded the death knell for our approach based on differential topology. There were no presentations of experimental work in this conference. The summer ended in shock for our whole group.

Our group reconvened at the Math Institute newly created at Warwick University by Chris Zeeman, for the academic year, and in a Summer conference, 1969. Experimental work (not chaotic) was presented by a single attendee, Minoru Urabe (1912–1975) of the Math Department, Kyoto University.[34] Ironically, he was a chaos disbeliever. In 1961, Ueda, then still a graduate student, had presented his chaotic data in Urabe's seminar. Urabe discouraged Ueda, whose results were not published until 1970. Chihiro Hayashi, Ueda's thesis advisor, also rejected the work and Ueda was forced to submit another thesis on electrical machines [35, pp. 23–80].

5. The Californian Period

In 1971–73, new people had arrived in my department, including John Guckenheimer, a recent Ph.D. with Stephen Smale, and veterans of our group, Dick Palais and Mike Shub. Dick Palais collaborated in creating a computational program using a digital mainframe and a primitive graphics terminal. We were able to recreate the attractors of Ueda, Lorenz, and Mira, with assistance of a talented group of undergraduates. After a couple years we also studied the Rössler attractor and other new developments. Extensive graphics of this work are on display in my website.[a] John Guckenheimer was also active in computational dynamics at UCSC in the 1970s.

Our experimental work with chaotic dynamical systems, begun in the summer of 1974, was soon matched by related but independent activity in the Physics Department of UCSC. And around 1974 I was invited to Stuttgart by Werner Guttinger to speak on dynamical systems theory. This year saw the first connection of the word chaos with dynamical systems theory, in a paper of Robert May,[36] beginning the chaos revolution. At this time, I abandoned the Smale program, and commenced a new career in experimental dynamics and chaotic behavior. And there in Stuttgart, I met Otto, already a professor at age 25, and also Gottfried Mayer-Kress, then a graduate student. All these — Yoshi, Christian, Otto, and Gottfried — became friends and collaborators. They were especially gifted in 3D visual

[a]www.visual-chaos.org.

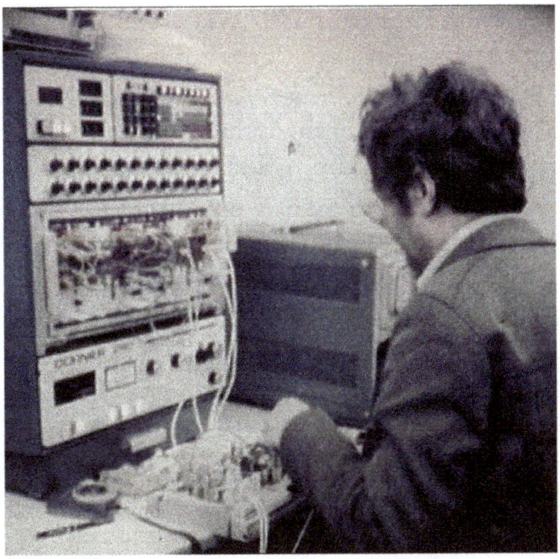

(a) Otto in the front of his computed in 1976

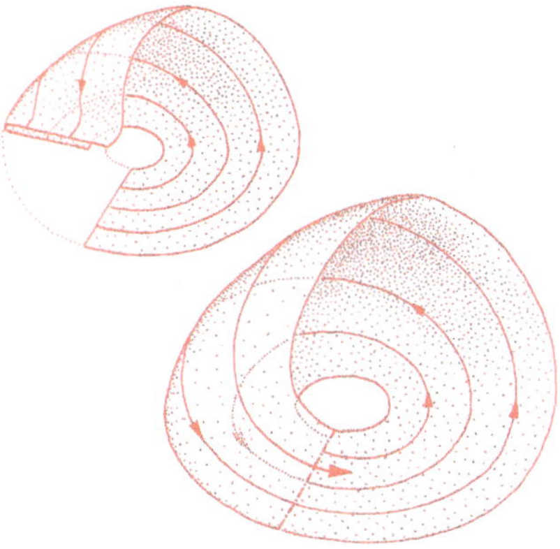

(b) Sketch of the Rössler attractor as drawn by Chris Shaw

Fig. 5. Otto E. Rössler found his eponym attractor by performing numerous simulation with his analog computer, a Dornier 240. The structure of his attractor was investigated by Abraham and Shaw.[37]

intuition. Rarely, people meet and become close friends instantly. This happened to Otto and me in 1974, before his famous attractor and the chaos revolution. Meeting Otto, and spending time with him in front of his analog computer (Fig. 5), excited me to change my line of work from topology to simulation in 1974.

Over the years since 1974 we have maintained contact. I have greatly admired him for his moral code, and his readiness to take action on behalf of a variety of issues, as well as his highly original new directions in mathematical physics. Otto has been extraordinarily prolific and innovative. More than 300 papers and 8 books in some 20 fields in less than 50 years. My own span is ordinarily narrow, just a small part of chaos theory. So I will comment here only an that small fraction of Otto's output. In chaos theory, Otto had a meteoric rise, from nowhere in 1975, just two years after his habilitation in theoretical biochemistry, to a commanding position in chaos theory the next year.

In 1974 or 1975, he received a crucial gift of chaos reprints from the late Art Winfree.[38] And in 1976 he published his wonderful papers on the attractor he discovered while searching for chaos in the simplest possible set of equations,[5] as well as a video illustrating its dynamics.[b] In the interim he must have spent uncountable hours in front of his analog computer. And by 1977, he had ascended to the leadership team of the historic conference on chaos theory organized by the New York Academy of Sciences.

Heinz Pagels, a far-sighted physicist and then president of the NY Academy of Sciences, arranged for a major conference in New York in 1977. Under the direction of Okan Gurel and Otto Rössler, a profoundly diverse collection of math, engineering, and science researchers were brought together for sharing new ideas from the radical frontier. Chaos, fractals, and complex systems were connected in new ways, and the chaos revolution was exploding. An entire section of the meeting was devoted to experimental results.

This meeting, entitled "*Bifurcation theory and applications in scientific disciplines*," October 31 to November 4, 1977, was jointly sponsored by the New York Academy of Sciences and the University of Tübingen. The proceedings were organized in eight parts[39]: Mathematics, Biology, Chemistry, Physics, Ecology, Economics, Engineering, and finally, Experimentation and Simulation. The last paper in the final section was the first announcement of my simulation of chaos using digital computer graphics. This work evolved into the graphic introduction for chaos theory in my book, written jointly with the artist, Christopher Shaw.[37] I met Otto again in in Tübingen in 1981, and perhaps also in Guelph (Canada) in 1981.

[b]https://www.youtube.com/watch?v=Tmmdg2P1RIM.

6. The Santa Cruz Chaos Cabal

Around 1978, a group of students (Fig. 6), primarily Rob Shaw, Doyne Farmer, Norman Packard, and Jim Crutchfield, later known as the *Santa Cruz Chaos Cabal* (after Gleick's best-seller[40]) began a literature seminar and chaos program on chaotic dynamics. This resulted in an audacious article in the *Scientific American* of December, 1986, in which chaos theory reached a wide popular audience for the first time.[41]

Some exchanges between Robert Shaw and James Crutchfield on January 1st, 2020[c]:

RS: "I do remember a *Physics Letter*, but how that came into my hands I don't know."

JC: "The PLA (1976) had Otto's simplest chaotic ODE.[5] Big deal for us, already running it in '78."

RS: "For me it meant (i) calculate all the Lyapunov exponents, (ii) couple two of them (each only requiring a single multiplier) on the Systron-Donner analog computer (and so look at sync and ∞-number of attractors at the period-doubling onset (eventually a project of Otto's and mine... thence to map lattices), and (iii) failed attempt to expand his ODEs around a single frequency to explain the phase coherence that we published on: "Power Spectral Analysis of a Dynamical System"[42] and (ii) "Power Spectra and Mixing Properties of Strange Attractors."[43]

JC: "When did the PLA show up exactly and how? Can't recall.

RS: "That said, I do recall that *Physics Letters A* was pretty much the only journal openly considering nonlinear dynamics. (Pre-*Physica D* and Joe Ford.) Point is, we were actively watching PLA's table of contents.

JC: "Information flow was a trickle pre-internet and pre-web."

[c]Communicated by James Crutchfield to Christophe Letellier.

Fig. 6. James Farmer, Robert Shaw and James Crutchfield at the University of California at Santa Cruz in 1978.
Source: Courtesy of James Crutchfield.

In the early 1980s, Rob Shaw, working with UCSC Physics Professor Peter Scott, began a careful experimental project on the chaotic dynamics of a dripping faucet. This was the subject of a short video in which the attractor reconstruction method was applied to physical data, as well as computer simulation of a dripping faucet. A complete report was published as a book.[44] Rob Shaw credits Otto with the idea for this project.

This 1977 conference was the occasion of first meetings of many from the chaos community. Ranking these attendees according to the number of entrees in the Author Index of the published proceedings, we find: Gurel (on singular points, 16 entries), Nicolis (chemical bifurcations, 14), May (bifurcations in ecology, 13), Marsden (bifurcations in engineering, 12), Yorke (bifurcations in fluid flow, 12), Haken (synergetics, 11), Rössler (with graphics of his famous equations, 11), and Ruelle (turbulence, 11). This meeting gave an enormous boost to chaos theory world-wide.

Thus, in only two years, Otto had ascended into the top 10% of the attendees, ahead of some of the pioneers of the field such as Smale (bifurcations in economics, 9), Joseph (eigenvalue branching, 7), Williams (Lorenz attractor, 6), Gollub (rotating and convecting fluids, 4), and Devaney (homoclinic orbits, 3).

In the following four decades Otto has continued to contribute to the field, comprising about 10% of his output. Chaos theory is greatly in his debt.

References

1. Y. Ueda, Some problems in the theory of nonlinear oscillations (Ph.D. Thesis, 1965). In *The Road to Chaos*, Aerial Press (1992).
2. E. N. Lorenz, Deterministic nonperiodic flow, *Journal of the Atmospheric Sciences.* **20**, 130–141 (1963). doi: 10.1175/1520-0469(1963)020⟨0130:DNF⟩ 2.0.CO;2.
3. I. Gumowski and C. Mira, *Dynamique chaotique: Transformations ponctuelles, transition ordre-désordre.* Cepadues Editions, Toulouse, France (1980).
4. O. E. Rössler, Chaotic behavior in simple reaction system, *Zeitschrift für Naturforschung A.* **31**, 259–264 (1976).
5. O. E. Rössler, An equation for continuous chaos, *Physics Letters A.* **57**(5), 397–398 (1976). doi: 10.1016/0375-9601(76)90101-8.
6. O. E. Rössler, *Continuous Chaos.* Springer Series in Synergetics, Vol. 2. Springer-Verlag, Berlin, Germany (1977), pp. 174–183.
7. S. Lefschetz, *Lectures on Differential Equations.* Princeton University Press (1946).
8. S. Smale, *The Mathematics of Time: Essays on Dynamical Systems, Economic Processes, and Related Topics.* Springer-Verlag, New York (1980).
9. M. M. Peixoto, Structural stability on two-dimensional manifolds, *Topology.* **1**(2), 101–120 (1962). doi: 10.1016/0040-9383(65)90018-2.
10. G. F. De Baggis, Dynamical systems with stable structure, *Uspekhi Matematicheskikh Nauk.* **10**(4), 101–126 (1955).
11. S. Smale, Generalized Poincaré's conjecture in dimensions greater than four, *Annals of Mathematics.* **74**(2), 391–406 (1961).
12. R. Thom, Un lemme sur les applications différentiables, *Boletín de la Sociedad Matemática Mexicana.* **1**, 59–71 (1956).
13. R. H. Abraham, Transversality in manifolds of mappings, *Bulletin of the American Mathematical Society.* **69**(4), 470–474 (1963).
14. S. Smale, Differentiable dynamical systems. I Diffeormorphisms, *Bulletin of American Mathematical Society.* **73**, 747–817 (1967).
15. G. D. Birkhoff and P. A. Smith, Structure analysis of surface transformations, *Journal de Mathématiques Pures et Appliquées.* **7**, 345–380 (1928).

16. H. Poincaré, Sur les courbes définies par les équations différentielles (3), *Journal de Mathématiques Pures et Appliquées, IV.* **1**, 167–244 (1885).
17. G. D. Birkhoff, Surface transformations and their dynamical applications, *Acta Mathematica.* **43**, 1–119 (1922). doi: 10.1007/BF02401754.
18. M. Morse, The critical points of a function of n variables, *Transactions of the American Mathematical Society.* **33**, 72–91 (1931). doi: 10.1090/S0002-9947-1931-1501576-4.
19. A. A. Andronov and L. S. Pontryagin, Rough systems, *Doklady Akademii Nauk SSSR.* **14**(5), 247–250 (1937) (in Russian).
20. R. Thom, Sur une partition en cellules associée à une function sur une variété, *Comptes-Rendus de l'Académie des Sciences.* **228**, 973–975 (1949).
21. L. E. El'sgol'ts, An estimate for the number of singular points of a dynamical system defined on a manifold, *Matematicheskii Sbornik.* **26**(68), 215–223 (1950). Translated for the American Mathematical Society, 1952.
22. G. Reeb, Sur certaines propriétés topologiques des trajectoires des systèmes dynamiques, *Mémoires de l'Académie Royale de Belgique: Classe des Sciences.* **27**, 9 (1952).
23. M. M. Peixoto, On structural stability, *Annals of Mathematics.* **69**(1), 199–222 (1959).
24. R. Abraham, Lectures of Smale on differential topology. Technical Report, Columbia University (1962). Available at https://www.math.cuhk.edu.hk/course_builder/1415/math5070/.
25. R. H. Abraham and J. E. Marsden, *Foundations of Mechanics: A Mathematical Exposition of Classical Mechanics.* W. A. Benjamin, Reading, MA (1967).
26. R. H. Abraham and J. W. Robbin, *Transversal Mappings and Flows.* W. A. Benjamin, Reading, MA (1967).
27. R. H. Abraham, L. Gardini, and C. Mira, *Chaos in Discrete Dynamical Systems: A Visual Introduction in 2 Dimensions.* Springer-Verlag, New York (1997).
28. R. Thom, *Structural Stability and Morphogenesis: An Outline of a General Theory of Models.* W. A. Benjamin, New York (1973). Translated from the French by David Fowler.
29. C. E. Zeeman, *Catastrophe Theory: Selected Papers, 1972-1977.* Addison-Wesley, Reading, Massachusetts (1977).
30. S.-S. Chern and S. Smale (eds.), *Global Analysis.* Number 14 in Proceedings of Symposia in Pure Mathematics. American Mathematical Society, Providence, RI (1970). (Berkeley, CA, 1–26 July 1968).
31. D. Ruelle, Strange attractors, *The Mathematical Intelligencer.* **2**, 126–137 (1980).
32. C. Mira, étude de la frontière de stabilité d'un point double stable d'une récurrence non linéaire autonome du deuxième ordre. In *Proceedings of the IFAC Symposium on Pulse-Rate and Pulse-Number Signals in Automatic Control* (1968), pp. D43–7II.

33. I. Gumowski and C. Mira, Sensitivity problems related to certain bifurcations in non-linear recurrence relations, *Automatica.* **5**, 303–317 (1969).
34. *Proceedings of the Symposium on Differential Equations and Dynamical Systems.* Lecture Notes in Mathematics, Vol. 206. Springer-Verlag (1971). Summer School, July 15–25, 1969.
35. R. H. Abraham and Y. Ueda, *The Chaos Avant-Garde: Memories of the Early Days of Chaos Theory.* World Scientific Publishing, Singapore (2000).
36. R. May, Biological populations with nonoverlapping generations; stable points, limit cycles, and chaos, *Science.* **186**, 645–647 (1974).
37. R. H. Abraham and C. D. Shaw, *Dynamics the Geometry of Behavior.* Aerial Press, Santa Cruz, CA (1981–1988).
38. C. Letellier and V. Messager, Influences on Otto E. Rössler's earliest paper on chaos, *International Journal of Bifurcation & Chaos.* **20**(11), 3585–3616 (2010).
39. O. Gurel and O. E. Rössler, *Bifurcation Theory and Appliations in Scientific Disciplines.* Annals of the New York Academy of Sciences, Vol. 316. New York Academy of Sciences (1979). Papers from the Conference Held on October 31–November 4, 1977.
40. J. Gleick, *Chaos: Making a New Science.* Viking Press, New York (1987).
41. J. P. Crutchfield, J. D. Farmer, N. H. Packard, and R. S. Shaw, Chaos, *Scientific American.* **255**(6), 46–57 (1986).
42. J. Crutchfield, D. Farmer, N. Packard, R. Shaw, G. Jones, and R. Donnelly, Power spectral analysis of a dynamical system, *Physics Letters A.* **76**(1), 1–4 (1980). doi: 10.1016/0375-9601(80)90130-9.
43. D. Farmer, J. Crutchfield, H. Froehling, N. Packard, and R. Shaw, Power spectra and mixing properties of strange attractors, *Annals of the New York Academy of Sciences.* **357**(1), 453–471 (1980). doi: 10.1111/j.1749-6632.1980.tb29710.x.
44. R. Shaw, *The Dripping Faucet as a Model Chaotic System.* Aerial Press, Santa Cruz, CA (1984).

https://doi.org/10.1142/9789819823130_0005

Chapter 5

BORIS CHIRIKOV — SPUTNIK OF CHAOS

Dima L. Shepelyansky

Université de Toulouse, CNRS, UPS, 31062 Toulouse, France
dima@irsamc.ups-tlse.fr

1. Introduction

In Russian, the word *Sputnik* means companion. But after the very first artificial satellite *Sputnik*, launched back in 1957, this word became also a metaphor for a pioneering, outstanding achievement. These meanings are appropriate for a portrait of Boris Chirikov (6 June 1928, Oryol — 12 February 2008, Akademgorodok, Novosibirsk, Russia), who was the founder of the physical theory of Hamiltonian chaos and made pioneering contributions to the theory of quantum chaos.[1,a] In 1959, he invented a simple analytical criterion, now known as the Chirikov criterion, which determines the conditions for emergence of deterministic chaos in dynamical Hamiltonian systems.[2,3] In the same paper, he applied this criterion to explain some puzzling experimental results on plasma confinement in open mirror traps, or magnetic bottles, that had just been obtained at the Kurchatov Institute.

There are various research directions launched by Boris Chirikov in the field of chaos. They include chaotic dynamics of particles in plasma magnetic traps and accelerators, chaos border for the Fermi acceleration

[a]D. L. Shepelyansky, Website dedicated to Boris Chirikov, Last access, August 2020: www.quantware.ups-tlse.fr/chirikov/.

model, emergence of chaos in various Hamiltonian systems including the Fermi–Pasta–Ulam problem border for the Fermi–Pasta–Ulam problem,[4,5] quantum chaos,[6,7] in dissipative dynamical systems,[8,9] and many others.

2. A Master's Touch

I joined Chirikov's group at the theory division of the Institute of Nuclear Physics (INP) in September 1976, at the beginning of my fourth year at the Novosibirsk State University. At the University, it was common practice to attach students of this year to specific research Laboratories or groups at the Academy Institutes at Akademgorodok. As many other students, I knew Chirikov from the course of Electrodynamics given by him and Igor Meshkov at our second year. But my choice was also significantly influenced by a recommendation of George Zaslavsky, whom I knew, who had worked with Chirikov and gave outstanding recommendations for his research. Also Chirikov was favourable to have a new student and had that possibility from the INP side.

Chirikov (see photo in Fig. 1) was head of the sector T3 of the theory division directed by Spartak Belyaev. The division was composed of three sectors, and there were about ten people in T3. However, the actual group

Fig. 1. Boris Chirikov, Toulouse, June 6, 1998.
Source: Photo by D. L. Shepelyansky.

working with Chirikov on nonlinear dynamics and *stochasticity* (now we say *chaos*) was rather small; it included essentially Felix Izrailev, Vitold Vecheslavov and Lida Hailo, who worked as a programmer. Two young researchers, Valery Tayursky and Oleg Zhirov, soon moved from T3 to other INP Labs.

I remember Chirikov's office in 1976–1978. It was a small room of 12 m^2, located on the fifth floor at the back yard of the main INP building. There were three desks of Chirikov, Izrailev, and Zhirov. The main focus of the room was a teletype terminal directly connected to a computer BESM-6 at the Computer Center of Siberian Division of Russian Academy of Sciences, located at about 1 km distance down along prospect Nauka. This was the most powerful soviet computer at that time. From the terminal it was possible to submit short runs on BESM-6, and even to work in interactive mode. The terminal produced a deafening rumble, but everybody in the group — especially Chirikov — was proud of it.

Chirikov defined the main scientific group aim as the investigation of fundamental laws of chaos and foundations of statistical mechanics for classical and quantum systems.

3. Chirikov Standard

In 1977, the now famous quantum kicked rotator model was invented. The model is the quantized version of the classical standard map, now known as the Chirikov standard map. This simple but generic symplectic map was essencially invented and analyzed by Boris Chirikov.[4,5] It has the form

$$\begin{cases} \bar{p} = p + K \sin x \\ \bar{x} = x + \bar{p} \end{cases} \tag{1}$$

where bars mark the new values of conjugated variables of momentum p and coordinate x, K is dimensionless parameter characterizing the kick strength. An example of the Poincaré phase space is shown in Fig. 2 for an eigenvector of the corresponding Perron–Frobenius operator of the Ulam method approximate.[10] In the quantum case these variables become the Heisenberg operators with the commutator $[p, x] = -i\hbar$. The properties and various applications of the Chirikov standard map are described in the *Scholarpedia*.[11]

The classical and quantum models became the corner-stones of the well established fields of classical and quantum chaos. But in 1977, many people looked at the quantum model with suspicion. I remember that the usually

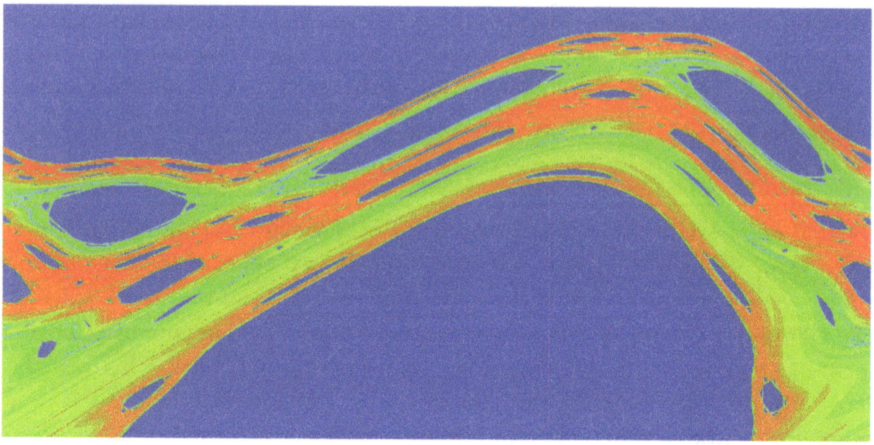

Fig. 2. Amplitude of the eigenstate of the Ulam approximate of Perron–Frobenius operator of the Chirikov standard map at $K = 0.971635406$; the number of cells and matrix size are $N_d = 127282$ and eigenvalue is $\lambda_2 = 0.99878108$; amplitude is proportional to color with maximum for red and zero for blue; upper part of phase plane is shown for the range $0 < \frac{x}{2\pi} \leq 1$, $0 < \frac{p}{2\pi} \leq 0.5$.

Source: From Frahm and Shepelyansky.[10]

heated atmosphere of the INP theory seminar, became a typhoon when Chirikov presented the results on kicked rotator for the first time:

> "Why you have here a delta-function? Why does your spectrum grow quadratically with the level number? Why does the energy of your quantum system grow slower than in the classical system, while the quantum theory is probabilistic?..."

Only Chirikov understood that the standard map describes a generic behaviour of chaotic systems, and since the correspondence principle, invented by Niels Bohr, should be valid at small dimensionless values of Planck constant, the kicked rotator model should demonstrate generic properties of quantum systems chaotic in the classical limit (now we say properties of quantum chaos). And indeed, the kicked rotator demonstrated a generic phenomenon now named dynamical localization. An analytical estimate of the number of populated quantum states, which is essentially the localization length, was obtained by Chirikov and his group at INP in 1980–1981.[6] However, the analogy with the well-known phenomenon of Anderson localization was established by the Maryland group of Shmuel Fishman, Grempel and Prange in 1982.[12]

Back in the late spring of 1977, Chirikov suggested that I work on the kicked rotator model, starting from the improvements of the computer code. Following his suggestions, I achieved a significant reduction of the CPU time, and I am still proud that the improved figures we obtained, were used in the Russian version of the kicked rotator paper written at INP and published as INP preprint in 1978. The English version published in 1979 was based on a presentation of this work given by foreign co-authors at a conference in Italy in 1977.[13] In fact, there were significant differences from the Russian version, since communications between the USSR and the West were rather slow at those times.

4. Our Chaos is the Best in the World

In the summer 1978 I graduated from the University, and became a stagier at INP, and soon a Ph.D. student under Chirikov's supervision. In the fall of 1978 the theory division moved to the top floor of new building behind the main one. The development of chaos theory had great potential for expansion, in the USSR and abroad. In Chirikov's group we were enthusiastic, and completely sure that our chaos was the best in the world.

Indeed, at those times even chaotic dynamics in nonlinear classical systems was a rather new and unusual subject for the world scientific community. For example, there wasn't any specialized journal in this field (now there are about ten), and often it wasn't easy to explain to an editor how it happens that, in spite of Laplace determinism, simple equations produce chaotic unpredictable behaviour. Quite often, editors blamed errors of numerical simulations, and rejected papers on chaos. The world wide circulation of research results was initiated by Joe Ford at Georgia Tech, Atlanta, who, every week, patiently collected the abstracts of new preprints on chaos and nonlinearity, with young collaborator Franco Vivaldi, and send them to colleagues and friends. Chirikov knew Joe Ford from their first meeting in Kiev in 1966, where Ford came as a tourist with a group of school pupils to visit the USSR. From that time they continued a regular correspondence (available at the archive of Boris Chirikov) and of course the world chaos news were regularly sent to Chirikov in Siberia. Finally, the number of publications in nonlinear systems became so large that the first specialized nonlinear journal, *Physica D*, was created in 1980. During many years Ford and Chirikov worked in the editorial board of this journal. It was Joe Ford who also stimulated Chirikov to write his famous chaos review article, in *Physics Reports* in 1979.[5] At 2008, this was the most quoted single author article of Russian scientific research.

This was the situation with classical chaotic dynamics. At that time, there were only few people in the world working in field now known as quantum chaos. Besides the Chirikov group, there were Martin Gutzwiller at New York and George Zaslavsky at Krasnoyarsk. But the interest in this field was rapidly growing all over the world. Also in the theory division Dodik Shuryak, Sam Heifets and Valya Sokolov obtained interesting results, and followed closely the development in the area.

5. Chaos Easy to Discuss

Classical and quantum chaos were attracting the interest of various people. The spectrum of visitors in Chirikov's office was very broad: there were Soviet scientists from many cities, jobless physicists, foreign researchers from many countries, including East and West Germany, France, Italy, UK, USA. Chirikov often invited his group to follow a discussion. He had the same attitude towards any visitor, and never blamed a speaker for not knowing some well-known things. He used to say: "*Criticism should be constructive*" and he always tried to extract some positive things from a talk. That's why people felt themselves at ease to ask him about any scientific problem; this is nicely illustrated in the reminiscences of Igor Meshkov.[14] Even Chirikov's office reflected his simple working attitude: it was very modest, with a plain old desk, old chairs, and book-shells.

What is even more surprisingly is that Chirikov was able to communicate not only with physicists, but also with mathematicians and philosophers. Thus, in 1958, he had meeting with Kolmogorov at Kolmogorov's home. After listening the explanations of Chirikov about the criterion,[2] Kolmogorov said: "*one should be a very brave young man to claim such things!*" Indeed, even now a mathematical proof of the criterion is still lacking and there are even known counterexamples of nonlinear systems with a hidden symmetry, where the dynamics remains integrable even in the case of strongly overlapped resonances. However, such systems with a hidden symmetry are quite rare and specific, while for generic Hamiltonian systems this physical criterion works nicely and determines very well the border for the onset of chaos.

After this meeting Chirikov maintained close contacts with Vladimir Arnold and Yakov Sinai and other members of this school. He was able to understand their formal theorems, tried always to extract their physical meaning, and applied them in his own research. I remember how Chirikov was telling me

> "Of course, it's usually very difficult for a physicist to read and understand a mathematical paper, but when you corner a good mathematician, like Arnold or Sinai@Sinaï, and discuss closely his results then, he will start to explain them to you as a physicist!"

At the celebration of Chirikov's 65th birthday, Sinai proposed a special toast for Chirikov and his respect for, and links to, mathematicians.

As to philosophy, it suffice to say that Chirikov published some articles in philosophy journals in German, English[15] and Russian, and was respected by philosophers both in the USSR and in the West, even if both sides usually did not respect each other. His talks at the philosophy seminar of INP were always attracting a full audience of the large conference hall of INP. I remember his aside note during such a talk:

> "The human mind is weak, and it needs a prompt to understand complex behaviour. Numerical simulations on a computer, or numerical experiments, give such a prompt. But computers have restricted abilities, and hence a researcher should find a good model, which on the one hand is sufficiently simple for simulations, but on the other captures the generic properties of the phenomenon being investigated."

I think that the very best example of such a model is the Chirikov standard map, which is still has puzzling phenomena in both classical and quantum gases, and which is still actively investigated in modern experiments with cold atoms and Bose–Einstein condensates.

6. Supercomputer Fervor

Computers were one of Chirikov's passions. At a very early stage, in the beginning of the 60s, he realized their importance for investigations of dynamical chaos, and since then used them extensively in his research. Back in 1979 Chirikov was allowed to make a return visit to Joe Ford at Atlanta for three months (Ford had visited Chirikov earlier in 1979). During this visit he got access to the most powerful computer of that time, the CRAY I. His work on CRAY I is lively described in the reminiscences of Franco Vivaldi,[14] while I know only how it was after his return back to INP. The first thing Chirikov showed to me was a white sheet of paper on which there was a hand-drawing of an american phone handset. He compared this image with a phone handset in his office, and concluded that in principle it was possible from Siberia to make computer runs on the CRAY in the USA. The attraction of CRAY was very strong, since Chirikov established that it was 500 times faster than BESM-6. It looked like a fantasy at that time. About two or three years later Jeff Tennyson, visiting INP, managed to establish a short connection with a CRAY in the USA, but it took about nine years before a real work on CRAY in Europe became possible during visits of foreign colleagues to the Chirikov group.

7. Chaos vs. Order

After Chirikov's return from the USA I continued with him our chaotic research. We discovered a slow algebraic decay of Poincaré recurrences in generic chaotic maps, presented at the International Conference in Kiev in 1981.[16] We also showed that homogeneous classical Yang–Mills fields have a chaotic dynamics, which shattered a dream of Vladimir E. Zakharov about the integrability of such all important equations in physics. Surely, a chaotic behaviour is generic while an integrable one is rare and exceptional. Chaos gains vs. Order but then again Order emerges from Chaos. But the main research line was linked to quantum chaos and the kicked rotator. A global picture of time scales in the regime of quantum chaos has been worked out and presented in the review of 1981 written by Chirikov, Izrailev and myself.[6] Soon after that, in 1982, I defended my Ph.D. and became a permanent researcher in Chirikov's group.

8. Chaos Summits

In 1983 Chirikov became a corresponding member of the Academy, and the development of chaos theory continued to expand to a higher level. During this year I developed a computer code which allowed to simulate an unexpectedly strong ionization of hydrogen atoms in Rydberg states by a microwave field, which was first observed in experiments of James Bayfield and Peter Koch at Yale in 1974.[17] In contrast to the classical numerical simulations of Ian Percival done in London in 1979 (in fact Percival visited Chirikov during summer 1983), my quantum simulations were showing, under certain conditions, the quantum suppression of classical chaotic excitation, which was rather similar to that seen in the kicked rotator. With a help of our american and italian colleagues, the following year this code was run on a CRAY at Livermore. The improved data were analyzed in detail in Siberia, and explained on the basis of the dynamical localization theory developed for the kicked rotator in 1981 and later. Finally, the analytical theory combined with numerical data was published in 1984.[18] This was a first application of quantum chaos theory to a real system studied experimentally. Further studies showed that this system is locally described by the quantum Chirikov standard map, and that the dynamical localization should be observable in experiments with a higher microwave frequency. During the workshop in Riga and the Vavilov Conference in Novosibirsk in 1987, Chirikov and I succeeded to convince Koch to make experiments in those conditions, and the theoretical predictions of dynamical localization theory were observed by his group at Stony Brook in 1988.[19]

In 1988 Chirikov and Vecheslavov, excited by a recent appearance of Halley's comet in 1986, performed an analysis of its 46 apparitions known from historical records and computer simulations and showed that the comet dynamics is described by a simple area-preserving map which is rather similar to the standard map. The dynamics of the comet was shown to be chaotic with a typical life time of 10 million years.[20] This was an amazing example of an enormously rich information extracted from only 46 numbers!

In 1988 INP (now Budker INP, BINP) celebrated Chirikov's 60th birthday with coffee, tea and songs at the round table, photo collage of theory division, directed now by Chirikov, and other festivities. At that time quantum chaos became a popular field of research, and a special Les Houches Summer School was organized in France in 1989 on that subject. Chirikov gave there a fundamental course on quantum chaos in time-dependent systems, Izrailev and I gave short courses. We met there many leading players in the field, whom we knew before only through their pioneering publications, including Oriol Bohigas, Martin Gutzwiller and many others. A few speakers, like Michael V. Berry and Ian Percival, were already known to us from their visits to Akademgorodok.

In the next two or three years quantum chaos flourished, with a variety of international workshops, schools and conferences organized in various countries. The manifestations of quantum chaos have been observed in various experiments in atomic and mesoscopic physics. But in spite of this experimental progress, it was becoming clear that the main effects of one-particle quantum chaos became understood at the end of second Millennium. It was a time to look for new ideas and lines of development of this field. In 1991 I got one year CNRS invitation to Toulouse and went there. A couple of months later the USSR disappeared. I got a research position at CNRS and continued to work in Toulouse. My visits to Akademgorodok persisted once a year.

9. French Connections

In 1973, Jean Bellissard, the leader of the theory group in Toulouse, and I succeeded in organizing a two month visit of Chirikov and his wife Olga Stepanovna Bashina to Toulouse. They arrived in December, 20 years after a virtual visit of Chirikov to Toulouse. Indeed, he was supposed to give an invited talk at a CNRS Conference on chaos theory in Toulouse 20 years earlier, but did not get a soviet permission as it was often the case at those times. His talk on dissipative dynamical chaos was presented by Joe Ford and was published in the Conference Proceedings.[8] Figure 3 shows an image of a strange attractor from this work. In fact Chirikov prefered to say "*chaotic attractor*", saying also "*strange for whom?*" (he was refering to the Anosov theorem about robustness which implies that a chaotic dynamics will remain if a small smooth dissipation is added). It appears that these results of Chirikov and Izrailev stimulated Michel Hénon, who was among conference participants, to invent the Hénon strange attractor.

Chirikov with wife came again to Toulouse in 1998 when Chirikov's 70th birthday was celebrated during the International Conference dedicated to him.[21] This was his last visit abroad.

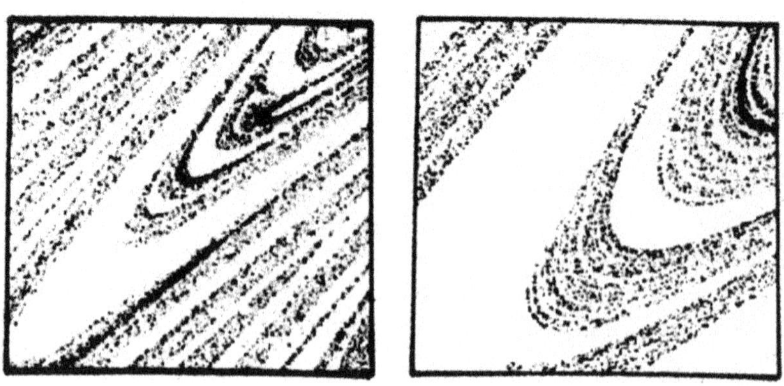

Fig. 3. An image of chaotic attractor (1973) from 8 (right panel gives 4 times magnification of left panel).

10. Land Siberia

During our meetings at INP (Fig. 4), and at his home, Chirikov told me many stories about Budker, Kurchatov Institute and INP history, his meetings with Kolmogorov, Ulam and other leading scientists.[1] Other stories were about the Second World War, the siege of Leningrad and his mother and him evacuation from Leningrad to the Krasnodar region, soon occupied by the German army around 1942–1943. A characteristic story of that time he told me:

> "A wounded Russian partisan hid in our home, and a german soldier came in for a search. I was alone at home, the situation was critical, and I was preparing to use one of the grenades I was hiding at home. The soldier came to a table in the center of the room, saw a soviet school manual of German language there and opened it. Then he spat, cursed, and went away. I looked at the opened page of the manual, it was the song text of the International!"

In May 2008 the Budker Institute of Nuclear Physics organized the Chirikov Memorial Seminar which demonstrated his pioneering contributions to the field of chaos. Indeed, it was a very early 1959,[2] when Chirikov freed the genie of Chaos, which spread the world over.

Fig. 4. Boris Chirikov and author at BINP office of Chirikov, September 2002.
Source: Photo by O. V. Zhirov.

References

1. D. Shepelyansky, Boris Valerianovich Chirikov, *Scholarpedia.* **3**, 6628 (2008).
2. B. V. Chirikov, Resonance processes in magnetic traps, *Atomnaya Energiya.* **6**, 630 (1959) (in Russian).
3. B. V. Chirikov, Resonance processes in magnetic traps, *The Soviet Journal of Atomic Energy.* **6**, 464–470 (1960). doi: 10.1007/BF01483352.
4. B. V. Chirikov, Research concerning the theory of non-linear resonance and stochasticity. Technical Report IYAF-267-TRANS-E, Institute of Nuclear Physics, Novosibirsk (1969). Translated at CERN from the Russian, CERN-Trans-71-40, 1971.
5. B. V. Chirikov, A universal instability of many-dimensional oscillator systems, *Physics Reports.* **52**(5), 263–379 (1979). doi: 10.1016/0370-1573(79) 90023-1.
6. B. V. Chirikov, F. M. Izrailev, and D. L. Shepelyansky, Dynamical stochasticity in classical and quantum mechanics, *Soviet Scientific Reviews C.* **2**(4), 209–267 (1981).
7. B. V. Chirikov, F. M. Izrailev, and D. L. Shepelyansky, Quantum chaos: Localization vs. ergodicity, *Physica D.* **33**(1), 77–88 (1988). doi: 10.1016/ S0167-2789(98)90011-2.
8. B. V. Chirikov and F. M. Izrailev, Some numerical experiments with a nonlinear mapping: stochastic component. In *Transformations ponctuelles et applications*, Vol. 229. Colloques Internationaux du CNRS, Toulouse (1973), pp. 409–428.
9. B. V. Chirikov and F. M. Izrailev, Degeneration of turbulence in simple systems, *Physica D.* **2**(1), 30–37 (1981). doi: 10.1016/0167-2789(81)90056-7.
10. K. Frahm and D. Shepelyansky, Ulam method for the Chirikov standard map, *European Physical Journal B.* **76**, 57–68 (2010). doi: 10.1140/epjb/ e2010-00190-6.
11. B. Chirikov and D. Shepelyansky, Chirikov standard map, *Scholarpedia.* **3**(3), 3550 (2008).
12. S. Fishman, D. R. Grempel, and R. E. Prange, Chaos, quantum recurrences, and Anderson localization, *Physical Review Letters.* **49**(8), 509–512 (1982). doi: 10.1103/PhysRevLett.49.509.
13. G. Casati, B. V. Chirikov, F. M. Izrailev, and J. Ford, Stochastic behavior of a quantum pendulum under a periodic perturbation, *Lecture Notes in Physics.* **93**, 334–352 (1979). doi: 10.1007/BFb0021757.
14. O. V. Zhirov, D. L. Shepelyansky, K. V. Epstein, and A. N. Skrinsky (eds.), *Boris Valerianovich Chirikov – Legislator of chaos.* Russian Academy of Sciences, Siberian Branch, Budker Institute of Nuclear Physics, Novosibirsk (2014). Partly in Russian. https://www.quantware.ups-tlse.fr/dima/myrefs/ chirikovbook2014.pdf.
15. B. Chirikov, Creating chaos and the life, 0503072 (2005). *arXiv:physics.*

16. B. V. Chirikov and D. L. Shepelyansky, Statistics of Poincaré recurrences and the structure of the stochastic layer of a nonlinear resonance. Technical Report PPPL-TRANS–133, Institute of Nuclear Physics, Novosibirsk (1981). Translated from the Russian, PPPL-TRANS-133, Plasma Physics Laboratory, Princeton University, 1983.
17. J. E. Bayfield and P. M. Koch, Multiphoton ionization of highly excited hydrogen atoms, *Physical Review Letters.* **33**(5), 258–261 (1974). doi: 10.1103/PhysRevLett.33.258.
18. G. Casati, B. V. Chirikov, and D. L. Shepelyansky, Quantum limitations for chaotic excitation of the hydrogen atom in a monochromatic field, *Physical Review Letters.* **53**(27), 2525–2528 (1984). doi: 10.1103/PhysRevLett.53.2525.
19. E. J. Galvez, B. E. Sauer, L. Moorman, P. M. Koch, and D. Richards, Microwave ionization of H atoms: Breakdown of classical dynamics for high frequencies, *Physical Review Letters.* **61**(18), 2011–2014 (1988). doi: 10.1103/PhysRevLett.61.2011.
20. B. V. Chirikov and V. V. Vecheslavov, Chaotic dynamics of comet Halley, *Astronomy & Astrophysics.* **221**(1), 146–154 (1989).
21. J. Bellissard, O. Bohigas, G. Casati, and D. L. Shepelyansky, A pioneer of chaos, *Physica D.* **131**(1), VIII–XV (1999). doi: 10.1016/S0167-2789(99)90007-6.

https://doi.org/10.1142/9789819823130_0006

Chapter 6

IN THE SHADOW OF CHAOS

Felix Izrailev

Universidad Autonoma de Puebla,
Puebla, Mexico
felix.izrailev@gmail.com

My recollections on the time I spent with Boris Chirikov, a pioneering Soviet physicist, who contributed to chaos theory. In particular, he introduced the so-called the Chirikov criterion that allows to track transition from regular to chaotic motion in nonlinear dynamical systems. He also contributed to the development of quantum chaos by introducing the standard map and the quantum kicked rotor.

1. Introduction

These somewhat chaotic memories begin with that distant day in 1964 when, I was a fourth-year student at the Faculty of Physics at Novosibirsk State University (NSU). I went to Boris Valerianovich Chirikov's room at the Institute of Nuclear Physics (INP) and said that I wanted to do my diploma in theoretical physics under his supervision.

First of all, Boris gave me a small task related to the equations of motion for a system of oscillators with cubic nonlinearity. As I later understood, this was a mathematical model from the Fermi–Pasta–Ulam preprint,[1] and I was to analyze the strange behavior of this system using Boris's new theory,[2] which predicted the emergence of chaotic oscillations in nonlinear systems.

At that time, Boris's main activity was related to experiments on the B-3M accelerator to create a stabilized high-density electron beam, one of the ideas of the Institute's director, Gersh I. Budker.[3] My first work with Boris coincided with his gradual transition into theoretical physics. However, Boris often emphasized that "we" were not theorists, referring to the small group that later formed in the theoretical department, whose main research tool was the so-called "numerical experiment." It should be noted that this term was not taken seriously by many physicists for a long time. Budker's critical attitude towards this type of activity is well-known; he believed that Boris should engage in more "serious" matters. Nevertheless, during Boris's doctoral dissertation defense in 1969, Budker acknowledged the validity of the method developed by Boris, which laid the foundations for the modern theory of dynamic chaos in deterministic systems.

2. Fermi–Pasta–Ulam Model: First "Numerical Experiment"

In essence, Chirikov's viewpoint was that besides studying the laws of Nature through experimental and theoretical physics, there is another equally effective approach based on "numerical experimentation." Unlike the standard use of computers as an auxiliary element for the theoretical analysis of a particular problem (for example, for calculating complex integrals or labor-intensive analytical transformations), numerical experimentation involves experimenting with the initial mathematical model itself. An essential element of this approach is that the final result is unknown in advance and may be obtained (or may not be!) during the research process by including or excluding certain terms in the equations of motion and changing the system's parameter values, followed by numerical integration of these equations.

This approach is very similar to real experimentation, where you can "turn the knobs" and conduct various measurements. On the other hand, numerical experimentation implies theoretical analysis, albeit not rigorous, without which simple "knob-turning" is practically useless. Moreover, it is absolutely necessary that the model in question is genuinely related to physical reality so that the newly discovered effects can then be confirmed in real experiments and studied using standard theoretical methods. It can be said that the history of physics development over the past 50 years has already confirmed the effectiveness of numerical experiments, which have led to the discovery of new physical phenomena such as dynamic chaos, the integrability of nonlinear equations, dynamic localization, and dissipative chaos (strange attractors), among others.

For those who closely interacted with Boris, it was obvious that he had a unique gift for understanding the essence of the most complex physical phenomena, intuitively sensing what was correct and what was not, and where truly interesting results could be expected. His ability to understand both theoretical physics and "high" mathematics, not to mention experimental physics, in which he successfully worked for some time, was remarkable. He could find errors in the works of world-class mathematicians and physicists, including Nobel laureates. The most derogatory epithet Boris used for the authors of incorrect works was "muddler."

The history of the Fermi–Pasta–Ulam problem is quite instructive, especially for those interested in nonlinear physics. A beautiful presentation of the physical and mathematical ideas associated with this problem and a historical analysis of the main results are presented in Weissert's book,[4] based on his dissertation on the history of nonlinear physics. A pleasant feature of this book is that it fairly fully presents the role of Soviet physicists and mathematicians in the development of nonlinear physics, unlike many other foreign publications. This is clearly due to the influence of the work's consultant, James D. Meiss, an American physicist — University of Colorado at Boulder — and expert in dynamical systems[5] who was well acquainted with the works of Soviet researchers and personally knew Boris.

In brief, the history of the FPU problem is as follows. In 1953, Enrico Fermi, having a brief respite from working on the atomic project at the Los Alamos Laboratory in the USA, decided to return to his long-time interest and check, using numerical modeling, the possibility of thermalization in a chain of nonlinearly coupled oscillators. It was expected that with a large number of oscillators (32 or 64), even weak nonlinearity would lead to an intense exchange of energy between linear modes, and consequently, to the system's relaxation to a statistically equilibrium state. In one of his earliest works (1923),[6] Fermi discussed the meaning of the ergodic hypothesis, which was then considered necessary to justify statistical physics, however, he did not specify the conditions under which the ergodicity emerges. Having temporarily at his disposal the most powerful computer at the time, he, along with the Polish mathematician, Stanislaw Ulam, and the Italian physicist John Pasta, conducted the first "numerical experiment" in history, leading to the discovery of new nonlinear phenomena. It turned out that instead of thermalization, the considered chain of oscillators exhibited periodic energy transfer from the initially excited harmonic to the others and back. There was no talk of thermalization, which made Fermi state that it was an incomprehensible but apparently very important effect. Unfortunately, these works were interrupted due to Fermi's illness and death in 1954, and the results were only formalized as an internal preprint of the Los Alamos Laboratory the following year.[1]

Attempts to explain the FPU paradox undertaken by various researchers over the next ten years did not yield significant results, and only in 1965 was there a breakthrough, largely due to Boris's ideas about the existence of a criterion according to which the system behaves either regularly or stochastically depending on the parameter values of the model.[2] Initially, Boris used the term "stochasticity" to denote the practically random behavior of the system, which arose despite the absence of any random parameters or perturbations in the equations of motion. Today, the term "dynamic chaos" or "deterministic chaos" or simply "chaos" is established in the literature. Jokingly or seriously, Boris used to say that studying "chaos" in such a stable country as the Soviet Union was somewhat provocative. It is now hard to imagine how unusual it was in the 1960s and 70s to talk about chaos in physical systems. Today, the scientific literature is "overflowing" with chaos to the extent that it is difficult to find a field of physics where the problem of chaotic behavior of a system does not arise.

Applying the "Chirikov criterion" to the FPU problem showed that under the parameters used in the numerical modeling in the work of Fermi, Pasta, and Ulam, the system's behavior should indeed be practically regular and characteristic of integrable systems. Theoretical analysis performed of the problem has been done in 1965 by Boris and myself. First, the results have been in the form of a preprint of the Institute of Nuclear physics and published as a regular paper next year, 1966.[7] This was my first paper and I was proud that my work under the guidence of Boris was successful. As a result, it was understood why the authors of the FPU preprint could not observe the equipartition of energy among linear modes, that was expected to see due to a nonlinear coupling between linear modes. The analytical estimated based on the overlap of nonlinear resonances have shown that the condition for the onset of strong chaos, depends not only on the interaction strength but also on the initial conditions in the dynamics of the nonlinear chain. Specifically, it was found that the thermalization is strongly suppressed if initially a few low frequency modes are excited. On the other hand, according to the obtained criterion for a strong sharing of energy beween the modes, if one exites initially high frequency modes, one can observe strong overap of resonance that could result in the statistical distribution of energy between all modes. This prediction has been confirmed in the following numerical experiments.[8,9] We thus effectively resolved the paradox that many physicists and mathematicians unsuccessfully tried to explain. One of the authors of the FPU work, Stanislaw Ulam, was so intrigued by the results achieved by Boris's group at the Institute of Nuclear Physics that one of the preprints of the Institute, in which the analytical predictions obtained by Boris and me (1965) were numerically confirmed, was translated into English and published in 1969 as a regular preprint of the Los Alamos Laboratory.[8] Moreover, when the International Mathematical Congress was held in Moscow in 1966, where Boris presented a report on the emergence of chaos in deterministic systems, Ulam specially flew to Moscow for a couple of days and spent most of his time in discussions with Boris. Apparently, Ulam faced some troubles due to this visit to Moscow, as he was one of the main developers of the hydrogen bomb. Later, Boris told me that after meeting Ulam and attempting to invite him to the Institute of Nuclear Physics for joint work for three months (a formal invitation had already been signed by the chairman of the Siberian Branch of the Academy of Sciences of the USSR, M. A. Lavrentyev), correspondence with Ulam ceased. After some time, an American physicist visiting the Institute wished to meet Boris. In the corridor, looking around, he quickly handed Boris a short note from Ulam, in which he apologized for his "silence," which was not his fault. When during Boris's three-month trip to the USA in 1979, he finally met Ulam, Ulam was still "under surveillance," as during their brief conversation after a seminar conducted by Boris in one of the American laboratories, Ulam was constantly accompanied (though without interfering with the conversation) by two "bodyguards" in plain clothes.

Apparently, Boris found in Ulam not only an excellent interlocutor but also a like-minded person, as Ulam was one of the first to understand the revolutionary significance of computers for the development of physics. In his memoirs,[10] he predicted that the future of physics would largely depend on the creative alliance (a kind of symbiosis) of a researcher-physicist and a computer. Combining analytical and numerical methods with great skill, Boris was effectively a pioneer in the practical implementation of this new direction in modern physics.

I was always amazed by Boris's absolute openness to perceiving new results and effects arising from studying specific problems. I remember how long ago, around 1968, I showed him a strange result produced by the M-20 (some might still remember this archaic vacuum-tube computer, which occupied a significant space in the Computing Center of the Siberian

Branch of the Academy of Sciences). The result was that in the numerical modeling of the FPU model's behavior (according to a program written by A. Khisamutdinov in machine code), even harmonics were excited over time in the wave packet dynamics, which should theoretically be absent due to some symmetry of the corresponding Hamiltonian. This was undoubtedly an effect related to rounding errors in the computer, but I had no idea why it quickly destroyed the correct dynamic picture of the system. When I told Boris that I didn't know how to suppress the arising instability, he immediately suggested not fighting it but using it as an indicator of chaos. The fact is that this instability of the solution was directly related to the emergence of chaos, and later we measured the degree of chaos through the exponential growth of even harmonics in the system's evolution.[8] Today, for specialists in nonlinear dynamics, this effect might seem trivial, but back then, everything was absolutely new and inexplicable. It is enough to say that the modern method of detecting chaos through measuring the so-called Lyapunov exponents was invented many years later.[11]

Unfortunately, in those distant times (1963–1974), chaos research received no support in scientific circles. All these "so-called" numerical experiments and the results associated with them were regarded, at best, as eccentricities unrelated to "real" physics. Although many physicists' personal attitudes toward Boris were always favorable, his scientific activity (I mean the study of chaos) was not properly appreciated. The situation was further complicated by the fact that Soviet physics journals, with rare exceptions, did not publish articles where numerical analysis was the basis of the research. Nevertheless, thanks to the Institute of Nuclear Physics having the right to publish its own preprints, many results from those times, obtained by Boris and his collaborators, were not lost.

The situation changed somewhat in the mid-1970s with the publication of the article by Ruelle and Takens,[12] which started the boom associated with the discovery of strange attractors — nonlinear dissipative systems with chaotic behavior together with an unusual phase space structure, described by ordinary differential equations. Boris immediately criticized this term, as there was nothing "strange" about the behavior of strange attractors to him. Interestingly, the original article by American mathematician and meteorologist Edward Lorenz, which served as the catalyst for Ruelle and Takens' work, was published much earlier, in 1963.[13] Initially, it was not accepted in a general physics journal (again due to the numerical results!) and was published in an atmospheric physics journal, where it was virtually buried until better times. This work carried out detailed numerical modeling of a simplified system in which turbulence arose, and this fact resonated with mathematicians working on dynamic systems and statistical physics.

From Boris's perspective, "strange attractors" were nothing more than chaotic behavior of a dissipative system (in the absence of random forces), which in some sense was analogous to the dynamic chaos well studied by that time (though little known to the general public). According to Boris, these strange attractors seem "strange" only to non-specialists. In English transcription, the phrase by Boris — "strange for strangers" — became quite popular, as noted by Lorenz himself in his book.[14] Moreover, a few years ago, I came across a small book[15] that aimed to popularize chaos theory and consisted entirely of comic illustrations. On one of the pages, the aforementioned phrase by Boris was depicted under two strange figures in dark coats and hats covering their eyes, representing Boris and me.

3. Tête-à-tête with the Computer

It should be noted that by 1975, the effect of the emergence of dissipative chaos had already been studied in Boris's group, and quite thoroughly. However, no one was interested in such things at the time, and the results were only published as reports presented at little-known conferences.[16] As a result, priority in the study of strange attractors in nonlinear systems described by discrete equations was lost. Boris's idea to "see" how dissipation affects strong dynamic chaos in a nonlinear Hamiltonian system described by a two-dimensional map almost immediately led to the discovery of an unknown effect, which later turned out to be the very "strange attractor." These and many other results of that time were obtained by Boris and me using a "display" constructed by the engineer B. Dolgovesov at the Institute of Automation and Electrometry of the Siberian Branch of the Academy of Sciences (SOAN). A very advantage of this device was that, with software written by L. Vasilieva,[17] it was directly connected to the high-speed BESM-6 computer at the Computing Center of the Siberian Branch of the Academy of Sciences, allowing for "on-line" cooperation with the computer. A distinctive feature of such a "tandem" was the ability to see the system's trajectory in real-time on the display screen, and most importantly, at any moment, the problem-solving process could be stopped and continued in time, with new parameter values changed using a "light pen," an analogue of the "mouse" in modern computers. The effectiveness of this kind of "experimenting" was astounding, and Boris participated in our visits to the computer center with great interest and enthusiasm, suggesting various methods of computation and recording the results to prepare for next "experiments."

Once, during one of our visits to the Computer Center (this was in 1974), a film crew shooting a popular science film about the use of computers in various applications became interested in our work on the display. Filming took place in various institutes in our Academgorodok which is near the Novosibirsk town. Seeing Boris and me by the display in the computer room, the film's director, Boris Zagryazhsky, and the cameraman, E. Antsis, wanted to include us in their film "*The Computer and Leonardo's Riddle*" (Kyivnauchfilm, 1974), especially since the display created at the Institute of Automation and Electrometry under the guidance of Y. E. Nesterikhin was indeed a unique development (unfortunately, this device was never put into mass production). Boris was asked to look into the camera, explain what we were doing, and demonstrate the work on the display. Later, B. Zagryazhsky told me several times that he was impressed by Boris's relaxed and graceful behavior in front of the camera. Thus, we were included in the film by the Kyiv studio named after Dovzhenko, or more precisely, its branch, which produced many excellent popular science films in those distant times. The film *The Computer and Leonardo's Riddle* was shown in Akademgorodok and at the Institute of Nuclear Physics, but unfortunately, it was not acquired by the Institute.

It should be noted that the film turned out to be very successful and won numerous awards, including international ones.[a] In one of the favorable reviews, in the Ukrainian magazine *Novyny Kinoekranu*, the cover featured photos from this film, where Boris and I demonstrate the properties of deterministic chaos to the audience (Fig. 1). For this unusual report on chaos research presented to the widest audience imaginable, Boris chose an example of how strong nonlinearity can "suppress" chaos and stabilize nonlinear oscillations.

By the way, the following story is related to this episode. Some time passed, and in a conversation with me, Boris thoughtfully said that the idea he presented on camera, and his speech, had not gone through the Institute expert commission. Thus, the "scientific information presented

[a]For instance, this movie received the Golden Seal at the XIII *Festival Internationale de Fantasienza*, Trieste, July 1975. This is a history of cybernetics from first machines to mankind's aspirations to create a being in its image, and about the computing power of machines that were able to detect Leonardo da Vinci's figure among the characters of his own paintings.

Fig. 1. Cover of the magazine *Cinema News* dated on August 1974, Kyiv, Ukraine. From left to right: Boris Zagryazhsky, Felix Izrailev and Boris Chirikov.

Source: Courtesy of Felix Izrailev.

to the public" had not been approved and could theoretically have "state secrets." Therefore, we decided to urgently write a corresponding short text of 4 pages, including the drawings shown in the film, and publish it as an Institute preprint (of course, after obtaining half a dozen signatures certifying that the text "*does not contain results representing state secrets*" and was, in general, of no use to anyone).[18] After that, we could sleep peacefully.

4. "Let the Reader Think!"

In the late 1970s, a new graduate student, Dima Shepelyansky, joined Boris's group. Practically, I co-authored only two articles with Dima;[19] in the rest, the main author was Boris. This happened as follows: Boris invited both of us to his office, seated us on either side of him, and we began creating the text of a given article. To be precise, Boris wrote the text, and Dima and I occasionally suggested our versions or asked questions. It must be said that such interaction with Boris was extremely beneficial for both of us, as numerous questions usually arose during the writing process, leading to discussions. It is impossible to describe how interesting these discussions were. I want to note that anyone who had scientific conversations with Boris would agree that his manner of discussion was extraordinarily engaging, and the discussions themselves were simultaneously deep and rich, stimulating further contemplation of the topic.

It is noteworthy that Boris was willing to spend as much time as necessary on discussions and explanations of the essence of the matter but did not like discussing such insignificant, in his view, details as the style of presentation. This trait also manifested itself when preparing texts jointly with our Italian partners, Giulio Casati and Italo Guarneri. Once, when Boris and me were visiting our Italian friends in Como (Italy), Boris prepared the initial draft of a joint article and brought the written text to us for reading and discussion. For us, this often began a period of "torment" if, in our opinion, it would have been better to write a particular paragraph differently to make it easier for the reader to understand the result or the logic of the presentation. To all our suggestions, Boris usually said, "*Never mind, let the reader think!*" Obviously, this attitude reflected his unique way of thinking, characterized by an extraordinarily high density of ideas and results in his articles' texts. Sam Haifetz Jr. once said that after the publication of Boris's famous 1979 review on dynamic chaos,[20] a small group of staff formed in the accelerator lab at Stanford University to study this review.[b] The goal of the regular seminars was a detailed analysis and discussion, step by step, of the numerous results obtained by Boris, which later formed the basis of the theory of dynamic chaos. This work by Boris[20] is considered by many physicists to be the "bible of chaos" — it is now among the top five most cited works by Russian physicists.

Boris's lectures at the Physics Department of Novosibirsk University were also distinguished by their material density. I still remember my unsuccessful attempts to take detailed notes on his lectures on electrodynamics, which Boris taught students at the Novosibirsk University in 1963. Once, after spending a couple of hours immediately after one of his lectures, trying to make my notes — before the details were forgotten — I unexpectedly discovered the complete useless of my naive attempt to write down "everything" that seemed important or simply interesting to me. Each of Boris's lectures evoked an association with a large intricate

[b]There is such a testimony by Celso Grebogi in chapter 19 regarding Allan Kauffman's group at Berkeley. Note from the editor.

Fig. 2. Boris Chirikov asking a question to Giulio Casati after a talk given at the Novosibirsk University in 1975.
Source: Courtesy of Giulio Casati.

structure, but one that extended not upwards but into a bottomless depth. Often, it was impossible to form a coherent and clear picture of what was presented, but each fragment revealed a vast number of interesting details, serving as a powerful stimulus for further study of the subject.

The uniqueness of Boris's lectures was especially evident in comparison with the excellent lectures of other great lecturers I was fortunate to listen to. First, I recall the lectures of Mikhail Vladimirovich Volkenstein at Leningrad University (molecular biology), as well as the lectures of Victor M. Galitsky (quantum mechanics) and Spartak Timofeyevich Belyaev (elementary particle physics) at Novosibirsk University. If Volkenstein, Galitsky, and Belyaev's lectures always had transparency and perspective, giving a sense of the subject's integrity, the distinguishing feature of Boris's lectures was the density of material, richness of detail, and many questions for independent contemplation.

5. Beginning of the International Collaborations

In 1975, a young Italian physicist, Giulio Casati, visited Boris's group at the Institute of Nuclear Physics in Novosibirsk (USSR). This marked the beginning of an extremely fruitful collaboration between Boris's team and other researchers, including Joe Ford from Georgia Tech University (USA),

Franco Vivaldi, and Italo Guarneri, who joined the group later. This collaboration laid the foundations for what would eventually become known as the study of quantum chaos.

Casati initially intended to explore how the KAM (Kolmogorov–Arnold–Moser) threshold is affected by increasing the number of particles in a system. However, this approach soon proved too complex. During the visit, Boris proposed a different line of inquiry: to investigate the role of chaos in quantum mechanics. This led to the development of the kicked rotor model, a quantum analogue of the classical standard map, which Boris had studied in great detail for his second doctoral degree in 1969. That earlier work later became the basis of his influential review paper — now considered foundational in applying classical chaos theory to physical systems. Casati remembers:

> This work on the Kicked Rotor was completed in 1977 by Casati, Chirikov, Ford, and Izrailev, and was presented at the very first conference dedicated to quantum chaos, titled "*Stochastic Behaviour in Classical and Quantum Hamiltonian Systems*," held in Como, June 1977. The paper was published in the conference proceedings.[21] We never submitted this paper (which remains my most cited publication) to any scientific journal. This is a significant detail because it reveals something essential about Boris's personality. In a world where everyone strives to publish and gain recognition, Boris had no interest in accolades or external validation — he was simply devoted to science itself. He considered our results — now known to be quite important — as only a preliminary step. He didn't want to submit the paper to a journal, believing that we needed to reflect more deeply on our findings. This speaks volumes about the kind of person he was, and I will never forget it.
>
> The second point is more of a curiosity, but still worth sharing. I vividly remember sitting side by side with Boris at a desk to write the first draft of the paper. I naturally expected him to begin writing, but instead, he handed me a blank sheet of paper and a pen, saying, "*Please, you write.*" I was surprised and asked why. He explained that since I was returning to Italy in a few days, he was worried that, at the airport, a border guard might see the manuscript written in someone else's handwriting and suspect me of smuggling sensitive scientific material out of the Soviet Union.

In 1983, the entire small group (Boris, Vitaly Vecheslavov, Valentin Davydovsky, Dima Shepelyansky, and I) in the theoretical department of IYAF[c] undertook the translation from English into Russian of the fundamental work by Allan Lichtenberg and Michael Lieberman[22] which compiled the most well-known results on nonlinear dynamics and chaos at that time. Boris was the chief editor, and we divided the material among ourselves, each translating several chapters as best we could. What does an editor usually do in such a situation? Usually — corrects and "polishes" the entire text. I do not know if there was ever another precedent when an editor (or translator) rechecked all the physical and mathematical results presented in the book on paper. This is exactly what Boris did, even though the book was an overview of a vast number of articles. As a result of such an incredible work, Boris discovered many significant inaccuracies, including simply incorrect results obtained by rather well-known physicists.

The book was published by MIR publishing house, but initially, the editors refused to accept our text because the number of editor's notes (made by Boris) exceeded the number of days in a year. They remarked, and I completely agreed, that it would have been better if we had written our own book. The situation was resolved by Boris — he translated all his "corrections" into English, sent them to the authors in the USA, and requested official permission to include them in the translation of the book (see one of the answers in Fig. 3[d]). Since the authors personally knew Boris and his works (many of which were discussed in detail in the book, even those published only as preprints), a scandal was avoided, and the book was successfully published in 1984 (Fig. 4).

6. From Classical Chaos to the Quantum One

By the way, many of Boris's colleagues repeatedly tried to persuade him to write his own book on dynamic chaos, but unsuccessfully, which is a pity. He always made excuses with various pretexts, the main one being "*there is work to do.*" Probably, the problem was that Boris did not like writing a book as a collection of his already published works (as many authors usually do). It is enough to remember his 1979 review — all the main results of the review were obtained during its writing, which took about a year. As a result, many of Boris's important results remained scattered across articles, preprints, and conference proceedings, and were often little known to a wide audience.

[c]Nuclear Physics Institute of the Siberian section of the USSR Academy of Science.
[d]https://www.quantware.ups-tlse.fr/chirikov/refs/chi1984lichlet.pdf.

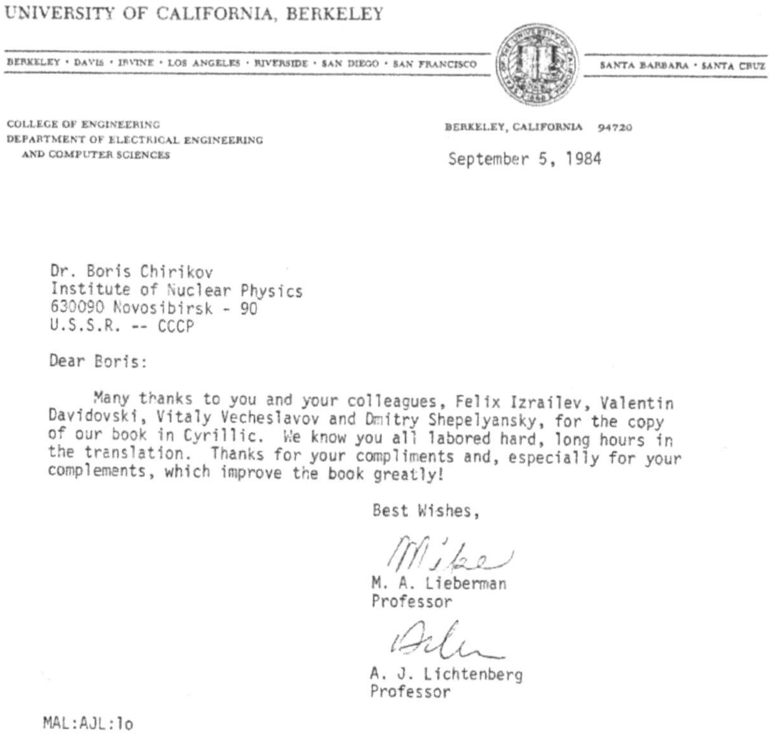

UNIVERSITY OF CALIFORNIA, BERKELEY

BERKELEY · DAVIS · IRVINE · LOS ANGELES · RIVERSIDE · SAN DIEGO · SAN FRANCISCO   SANTA BARBARA · SANTA CRUZ

COLLEGE OF ENGINEERING
DEPARTMENT OF ELECTRICAL ENGINEERING
AND COMPUTER SCIENCES

BERKELEY, CALIFORNIA 94720

September 5, 1984

Dr. Boris Chirikov
Institute of Nuclear Physics
630090 Novosibirsk - 90
U.S.S.R. -- CCCP

Dear Boris:

Many thanks to you and your colleagues, Felix Izrailev, Valentin Davidovski, Vitaly Vecheslavov and Dmitry Shepelyansky, for the copy of our book in Cyrillic. We know you all labored hard, long hours in the translation. Thanks for your compliments and, especially for your complements, which improve the book greatly!

Best Wishes,

Mike
M. A. Lieberman
Professor

A. J. Lichtenberg
Professor

MAL:AJL:lo

Fig. 3. Letter from Liberman and Lichtenberg to Chirikov on September 5, 1984.

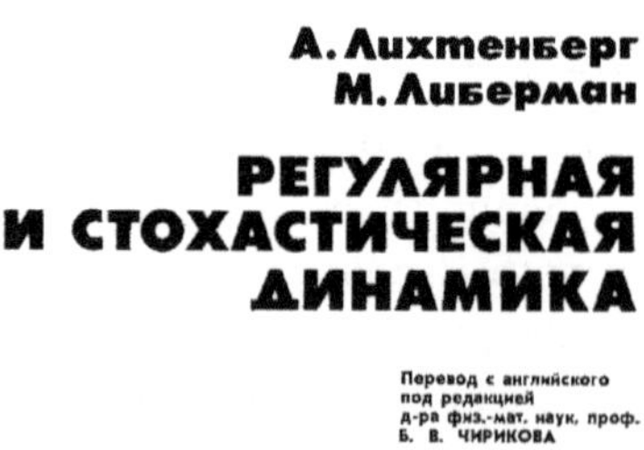

А. Лихтенберг
М. Либерман

РЕГУЛЯРНАЯ
И СТОХАСТИЧЕСКАЯ
ДИНАМИКА

Перевод с английского
под редакцией
д-ра физ.-мат. наук, проф.
Б. В. ЧИРИКОВА

Fig. 4. Russian translation of the Lieberman and Lichtenberg's book. MIR editions, 1984.

Boris's 1979 review[20] in some way marked the conclusion of research in the field of dynamic chaos in classical mechanics and the transition to chaos studies in quantum systems. The first work in this direction[21,23] brought unexpected results and later turned out to be a fundamental work in the theory of quantum chaos. This paper has been cited over 5500 times and widely assumed to be the cornerstone in the theory of quantum chaos.

By that time, Boris had established a collaboration with Giulio Casati (1975), which Joseph Ford joined in 1977, followed later by Franco Vivaldi and Italo Guarneri in the 1980s. This international cooperation proved to be extremely fruitful for all participants. Besides the scientific aspect, this collaboration was also important for our group (Boris, Dima Shepelyansky, and I) because it facilitated the publication of our works in foreign scientific journals. It should be reminded that in Soviet times, to publish an article abroad, it had to be previously published in "domestic press." This requirement naturally contradicted the practice of foreign journals, which accepted only original works. Given that the publication of our works, many of which were based on numerical calculations, met with active resistance from the editors of Soviet physics journals, publishing articles in foreign journals was undoubtedly beneficial. Furthermore, the joint work provided an opportunity to use the powerful CRAY supercomputer, first deployed in the USA and later in Italy.

The key participant in this fruitful collaboration was, of course, Boris. His initial question related to quantum chaos was — what happens to dynamic chaos when the system is described not by classical Newtonian equations but by the quantum Schrödinger equation? Initially, practically everything was unclear, and I remember numerous discussions between Boris and Joe Ford, who repeatedly visited the Institute of Nuclear Physics for brief visits in the late 1970s. As a result of these discussions, a new approach was developed, which now forms the basis of the modern theory of quantum chaos. The key element of this approach is the concept of quantum "pseudo-chaos," as opposed to the chaos that occurs in classical systems. The fundamental difference of quantum chaos is that it is not based on the instability of motion that arises in classical nonlinear systems, but on a completely different linear mechanism associated with a vast number of independent frequencies that determine the temporal evolution of a quantum system.

The estimation of the time during which an approximate correspondence between the global properties of quantum and classical systems is observed was proposed by Boris and turned out to be far from trivial, though seemingly simple. This showcased Boris's renowned skill in deriving important physical estimates "by fingers." By the way, this well-known unique ability of Boris to obtain unexpected estimates "from nothing" was humorously

Fig. 5. Cartoon prefacing Boris Chirikov's lectures (Course 8) given at the summer school *Chaos and quantum physics* at Les Houches (France), August 1–31, 1989.[24]

depicted in a friendly cartoon (Fig. 5) that prefaced the text of lectures given by him at an international school in France in 1989.[24]

A fundamental moment in understanding the properties of quantum chaos was the discovery (again, based on numerical experiments conducted under the guidance of Boris) of the mechanism of so-called dynamic localization, according to which quantum effects accumulate over time and lead to the freezing of the system's dynamics.[25] It was later shown that, in some sense, dynamic localization is analogous to the well-studied Anderson localization in the theory of disordered quantum systems.[26] However, it is essential to note the fundamental difference between dynamic localization and Anderson localization since the latter occurs in quantum systems with random potentials, whereas dynamic localization pertains to deterministic systems without any random parameters, to which a statistical approach is inherently inapplicable.

Several memorable moments are associated with the aforementioned school in France, more precisely, in the resort area of the French Alps. The first is when, in response to a question from Dima and me about what to do with the invitations we all (including Boris) received from the organizers, Boris unexpectedly told us to start processing the documents for all three of us. Until then, Boris's standard advice was to throw the invitation letters in the trash and send a reply saying that, unfortunately, it was impossible to attend such and such a conference (any reason could be given, but the rules of the game did not allow mentioning the true reason — "*not allowed to go!*"). But Boris already knew that "perestroika" times were approaching, and indeed, all of us were "let out," as they used to say at the time, and not just to some "near abroad," but straight to France. When the three of us appeared before the conference organizers, they were taken aback — they did not expect three participants from one institute in Novosibirsk, still thinking that maybe one might be "let out," but even that was unlikely. Boris promptly remarked that we were undergoing perestroika — "*And you haven't restructured yet!*"

7. "No One is a Prophet in There Own Land"

Boris had remarkable intuition — it seemed he could foresee where and in what situations new physical effects might appear. This was especially true in his choice of mathematical models for further research. His approach was to select a model that was as simple as possible for analysis, yet far from trivial and highly substantive. The most striking example of such a model is the relatively simple system of two discrete equations, known in the literature as the Chirikov standard map.[20] In the theory of classical chaos, this map plays a unique role. On one hand, it allows one to understand the basic properties of conservative chaos and, on the other hand, it serves as a touchstone for testing various mathematical methods used to study real physical systems. The number of published works investigating the standard map is enormous, and entire chapters in textbooks on nonlinear mechanics and dynamic chaos are devoted to it.

Another equally well-known model proposed by Boris is the "quantum rotator" (Kicked Rotator) model, which forms the basis of modern quantum chaos theory.[21,27] Essentially, a quantum analog of the standard map, this model describes the behavior of a quantum pendulum under periodic kicks from an external nonlinear perturbation. The number of works investigating this model is beyond estimation, with many using the term "quantum rotator" in their titles. The model's popularity stems from its relative simplicity and the multitude of non-trivial quantum effects it describes. A few years ago, in the USA, the quantum rotator was realized through the interaction of cooled atoms with an optical laser field, and the effect of dynamic localization was experimentally confirmed.[28]

Unfortunately, the contribution of Boris in the development of nonlinear mechanics and, especially, in the discovery of dynamic chaos and the study of its properties, is greatly underestimated by the physical community. One reason is that many of his results were published in preprints and conference proceedings rather than in prestigious scientific journals. Another reason

Fig. 6. Boris Chirikov working (without computer) with Giulio Casati at Como in 1989. *Source*: Courtesy of Felix Izrailev.

is that Boris did not pay due attention to the publicity of his results, apparently believing that sooner or later his achievements themselves would be recognized as outstanding. He was always alien to marketing activities aimed at gaining recognition for his work. Even in the USSR, for a long time, Boris's work on dynamic chaos did not receive due attention from the physics community, although the results obtained by Boris and his colleagues in the field of classical chaos were well known to mathematicians in Moscow and Gorky (now Nizhny Novgorod), specialists in the theory of dynamical systems. Close communication with renowned mathematicians such as Vladimir I. Arnold, Vasiliy M. Alekseev, Ya G. Sinai, and others, served both as support and a catalyst for Boris's scientific inquiries. Importantly, the interest in this communication was mutual — in many works by Ya G. Sinai and his students from the 1970s and 1980s, the influence of this interaction is evident.

It should also be noted that at the Institute of Nuclear Physics, Boris's scientific pursuits initially did not receive a positive response. This was likely due to Gersh I. Budker's high regard for Boris as a top-class experimentalist, and his reluctance to see him requalify as a "pure" theorist. Additionally, Budker had a negative view of "numerical experiments" and did not accept Boris's thesis on the special role of computers in modern physics. Despite this, Boris always held Budker in the highest esteem, appreciating his talent and considering him his teacher.

Nevertheless, attempts to engage Boris in "business" were not successful, and by 1976, Boris shifted his focus to studying the problem of dynamical chaos in quantum systems. Despite the forced interest in the dynamics of colliding beams, my collaboration with Boris continued uninterrupted — we successfully continued researching new problems such as Arnold diffusion[29] and strange attractors in nonlinear systems described by discrete mappings.[16] By the late 1970s, Boris had almost entirely switched to the problem of quantum chaos, and the activity of our group sharply increased with the start of long-term international cooperation (with Giulio Casati, Italo Guarneri, Franco Vivaldi and Joe Ford).

8. Conclusion

For Boris' 75th birthday, a conference was organized at the Institute of Nuclear Physics in 2003. Giulio Casati remembers.

> In 2003 a large, one week, conference was organized in his Institute in honor of his 75th birthday. A lot of people came from all over the world. However, Boris spent all the time in his office, discussing with Yu Orlov his results[30–32] that, on the first glance, where no relation with the quantum chaos problems. He attended only the Orlov's talk, asking many questions in view of their discussion. It looks as Boris lost an interest to standard physics and wanted very much to understand how the brain work. He wanted to understand how creative process is possible and his idea was that the origin of this was chaos, because chaos is the only possibility to produce new information (not simple rational activity). So, chaos must be at the root of creativity of the human being. The problem is that in order to make progress in this direction, he needed to know how our brain works, and it was difficult for him to find the books to learn about the brain. Now, at this conference was present Orlov (the former russian dissident who was by then professor in US University) who spent much time working on the problem of "free will" and its conection to physical determinism (Fig. 7). Orlov was supposed to stay in Novosibirsk only for the period of the conference, and Boris wanted to know as much as possible from him while he was there. For this reason, Boris spent the entire week discussing with Orlov and completely ignored the conference in his honor. He only attended official dinner and celebration. I strongly admired him for this. I would never have the courage to do what he did! About hundred people came to Novosibirsk for him and he did not attend any talk! Great!

Fig. 7. Yuri Orlov with Boris Chirikov and his wife, Olga, during the International Conference on *Dynamical Chaos in Classical and Quantum Physics* dedicated to his 75th Birthday. Institute of Nuclear Physics, Novosibirsk, 4–9 August 2003.
Source: Courtesy of Felix Izrailev.

In conclusion, returning to the years of my joint work with Boris Valerianovich, I would like to quote the words once said by Academician Dmitry N. Pryanishnikov about the Soviet geneticist of world-class, Nikolai I. Vavilov — "*Nikolai Ivanovich is a genius, and we fail to recognize it only because he is our contemporary.*" To some extent, these words can also be applied to Boris Valerianovich Chirikov.

References

1. E. Fermi, P. Pasta, S. Ulam, and M. Tsingou, Studies of the nonlinear problems. Technical Report LA-1940, Los Alamos National Laboratory, Los Alamos, NM (1955).
2. B. V. Chirikov, Resonance processes in magnetic traps, *Atomnaya Energiya.* **6**, 630 (1959) (in Russian).
3. G. I. Budker, A. V. Kiselev, N. G. Kon'kov, A. A. Naumov, V. I. Nifontov, G. N. Ostreiko, V. S. Panasyuk, V. V. Petrov, L. I. Yudin, and G. I. Yasnov, Start-up of a b-3m synchrotron injector for a positron-electron storage ring, *Journal of Nuclear Energy C.* **8**(6), 727 (1966). doi: 10.1088/0368-3281/8/6/314.

4. T. P. Weissert, *The Genesis of Simulation in Dynamics: Pursuing the Fermi-Pasta-Ulam Problem*. Springer, New York, NY (1997). doi: 10.1007/978-1-4612-1956-9.
5. J. D. Meiss, *Differential Dynamical Systems*. Society for Industrial and Applied Mathematics, Philadelphia, PA (2017). doi: 10.1137/1.9781611974645.
6. E. Fermi, Beweis dass ein mechanisches Normalsystem im allgemeinen quasi-ergodisch ist, *Physikalische Zeitschrift*. **24**, 261–265 (1923).
7. F. M. Izrailev and B. V. Chirikov, Statistical properties of a nonlinear string, *Doklady Akademii Nauk SSSR*. **166**, 57–59 (1966).
8. F. M. Izrailev, A. I. Khisamutdinov, and B. V. Chirikov, Numerical experiments with a chain of coupled anharmonic oscillators. Technical Report Technical Report 252, Institute of Nuclear Physics, Novosibirsk, USSR (1968). English translation: Los Alamos Scientific Laboratory, LA-4440-TR, 1970.
9. B. Chirikov, F. Izrailev, and V. Tayursky, Numerical experiments on the statistical behaviour of dynamical systems with a few degrees of freedom, *Computer Physics Communications*. **5**(1), 11–16 (1973). doi: 10.1016/0010-4655(73)90003-9.
10. S. Ulam, *Adventures of a Mathematician*. Charles Scribner's Sons, New York (1983).
11. G. Benettin, L. Galgani, and J.-M. Strelcyn, Kolmogorov entropy and numerical experiments, *Physical Review A*. **14**, 2338–2345 (1976). doi: 10.1103/PhysRevA.14.2338.
12. D. Ruelle and F. Takens, On the nature of turbulence, *Communications in Mathematical Physics*. **20**, 167–192 (1971).
13. E. N. Lorenz, Deterministic nonperiodic flow, *Journal of the Atmospheric Sciences*. **20**, 130–141 (1963). doi: 10.1175/1520-0469(1963)020⟨0130:DNF⟩2.0.CO;2.
14. E. N. Lorenz, *The Essence of Chaos*. University of Washington Press (1993).
15. Z. Sardar and I. Abrams, *Introducing Chaos :A Graphic Guide*. Icon Books, Zittau, Germany.
16. B. V. Chirikov and F. M. Izrailev, Some numerical experiments with a nonlinear mapping: Stochastic component. In *Transformations ponctuelles et applications*, Vol. 229. Colloques Internationaux du CNRS, Toulouse (1973), pp. 409–428.
17. L. V. Vasilyeva, F. M. Izrailev, and P. A. Kim, Computer graphics and its applications. Technical Report, Novosibirsk Computer Center, Novosibirsk, USSR (1974).
18. B. V. Chirikov and F. M. Izrailev, Numerical experiments on stabilization of stochastic instability using a display in dialog mode. Technical Report 74-13, Institute of Nuclear Physics, Novosibirsk, USSR (1974) (in Russian).
19. F. M. Izrailev and D. L. Shepelyansky, Quantum resonance for the rotator in a non-linear periodic field, *Doklady Akademii Nauk SSSR*. **249**, 1103–1107 (1979).

20. B. V. Chirikov, A universal instability of many-dimensional oscillator systems, *Physics Reports.* **52**(5), 263–379 (1979). doi: 10.1016/0370-1573(79) 90023-1.
21. G. Casati, B. V. Chirikov, F. M. Izrailev, and J. Ford, Stochastic behavior of a quantum pendulum under a periodic perturbation, *Lecture Notes in Physics.* **93**, 334–352 (1979). doi: 10.1007/BFb0021757.
22. A. Lichtenberg and M. Lieberman, *Regular and Chaotic Dynamics. Applied Mathematical Sciences*, Vol. 38. Springer-Verlag (1992). doi: 10.1007/ 978-1-4757-2184-3. First edition in 1983.
23. B. V. Chirikov, F. M. Izrailev, and D. L. Shepelyansky, Dynamical stochasticity in classical and quantum mechanics, *Soviet Scientific Reviews C.* **2**(4), 209–267 (1981).
24. B. V. Chirikov, Time-dependent quantum systems. In M.-J. Giannoni, A. Voros, and J. Zinn-Justin (eds.), *Chaos and Quantum Physics*, Vol. LII. Elsevier (1991), pp. 443–545.
25. B. Chirikov, F. Izrailev, and D. Shepelyansky, Quantum chaos: Localization vs. ergodicity, *Physica D.* **33**(1), 77–88 (1988). doi: 10.1016/S0167-2789(98) 90011-2.
26. S. Fishman, D. R. Grempel, and R. E. Prange, Chaos, quantum recurrences, and Anderson localization, *Physical Review Letters.* **49** (8), 509–512 (1982). doi: 10.1103/PhysRevLett.49.509.
27. F. M. Izrailev, Simple models of quantum chaos: Spectrum and eigenfunctions, *Physics Reports.* **196**(5), 299–392 (1990). doi: 10.1016/0370-1573(90) 90067-C.
28. F. L. Moore, J. C. Robinson, C. F. Bharucha, B. Sundaram, and M. G. Raizen, Atom optics realization of the quantum δ-kicked rotor, *Physical Review Letters.* **75**, 4598–4601 (1995). doi: 10.1103/PhysRevLett.75.4598.
29. G. V. Gadiyak, F. M. Izrailev, and B. V. Chirikov, Numerical experiments on a universal instability in nonlinear oscillator systems (the Arnold diffusion). In O. B. Lykova, A. M. Samoilenko, and A. N. Sharkovskii (eds.), *Proceedings of 7th International Conference on Nonlinear Oscillations*, Vol. 2. Berlin (1975), p. 315.
30. Y. F. Orlov, A quantum model of doubt, *Annals of the New York Academy of Sciences.* **373**(1), 84–92 (1981). doi: 10.1111/j.1749-6632.1981.tb51134.x.
31. Y. F. Orlov, The wave logic of consciousness: A hypothesis, *International Journal of Theoretical Physics.* **21**, 37–53 (1982). doi: 10.1007/BF01880263.
32. Y. F. Orlov, Origin of quantum indeterminism and irreversibility of measurements, *Physical Review Letters.* **82**, 243–246 (1999). doi: 10.1103/ PhysRevLett.82.243.

https://doi.org/10.1142/9789819823130_0007

Chapter 7

MY DANCE WITH CHAOS

Otto E. Rössler

University of Tübingen, Tübingen, Germany
oeross00@yahoo.com

Following my original attempt at finding equations for the origin of life, I developed more manageable problems by looking at elementary chemical reactions as components of an electronic circuit. I so arrived at chemical reactions producing periodic oscillations. Then I experienced a stimulating freedom in discovering the beauty of chaotic attractors. Exploiting the possibilities offered by the analog computer enabled new results.

1. My First Steps

The first skill that I developed and which was significant for my contribution to chaos was related to my interest in telephones. When I was 13, I built a connection between the floor on which my parents lived with my room under the roof, so that I could listen while the other side could not (I used three wires instead of two, which could have been more elegant). It gradually led to repairing and then building radios and radio emitters. I got my radio amateur's licence DL9KF when I was 17. I still own it. I never forgot my first interest in electronics and the little book entitled *Measuring-signal generators, frequency measuring devices and multivibrator* from the radio-amateur library [Fig. 1(a)].[1]

Stimulated by the Franciscan monk Pater Guido, a friend of my family, I decided to become a begging monk but my father forbade it. I was half-relieved because I did not know whether I would be strong enough for a life on the streets as a begging monk. To overcome my bad conscience,

(a) Sutaner, 1952 (b) Richter, 1953 (c) Richter, 1957

Fig. 1. The books in electronics used by Otto E. Rössler during his teenager years.

and pushed by my father, I enrolled in medicine, out of the same idealistic motivation. During the third year, I got attracted by scientific medical books (histology, biochemistry) and began working with mice in a long-term immunization study under the supervision of one of the most renown German immunopathologists, Erich Letterer (1895–1982). He was known for the Letterer–Siwe disease[2] and was the head of the Institute of Pathology at the University of Tübingen. Unexpectedly, my mice became immune-tolerant to the bovine gamma globulin used in the immunization, a result which was not expected. In fact, the Australian Frank MacFarlane Burnet (1899–1985)[3] and the British biologist Peter Medawar (1915–1987)[4] received in 1960 the Nobel prize for the discovery of acquired immunological tolerance. I developed weak experimental skills and my thesis entitled *Long-term immunization of albino mice with bovine-γ-globulin*[5] has only 24 pages.

During the last years of my medical studies, I read the book *The History of Nature* published in 1948 by Carl Friedrich von Weizsäcker.[6] I was impressed by Weizsäcker's ideas. Weizsäcker (1912–2007) had made major contributions in nuclear physics as well as in philosophy. To give an impression of the type of ideas I enjoyed in his book:

> A living organism, understood in physical terms, is a certain kind of material system. The organism, though engaged in a constant interchange of substances with its surroundings, can be distinguished from them with a certain degree of accuracy. Living organisms consist of some of the same chemical elements that make up their inorganic surroundings. We have succeeded, in the laboratory, in producing a large measure of the specific 'organic' combinations of elements that occur in living things.

> This suggests the idea that, given sufficient knowledge of the internal structure of organisms, we should be able to explain the origin and behavior of organisms entirely in physical and chemical terms.

I then attempted to get scientific contacts (and perhaps recognition) by writing him quite long letters (one of them had 25 pages).[a]

> I understand that you have clearly expressed the insight that a scientist needs to be a physician, and that science wherever it arises must be medicine. That is, that the natural scientist, without having wished so and without being aware of it, always has been a physician, and that science has always been a branch of medicine..

After a few exchanges by letters, I was invited to meet him at his mother's home (Fig. 2). Indeed, Weizsäcker wrote[b]:

> Indeed, according to the content of your letter, it seems to me that it would be easier for us to communicate even in a relatively short conversation than in an exchange of letters. The simultaneous presence in the same room is probably much more suitable than a telephone call. I would only resort to the latter if the former is technically too difficult to achieve. On the other hand, I have to warn you that my time is always very limited and that I can only offer you about half an hour for direct discussions. The week after Pentecost is not suitable for me. If you want to come to Hamburg after your current stay in Marburg, I could suggest to you at the earliest the late afternoon of June 14. Otherwise, we could postpone the conversation until I am near you again. I will be in Hamburg until the end of July, vacationing in August, I cannot not overlook exactly what I will do in the first half of September, but will be in southern Germany towards the end of September and spend the first half of October in Munich.

We finally met on September 11, 1965.[c] During this meeting, we visited Weizsäcker's mother who thanked me for my visit [Fig. 2(b)].[d]

---

[a] Letter from Otto to Carl F. Weizsäcker on May 31, 1965.

[b] Letter from von Weizsäcker to Rössler, June 4, 1965.

[c] Letter from Weizsäcker to Otto on August 4, 1965.

[d] "Dear Mr Rössler,

What lovely roses you gave my son and me! Thank you very much for that: also for your lovely, subtle accompanying words. We both enjoyed your visit very much and even continued talking about the topics raised. Warm wishes and many greetings, also from my son, sends you the old grandmother.

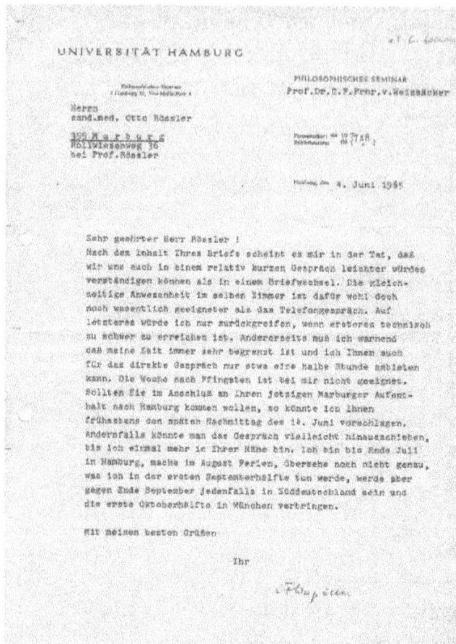

UNIVERSITÄT HAMBURG

PHILOSOPHISCHES SEMINAR
Prof.Dr.C.F.Frhr.v.Weizsäcker

Herrn
cand.med. Otto Rössler

355 Marburg
Rollwiesenweg 36
bei Prof.Rössler

4. Juni 1965

Sehr geehrter Herr Rössler !
Nach dem Inhalt Ihres Briefs scheint es mir in der Tat, daß wir uns auch in einem relativ kurzen Gespräch leichter würden verständigen können als in einem Briefwechsel. Die gleichzeitige Anwesenheit im selben Zimmer ist dafür wohl doch noch wesentlich geeigneter als das Telefongespräch. Auf letzteres würde ich nur zurückgreifen, wenn ersteres technisch zu schwer zu erreichen ist. Andererseits muß ich warnend daß meine Zeit immer sehr begrenzt ist und ich Ihnen auch für das direkte Gespräch nur etwa eine halbe Stunde anbieten kann. Die Woche nach Pfingsten ist bei mir nicht geeignet. Sollten Sie im Anschluß an Ihren jetzigen Marburger Aufenthalt nach Hamburg kommen wollen, so könnte ich Ihnen frühestens den späten Nachmittag des 14. Juni vorschlagen. Andernfalls könnte man das Gespräch vielleicht hinausschieben, bis ich einmal mehr in Ihrer Nähe bin. Ich bin bis Ende Juli in Hamburg, mache im August Ferien, übersehe noch nicht genau, was ich in der ersten Septemberhälfte tun werde, werde aber gegen Ende September jedenfalls in Süddeutschland sein und die erste Oktoberhälfte in München verbringen.

Mit meinen besten Grüßen

Ihr

(a)

899 Lindau (B), Mozacherweg 77, 15.9.65
Lieber Herr Rössler,
mit was für herrlichen Rosen haben Sie meinen Sohn und mich erfreut! Seien Sie aufs Herzlichste dafür bedankt; ebenso auch für Ihre lieben, feinsinnigen Begleitworte. Wir haben Beide viele Freude an Ihrem Besuch gehabt und manches angeschnittene Thema noch weiter erörtert. Herzliche Wünsche und viele Grüße, auch von meinem Sohn, sendet Ihnen die alte Großmama

(b)

Fig. 2. (a) Letter from Carl F. Weizsäcker to Otto E. Rössler, dated on June 4, 1965. (b) Letter from Weizsäcker's mother to Otto, September 9, 1965.

Source: The picture is from her house, Motzacher Weg, Lindau.

A few day later, I was invited by the German biologist and cybernetician Horst Mittelstaedt (1923–2016) to attend a scientific Colloquium organized at Seewiesen[e]:

> Mr. von Weizsäcker told me about your interest in cybernetics and suggested inviting you to this year's Mind-Body Colloquium, which will take place here in Seewiesen on October 14th and 15th, 1965. [...] This year's program includes a presentation by Mr. von Weizsäcker, one by Mr. Schmitz (about "awareness of something as a bodily occurrence"), one by Mr. Marko (about "subjective information") and one by Mr. Riemann (on a psychopathological topic).

Weizsäcker recommended to me to complete my medical studies and to work as a scientist afterwards. He even secured for me a first scientific position at the Max Planck Institute for Behavioural Physiology in Biocybernetics at

[e]Letter from H. Mittelstaedt to Otto on September 15, 1965.

Seewiesen. During that year (from April 1st, 1966 to February 1st, 1967), I developed a big friendship in very long discussions with Konrad Lorenz (1903–1989), one of the founders of ethology. He took seriously my ideas about biogenesis and the ensuing deductive biological approach.[7–9] At that time, I discovered the work by the British developmental biologist C.H. Waddington (1905–1975), in particular his book on Genes,[10] and his book on the Nature of Life (chapter 4 is about evolution).[11]

At Seewiesen, I started to write a paper initially entitled *General meta-atomic theory of evolution*, inspired by von Weizsäcker's book.[6] Using the assumption that the development of a given system is going on under a selection pressure, it passes through universal evolutionary stages. I gave a talk on that topic at the Max-Planck Institute für Verhaltensphysiologie in March, 1967.

I then spent one year as an medical intern at the University of Marburg partly under the supervision of Reimara Waible who became my wife, one year later. Mittelstaedt recommended me to the German mathematician and computer scientist, Carl Adam Petri (1926–2010), who was working in Bonn-Beuel at an institute for applied mathematics, computer issues, artificial intelligence, etc.[f] Petri was according to Mittelstaedt one of the few really important mathematicians in Germany. His Ph.D. thesis was devoted to communication with asynchronous automata. This theory provides a means of representing complex organizational processes of any kind.[12] He describes his automata with a sort of graph.

During that period, I wrote — in German — a first paper entitled "*Contributions to the theory of spontaneously evolving systems I: a simple model class*" (see Chapter 8, p. 205). I wrote many drafts during nearly two years, and I finally submitted one to the *Journal of Theoretical Biology*. In February 1970, I started working with Friedrich-Franz Seelig at the Division for Theoretical Chemistry at the University of Tübingen on modeling chemical systems.

2. My First Written Paper

In 1967, when I was still in Seewiesen, I tried to put together a longer publication for the *Journal of Theoretical Biology*. The "instructions for authors" (which allowed submission of German-language manuscripts)

[f]Letter from Mittelstaedt to Otto on August 26, 1968.

were encouraging. When we had found the Advertisement for a "visiting appointment" at the Center for Theoretical Biology in the *Journal of Theoretical Biology*, this (together with a positive response) added to the fun of writing the paper. In February of 1967, I returned to Marburg to complete my medical education, with clinical work. At the same time I was finishing my first manuscript. A correspondence with the German microbiologist Reinhard Walter Kaplan (1912–2003) about his book *The Origin of Life*,[13] and with Waddington about his contribution related to theoretical biology[14,15] — who stayed in Buffalo — also contributed to the motivation.

Whether a bit of adrenaline also contributed I cannot exclude in hindsight. There were rumors that a famous colleague was preparing a paper on an alternative approach. But if any such "push" existed, the "pull" by the appeal of the "second arrow" in nature was certainly greater.

I took some references about theoretical physics from the book by Sybren Ruunds de Groot and Peter Mazur:[16] for instance, the principle of dynamical reversibility introduced by Lars Onsager.[17,18] One of the most influential contributors was certainly the Japanese thermodynamicist and cybernetician, Motoyosi Sugita (1905–1990), who led me to discover the papers by Britton Chance,[19,20] Harold Garfinkel,[21] and Ilya Prigogine.[22,23] At that time, I was regularly browsing the *Journal of Theoretical Biology.* Most of the references were quoted to justify my approach or to confirm my point of view. However, the papers by Sugita and Nobuo Fukuda, the Scottish electrical engineer Alistair MacFarlane (1931–2021), the founder of general system theory, Ludwig von Bertalanffy (1901–1972) and those by Reinhard W. Kaplan had a strong influence on my ideas. I was aware of Bertalanffy's work since 1965.[g] In his book,[24] I discovered Ross Ashby's contribution.[25,26]

Prigogine intimidated me ever since his name was mentioned to me after my talk entitled "*Einige verhaltensphysiologisch interessante Aspekte einer allgemeinen meta-atomaren Evolutionstheorie*" on March 31, 1967 in Seewiesen. Prigogine had an early 1949 law on the decreasing rate of change of energy dissipation in an open system under linear conditions.[27] This seemed to be related to my principle of self-inhibition of energy dissipation, but I lacked the formal techniques even to attempt a comparison. I always knew that someday I would have to study his work in detail.

[g]There is a letter from Otto to Bertalanffy, on February 25, 1965.

Prigogine was very generous. He did me many favors and always acted as a friend. To me, he is with his radiating and inspiring personality something like an incarnation of the second arrow. Nevertheless, I never investigated Prigogine's works in depth. This was a kind of "self-protection" to be able to continue my research without any pressure. He was a master in the field who apparently had said very similar things as I was trying to say. Would it not have been natural to go to this source? At this time, I had no background in statistical mechanics, not to mention its new far-from-equilibrium branch, knowing not even what a differential equation is, I simply was not up to this task.

A second point which is important is that it was quite natural for me to make the analogy between electronic circuits and chemical reactions. I had got my background in electronics from the books (Fig. 1) by Hans Sutaner,[28] and by Heinz Richter.[29,30]

How did the idea of a distorsion-free amplitude modulation with direct carrier modulation arise? In my home-made shotwave transmitter of 1957 I could not afford an expensive high-power-output audio frequency amplifier, as would have been necessary for a high-output "anode modulation". Only a "screening-grid" low-power modulation was affordable. Here I empirically "overmodulated", with the consequence of severe distorsions. Then I tried to get rid of the distorsions by introducing a negative feedback, from the rectified transmitted output to the input side of the audio frequency amplifier. Amazingly, the negative feedback worked marvellously. All the audible distorsions disappeared. Thus, much less power was needed on the part of the transmitter to transport an undistorted audio signal by a modulated rather than constant carrier wave.

This gave me much confidence in the reliability of the principle of negative feedback and the intuitive accessibility of strongly nonlinear "dynamical systems"; I remember reading the term "dynamical systems" for the first time in 1967 in Erich Kamke's book.[31]

Such a connection had already been made by the German chemist Karl Friedrich Bonhöffer (1899–1957) in 1948[32] but I was not aware of it. For me, it was natural to make direct links between the properties a chemical reaction should have and the dynamical properties which can be observed in electronic oscillating circuits.[33]

The connection to me always was via water flowing in narrow transparent pipes between sperical vessels, See Fig. 2 in the chapter which follows (p. 209). So the substances were connected: by reaction lines like by transparent pipes. The connection to electronics was secondary: the pipes became wires and the spherical vessels became capacitors. Unlike in electronics, there exist also nonlinear elements in chemistry. Only the amplifier took a longer way to identify. And the self-induction (coil) has still not found its counterpart in chemical reaction networks. Later, I learned from Bob Rosen (1934–1998) of Alfred Lotka's work of the 1920s.[34,35]

Except for the iron wire-model of the German chemist Wilhelm Ostwald (1875–1952)[36,37] and the Canadian biologist Ralph Stayner Lillie (1896–1979),[38] I learned of Ostwald's chaotic measured curves, only in 1999 from Peter Plath at his Ostwald conference, in Ostwald's house.[39] In fact, Alfred Lotka was had already a "dynamical system's mind" in the sense that he understood, as did van der Pol, Andronov and co-workers, that the most important fact is how the elements of a system are interacting, the nature of the elements themselves being irrelevant for the underlying dynamics. In other words, what is important is the algebraic form of the equations studied and not where the equations are coming from. Thus you may have similar equations for a chemical reaction[34] and a food chain[35] as Lotka investigated in the 1920's. By the time I wrote my first long paper, I was not aware of the existence of oscillating chemical reactions as observed by the American chemist William Bray (1879–1946),[40] for instance.

Thus, I considered a chemical soup, with a large number of chemical species, as a network of elementary chemical reactions viewed as simple ordinary differential equations. This is an approach quite common for designing electronic circuits. The network was viewed as a directed graph and eventually nonlinear effects could emerge from the topology of the network. An elementary chemical reaction was considered as a small subgraph of the whole graph and can be viewed as a "molecular automaton" which is a kind of switching circuit with a certain logical function.[41] The purpose was then to determine the different types of dynamics which can be encountered and which types of reaction among the known ones were involved for producing a given dynamics. Identifying the element responsible for the growth of the system was one question addressed in this first work. The approach was qualitative and the transient regime was seen as an evolutionary process of the chemical soup toward a "stable" stationary dynamics. Thinking in terms of control theory and feedback systems, I learned from Bob Rosen.[42]

In fact, Sugita proposed in his 1966 paper[43] a system which illustrates quite accurately what I described in my first paper. The model describes cell size regulation[44] and reads as

$$\begin{cases} \dot{x} = \overline{T}_{\rm M} - x \\ \dot{y} = \overline{T}_{\rm M} T_{\rm c} - y \\ \dot{z} = \alpha y - \beta z \end{cases} \tag{1}$$

where $\overline{T}_{\rm M} = 0$ during mitosis and 1 otherwise. The second parameter $T_{\rm c}$ switches between two values as

$$T_{\rm c} = \left| \begin{array}{ll} 1 & \text{if } x \geqslant z \\ 0 & \text{otherwise.} \end{array} \right. \tag{2}$$

The system (1) is hybrid, that is, it is a combination of a continuous system and a logical circuit with a binary switching mechanism. Sugita was interested in using oversimplified mathematical models for biological processes. Here, z is the concentration of a protein (RNA); x and y are the concentrations in some regulatory metabolites as RNA or similar compounds. Sugita determined empirically the parameter values α and $\beta \ll 1$ for getting a curve reasonably similar to those experimentally obtained. He thus obtained time series similar to those plotted in Fig. 3. The synthesis of the second metabolite starts shortly after t_0 since $x < z$. Then, y tends to reach its saturation value before strongly decaying up to $t = t_1$, and then oscillating with a small amplitude.

Indeed, the first phase ($0 \leqslant t \leqslant t_1$) corresponds to what I called the *transition phase* during which the system presents an *evolution*. Thereafter, it reaches a *stationary phase* during which it oscillates. In fact, the first equation is independent from the two others, and quickly tends to the limit value $x_\infty = 1$ which corresponds to $\dot{x} = 0$. The system (1) is thus reduced to the two-dimensional system

$$\begin{cases} \dot{y} = \overline{T}_{\rm M} T_{\rm c} - y \\ \dot{z} = \alpha y - \beta z \end{cases} \tag{3}$$

which has two singular points, $\mathrm{S}_0 = (0, 0)$ and $\mathrm{S}_1 = (1, \alpha/\beta)$ which are both stable nodes. The dynamics can be plotted in the two-dimensional state space $\mathbb{R}^2(y, z)$; it depends on the time step δt used in the fourth-order Runge-Kutta algorithm for integrating the system (3); particularly the amplitude of the fluctuations. In fact, the parameter $T_{\rm c}$ depends on time and, consequently, is not a parameter but a variable. Rigorously speaking, this system is three-dimensional, thus explaining how a period-52 orbit can be produced.

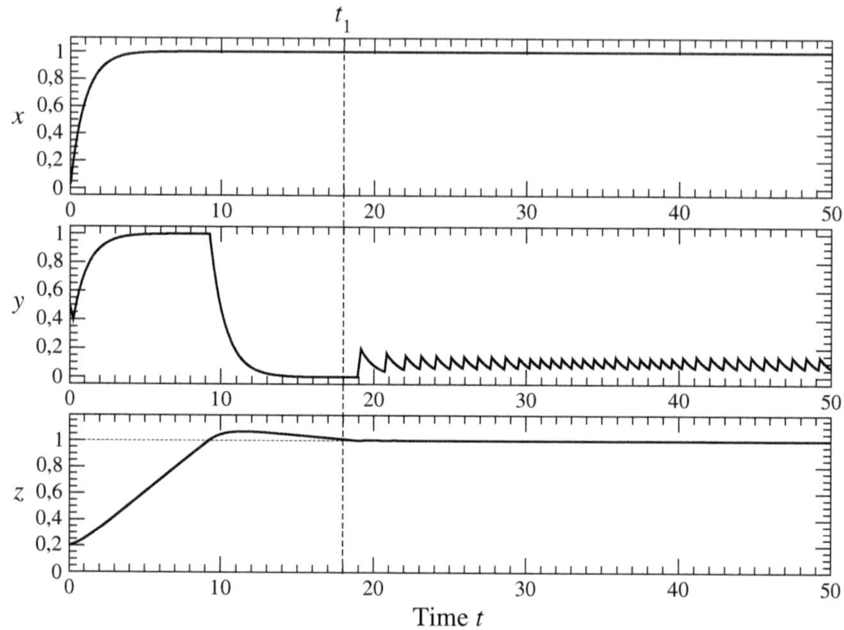

Fig. 3. Time series of the regulatory metabolites x and y, and the protein z produced by the Sugita model (1).

Notes: Parameter values: $\alpha = 0.1$, $\beta = 0.01$, and $\overline{T}_M = 1$. Initial conditions: $x_0 = 0.01$, $y_0 = 0.5$, and $z_0 = 0.2$. Time step: $\delta t = 0.01$ s.

After submitting my manuscript in July 1969, I continued to investigate the problem of modelling an evolutionary chemistry, by corresponding with Reinhard Kaplan (Institut für Mikrobiologie der Goethe Universität, Frankfurt). I provided him with a copy of my manuscript. Kaplan answered as follows.[h]

> I recently came across your essay, "*Contributions to the Theory of Spontaneously Evolutionary Systems I*". I am very interested in the issues you have discussed there. Unfortunately, since I was only able to keep the copy for a short time, I did not read it intensively. I would be very grateful if you would let me have a copy or an offprint. It seems to me particularly important from the biologists' point of view that you include the evolutionary ability of living material systems in your considerations. So far, the "skillfulness" of living systems as

[h]Letter from Kaplan to Otto on June 27, 1969.

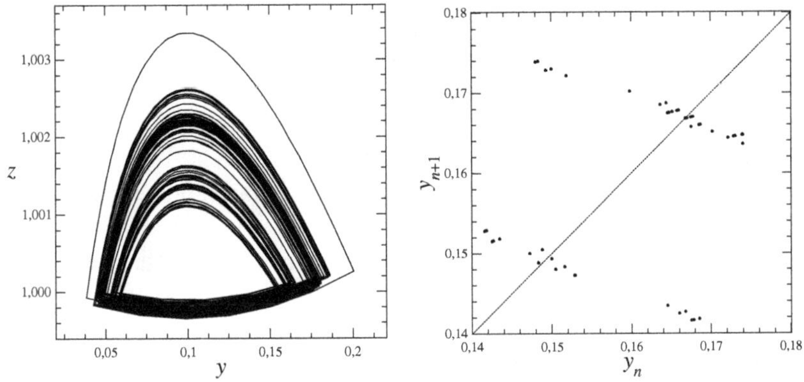

(a) State portrait (b) First-return map

Fig. 4. State portrait and first-return map to a Poincaré section of the solution produced by the two-dimensional Sugita model (3). This is a period-52 orbit. Same parameter values as in Fig. 3 but the time step is $\delta t = 0.03$ s for esthetic arguments.

a result of hereditary change has not really come into its own in the papers I am aware of, as for instance by Bertalanffy, compared to reproduction (= autocatalysis) and regulation (= equifinality), although it is a key property of organisms. I wish you every success in your future work; I would be very grateful if you would also send reprints of your future work.

I replied to Kaplan[i]

The question of the physico-chemical feasibility of complex abstract systems is today a controversial area (Wigner, Petri,[12] Trintscher[45]). Conversely, many biochemical phenomena cannot yet be satisfactorily imagined physically (Pattee[46]).

Since the end of the era of classical physics, the fact that "historical" systems of the most complex precision are possible in mathematics has been largely forgotten. However, the disadvantage of quantum physics (the fuzzy relationship) is coupled with the advantage that, at the same time, Democrit's demand for a basis that does not sink into the infinitesimal can be fulfilled in an ideal way. Thus, the mathematically ideal "discreteness of the state variables" required for abstract evolutionary systems is found again as a part of the physical world. However, the problems that arise here have not yet been formulated. At least the path seems passable.

[i]Letter from Otto to Reinhard Kaplan, July 17, 1969.

The founding editor of the *Journal of Theoretical Biology*, the English biologist James Danielli (1911–1984) in October 6, 1969 asked for a *major revision* based on the referee's comments which read:

(1) the paper contains some interesting ideas,
(2) these ideas are not presented in a sufficiently sharp and concise manner,
(3) the ideas must be carried to some kind of conclusion or prediction or some other kind of tangible consequence, to further reduce the speculative character.

Looking back at the unpublished paper after four decades, it is hard to tell what it still means to me. It is imperfect — the delay caused by reaction chains becoming slowly dripping, so to speak — was not made explicit. And the fact that evolution is a growing automaton was not yet explicitly seen. The overwhelming power of autonomous growth and self-maintenance, the "drive" in evolution, was fascinating to me. The fun of riding on an invincibly thrusting rocket ("riding a cannon ball" in Maxwell's words) was still fresh. The nonlinearities considered — Michaelis-Menten-type "saturations" — were naive in the exclusiveness they were focussed on (as thresholding devices). Still, I was very much encouraged in restrospect when in 1977, the American biologist Michael Conrad (1941–2001) in a dialog in our home seriously proposed to run the Belousov–Zhabotinsky reaction in a continuous stirred tank reactor, not for hours (as we had done), but for months and years. He saw what I had seen, and it was an independent intuition. And it was so concrete that I had not even dared thinking about it.

Even today, his proposal remains unconsummated. It would not cost very muchto do so. It would be an analog to the Urey–Miller experiment (about which we both knew),[47,48] yet well-stirred. Also the great acidity would cause most chemists to question that anything of combinatorial complexity could come out. But if 10^{100} compounds are possible at that acidity (which no one can exclude for sure), a path toward opening up that universal library would, presumably, be found. Of course, half a year of running time may still be ridiculously short. But then I read Robert Forward's book ("*The Dragon's egg*") on evolution on a neutron star — an idea that had also come to me as an illustration of the principles of the 1969 paper in a nuclear-reaction rather than chemical reactions setting —, I realized that it would indeed be worthwhile to look for life on Jupiter (not-carbon-chains, but borane-chains-based).

I presume that the British cybernetician Gordon Pask (1928–1996) — whom I once met at a conference and who had an infectious smile — may have uttered similar things early on. For some reason we did not come together at the time.

3. Visiting the Centre for Theoretical Biology

On October 3, 1968, I applied for a visiting appointment at the Center for Theoretical Biology (CTB) at the State University of New York at Buffalo. My application was triggered by an announcement published in the *Journal of Theoretical Biology* by the CTB. They encouraged applications from scientists working on "the theoretical basis of life processes." My letter read as follows.

> In response to your "Announcement" in the *Journal of Theoretical Biology* of 1968, August, I should like to offer myself as an applicant for one of your "visiting appointments."
>
> I am a "Dr. med." who is just leaving his "internship". My doctoral thesis dealt with an immunological problem (amyloidosis) in some experimental way (under Prof. E. Letterer, Tübingen, Germany).
>
> Interested in the theoretical foundations of Medicine, I have developed a cybernetical approach to biogenesis and evolution perhaps capable of delivering an invariant theory for a good deal of all possible "Biologies". This idea, partially worked out at the Max-Planck-Institut für Verhaltensphysiologie in Seewiesen, Bavaria, in contact with the heads of the institute, K. Lorenz and H. Mittelstaedt, has not been published so far, the first paper being in preparation. Some of the fundamentals of the theory can be found in papers by M. Sugita[49] and by R. Rosen.[50]
>
> I must say, that I am not a mathematician, nor a thermodynamicist, and not an information theorist, either. From the viewpointof chemistry, the model could be called pure "kinetics", being essentially a cybernetical approach. At the moment, the main problem for me is relating this approach to non-equilibrium statistical mechanics and to information theory.
>
> My English is even worse than that of this letter because writing leaves me more time for the selection of words. In the case you should want me to visit you, I should like to come at the same time my wife can follow a research fellowship in your country , continuing her experimental work on lymphocyte stimulation in liver diseases (in an hospital in New York).

Intrigued by the manuscript that I submitted to the *Journal of Theoretical Biology*, Robert Rosen — professor at the Center for Theoretical Biology and, most likely, one of the referees — supported my application for a position at the CTB. I got my visa for the US on August 28, 1969 and flew a few days later. I returned to Germany in February 1970. At the CTB, I met Robert Rosen, James Danielli, C.H. Waddington and Ludwig von Bertalanffy. A series of seminars on "Systems in Science" given at the CTB, organized by Bertalanffy, covered the following topics:

- September 30, 1969. Robert Rosen presented his forthcoming book entitled *Dynamical System Theory in Biology* (this was the 1st of three seminars);
- October 10, 1969. Robert Rosen presented his forthcoming book entitled *Dynamical System Theory*;
- October 17, 1969. Robert Rosen, *Dynamical System Theory*, 3rd part;
- October 31st, 1969. Gabor T. Herman, from the Computer Science Department at Buffalo State College, discussed a *theory of automata and related topics*;
- November 7, 1969. Ludwig von Bertalanffy spoke about the *theory of Open Systems: particularly in application to biology*, and discussed a dynamic model of life phenomena;
- November 13, 1969. Stuart Kauffman, *Metabolic stability and epigenesis* based on genetic control circuitry models;
- November 21, 1969. Ludwig von Bertalanffy gave a talk on *Open Systems: thermodynamic considerations* with a discussion on the evolution and the origin of life;
- January 29, 1970. Ludwig von Bertalanffy gave another talk on the *Theory of Open Systems*;
- February 5, 1970. C. H. Waddington gave a talk on *Gene Reiteration and Amplification in Higher Organisms.*

Rosen had obtained his Ph.D. at the University of Chicago under the supervision of Nicholas Rashevsky (1899–1972), the pioneer of mathematical biology.[51–53] In continuation of Rashevsky, Rosen developed a bridge between dynamical systems theory and biology.[54] Bob and I had an immediate resonance. We could talk at long lengths. We both liked Gene Roddenberry's Sci-Fi movies of the *Star Trek* series. Rosen promoted the fact that Life cannot be described by "closed systems", that is, by systems whose solutions can be analytically written down. He stressed the importance of organization and the relations between the functional parts of a whole organism over the underlying physicochemical structures of its components.[55] A key aspect of living organisms is that they are capable of both Metabolism and Repair — (M,R)-systems — in their complex organization.[50,56] From the draft of his book *Dynamical System Theory in Biology*, I discovered Andronov, Khaikin and Vitt's textbook[57] which later lead me to building a three- and then a two-variable chemical multivibrator.[33] Bob's book made me firm in dynamical systems thinking.

4. Getting a Position at Tübingen

A few years before, I had met Friedrich-Franz Seelig who got a chair for Theoretical Chemistry at the University of Tübingen while I was in Buffalo. Seelig offered me to join his new group with a stipend from the Deutsche Forschungsgemeinschaft (DFG). In the early 1960s, Seelig had done his diploma work with the Swiss Hans Kuhn (1919–2012) and the German physicist Fritz-Peter Schäfer (1931–2011) to build an analog computer consisting of a network of electrical oscillators, connected by capacitors to solve the two-dimensional Schrödinger equation.[58] This system was triggered by means of a radio frequency generator. In 1965, Seelig solved a two-dimensional Schrödinger equation with a digital computer (IBM 7090).[59] I had met Seelig via Hans Kuhn who was working on the origin of life, my first research topic. Kuhn handed me over to Seelig as it were. Sharing an interest for the origin of life, in differential equations and electronics (computers), we agreed that nonlinear systems like my evolutionary soup and electronic systems (without self-induction and without coupling condensers) were virtually isomorphic. This triggered a cooperation project between Seelig — a quantum chemist — and me — a medical physiologist — to look for reaction-kinetic analogs to electronic circuits.

It was a very stimulating environment at Buffalo. It enriched my knowledge for developing models and finding concrete examples providing a link between dynamical systems theory and chemistry. The following letter gives an idea of the context[j]:

> The reason for this second letter is that my DFG grant application has been fully accepted by the DFG for two years taking effect from October 1st. So I no longer need to give you any headaches about this when I move into your institute. This increases the anticipation.
>
> For the research grant (a total of 10,000 DM, 9,000 for computing costs, and 1,000 for material expenses) [...]

[j] Letter from Otto to Seelig on October 1st, 1969.

I imagine by now you've had your inaugural lecture. From here I can hardly predict the current size of the avalanche that was set in motion. Theoretically, the little fire lit in Tübingen should find rich nourishment.

Yesterday, Rosen gave a seminar on "Biochemical Regulations" to pharmacologists. After a very transparent presentation of the possible mathematical behavior of 2-dimensional dynamical systems (e.g. the Volterra system

$$\begin{cases} \dot{x}_1 = x_1 - x_1 x_2 \\ \dot{x}_2 = -x_2 + x_1 x_2 \end{cases} \tag{4}$$

which has two singular points, a "neutrally stable" = Center, and a "conditionally stable = Saddle point), and after the obligatory review of the equivalence of the realization with chemical or hydraulic means (e.g. Volterra's system), he did not fail to point out that the "transparent" (most conspicuous) state variables which most rationally determine the behavior of a given system describe, do not need to exist in the substances themselves, but could just as well be in linear combinations of such, whichwould then not be experimentally detectable. He answered an intelligent student's question as to whether the design of complicated networks of mutually activating and inhibiting enzymes would really be as simple as the one of electronic networks: that this would be the case as the equally good suitability of the analog computer for both tasks proves. I found that the way the question was asked (in disbelief) and the way it was answered (generally, Garfinkel[21] was of course cited as an example) are typical of the current situation. All actual possibilities are yet to be explored. By the way, Danielli, the discoverer of the double-layer membrane as the theoretically simplest and most stable possibility of membrane formation,[60] hopes that his results will be industrially applicable: A hint to the second great "no man's land" that exists.

Incidentally, I still have not spoken to Danielli, but I have an appointment with him now. I actually saw him at a conference of experimental biologists here in the house (*control mechanisms in genetic expression of cellular phenotype*) and had the impression of a very witty person.

On the news here, I still have two mere possibilities to report, both of which depend on your consent, in the second case, even only on your initiative. The first case says briefly that Rosen would like to keep me here a little longer (three months at most), but that the financial side and the visa have not yet been clarified. I declared my basic readiness because I do not think my "Outpost" mission is pointless. Nevertheless, this preliminary decision was not easy for me: not because I would not have expected your consenting , but because I am actually looking forward to Tübingen.

The second news maybe starts with Mr. Rosen wanting to spend his academic year, next year (that happens every four years here), with his family in Europe, namely in Western Europe, although he has more acquaintances in Eastern countries. (Despite or even daring to have limited material resources, East Germany, Czechoslovakia, Romania — would work very well theoretically there. But one would not like to be there with family.) He intends to go to Vienna, once because of the Czech emigrants who were there. On the other hand, because his well-known Locker (Bertalanffy student, publisher of the Heligoland symposia on quantitative biology of metabolism) has been granted a position there in theoretical biology (at least the first from Europe).

However, it can take up to two years for this to go through the last administrative instances. I told him a bit about Tübingen, which certainly has the most stimulating atmosphere in Germany to justify a visit. I also told him a bit about you (that you are also still young and that you come from quantum theoretical chemistry). He was very enthusiastic about the possibility of talking to someone about his quantum theoretical-genetic work (which I do not exactly know) to be able to talk. I have not given the slightest hint about the possibility of a visit to Tübingen, as I can neither decide nor arrange it. I do not know at all whether it is something like a visiting professorship, or some other activity which allows visitor status in Tübingen, but I suspect Rosen would take advantage of such an opportunity if offered.

Personally, of course, I think that the presence of a well-known mathematical biologist with basically the same purpose could only be stimulating for us. Rosen is also thoroughly aware of the almost limitless possibilities of systems-theoretical chemistry. With his M,R systems,[k] he has also described a mathematical model class for enzymatic reaction systems. Possibly, he set his modeling one step too high, namely on the biological level. As a result, he cut off the bridge to chemistry. and in all the troubles (mathematically perhaps interesting for later developments) he has to start with the "natural" ones. Extracting them secondarily from biological models is perhaps in fact hardly possible. It is a pity, however, that this experience has led to the admitted masses of "also" suitable chemical variables being attributed less "reality" than the postulated mathematically simplest possible description (knowing which is of course still a great advantage in individual cases can; however, it then remains an aid). I keep searching for the "generalized observation space" (which is no longer continuous, even with the simplest systems, and about which one can only hope for very general statements, even in the best case, once the theory is developed according to the life's work of a mathematician will be, where it may be true that these few general statements will then have just as general applicability as those of set and category theory) for a risky detour when it comes to achieving the simple results that can be achieved taking into account the actual existing chemical constraints. (The same special constraints can, of course, also be fed later into the finished "generation space" theory, and then lead to results that were not accessible in the direct way. But the reverse is also true.)

[k]Metabolism-Repair (M,R) system is an abstract biological network that is not reducible to sub-models and that is not computable on a Turing machine.[5,61,62] The existence of such a system serves as an example for non- computable biological systems.

Starting with chemistry, which are models that apply to much larger classes of systems than one really wants to describe, instead of biology saves the detour via "metaphors", and their reduction to the real "model" is just as difficult.

While preparing my coming back to Tübingen from Buffalo, framing my topic of research, that is, toward theoretical considerations, far from my initial medicine, m Seelig replied as follows.[l]

Dear Mr. Rössler,

Many thanks for your two letters of October 12th and November 1st. I owe you a great deal for that. I am very pleased that the DFG has approved everything you requested. (I don't know, however, whether you also applied for the travel costs; you rarely write anything about the approval of these costs).

I am very honored by your assumption that I have now made significant progress in the kinetic theory, but you overestimate my work mania. I have had a lot of administrative work to do lately (in connection with setting up the chair), and I have also got two lectures at the moment, of which I am preparing lecture notes together with the lecture from the previous semester, which I am very interested in claim. I have not even finished my inaugural lecture, nor even made an appointment for it.

Except for the moment, only Mr. [Friedrich] Göbber is working in the field of kinetic theory, but he is still quite busy with the technical problems (simulation programs for the Tübingen computer system).[m] I have so little time at the moment that I cannot even go into detail about your very numerous scientific comments, I think we will have to discuss this complex questions when you will be here in Tübingen.

As far as your possible further stay in America is concerned, it would of course be somewhat regrettable for me, although your scientific advancement is of course paramount. Moreover, as you write, you also see your activity there as an outpost for us. So I would say if you can stay longer, go ahead, your spot here will be reserved anyway.

As far as the other project (Mr. Rosen's possible stay with us) is concerned, at the moment I cannot oversee the possibilities. In terms of space, of course, it would not be a problem for us right now, but as far as the financial side is concerned (guest professorship), I do not see any concrete possibilities at the moment. In any case, it would be worthwhile if Mr. Rosen could give a lecture with us during his stay in Europe, the financing of which would not cause any difficulties.

[l]Letter from Seelig to Otto, November 11, 1969.

[m]In fact, Seelig was preparing with Göbber two papers devoted to simulations of two chemical reactions: Göbber was writing the code.[63,64]

Incidentally, I found in the course catalog that there is a "theoretical" biologist (whose topic is biological cybernetics) here in Tübingen, Prof. [Dezsö] Varjú.[n] I have not contacted him yet though, I just wanted to wait until my paper is published in the *Journal of Theoretical Biology* so I can demonstrate more concretely what I am working on.[o]

I am afraid I have to close my letter now, otherwise it will lie there for another week. I think we can talk more about it in a bit. At the moment I greet you and your wife warmly.

Your F. F. Seelig

On February 9, 1970, I therefore joined Seelig in Tübingen returning from Buffalo but I was still hoping for a semi-permanent arrangement to spend some time at Buffalo each year as Rosen tried to arrange.[p] Nevertheless, in spite of Rosen's effort, I never returned to Buffalo. Rosen visited us and gave a talk at the Lehrstuhl für Theoretische Chemie der Universität Tübingen on March 3, 1970 about *Dynamical systems in biology and chemistry.* Rosen returned six times to Tübingen and gave talks there, namely (i) on April 26, 1976, (ii) June 6, 1983, (iii) May 7, 1984, (iv) May 9, 1988, (v) December 3, 1990 (see Fig. 5) and (vi) May 17, 1993.

Fig. 5. Robert Rosen and Otto E. Rössler in December 1990 when visiting Tübingen for a talk.

[n]In 1968, he was offered a chair for zoology which was renamed into Department of Biological Cybernetics.[65]

[o]Seelig's paper was submitted on June 2, 1969, and finally published in 1970.[66] It deals with a chemical analogue to the well-known electronic RC phase-shift oscillator producing periodic oscillations. Sugita's work is also mentioned.

[p]Letter from Rosen to Otto, January 10, 1970.

With the founding money obtained with his new position at Tübingen, Seelig bought (with 80 000 DM) an analog computer — a Dornier 240 (Fig. 5) — equipped with potentiometers, a digital clock and two function generators... With this computer, Seelig obtained oscillations — plotted in the state space — in the simulation of a linear chemical model[64] and a limit cycle in a model of a spike oscillator.[67] It is worth noting that oscillatory behavior of a two-dimensional system in the state space was often obtained by theoretical biologists playing with computers. I started at the University of Tübingen lecturing on analog computing. Just before, Seelig had sent me from the Division of Theoretical Chemistry to attend an EAI (Enterprise Application Integration) course on analog computing. As a Stipend-holder" of the DFG, I was free in my research andI started to study few-variable systems. Due to the detailed teaching efforts of my decade-long school friend Roland Wais who had become a mathematician working with the German mathematician Lothar Collatz (1910–1990) in Hamburg,[68] I could write a numerical differential-equations solver. This was the ideal complement to the Dornier analog computer I had available.

Morphogenetic reaction systems, devised by Rashevsky[69] and Turing,[70] fitted in, enabling the design of a chemical oscillator based on a chemical flip-flop, that is, a bistable multivibrator that has two stable states and hence can be used as a one-bit memory. The latter had been invented by the British physicists William Eccles (1875–1966) and Frank Jordan (1881–1941).[71,72] In this vein, with Seelig, I jointly proposed a two-cellular homogeneous chemical multivibrator[73,74]

$$\left\{\begin{aligned} \dot{x}_1 &= (k_1 z - k_4 y_1 - k_5)\, x_1 - k_2 x_1 x_2 + k_3 \rho \\ \dot{y}_1 &= k_5 x_1 - (k_4 x_1 + k_6)\, y_1 \\ \dot{x}_2 &= (k_1 z - k_4 y_2 - k_5)\, x_2 - k_2 x_1 x_2 + k_3 \rho \\ \dot{y}_2 &= k_5 x_2 - (k_4 x_2 + k_6)\, y_2 \\ \dot{z} &= j - k_1 z\, (x_1 + x_2) \end{aligned}\right. \tag{5}$$

which is still quoted as one of the very first chemical reaction systems designed to implement logic circuits.[75] An example of a homogeneous system involving a two-variable bistable system (switch) as a subsystem that I found on the analog computer was

$$\left\{\begin{aligned} \dot{x} &= k_1 + \beta_x - k_2 x - k_3 \frac{xy}{K + x} + k_6 z \\ \dot{y} &= k_1 + \beta_y - k_2 y - k_3 \frac{xy}{K + y} \\ \dot{z} &= k_4 y - k_5 z. \end{aligned}\right. \tag{6}$$

Computer outputs of this abtract chemical reaction (Fig. 2 from my 1975 paper[76]) were equivalent to those produced by the well-known electronic multivibrating devices[77]: it is a slightly modified version of a system proposed in 1972.[78] I investigated the synchronization with periodic oscillations which were replacing the parameter value β_y. I also found that this system was producing oscillations resembling the one produced by Selkov's model for glycolysis.[79] Remembering my early studies in electronics, I introduced *dynamical automata* as components for building up complex chemical reaction systems: in other words, I had in mind to build chemical reaction systems as complex as electronic circuits can be.[33] This system was an example of an Abraham-Bloch multivibrator.[80]

In fact, in 1972, I already proposed a list of dynamical automata of various types: non-resettable switch, resettable switch, conditional switch, multivibrator, monoflop, T-flipflop, and more predictable oscillators.[33] The dimension d of the proposed systems is such that $1 \leqslant d \leqslant 4$. By "interconnecting these elementary systems, switching networks of in principle unlimited complexity can be set up." The building block principle inherited from electronics was thus transferred to chemistry.

My interest for the origin of life was not totally lost. I showed that the problem of self-reproduction can be reduced to that of a multiple step serial message transfer, at least for structure-determined machines driven by a standardized, exogenous message.[78] This contribution stemmed from what I had learned during my stay.

5. Art Winfree's Invitation to Chaos

Chaos theory is fun. While one variable can only attain a stable state, two can mutually oscillate, and three can do more things with each other: Birkhoff's recurrent motions can fall into a dance that is beautiful and non-repeating and surprising at every round. This dance is chaos. The American biologist Art Winfree (1942–2002) stimulated and paved my way into the fascinating field of chaos theory.

In 1972, Art Winfree had invited me for a talk on chemical automata at Purdue University — "well-stirred fluid automata," something new at the time.[33,81] He showed me his subsequntly well known beautiful experiments with the BelousovZhabotinsky reaction.[82,83] We started to exchange letters about interpreting chemical reactions in terms of dynamical system theory. In 1975, we met again at a *Chronobiology Meeting* held in Vienna where I gave a talk on biological clocks. Art attended my talk and told me afterwards that he had found it a little bit boring. *Could I not do something more interesting in the context of my liquid automata?*[84] I told him that I was thinking about a three-variable limit cycle that looks like a knot and hence cannot be flattened into a planar circle-like thing for being irreducibly three-dimensional. He replied that this sounded to him like chaos. I asked: *What do you mean with this word, I know the term only from traffic jams.* He told me that he had just attended a conference in Aspen, Colorado, as usual without proceedings, on "chaos" and that he had collected all the papers written on the subject and he would send me a folder containing them. Lo and behold, four weeks later I received a big folder with all the papers that had been written on chaos up to that time, most still in preprint form.

The folder included Edward Lorenz's paper of 1963[85] and more recent ones like those by Jim Yorke,[86] Bob May and George Oster. Art wrote me explicitly that I should do him the favor of finding a chemical version of the Lorenz attractor.[87]

So I had to do him the favor of finding chemical rate equations for chaos, especially a chemical Lorenz attractor. Of course, I didn't succeed, but out of the obligation not to disappoint him I came up with the idea of a rope wrapped about my nose several times before falling down and then coming back up again in a loop. This mental picture was the origin.

If you draw an expanding spiral on a sheet of paper and lay a second copy of it made on a transparent sheet upon it, you can switch at the border of one sheet into the flow pattern on the other sheet, and back from another threshold downstairs. For certain relative positions of the two spirals and their switching thresholds, you get all kinds of chaotic motion. This was the main step.

Next came a characterless turning of the knobs on the analogue computer. The sharp thresholds of the paper game meant complicated multiplication and division operations on the machine. Simplifying and reducing the number of such elements then was sufficient to arrive at the desired attractor.[88] Making it simpler to have fewer terms was just a trial-and-error procedure. So I slowly reduced the number of multipliers and wires you had to plug in, and so on.[89]

Since I was convinced that the principle should work also with non-negative concentrations to make it realistic, I tried to find parameters that allow to implement this. In Mathematics, the method I used is called *singular perturbation.* The same method was independently employed by Christian Mira at the University of Toulouse. The underlying two two-dimensional equations to be connected via thresholds

$$\begin{cases} \dot{x} = -y + 0.95z \\ \dot{y} = x + 0.15y \\ \epsilon\dot{z} = (1 + z^2)(z - 1 + x) - \delta z \end{cases} \tag{7}$$

were easy to write down. Computers do not like such *stiff* differential equations, however. Describing sharp transitions at the two thresholds [Fig. 6(a)] where the trajectories suddenly-switch from one two-dimensional plane onto the other, is a bit demanding numerically. I felt it might be interesting to simplify the equations further to get rid of any *unnecessary* parts in the equation. Even the features which made the equation non-negative (so as to represent realistic chemical concentrations) turned out to be redundant for retaining the chaotic behavior. So I could omit quite a few nonlinearities altogether. After simplifying the equation as much as possible by trial and error, the *miracle* happened that the two formerly overlaid simple linear two-dimensional flows gave rise to a new everywhere smooth three-dimensional flow[88]

$$\begin{cases} \dot{x} = k_1 + k_2 x - (k_3 y + k_4 z)\dfrac{x}{x + K} \\ \dot{y} = k_5 x - k_6 y \\ \mu\dot{z} = k_7 x + k_8 z - k_9 z^2 - \dfrac{k_{10} z}{z + K'} \end{cases} \tag{8}$$

of beautiful appearance (Fig. 6). In other words, it was possible to keep the esthetics of the flow while simultaneously making the equations much simpler. So by varying parameters systematically and putting as many of them as possible equal to zero, it was possible to find a much simpler equation

$$\begin{cases} \dot{x} = -y - z \\ \dot{y} = x + ay \\ \dot{z} = b + z(x - c). \end{cases} \tag{9}$$

It is not actually the really simplest one, by the way, since I could later find a still simpler one[90]

$$\begin{cases} \dot{x} = -y - z \\ \dot{y} = x \\ \dot{z} = b(1 - y^2) - cz \end{cases} \tag{10}$$

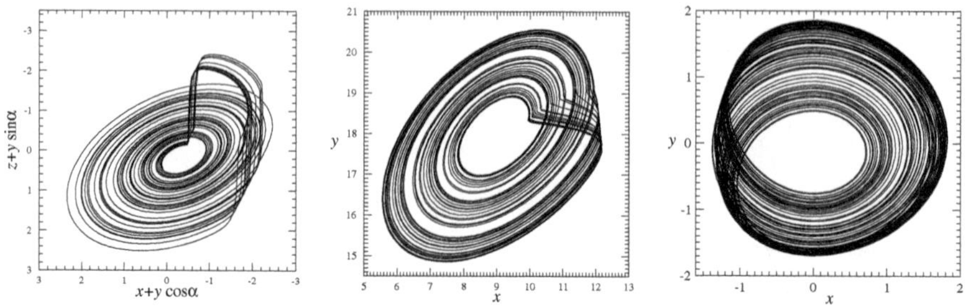

(a) System (7) (b) System (9) (c) System (10)

Fig. 6. The first beautiful chaotic attractor obtained by a trial-and-error procedure, after some simplifications of the original equations. Parameter values for system (9): $a = 0.2$, $b = 0.2$, $c = 5.7$. For system (10) $b = 0.275$ and $c = 0.2$ and initial conditions are $x_0 = 1$, $y_0 = -1.4$ and $z_0 = -0.4$.

which did not become well known, because I published it with a misprint (with x^2 in the third equation in place of y^2). I should add here that the stimulation I had obtained from Ralph Abraham and his school — the "Santa-Cruz kids" — was crucial. Norman Packard, Rob Shaw, Jim Crutchfield and Doyne Farmer then jointly coined the name for the attractor.[91]

I had invited Stephen Smale to Tübingen in 1972. At that time, I showed him a simpler (lower-dimensional) analog[84] to his complicated 8-variable bistable reaction-diffusion system,[92] which he accepted. In September of 1976, I visited him at Yale. He had a preprint of my simplest 3-variable chaos equation lying before him on his desk.[88] The mathematical Smale horseshoe is implicit in the 3-variable chaos equation (Fig. 7). I deeply admired his math but the visible and audible reality that I had stumbled upon seemed to represent a realistic, actually palpable reality of its own. I felt like the naive kid in a high-brow community — and like a farmer who had hit upon a previously only dreamed-of golden treasure in the ground.

6. Spiral Chaos and Beyond

The movie with the "*sound of chaos*" (named after Simon and Garfinkel's "*Sound of Silence*") was obtained on the analog computer jointly with Reimara.[q] The machine had the big advantage to it that one could choose, by throwing a switch, between different speeds of calculation — such that

[q]https://www.youtube.com/watch?v=Tmmdg2P1RIM.

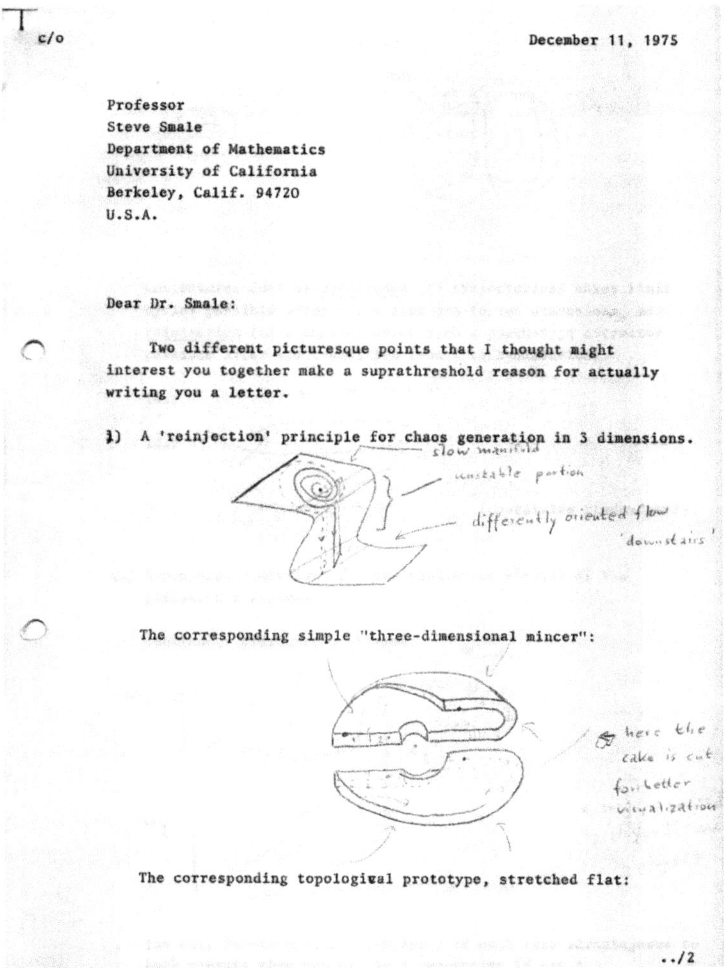

c/o

December 11, 1975

Professor
Steve Smale
Department of Mathematics
University of California
Berkeley, Calif. 94720
U.S.A.

Dear Dr. Smale:

Two different picturesque points that I thought might interest you together make a suprathreshold reason for actually writing you a letter.

1) A 'reinjection' principle for chaos generation in 3 dimensions.

The corresponding simple "three-dimensional mincer":

The corresponding topologisal prototype, stretched flat:

../2

Fig. 7. Letter from Otto to Stephen Smale, on December 11, 1975. Otto's paper model, exhibiting the presence of a horseshoe map in its chaotic attractor.

one could listen to the watched chaos either when it was running very slowly or else when it was ten times or a hundred or a thousand times faster. The sound produced by this simulational reality proved to be familiar to the ear. Hoarseness is exactly the sound of chaos! Thus, everybody has a perfect ear for chaos. If we then switched the analog computer to a lower speed but with the same motion — the same chaotic trajectory made a sound that proved familiar, too. In Tübingen, our flat lies close to

Fig. 8. Okan Gurel, Stephen Smale and Otto E. Rössler at the Conference on Bifurcation Theory and Application in Scientific Disciplines, New York, 1977.

a yellow postal mailbox where people would stop with their motorcycles from time to time to put a letter in and drive off again. It turned out that I had often heard this very sound before — that of an idling motorcycle. It is just low-frequency chaos. Or else when you are stopping with your car at a red light and you look at your gear shift, you realize that the whole car is vibrating — not completely periodically but the motor when idling is invariably chaotic. So chaos is something that is very close to everyday life. Everyone is familiar with periodic motions and periodic sounds, but the chaotic sounds are just as characteristic and ubiquitous. For example, Christophe Letellier successfully employed chaos control methods to diagnose and monitor patients suffering from respiratory and circulation problems.[93]

Just as chaos comes as a surprise in three dimensions, so hyperchaos comes in four.[94] Hyperchaos is closely related to chaos, which fact is not really surprising: one more direction of exponential divergence between neighboring trajectories. But there ought to exist also something radically new in the 4th dimension that is just as surprising as chaos is in the third dimension. This unknown behavior I called the X-attractor in 1976.

In 1978, I contributed to the 10th Summer Seminar on Applied Mathematics, organized by the American mathematician Frank Hoppensteadt at the University of Utah. My talk, entitled *Chaotic oscillations*, was my first attempt to summarize my views on the various types of chaos possible. I started with a brief historical tour as follows.

(1) Poincaré's contribution to the three-body problem in 1899,[95]started nonlinear dynamics, followed by George Birkhoff,[96] Eberhard Hopf,[97,98] Jürgen Moser,[99] Michel Hénon and Carl Heiles,[100] Vladimir Arnol'd,[101] Dmitri Anosov,[102] Martin Gutzwiller;[103,104]
(2) Gaston Julia's contribution[105] to the properties of one-dimensional maps was followed by Pierre Fatou,[106–108] Pekka Myrberg,[109,110] Oleksandr Sharkovsky,[111] Christian Mira,[112,113] and crowned by the Li-Yorke theorem;[86]
(3) Forced oscillators as investigated by Mary-Lucy Cartwright & John Littlewood,[114] Norman Levinson,[115] and Frank Hoppensteadt[116] were the next step.
(4) Diffeomorphisms as developed by Stephen Smale[117] in the spirit Of Poincaré, Birhoff and Moser followed suit.
(5) The Lorenz equation introduced by Edward Lorenz in 1963 was followed by Derek W. Moore & Edward A. Spiegel,[118] A. E. Cook & P. H. Roberts[119]).
(6) Generic structural instability was studied by Stephen Smale,[120] Jacob Palis & Sheldon Newhouse,[121,122] and Robert Williams.[123,124]
(7) The turbulence described by Ruelle and Takens[125]followed next.

I then introduced some properties related to chaos: how to design chaotic flows. To classify all the possible types of chaotic attractors, I proposed the *hierarchy theorem* stating that every additional dimension carries a new non-trivial type of attractor. This is related to the concept of "maximally extended attracting invariant sets". That is, the behavior characterized by a single negative Lyapunov exponent. In one-dimensional space, you have only point attractors, in two-dimensional space, periodic attractors, in three-dimensional space, chaotic attractors, then hyperchaos in four-dimensional space, and so on. Hence in dimension $d \geqslant 3$, we have maximally $d - 2$ positive Lyapunov exponents. This led to the 1983 paper entitled "hierarchy of chaos".[126] Toroidal chaos got overlooked at the time. A second attempt was completed with explicit examples, including toroidal chaos in 1990 by Michael Klein and Gerold Baier.[127] Since, Christophe Letellier, has turned this hierarchy into a taxonomy, providing a more complete description of the different types of maximally extended invariant sets observed in d-dimensional space.[128] Thus, according to this taxonomy, in four dimension, the two new types of behavior are: the hyperchaotic one that I found in 1979, and toroidal chaos, which was found by Ueda (see p. 113).

In 1979, an important conference was organized in Aachen by Ulrich Franck (1915–1996) and Ewald Wicke, entitled "*Kinetics of Physicochemical Oscillations*". After the first announcement, they received positive answers from 200 chemists: they added one day (September 19–22, 1979). This success confirms that the topic was mature: the 1960s and the 1970s stabilized the concept of oscillations in chemical reactions. Indeed, the key word was "oscillation". This conference was the culmination point of a rather long story about the production of oscillations in chemical, electrochemical, biological or physiological systems.

The main organizer, Franck, was involved in oscillating chemical reactions since the early 1950s. As Christophe Letellier found out all of this started with some oscillations observed in chemical reactions by Wilhelm Ostwald[36] and William Bray,[40] and in nerve conduction by Ralph Lillie,[38] and in physiology by Albrecht Bethe.[129–131] The description of relaxation oscillations in 1926 by Balthazar van der Pol[132,133] was a key step. The bridge between electronics and chemical and biological processes was built by Karl Friedrich Bonhöffer[134]: so in particular with the modified van der Pol system as a nerve model.[32] Bonhöffer, since his papers from the early 1940s, knew the works of Ostwald, Lillie and Bethe as well as van der Pol. Bonhöffer worked at the university of Göttingen, the university where Franck started his career. In one of his early papers, Franck thanks Bonhöffer for valuable discussions and his interest in his work.[135] Franck was comparing the dynamics of an iron electrodissolution in sulphuric acid with the Ostwald-Lillie nerve model. Franck was already working in the spirit of Bonhöffer's approach. Two years later, these two scientists published together;[136] they were investigating the conditions required for producing oscillations in chemical reactions. Franck got from Bonhöffer the references to these early contributions to the world of oscillations. Franck then continued to investigate periodic chemical reactions with period-2 and relaxation oscillations.[137]

Around 1950, Franck visited the National Institute of Health (Bethesada, Maryland) where Richard FitzHugh (1922–2007) was working. FitzHugh was already aware of the work of Bonhöffer.[138] They published together a paper in March 1961 about a mathematical model for describing periodic electrode processes[139]: they both worked in continuation of Bonhöffer's work,[32] that is, in exploiting the analogy between electronic circuits and chemical reactions. Their joint paper was published in March 1961, that is, the month during which FitzHugh submitted his well-known paper (received by the *Biophysical Journal* on March 9) in which he used a modified van der Pol model as a nerve model.[140] Compared to Bonhöffer's contribution,[32] FitzHugh provided the explicit expression of the cubic nonlinearity used: results became reproducible! Moreover, FitzHugh had a background in nonlinear dynamics, quoting Solomon Lefschetz,[141] Alfred Lotka,[142] and Nicolai Minorsky.[143] He so provided a careful and rigorous analysis of the dynamics in state space.

Franck was therefore in the right stream. He then proposed, for instance, an electronic circuit made of an electrolytic system with a non-monotonic current-voltage curve.[144] He thus obtained periodic oscillations with various shapes, some of them being clearly relaxation oscillations. He was aware of the structural properties a system should have for producing oscillations. He demonstrated the existence of a feedback loop in such systems to produce oscillations.[145] The difficulties encountered in formulating this bridge were summarized by Frank Jenik[146]:

> The investigation of nervous systems is one of the tasks of biologists, and Engineers hope that their questions can be answered by the biologists. However two difficulties arise: firstly, many biologists are not interested in engineering problems, and secondly, biologists and engineers do not approach problems from the same angle. The evaluation of biological work is made more difficult for the engineer because details are missing which are important for him, though insignificant to the biologists. Cooperation between the two sciences would therefore be advantageous...

He stated the problem as follows.

(1) What are the input-output relationships of neurons and neuronal networks?
(2) How do input-output relationships depend on the properties of neurons and neuronal networks?
(3) What properties of neurons and neuronal networks are essential for information processing?
(4) What input-output relationships do electronic devices have which exhibit several basic properties of a neuron?
(5) How do networks of such electronic devices behave?

According to this, "*neuron models are electronic devices with the same basic properties as neurons.*" He then proposed a review of some models for a neuron. Finding a good model for nerve activity was thus a problem quite frequently addressed in the 1960s.

By the end of the 1970s, Ulrich Franck stated[147]

> It is a remarkable fact that in spite of the intensive research activities in the field of periodic phenomena all over the world all physicochemical oscillations known so far were found exclusively by accident and no oscillatory reaction system has been "synthetized" by well-known elementary reactions up until now. For this reason the unbiased phenomenology of oscillatory systems shall be the starting position of this text.
>
> The lines of approach to the elucidation of physicochemical oscillations at this meeting vary in a striking way, obviously because there are many different disciplines of natural and mathematical science interested in the spontaneous occurrence of periodic phenomena, although, equipped with different epistemological tools. Another reason may be the fact that spontaneous oscillations are observed in physicochemical systems of a completely different substantial nature.

At the time, Richard Noyes — one of the authors of the *Oregonator* [148,149] for reproducing periodic oscillations in the Belousov–Zhabotinsky reaction — remained skeptical towards the emergence of chaotic systems and their implications for the interpretation of certain time series[150]: "*Theoretical aspects of oscillators are ignored completely, and discussion will be restricted to known experimental systems*". He added: "*Most of my references will be to systems with which I have had personal experience or about which I believe some mechanistic conclusions can be drawn*". Consequently, Rössler and Wegmann's experimental results about a chaotic Belousov–Zhabotinsky reaction[151] were also ignored by Noyes. The Experimental evidences for chaotic time series obtained by Jack Hudson's group,[152] were reduced to "*complex limit cycles*", in spite of the connection established in this work with the 1976 paper.[88] Nevertheless, Noyes quoted Milos Marek and Ivan Stuchl's Belousov–Zhabotinsky experiments in which they observed aperiodic (chaotic) oscillations.[153] and cited a paper where Ruelle announced that "*a periodic oscillation may bifurcate to a nonperiodic oscillation of "turbulent" character*" (at that time, Ruelle was using the word turbulent for what became chaotic later),[154] As he admitted, Noyes was focused on the chemical mechanisms which can produce periodic oscillations. Benno Hess was equally reluctant[155]: "*only stable periodic solutions guarantee the possibility of experimental observation*". This is particularly surprising since his co-author, John Tyson, observed a quasi-periodic regime in two coupled Brusselators[156]

$$\begin{cases} \dot{x} = a - bx + x^2 - x \\ \dot{y} = bx - x^2 y \\ \dot{u} = x - uv \\ \dot{v} = cx - uv + v^2 w - v \\ \dot{w} = uv - v^2 w \end{cases} \tag{11}$$

with $a = 0.4$, $b = 1.2$, and $c = 1.13$ (Fig. 9).

There are other examples of misinterpretation. For instance, A. Vaskelis and co-workers presented a time series that is obviously aperiodic (see, for instance, Figs. 1(b) and 1(d) of their paper, p. 715) but was designated as "normal" and "irregular", respectively, only periodic oscillations were in fact considered.[157] Graae Sorensen assumes that the aperiodic nature of the oscillations observed in a stirred flow reactor results from the turbulent flow and not from the Belousov–Zhabotinsky reaction.[158]

Contrary to the previous examples, Anatol Zhabotinsky recognized the existence of chaotic solutions,[159] thus quoting works by Hudson's group,[152, 160] Rössler and Wegmann,[151] and also Marek and Svobodova.[161]

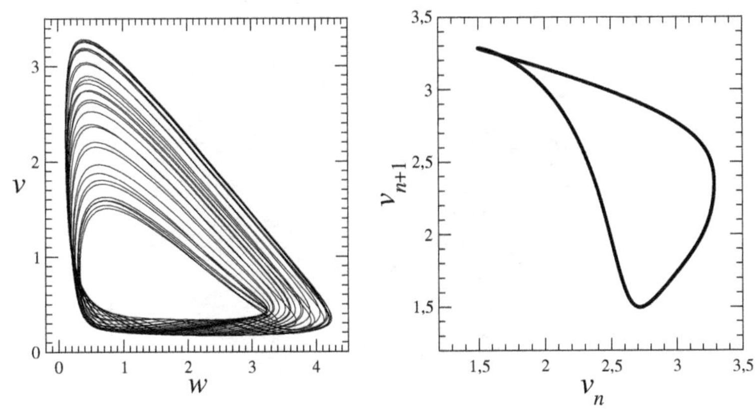

Fig. 9. Quasi-periodic regimes observed in the two coupled Brusselators by Tyson. The annular first-return map to a Poincaré section (right) confirmed the toroidal structure of the attractor. This solution was designated as being "almost periodic" by Tyson.

All these works show at least, evidence of aperiodic oscillation. The main difficulty for Zhabotinsky was, that all the proposed models were unable, at the time, to produce these complex oscillations. Lars Olsen presented his model for the peroxidase-oxidase reaction[162]: most likely the state portrait of his Fig. 6 corresponds to a quasi-periodic regime, but he clearly favored the possibility of a chaotic chemical reaction (see Chapter 16, p. 385). Another scientist contemplating chaotic oscillations in chemical reactions was Okan Gurel.[163] He proposed toroidal oscillations using a three-dimensional mathematical model[164]:

$$\begin{cases} \dot{x} = x - ay - xz \\ \dot{y} = bx + y - yz \\ \dot{z} = x^2 + y^2 + cz - \dfrac{z}{z+d}. \end{cases} \tag{12}$$

The resulting behavior is a quasi-periodic regime (Fig. 10). Of course this was expected since Gurel co-organized our workshop held at the New York Academy of Sciences, one month later.[163] Another interesting example was provided by K. F. Jensen and W. H. Ray presenting chaotic time

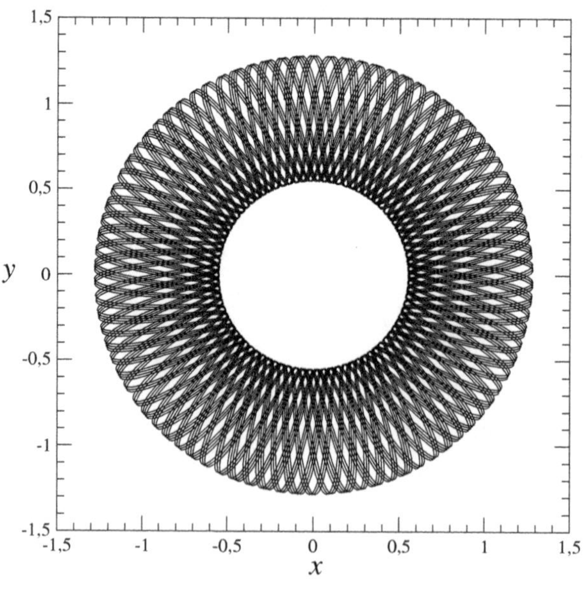

Fig. 10. Quasi-periodic regime produced by the three-dimensional system (12). *Notes*: Parameter values: $a = b = c = d = 0.1$.

Fig. 11. 1979 Aachen-Group at the conference entitled *Kinetics of Physicochemical Oscillations* held at Aachen, Germany, September 19–22, 1979.[r]

[r]From the site *History of the Belousov–Zhabotinsky Reaction* (accessed March 9, 2024), https://woosterdigital.org/BZ-history/items/show/57.

series observed on a catalytic surface.[165] Roger A. Schmitz and coworkers provided a state portrait of a chaotic attractor produced by a model for the catalytic oxidation of hydrogen on a nickel foil.[166] Roger Schmitz's former supervisor, Jack Hudson, proposed some experimental time series from a Belousov–Zhabotinsky reaction with an obvious aperiodic nature: chaos was suspected.[167] Chaotic evidence was found in a five-dimensional model of the Belousov–Zhabotinsky reaction by Jack Turner.[168]

If we return to an ordinary chaotic flow in three dimensions where everything is absolutely intuitive and locally parallel, we do nevertheless get this divergence of neighboring points. When we now punch two holes into the chaotic flow, as is easy to do by playing with the three free parameters in the differential equation, one gets a layered fractal dividing line between two neighboring state points — initial conditions — that either go into the one drain (point attractor) in the long run or else into the other. If these holes are now given different colors, the striped structure that results consists of neighboring colored layers in all sizes and sub-sizes: a product of

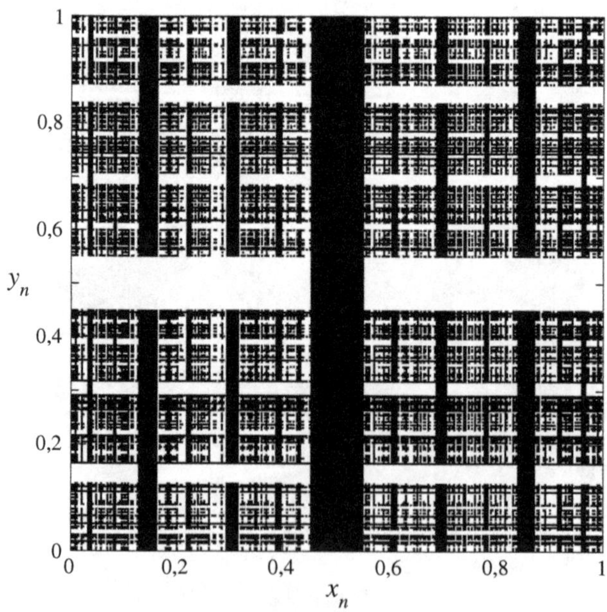

Fig. 12. Julia sets constructed with two Logistic equations $x_{n+1} = 4.04x_n(1 - x_n)$, and $y_{n+1} = 4.04y_n(1 - y_n)$ from a grid of initial conditions marked in black (white) when the trajectory is ejected with $x_n < 0$ ($y_n < 0$).

a Cantor set and a line — or a "one-dimensional fractal" for short (Fig. 12). The Gumowski School — Igor Gumowski and Christian Mira — saw this early on.[169] As stated, any trajectory in the chaotic cross section goes eventually either to the one hole or to the other. By systematically scanning the initial conditions in the area in question by using a computer, you can color them according to where they eventually go. What then remains is the boundary, a dividing line between red and blue. This fractal if generated by ordinary chaos consists of locally parallel stripes: a layered Cantor set. Even though it is mathematically most intricate, it is not very impressive when you look at it: stripes of all sizes. But if we have hyperchaos from the beginning in a four-dimensional flow, then the boundary between the two colored domains is a beautiful self-similar or self-affine fractal of the type made famous by Mandelbrot.[170] The striking beauty of the famous Mandelbrot set is the consequence of the existence of a special kind of hyperchaotic flow. The flow in question is a symplectic flow which has an analytic map for its cross section. The flow that underlies the famous Mandelbrot set has actually never been indicated. These beautiful sets are "eternally self-similar" since there is no end in the smallest domain at arbitrarily high resolution. This finding is a living illustration of the infinitely powerful creative mind first conceived by Anaxagoras around 500 B.C. which we find here in the Mandelbrot set.

In the early 1980s, I wrote the draft of a book devoted to chaos theory. I submitted it to Princeton University Press and then to Rowohlt Verlag. It got rejected and I left the manuscript in a drawer. It was only 30 years later, when Christophe was visiting us and showed his interest for my archives that Reimara reminded me of the existence of this old manuscript. After a few months, she finally recovered the manuscript which is now published with the help of Christophe with Springer.[171] It contains the viewwhich I held at the time about chaos and its role in the sciences of nature. At that time my views on the hierarchy of chaos that I published in 1983 had not yet matured.[126] An updated version of this hierarchy was subsequentlyprovided by Michael Klein and Gerold Baier.[127] Jointly with Christophe, a third updated version was offered in 2022.[128] This hierarchy more or less closed my research program on chaos.

7. Conclusion

In the early 1970s, I developed a bridge between electronic circuits and oscillating chemical reactions. Rosen initiated me to differential equations.

Seelig enabled their study on an analog computer, and Winfree pushed me into the chaotic world. Fascinated by the structure of chaotic solutions in state space, I described "their topology", I in the form of *paper models*. In this way, I quickly switched to the problem of finding the governing equations generating different attractors. After two years of intensive work, my zoo was rich enough to make an attempt at the "hierarchy of chaos." It took a few additional years to complete my hierarchy of chaos.[126]

In parallel, the 1979 Aachen conference was the confirmation that oscillating chemical reactions were actually accepted, although some timidity in recognizing their chaotic nature was still palpable. I exploited most of the abstract concepts available for producing a rich zoology of chaotic attractors. In resonance with Joe Ford's easiness to extrapolate the consequences of chaos theory to many fields, with a holistic approach, I started to be more active in other domains like endophysics and cosmology.

The mind is indeed capable of un-mixing an infinitely fine mixture, as claimed by Anaxagoras. The mind can produce simplicity from the complex. So the mind has a fineness to it which is absolutely unfathomable. I would think this was basically a religious idea on the part of Anaxagoras. It is described in his surviving fragment number 12. He speaks there explicitly of the "mind." Even though the word "chaos" does not appear in this fragment, the word "Tohu-wa-bohu" of the bible later took on the meaning of *chaos* — no doubt under the influence of Anaxagoras' school.

Acknowledgements

I thank Reimara and Christophe for their help in producing this chapter.

References

1. H. Sutaner, *Meßsender Frequenzmesser und Multivibratoren.* Radio-Praktiker Bucherei, Vol. 128/130. Franzis Verlag, Munich (1966).
2. E. Letterer, Aleukämische Retikulose. Ein Beitrag zu den proliferativen Erkrankungen des Retikuloendothelialapparates, *Frankfurter Zeitschrift für Pathologie.* **30**, 377–394 (1924).
3. F. M. Burnet and F. Fenner, *The Production of Antibodies*, 2nd edn. Macmillan, Melbourne, Australia (1949).
4. R. E. Billingham, L. Brent, and P. B. Medawar, 'Actively acquired tolerance' of foreign cells, *Nature.* **172**, 603–606 (1953). doi: 10.1038/172603a0.
5. O. E. Rössler, Dauerimmunisierung von Albinomäusen mit Rinder-γ-Globulin. Ph.D. Thesis, University of Tübingen (1966).
6. C. F. von Weizsäcker, *Die Geschichte der Natur.* Zwölf Vorlesungen, Göttingen (1946). *The History of Nature.* Routledge & Kegan Paul, London (1951).

7. O. E. Rössler, Adequate locomotion strategies for an abstract organism in an abstract environment — A relational approach to brain function, *Lecture Notes in Biomathematics.* **4**, 342–369 (1974).
8. O. E. Rössler, An artificial cognitive-plus-motivational system. In R. Rosen (ed.), *Progress in Theoretical Biology.* Academic Press (1981), pp. 147–160. doi: 10.1016/B978-0-12-543106-4.50012-X.
9. O. E. Rössler, L.-R. Vial, F. Kuske, A. Nitschke, T. Ikegami, and A. Ujica, Brain equation and personogenesis, *Clinics in Pediatrics.* **2**, 1–11 (2019).
10. C. H. Waddington, *The Strategy of The Genes.* Allan & Unwin, London (1957).
11. C. H. Waddington, *The Nature of Life.* Allen & Unwin (1961).
12. C. A. Petri, Kommunikation mit Automaten. Ph.D. Thesis, Fakultät für Mathematik und Physik der Technischen Hochschule Darmstadt, Schriften des Rheinisch-Westfälischen Institutes für Instrumentelle Mathematik an der Universität Bonn (June, 1962).
13. R. Kaplan, *Der Ursprung des Lebens. Biogenetik, ein Forschungsgebiet heutiger Naturwissenschaft.* Georg Thieme Verlag, Stuttgart, Germany (1972).
14. C. H. Waddington, Towards a theoretical biology, *Nature.* **218**, 525–527 (1968). doi: 10.1038/218525a0.
15. C. H. Waddington, Theoretical biology and molecular biology. In C. H. Waddington (ed.), *Towards a Theoretical Biology, Prolegomena*, Vol. 1. International Union of Biological Sciences & Edinburgh University Press (1968), pp. 103–108.
16. S. R. de Groot and P. Mazur, *Non-Equilibrium Thermodynamics.* North-Holland Publishing Company (1962). Reprinted by Dover in 1984.
17. L. Onsager, Reciprocal relations in irreversible processes I, *Physical Review.* **37**, 405–426 (1931).
18. L. Onsager, Reciprocal relations in irreversible processes II, *Physical Review.* **38**, 2265–2279 (1931).
19. B. Chance, The kinetics of the enzyme-subrate compound of peroxidase, *Journal of Biological Chemistry.* **151**(2), 553–577 (1943). doi: 10.1016/S0021-9258(18)44929-0.
20. B. Chance, Analogue and digital representations of enzyme kinetics, *Journal of Biological Chemistry.* **235**, 2440–2443 (1960). doi: 10.1016/S0021-9258(18)64639-3.
21. D. Garfinkel, Computer simulation in biochemistry and ecology. In T. Waterman and H. Morowitz (eds.), *Theoretical and Mathematical Biology.* Blaisdell (Ginn), New York (1965), pp. 292–310.
22. I. Prigogine and J. M. Wiame, Biologie et thermodynamique des phénomènes irréversibles, *Experientia (Basel).* **2**, 451–453 (1946).
23. I. Prigogine and R. Balescu, Sur les propriétés différentielles de la production d'entropie II, *Bulletin de l'Académie Royale de Belgique.* **41**, 917–928 (1955).
24. L. von Bertalanffy, *General System Theory: Foundations, Development, Applications.* George Braziller, New York (1968).
25. W. R. Ashby, *Design for a Brain.* Chapman and Hall, London (1952).

26. W. R. Ashby, *An Introduction to Cybernetics*. Chapman and Hall, London (1956).
27. I. Prigogine, V. Mathot, and A. Desmyter, Propriétés thermodynamiques des solutions associées, *Bulletin des Sociétés Chimiques Belges*. **58**(10–12), 547–565 (1949). doi: 10.1002/bscb.19490581014.
28. H. Sutaner, *Das Spulenbuch Hochfrequenzspulen*. Franzis Verlag, Munich (1952).
29. H. Richter, *Radiotechnik für Alle*. Franckh'sche Verlagsdhandlung (1953).
30. H. Richter, *Neues Bastelbuch für Radio und Elektronik: Selbstbaugeräte und Versuche*. Franckh'sche Verlagsdhandlung (Franckh, 1959).
31. E. Kamke, *Differentialgleichungen I: Gewöhnliche Differentialgleichungen*. Leipzig Akademischen Verlagsgesellschaft, Leipzig (1944).
32. K. F. Bonhöffer, Activation of passive iron as a model for the excitation of nerve, *Journal of General Physiology*. **32**(1), 69–91 (1948).
33. O. E. Rössler, Grundschaltungen von flüssigen Automaten und Relaxationssystemen, *Zeitschrift für Naturforschung B*. **27**, 333–343 (1972).
34. A. J. Lotka, Contribution to the theory of periodic reactions, *Journal of Physical Chemistry*. **14**, 271–274 (1910). doi: 10.1021/j150111a004.
35. A. J. Lotka, Analytical note on certain rhythmic relations in organic systems, *Proceedings of the National Academy of Sciences*. **6**(7), 410–415 (1920). doi: 10.1073/pnas.6.7.410.
36. W. Ostwald, Chemische Fernewirkung, *Zeitschrift für Physikalische Chemie*. **9**, 540–552 (1892). doi: 10.1515/zpch-1892-0944.
37. W. Ostwald, Periodische Erscheinungen bei der Auflösung des Chroms in Säuren, *Zeitschrift für Physikalische Chemie*. **35**(1), 204–256 (1900). doi: 10.1515/zpch-1900-3515.
38. R. S. Lillie, The passive iron wire model of proto-plasmic and nervous transmission and its physiological analogues, *Biological Reviews*. **11**(2), 181–209 (1936). doi: 10.1111/j.1469-185X.1936.tb00501.x.
39. P. Plath, Synergetic impact of synergetics — remembrances of a chemist. In G. Wunner and A. Pelster (eds.), *Selforganization in Complex Systems: The Past, Present, and Future of Synergetics*. Springer International Publishing (2016), pp. 59–74.
40. W. C. Bray, A periodic reaction in homogeneous solution and its relation to catalysis, *Journal of the American Chemical Society*. **43**, 1262–1267 (1921).
41. M. Sugita, Functional analysis of chemical systems in vivo using a logical circuit equivalent. II. The idea of a molecular automaton, *Journal of Theoretical Biology*. **4**(2), 179–192 (1963). doi: 10.1016/0022-5193(63)90027-4.
42. R. Rosen, Turing's morphogens, two-factor systems and active transport, *Bulletin of Mathematical Biophysics*. **30**, 493–499 (1968). doi: 10.1007/BF02476609.
43. M. Sugita, Kinetic theory and mathematical model of cell metabolism, *Helgoländer wissenschaftliche Meeresuntersuchungen*. **14**, 78–109 (1966). doi: 10.1007/BF01611615.

44. M. Yas, M. Sugita, and A. Bensam, A model of cell size regulation, *Journal of Theoretical Biology.* **9**(3), 444–470 (1965). doi: 10.1016/0022-5193(65) 90043-3.
45. K. S. Trintscher, Biology and information: A discussion of problems of biological thermodynamics (1966) (in German). Original in Russian, Moscow, 1964.
46. H. H. Pattee, The recognition of hereditary order in primitive chemical systems and their molecular matrices. In S. W. Fox (ed.), *Proceedings of the Conference Wakulla Springs Florida 1963* (1965), pp. 385–402.
47. S. L. Miller, Production of amino acids under possible primitive earth conditions, *Science.* **117**, 528–529 (1953). doi: 10.1126/science.117.3046.528.
48. S. L. Miller and H. C. Urey, Organic compound synthesis on the primitive earth, *Science.* **130**, 245–251 (1959). doi: 10.1126/science.130.3370.245.
49. M. Sugita, Functional analysis of chemical systems in vivo using a logical circuit equivalent. III. Analysis using a digital circuit combined with an analogue computer, *Journal of Theoretical Biology.* **4**(2), 412–425 (1963). doi: 10.1016/0022-5193(63)90027-4.
50. R. Rosen, A relational theory of biological systems, *Bulletin of Mathematical Biophysics.* **20**(3), 245–260 (1958). doi: 10.1007/bf02478302.
51. N. Rashevsky, *Mathematical Biophysics: Physico-Mathematical Foundations of Biology.* University of Chicago Press (1938).
52. N. Rashevsky, *Some Medical Aspects of Mathematical Biology.* C. C. Thomas, Springfield, IL (1964).
53. N. Rashevsky, Outline of a unified approach to physics, biology and sociology, *Bulletin of Mathematical Biophysics.* **31**(1), 159–198 (1969). doi: 10.1007/bf02478215.
54. R. Rosen, *Dynamical Systems Theory in Biology.* Wiley Interscience, New York (1970).
55. I. C. Baianu, Robert Rosen's work and complex systems biology, *Axiomathes.* **16**(1–2), 25–34 (2006). doi: 10.1007/s10516-005-4204-z.
56. R. Rosen, The representation of biological systems from the standpoint of the theory of categories, *Bulletin of Mathematical Biophysics.* **20**(4), 317–341 (1958). doi: 10.1007/bf02477890.
57. A. A. Andronov, A. A. Vitt, and S. E. Khaikin, *Teoriya kolebanii.* Fizmatgiz, Moscow (1959). *Theory of Oscillators*, Pergamon Press (1966).
58. W. Huber, G. Simon, and H. Kuhn, Analogiebetrachtungen und Analogrechner zur Behandlung der Korrelation von π-Elektronen (1), *Zeitschrift für Naturforschung A.* **17**(2), 99–114 (1962). doi: 10.1515/zna-1962-0201.
59. F. F. Seelig, Numerical solution of 2- and 3-dimensional Schrödinger equations for any molecular potential by iterative variation of numerical test functions with a digital computer — I. Theoretical principles: Description of a computer program for solution of 2-dimensional Schrödinger equation, *Zeitschrift für Naturforschung A.* **20**, 416–427 (1965).

60. J. F. Danielli and H. Davson, A contribution to the theory of permeability of thin films, *Journal of Cellular and Comparative Physiology.* **5**(4), 495–508 (1935). doi: 10.1002/jcp.1030050409.
61. R. Rosen, Abstract biological systems as sequential machines, *Bulletin of Mathematical Biophysics.* **26**, 103–111 (1964). doi: 10.1007/BF02476831.
62. R. Rosen, Abstract biological systems as sequential machines II: Strong connectedness and reversibility, *Bulletin of Mathematical Biophysics.* **26**, 239–246 (1964). doi: 10.1007/BF02479045.
63. F. F. Seelig and F. Göbber, Stable linear reaction oscillator, *Journal of Theoretical Biology.* **30**(3), 485–496 (1971). doi: 10.1016/0022-5193(71)90003-8.
64. F. F. Seelig, Activated enzyme catalysis as a possible realization of the stable linear chemical oscillator model, *Journal of Theoretical Biology.* **30**(3), 497–514 (1971). doi: 10.1016/0022-5193(71)90004-X.
65. D. Varju and W. Reichardt, übertragungseigenschaften im Auswertesystem für das Bewegungssehen II. Folgerungen aus Experimenten an dem Rüsselkäfer Chlorophanus viridis, *Zeitschrift für Naturforschung B.* **22**(12), 1343–1351 (1967). doi: doi:10.1515/znb-1967-1221.
66. F. F. Seelig, Undamped sinusoidal oscillations in linear chemical reaction systems, *Journal of Theoretical Biology.* **27**(2), 197–206 (1970). doi: 10.1016/0022-5193(70)90137-2.
67. H. R. Karfunkel and F. F. Seelig, Reversal of inhibition of enzymes and the model of a spike oscillator, *Journal of Theoretical Biology.* **36**(2), 237–253 (1972). doi: 10.1016/0022-5193(72)90095-1.
68. L. Collatz, *Funktionanalysis und numerische Mathematik.* Springer-Verlag, Berlin (1964). *Functional Analysis and Numerical Mathematics.* Academic Press, New York (1966).
69. N. Rashevsky, An approach to the mathematical biophysics of biological self-organization and a cell polarity, *Bulletin in Mathematical Biophysics.* **2**, 65–67 (1940).
70. A. M. Turing, The chemical basis of morphogenesis, *Philosophical Transactions of the Royal Society of London B.* **237**(641), 37–72 (1952). doi: 10.1098/rstb.1952.0012.
71. W. H. Eccles and F. Jordan, Improvements in ionic relays, British Patent number: GB 148582 (filed June 21, 1918 and published August 5, 1920).
72. W. H. Eccles and F. Jordan, Sustaining the vibration of a tuning fork by a triode valve, *The Electrician.* **82**, 704 (1919).
73. F. F. Seelig and O. E. Rössler, Model of a chemical reaction flip-flop with one unique switching input, *Zeitschrift für Naturforschung B.* **27**(12), 1441–1444 (1972). doi: 10.1515/znb-1972-1201.
74. O. E. Rössler and F. F. Seelig, A Rashevsky-Turing system as a two-cellular flip-flop, *Zeitschrift für Naturforschung B.* **27**(12), 1444–1448 (1972). doi: 10.1515/znb-1972-1202.
75. K.-P. Zauner, Molecular information technology, *Critical Reviews in Solid State and Materials Sciences.* **30**(1), 33–69 (2005). doi: 10.1080/10408430590918387.

76. O. E. Rössler, A multivibrating switching network in homogeneous kinetics, *Bulletin of Mathematical Biology.* **37**, 181–192 (1975). doi: 10.1016/S0092-8240(75)80024-3.
77. P. M. Chirlian, *Electronic Circuits: Physical Principles, Analysis and Design.* McGraw-Hill, New York (1971).
78. O. E. Rössler, Self-reproduction and serial message transfer: Two related problems, *Bulletin of Mathematical Biophysics.* **34**, 343–353 (1972). doi: 10.1007/BF02476447.
79. E. E. Sel'kov, Self-oscillations in glycolysis 1. A simple kinetic model, *European Journal of Biochemistry.* **4**(1), 79–86 (1968). doi: 10.1111/j.1432-1033.1968.tb00175.x.
80. H. Abraham and E. Bloch, Mesure en valeur absolue des périodes des oscillations électriques de haute fréquence, *Annales de Physique.* **9**(12), 237–302 (1919). doi: 10.1051/anphys/191909120237.
81. O. E. Rössler and D. Hoffmann, Repetitive hard bifurcation in a homogeneous reaction system. In H. C. Hemkers and B. Hess (eds.), *Analysis and Simulations of Biochemical Systems.* North-Holland, Amsterdam (1972), pp. 91–102. *Proceedings of the 8th FEBS Meeting.*
82. A. T. Winfree, Spiral waves of chemical activity, *Science.* **175**(4022), 634–636 (1972). doi: 10.1126/science.175.4022.634.
83. A. T. Winfree, Scroll-shaped waves of chemical activity in three dimensions, *Science.* **181**(4103), 937–939 (1973). doi: 10.1126/science.181.4103.937.
84. O. E. Rössler, Chemical automata in homogeneous and reaction-diffusion kinetics, *Lecture Notes in Biomathematics.* **4**, 399–418 (1974).
85. E. N. Lorenz, Deterministic nonperiodic flow, *Journal of the Atmospheric Sciences.* **20**, 130–141 (1963). doi: 10.1175/1520-0469(1963)020⟨0130:DNF⟩2.0.CO;2.
86. T.-Y. Li and J. A. Yorke, Period three implies chaos, *The American Mathematical Monthly.* **82**(10), 985–992 (1975).
87. C. Letellier and L. A. Aguirre, On the interplay among synchronization, observability and dynamics, *Physical Review E.* **82**, 016204 (2010). doi: 10.1103/PhysRevE.82.016204.
88. O. E. Rössler, Chaotic behavior in simple reaction system, *Zeitschrift für Naturforschung A.* **31**, 259–264 (1976).
89. O. E. Rössler, An equation for continuous chaos, *Physics Letters A.* **57**(5), 397–398 (1976). doi: 10.1016/0375-9601(76)90101-8.
90. O. E. Rössler, *Continuous Chaos.* Springer Series in Synergetics, Vol. 2. Springer-Verlag, Berlin, Germany (1977), pp. 174–183.
91. J. Crutchfield, D. Farmer, N. Packard, R. Shaw, G. Jones, and R. Donnelly, Power spectral analysis of a dynamical system, *Physics Letters A.* **76**(1), 1–4 (1980). doi: 10.1016/0375-9601(80)90130-9.
92. S. Smale, The mathematical model of two cells via Turing's equation, *Lectures on Mathematics in the Life Sciences.* **6**, 17–26 (1974).
93. C. Letellier, H. Rabarimanantsoa, L. Achour, A. Cuvelier, and J.-F. Muir, Recurrence plots for dynamical analysis of non-invasive mechanical ventilation, *Philosophical Transactions of the Royal Society A.* **366**(1865), 621–634 (2008). doi: 10.1098/rsta.2007.2114.

94. O. E. Rössler, An equation for hyperchaos, *Physics Letters A.* **71**, 155–157 (1979).
95. H. Poincaré, *Les méthodes nouvelles de la mécanique céleste III.* Gauthier-Vilard, Paris (1899).
96. G. D. Birkhoff, *Dynamical Systems.* American Mathematical Society, New York (1927).
97. E. Hopf, *Ergodentheorie.* Springer-Verlag, Berlin (1937). doi: 10.1007/978-3-642-86630-2.
98. E. Hopf, Abzweigung einer periodischen Lösung von einer stationären Losung eines Differentialsystems, *Berichten der Mathematisch-Physischen Klasse der Sächsischen Akademie der Wissenschaften zu Leipzig.* **94**, 1–22 (1942).
99. J. Moser, On invariant curves of area-preserving mapping of an annulus, *Matematika.* **6**, 51–68 (1962).
100. M. Hénon and C. Heiles, The applicability of the third integral of motion: Some numerical experiments, *The Astronomical Journal.* **69**(1), 73–79 (1964).
101. V. I. Arnol'd, Small denominators and problems of stability of motion in classical and celestial mechanics, *Russian Mathematical Surveys.* **18**, 85–191 (1963). doi: 10.1070/RM1963v018n06ABEH001143.
102. D. V. Anosov, Ergodic properties of geodesic flows on closed Riemannian manifolds of negative curvature, *Doklady Akademii Nauk SSSR.* **151**(6), 1250–1252 (1963).
103. M. C. Gutzwiller, Periodic orbits and classical quantization conditions, *Journal of Mathematical Physics.* **12**, 343–358 (1971).
104. M. C. Gutzwiller, The anistropic Kepler problem in two dimensions, *Journal of Mathematical Physics.* **14**, 139–152 (1973).
105. G. Julia, Mémoire sur l'itération des fonctions rationnelles, *Journal de Mathématiques Pures et Appliquées.* **8**, 47–245 (1918).
106. P. Fatou, Sur les substitutions rationnelles, *Comptes-Rendus de l'Académie des Sciences.* **164**, 806–808 (1917).
107. P. Fatou, Sur les substitutions rationnelles, *Comptes-Rendus de l'Académie des Sciences.* **165**, 992–995 (1917).
108. P. Fatou, Sur les équations fonctionnelles, *Bulletin de la Société Mathématique de France.* **47**, 161–271 (1919).
109. P. J. Myrberg, Sur l'itération des polynômes réels quadratiques, *Journal de Mathématiques Pures et Appliquées.* **41**, 339–351 (1962).
110. P. J. Myrberg, Iteration der reellen Polynome zweiten Grades III, *Annales Academiae Scientarum Fennicae A.* **336**, 2–18 (1963).
111. O. M. Sarkovskii, Coexistence of cycles of a continuous map of a line into itself, *Ukrains'kyi Matematychnyi Zhurnal.* **16**, 61–71 (1964). doi: 10.3842/umzh.v76i1.8026.
112. J. Bernussou, L. Hsu, and C. Mira, Quelques exemples de solutions stochastiques bornées dans les récurrences autonomes du 2nd ordre. In *Transformations ponctuelles et leurs applications*, Vol. 229. Colloques Internationaux du CNRS, Toulouse (1973), pp. 195–220.

113. C. Mira, Dynamique complexe engendrée par une équation différentielle d'ordre 3. In R. Conti, G. Sestini, and G. Villari (eds.), *Proceedings Equadiff'78* (1978), pp. 25–37.
114. M. L. Cartwright and J. E. Littlewood, On non-linear differential equations of the second order, I, *Journal of the London Mathematical Society.* **20**(3), 180 (1945).
115. N. Levinson, A second-order differential equation with singular solutions, *Annals of Mathematics.* **50**(1), 127–153 (1949). doi: 10.2307/1969357.
116. J. E. Flaherty and F. C. Hoppensteadt, Frequency entrainment of a forced van der Pol oscillator, *Studies in Applied Mathematics.* **58**(1), 5–15 (1978). doi: 10.1002/sapm19785815.
117. S. Smale, Differentiable dynamical systems. I diffeormorphisms, *Bulletin of American Mathematical Society.* **73**, 747–817 (1967).
118. D. W. Moore and E. A. Spiegel, A thermally excited non-linear oscillator, *Astrophysical Journal.* **143**, 871–887 (1966).
119. A. E. Cook and P. H. Roberts, The Rikitake two-disc dynamo system, *Mathematical Proceedings of the Cambridge Philosophical Society.* **68**(2), 547–569 (1970). doi: 10.1017/S0305004100046338.
120. S. Smale, Generalized Poincaré's conjecture in dimensions greater than four, *Annals of Mathematics.* **74**(2), 391–406 (1961).
121. S. Newhouse and J. Palis, Hyperbolic nonwandering sets on two-dimensional manifolds. In M. M. Peixoto (ed.), *Dynamical Systems.* Academic Press (1973), pp. 293–301. Proceedings of a Symposium Held at the University of Bahia, Salvador, Brasil (1971).
122. S. Newhouse and J. Palis, Bifurcations of Morse-Smale dynamical systems. In M. M. Peixoto (ed.), *Dynamical Systems.* Academic Press (1973), pp. 303–366. doi: 10.1016/B978-0-12-550350-1.50030-8. Proceedings of a Symposium Held at the University of Bahia, Salvador, Brasil (1971).
123. R. F. Williams, Expanding attractors, *Publications Mathématiques de l'Institut des Hautes Études Scientifiques.* **43**(1), 169–203 (1974). doi: 10.1007/BF02684369.
124. R. F. Williams, The Lorenz attractor, *Lecture Notes in Mathematics.* **615**, 94–112 (1977).
125. D. Ruelle and F. Takens, On the nature of turbulence, *Communications in Mathematical Physics.* **20**, 167–192 (1971).
126. O. E. Rössler, The chaotic hierarchy, *Zeitschrift für Naturforschung A.* **38**, 788–801 (1983).
127. M. Klein and G. Baier, Hierarchies of dynamical systems. In *A Chaotic Hierarchy.* World Scientific Publishing (1991), pp. 1–24.
128. C. Letellier, N. Stankevich, and O. E. Rössler, Dynamical taxonomy: Some taxonomic ranks to systematically classify every chaotic attractor, *International Journal of Bifurcation & Chaos.* **32**, 2230004 (2022). doi: 10.1142/S021812742230004X.
129. A. Bethe, Die biologischen Rhythmus-Phänomene als selbständige bzw. erzwungene Kippvorgänge betrachtet, *Pflüger's Archiv für die gesamte Physiologie des Menschen und der Tiere.* **244**, 1–42 (1940).

130. A. Bethe, Teilrhythmus, Alternans, Amplitude und die Grenzen des Alles-oder-Nichts-Gesetzes, *Pflüger's Archiv für die gesamte Physiologie des Menschen und der Tiere.* **244**, 43–49 (1940).
131. A. Bethe, Analogien zwischen Rhytmusstörungen des Herzens und Vorgängen in konstlichen Kippschwingungsystemen, *Klinische Wochenschrift.* **20**, 33–36 (1941).
132. B. van der Pol, On "relaxation-oscillations", *Philosophical Magazine.* **2**(11), 978–992 (1926).
133. B. van der Pol and J. van der Mark, Frequency demultiplication, *Nature.* **120**, 363–364 (1927). doi: 10.1038/120363a0.
134. K. F. Bonhöffer, Zur Theorie des elektrischen Reizes, *Naturwissenschaften.* **31**, 270–275 (1943). doi: 10.1007/BF01475687.
135. U. F. Franck, Über das anodische Verhalten des Eisens in Schwefelsäure. Ein Beitrag zur Kenntnis des Ostwald-Lilliescben Nervenmodells, *Zeitschrift für Naturforschung A.* **4**(5), 378–391 (1949). doi: doi:10.1515/zna-1949-0510.
136. K. F. Bonhoeffer and U. F. Franck, Über die elektrolytische und die chemische Passivierung und Aktivierung von Eisen, *Zeitschrift für Elektrochemie und angewandte physikalische Chemie.* **55**(3), 180–184 (1951). doi: 10.1002/bbpc.19510550303.
137. U. F. Franck, Instabilitätserscheinungen an passivierbaren Metallen, *Berichte der Bunsengesellschaft für physikalische Chemie.* **62**(6–7), 649–655 (1958). doi: 10.1002/bbpc.19580620607.
138. R. FitzHugh, Mathematical models of threshold phenomena in the nerve membrane, *Bulletin of Mathematical Biophysics.* **17**, 257–278 (1955). doi: 10.1007/BF02477753.
139. U. F. Franck and R. Fitzhugh, Periodische Elektrodenprozesse und ihre Beschreibung durch ein mathematisches Modell, *Berichte der Bunsengesellschaft für physikalische Chemie.* **65**(2), 156–168 (1961). doi: 10.1002/bbpc.19610650210.
140. R. FitzHugh, Impulses and physiological states in theoretical models of nerve membrane, *Biophysical Journal.* **1**, 445–466 (1961). doi: 10.1016/S0006-3495(61)86902-6.
141. S. Lefschetz, *Lectures on Differential Equations.* Princeton University Press (1948).
142. A. J. Lotka, *Elements of Physical Biology.* Williams and Wilkins Co, Baltimore (1925).
143. N. Minorsky, *Introduction to Non-Linear Mechanics* (1947).
144. U. F. Franck and F. Kettner, Zur Kinetik erregbarer Systeme, *Berichte der Bunsengesellschaft für physikalische Chemie.* **68**(8–9), 875–882 (1964). doi: 10.1002/bbpc.19640680845.
145. U. F. Franck, Chemical oscillations, *Angewandte Chemie International Edition in English.* **17**(1), 1–15 (1978). doi: 10.1002/anie.197800013.
146. F. Jenik, Electronic neuron models as an aid to neurophysiological research. In H. Autrum and *et al.* (eds.), *Ergebnisse der Biologie.* Springer, Berlin (1962), pp. 206–245.

147. U. F. Franck, Feedback kinetics in physicochemical oscillators, *Berichte der Bunsengesellschaft für physikalische Chemie.* **84**(4), 334–341 (1980). doi: 10.1002/bbpc.19800840407.
148. R. J. Field and R. M. Noyes, Oscillations in chemical systems. IV. Limit cycle behavior in a model of a real chemical reaction, *The Journal of Chemical Physics.* **60**(5), 1877–1884 (1974). doi: 10.1063/1.1681288.
149. D. Edelson, R. J. Field, and R. M. Noyes, Mechanistic details of the Belousov-Zhabotinskii oscillations, *International Journal of Chemical Kinetics.* **7**(3), 417–432 (1975). doi: 10.1002/kin.550070309.
150. R. Noyes, Oscillations in homogeneous systems. In U. F. Franck, B. Hess, F. Schögl, and E. Wicke (eds.), *Kinetics of Physicochemical Oscillations*, Vol. 3. (1979), pp. 591–605.
151. O. E. Rössler and K. Wegmann, Chaos in the Zhabotinskii reaction, *Nature.* **271**, 89–90 (1978). doi: 10.1038/271089a0.
152. R. A. Schmitz, K. R. Graziani, and J. L. Hudson, Experimental evidence of chaotic states in the Belousov–Zhabotinskii reaction, *The Journal of Chemical Physics.* **67**(7), 3040–3044 (1977). doi: 10.1063/1.435267.
153. M. Marek and I. Stuchl, Synchronization in two interacting oscillatory systems, *Biophysical Chemistry.* **3**, 241–248 (1975). doi: 10.1016/0301-4622(75)80016-0.
154. D. Ruelle, Some comments on chemical oscillations, *Transactions of the New York Academy of Sciences.* **35**(1 Series II), 66–71 (1973). doi: 10.1111/j.2164-0947.1973.tb01505.x.
155. B. Hess, Oscillations in biochemical systems. In U. F. Franck, B. Hess, F. Schögl, and E. Wicke (eds.), *Kinetics of Physicochemical Oscillations*, Vol. 3 (1979), pp. 606–620.
156. J. J. Tyson, Some further studies of nonlinear oscillations in chemical systems, *Journal of Chemical Physics.* **58**, 3919–3930 (1973). doi: 10.1063/1.1679748.
157. A. Vaskelis, N. Salkausaks, and A. Prokoptchik, Periodic oscillations in electroless plating processes. In U. F. Franck, B. Hess, F. Schögl, and E. Wicke (eds.), *Kinetics of Physicochemical Oscillations*, Vol. 3 (1979), pp. 714–723.
158. G. Sorensen, Heterogeneity properties of the Belousov-Zhabotinsky reaction for operating conditions close to homogeneity. In U. F. Franck, B. Hess, F. Schögl, and E. Wicke (eds.), *Kinetics of Physicochemical Oscillations*, Vol. 1 (1979), pp. 41–50.
159. A. Zhabotinsky, Oscillating bromate oxydative reactions. In U. F. Franck, B. Hess, F. Schögl, and E. Wicke (eds.), *Kinetics of Physicochemical Oscillations*, Vol. 3 (1979), pp. 657–671.
160. K. R. Graziani, J. L. Hudson, and R. A. Schmitz, The Belousov-Zhabotinskii reaction in a continuous flow reactor, *The Chemical Engineering Journal.* **12**(1), 9–21 (1976). doi: 10.1016/0300-9467(76)80013-5.
161. M. Marek and E. Svobodová, Nonlinear phenomena in oscillatory systems of homogeneous reactions — Experimental observations, *Biophysical Chemistry.* **3**(3), 263–273 (1975). doi: 10.1016/0301-4622(75)80019-6.

162. L. F. Olsen, Models of the single enzyme oscillating reaction catalysed by peroxidase. In U. F. Franck, B. Hess, F. Schögl, and E. Wicke (eds.), *Kinetics of Physicochemical Oscillations*, Vol. 3 (1979), pp. 704–713.
163. O. Gurel and O. E. Rössler, *Bifurcation Theory and Appliations in Scientific Disciplines.* Annals of the New York Academy of Sciences, Vol. 316. New York Academy of Sciences (1979). Papers from the Conference Held on October 31–November 4, 1977.
164. O. Gurel and O. E. Rössler, Bifurcation to toroidal surfaces, *Mathematica Japonica.* **23**, 491–507 (1979).
165. K. F. Jensen and W. H. Ray, A new approach to modelling the dynamics of catalysis surface. In U. F. Franck, B. Hess, F. Schögl, and E. Wicke (eds.), *Kinetics of Physicochemical Oscillations*, Vol. 1 (1979), pp. 243–253.
166. R. A. Schmitz, G. T. Renola, and A. P. Zioudas, Dynamic complexities in the catalytic oxidation of hydrogen on a nickel foil. In U. F. Franck, B. Hess, F. Schögl, and E. Wicke (eds.), *Kinetics of Physicochemical Oscillations*, Vol. 1 (1979), pp. 221–230.
167. J. L. Hudson, D. Marinko, and C. Dove, Experiments on complex oscillations and chaos in the belousov-zhabotinsky reaction. In U. F. Franck, B. Hess, F. Schögl, and E. Wicke (eds.), *Kinetics of Physicochemical Oscillations*, Vol. 1 (1979), pp. 71–80.
168. J. S. Turner, Periodic and nonperiodic oscillations in the belousov-zhabotinsky reaction. In U. F. Franck, B. Hess, F. Schögl, and E. Wicke (eds.), *Kinetics of Physicochemical Oscillations*, Vol. 1 (1979), pp. 61–70.
169. I. Gumowski and C. Mira, *Dynamique chaotique: Transformations ponctuelles, transition ordre-désordre.* Cepadues Editions, Toulouse, France (1980).
170. B. Mandelbrot, *The Fractal Geometry of Nature.* Freeman, San Francisco (1977).
171. O. E. Rössler and C. Letellier, *Chaos: The World of Nonperiodic Oscillations.* Springer Nature, Switzerland AG (2020). doi: 10.1007/978-3-030-44305-4.

https://doi.org/10.1142/9789819823130_0008

Chapter 8

CONTRIBUTIONS TO THE THEORY OF SPONTANEOUSLY EVOLVING SYSTEMS I: A SIMPLE MODEL CLASS

Otto E. Rössler

University of Tübingen, Tübingen, Germany
oeross00@yahoo.com

German-language version received May 19, 1969 by the *Journal of Theoretical Biology*, Buffalo, New York.[a]

There exist both chemical and mathematical systems whose components change through some kind of "evolution." Dynamical systems with a highly complex structure can be realized by chemical means. The behavior of but a few "functional elements" that can arise in chemical systems is known. Moreover, the astronomically high number of interactions which arise in some reaction systems ("evolutionary soups") enforces the analysis of strongly simplified models. Starting out from the principle of parametric action as formulated by Sugita, a class of chemical systems is arrived at which has some features in common with linear compartment models. The class is mathematically described by a very large system of ordinary differential equations of the nonnegative type that are "linear for the most part." The new class constitutes the prototype in a whole series of less and less idealized models. A "biopaleontological question" is then applied to the class: Which types of subsystems are prevalent, in the majority of cases, during successive epochs of their time evolution? Several successive epochs characteristic

[a]This paper was accepted for publication by the *Journal of Theoretical Biology* but never published because the editor asked O. Rössler for an English translation that Otto never produced. It was only for this book that he did the translation.

for the "early transient phase" of the systems' development, when a combination of graph-theoretic, statistical and analytical methods is employed. In the first model class, the emergent subsystems do not show the "genome-phenome differentiation" characteristic of biological systems. However, this seems to be the case already in the next, slightly more realistic, model class so that a first-principles approach to chemical evolution becomes possible.[b]

1. Introduction

Any spontaneous change of state in a macroscopic physical system requires the presence of a potential of free energy that can decrease during the process. However, the time course adopted by the variables of the system depends also on other properties of the energy-consuming variables. Frequently a nonlinear dependence holds true between the change of state of a variable ("flux") and the potential difference ("force") that makes the change possible. The second pure case is a complete lack of a numerical dependence of the fluxes on the forces — so that the latter acquire the role of mere "preconditions." This is equally frequent. Examples would be the spring tension in a watch which does not appreciably determine its rate, the pressure in an oxygen flask that is equipped with a pressure-reducing valve, or the chemical potential of a reaction whose rate is determined by a third substance that acts as a "catalyst" while being present in a low, rate-limiting concentration.

Prigogine and Wiame have indicated a law,[1] valid for a large group of "open systems" in the sense of von Bertalanffy,[2] which shows they are "spontaneously evolving" in a certain sense. Specifically, the law asserts a spontaneous decrease of entropy-production during the formation of "dynamical structures" in such systems.[3,4] The validity of this evolutionary law is, however, contingent on a (essentially) linear dependence of the "fluxes" on the "forces" that make the latter possible in the first place.[5–7] Some attempts at broadening the scope of the principle have been made.[8] They correspond to a transition into the range of overlap with an in the limit no longer energetically interpretable, second type of determination. The latter was called "parametric causation" by Sugita.[9,c]

[b]Supported in part by the Deutsche Forschungsgemeinschaft.

[c]Robert Mayer[10] very early recognized the importance of a control by mere signals, coining the term "releasability" (Auslösbarkeit) for this second principle of causal determination.

Thus the question poses itself of, whether it might be possible to, by starting out from this second (parametric rather than energetic) type of causation, arrive at equally general evolutionary properties in certain dynamical system models. If the latter retain their validity in the range of overlap mentioned above, a synthesis of both approaches will become possible. As a first case in point, the system models employed in the analysis of engineering systems [11] must be mentioned, in which "control by energy" and "control by signals" are routinely treated on an equal footing. The same is expected to hold true for the reaction schemes of chemical kinetics.[12–14]

2. Cybernetic Systems Endowed with an Evolutionary Character

Some cybernetic system models with a largely lacking energetic control do have certain evolutionary properties in common with the evolutionary system of biology. The simplest example, perhaps, is the "positive-integral feedback" which spontaneously occurs in many nonliving physical systems. Examples would be a snow avalanche, a forest fire, a chemical explosion, a rusting process on a piece of iron or any other example of a so-called "autocatalytic" process. Other frequently used names are "nucleation process," "vicious circle," "spiral," "escalation" or "divergent chain reaction." All these processes thrive on free energy but, at the same time, possess another type of control so that they appear to possess an autonomous "life" of their own.

The "signal flow chart" of such a process differs from that of an integral-type (negative) feedback circuit in the polarity (sign) of the returned signal. This fact has been illustrated in Fig. 1.

(a) Positive integral feedback (b) Negative integral feedback

Fig. 1. Block diagram of a positive integral feedback (a) compared with a "negative integral feedback control" (b). The "soll value" (value-to-be-achieved) is equal to zero in either case, once in negative time (a), once in positive time (b).

Notes: I = integrator; k = feedback factor; I.C. = initial condition.

The behavior of these systems is, in the simplest case, described by a single ordinary differential equation of the first order, as is well known:

$$\dot{x} = kx \tag{1}$$

It has the solution $x(t) = x(0)e^{kt}$, where x is the output value of the circuit, k the coupling factor, $\dot{x} = dx/dt$ the rate and $x(0)$ the positive initial value of the state variable x. This prototype model of a progressively growing system is in use since the 18th century to illustrate the behavior of complicated systems with an intrinsic evolutionary character.[15,16]

As can be demonstrated by using any technical illustration of a positive integral-feedback-loop, these systems show the assumed potential-independent behavior only within certain limits. In the case of Fig. 2, a more complicated rate equation holds true more realistically which has the following form:

$$\dot{x} = k_1 x - k_2 x^2, \tag{2}$$

with $k_1 \gg k_2$. It has the solution $x(t) = \frac{k_1 C}{e^{k_1 t} + k_2 C}$, with $x_{\text{max}} = \lim_{t\to\infty} x = \frac{k_1}{k_2}$, $C = \frac{x(0)}{k_1 - k_2 x(0)}$. This solution is related to the equation for the slanted-S shaped so-called autocatalytic curve of reaction kinetics.[17] In case the controlling valve V in Fig. 2(a) is of the "pressure-reducing" type, the slanted-S curve has a solution that, below the threshold thereby generated, looks like that of Eq. (2). As already mentioned, chemical catalysts do often possess such pressure-reducing properties.

We now make the transition to several autocatalytic systems that coexist in the same reaction medium. To an external observer (equipped with the right measuring instruments), they can present a rather interesting behavior. If the "feedback factors" k are different and so are the initial concentrations $x(0)$, the partial system showing the greatest potential for growth (the largest k) is not necessarily the one that grows ahead of the others (as would be the case with equal initial concentrations). Rather, the observer now witnesses so-called take-over phenomena on. The latter can occur rather suddenly — by a sharp qualitative jump in the composition of the system. Much as in any other race, the terminology of "competition" and "selection" offers itself. Indeed, an outside bystander could hardly avoid the impression of witnessing a selection process going on in which,

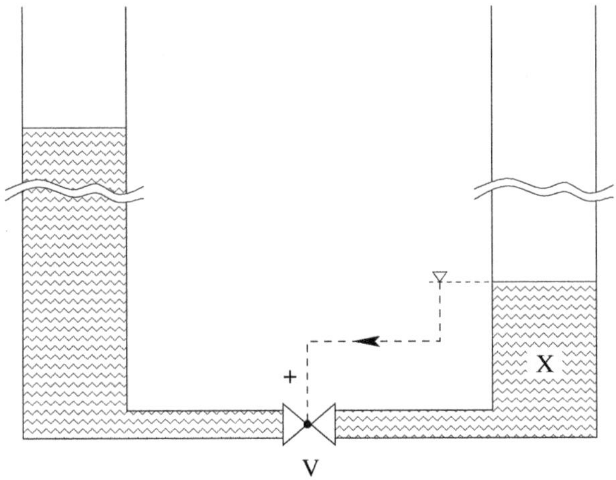

Fig. 2. Schematic of a continuous water-level control with value-to-be zero that is "fed back with the wrong sign" — as an example of a realistic positive integral feedback system.

Note: V = valve.

after a longer or shorter period of time, the most strongly positively fed-back partial system prevails.

A second example, which was already indicated by Darwin, can be interpreted as a modification of the first. In his well-known letter of 1871, Darwin endows a "protein compound" (which was obviously thought to possess autocatalytic properties) with the readiness to transform itself: "ready to undergo more complex changes." This model of biogenesis — if one may call it so — is still widely quoted today. Ashby[18] adopted it in his "*Introduction to cybernetics*" and Schramm *et al* use it in their interpretation of experiments designed to explain the spontaneous origin of biological key substances.[19] However, according to Kaplan,[20] this model, owing to its lack of a phenome-genome distinction (which according to Waddington[21] is of fundamental importance for biological evolution), has to be considered as a model of a pure "gene-life" as he calls it. The growth of copolymers would provide a simple example.[22]

The phenomenological characteristics of this more complicated second class of systems (endowed with mutability) also deserve to be put into perspective briefly. It is characterized by the existence of pre-formed "mutation bridges" between the individual autocatalytic subsystems. In this way, also an initial concentration of zero no longer constitutes an obstacle to a later "predominance" of a partial system if only its feedback factor is large enough, for it can arise through mutation out of another partial system. The behavior of this second type of system is, therefore, determined by the matrix of mutational connections between the individual partial systems. The evolutionary "emergence" of new partial systems can then be registered in the manner of a paleontological temporal scheme (see below). However, an argument against the early biohistorical importance of this second class of a pure gene life is, apart from its chemically speaking rather small probability of spontaneous occurrence, the high specificity of the involved catalysts that is assumed from the beginning.

The prototype of a third class of models that meets some of these objections is a system indicated by Cairns-Smith.[23] It admits substances into the medium that are incapable of autocatalysis themselves but take influence on the growth rate of other substances that are capable of self-amplification. Part of the latter, in turn, possess besides their autocatalytic properties also heterocatalytic ones. The combination of substances from both groups in a single system leads according to Cairns-Smith (given the right supplies or, in Darwin's words, "with all sorts of ammonia and phosphoric salts, light, heat, electricity etc. present") to the spontaneous occurrence of a genome-phenome differentiation in a larger self-multiplying unit. Moreover, autocatalytic "gene substances" that had been active at the beginning because of their modest requirements for auxiliary substances, are proposed to later give way to a second-generation gene substance that while more demanding is also more versatile in its producing heterocatalytic effects. In the history of the total system, three stages can be distinguished: A "homogenetic" initial stage is followed by a long "heterogenetic" intermediary stage in which several heterocatalytic substances coexist reciprocally facilitating each other's multiplication via the auxiliary substances they catalyze; then in a "competitive metamorphosis," again a largely homogenetic "third stage" is postulated to follow. With Cairns-Smith's model, for the first time an evolutionary system appears to have been indicated which makes only parsimonious use of highly specialized (*ad hoc*) functional and chemical assumptions. However, it of course still depends on the good will of the reader whether he or she finds plausible the spontaneous occurrence of the postulated, combined autocatalytic and heterocatalytic features in a few types of primordially assumed "clay-like" substances.

3. Basic Properties of a General Chemical Evolutionary System

If one views the above-described chemical systems as examples of "cybernetic" evolutionary systems, it becomes possible to arrive at a completely "non-specialized" evolutionary chemical reaction system. It does involve a host of different chemical transformations and catalytic and modifying activities. However, the overwhelming majority of reaction products possess a very low spontaneous production rate. Similarly, the overwhelming majority of catalytic effects possess a very limited activity. Strong catalytic and catalysis-modifying effects are confined to only a few constellations of variables.

In this way, we need not discuss which special physical and chemical conditions are required in order for these postulates to be heeded. This concerns the temperature range, the medium of choice, the requirements on the atomic constituents involved — like a capability to form a so-called "universal library" of molecular species — or the facilitatory presence of certain metal ions and crystals of argillaceaous earth (for example), in the case the medium is watery and carbon is the backbon former. None or any of these specific assumptions are necessary.

As is well known, every reaction system undergoes, once constituted, a transition phase on the way toward a steady state regime and, from there on, to the final thermodynamic equilibrium. Our conjecture is that the transition phase in this evolution will, in a system of the present very large type, be not only very long and complex, but also exhibiting further properties that up until now were considered to be typical of the biological evolutionary process. A continuity between the biological example and simpler thermodynamic and cybernetic systems has also been postulated by Waddington.[24] Thus, the question at this point is whether the evolutionary potential *already implicit* in certain simple chemical soups[d] may not be sufficient already to define a whole class of evolutionary cybernetic systems. Let us follow a few steps further in this direction.

All chemical systems with an evolutionary character owe this character to the property of "catalysis." Functionally speaking, this notion — which goes back to Berzelius (1836, quoted in Ref. 28) — refers to the control of the formation of one chemical species by the amount (or concentration)

[d]25 spoke of a "*warm little pond,*"[26] of a "*hot diluted soup.*" Thus, the term "*Darwin's soup*" suggests itself, cf. also.[27]

in which another is present. Under certain chemical preconditions (like a moderate excess of substrate, an approximate constancy of the chemical potential, an absence, or presence in saturation, of modifiers, etc.), this control turns out to be linear: $\dot{x}_a = \pm x_b$. The mentioned preconditions are all of the necessary (rather than sufficient) type; that is, if their limits are surpassed, other (more or less nonlinear) functions take the role of the previous linear ones. The previously linear catalytic effect can then be described as being either "modulated" or "cut off."

In isothermic homogeneous reaction systems, the occurrence of nonlinear functions depends only on the quantitative relations between the individual substances (since functional relations other than those between two reaction partners, or precursor and product, or substrate and catalyst, or catalyst and modifier — and combinations thereof — are absent in such systems). Since the linear catalytic function holds true again and again over large concentration ranges, it can perhaps be considered as a kind of natural bottleneck. This is the "principle of flexible throttle control" of Sugita.[14] It apples to a large subset of the processes that occur spontaneously in complex reaction systems rich in catalytic relations. A description of the qualitative behavior of these systems can, therefore, be attempted from the viewpoint of the linear limiting function of the catalysts that are present within them. This is the proposal made in the following.

To date, computer simulations of chemical and biochemical reaction systems represent a recognized field of instrumental mathematics.[14, 29–33] In spite of these developments, however, the exact simulation of a not too limited number of chemical variables possessing the properties assumed above appears to be hard to achieve. For the requisite data sets are not yet available to a sufficient degree of accuracy and completeness — neither empirically nor according to deductive quantum-physical arguments and calculations. Moreover, reaction-kinetic approaches mostly presuppose isothermy, homogeneity (at least within compartments) as well as mathematical continuity and determinism — idealizations that can greatly restrict the range of validity of the results obtained. Here, the cybernetic (system-theoretic) method can be of help since it allows one to derive the possible qualitative implications of one's combining a few already known types of (reaction) elements.

Surprisingly, applied systems theory in the above sense proves to be dependent, on the results of chemistry itself. Only chemistry can provide the means for the realization of very complex dynamical systems. For only the quantum-mechanically explicable, so-called "identity" of the basic chemical constituents allows one to span up a phase space with an arbitrarily sharp resolution power. While giant molecules often represent pseudo-continuously smeared-out entities in the sense of quantum mechanics (and hence cannot be assumed to be completely identical), the same drawback holds true to a lesser degree for small molecules whose vibrational states may be "frozen".[34] Only by taking the roundabout way of looking at the chains of synthesis that start out from small molecules (think of the crystallization process as a simple example), it is possible to maintain a constant product quality over an arbitrary number of steps, in the realm of chemistry. In the ordinary thermodynamic machines,[35] the same goal can only be achieved through the repetitive application of quantum-mechanical resettings of the calibration-point combined with special error-correction methods.

4. Formulation of a Strongly Simplified First Model

To describe the behavior of the functional elements that occur in large chemical reaction systems, it appears possible to start out from the above-mentioned "bottleneck behavior" of the linear catalytic function, all other functions being interpreted as functional pre-conditions for the latter. In this way, the functional structure of large reaction systems rich in catalytic connections becomes a network of linear, positive and negative controlling relations. The change of concentration of each chemical depends either positively or negatively on the current concentration of all other substances. Such a feedback relation was, under the name "structure-activity feedback," already proposed by Pask[36] to be of fundamental importance for the understanding of growing systems. However, since concentrations cannot assume negative values, all negative controlling influences affecting a chemical species, necessarily cease to apply when the concentration of the species in question approaches the value of zero. The seemingly "linear network" proposed above therefore actually includes well-defined nonlinearities. Figure 3 shows a cybernetic "configuration of effects" (*Wirkungsgefüge* in the sense of Mittelstaedt[37]) that reflects this basic structure.

Superficially similar models are often called "compartmental models" in the chemical and biological literature. The function of these systems is understood in terms of the mutual influences occuring between the different "compartments" of a system of fluxes.[38] However, linear multicompartment models by definition never contain any catalytic controlling relations if the

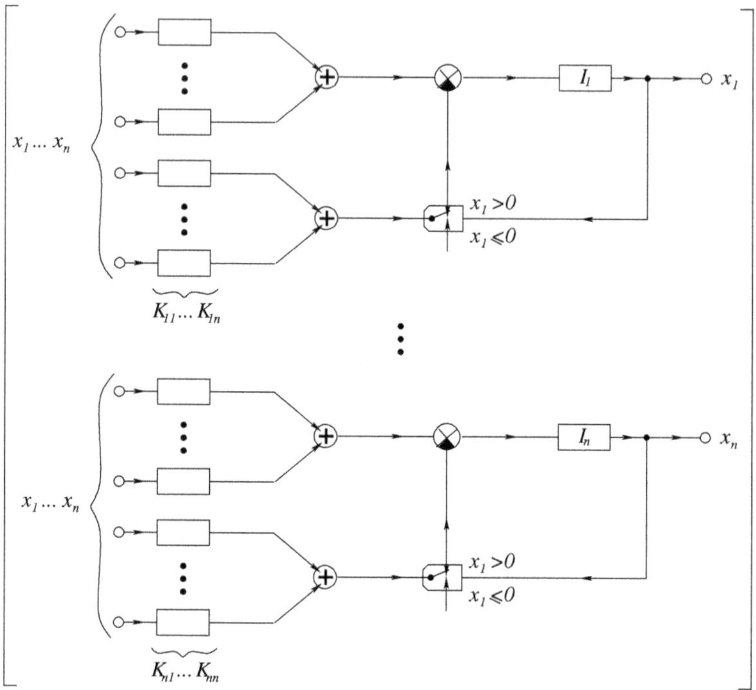

Fig. 3. Configuration of effects (block diagram) of a simple spontaneously evolving system derived from the chemical assumptions described in the text.
Note: $I_1, I_2, \ldots, I_n$ = integrators.

law of mass conservation is taken into account.[39] The here described system, in contrast, does contain catalytic effects and hence fluxes that are not contingent on an equal consumption of the controlling substance. In spite of the nonlinear switching functions contained, however, the present class of systems appears to form a prototype of its own.

Any network of topological relations stands in algebraic correspondence with a so-called connection matrix. If the functions connected by the matrix are those of a switching relay, one speaks of a "finite automaton." In case algebraic functions are employed, the name "self-organizing system" is customary.[40] If differential equations are involved, one speaks of a "dynamical system" in the mathematical sense.[41] The system model of Fig. 4 belongs into the latter class.

Specifically, the system of Fig. 4 is mathematically described by a system of ordinary homogeneous differential equations "with conditionally constant

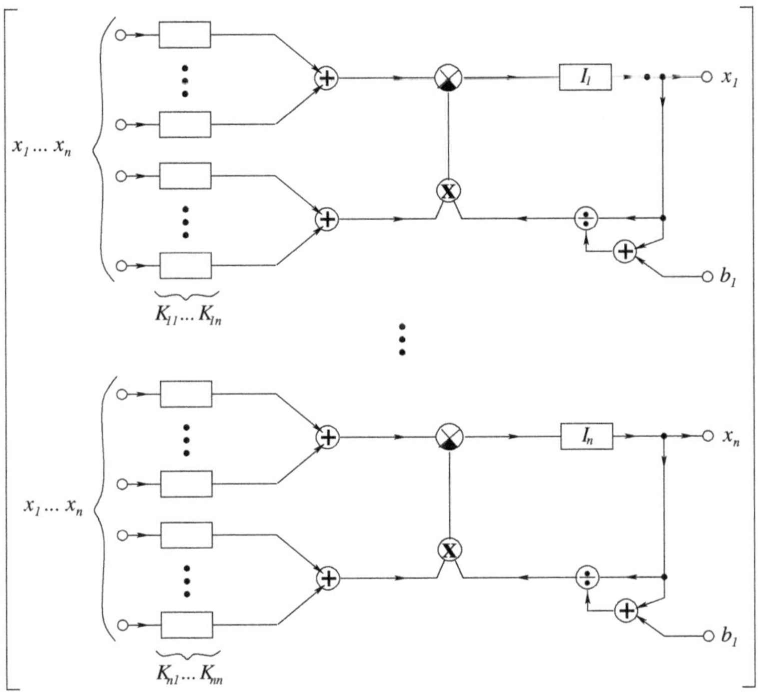

Fig. 4. A variant of the action pattern of Fig. 3: The more fundamental continuous case.

cœfficients." In matrix notation, it can be written

$$\dot{x} = F \cdot x, \tag{3}$$

with matrix elements $f_{ij} = k_{ij} \cdot a_{ij}$. Here the

$$a_{ij} = \begin{cases} 1 \\ 0 \end{cases} \text{if} \begin{array}{lll} k_{ij} \geq 0 & \text{or} & (k_{ij} < 0 \text{ and } x_i > 0) \\ k_{ij} < 0 & \text{and} & x_i < 0) \end{array} \tag{4}$$

with the k_{ij} being either positive or negative constants ($i = 1, 2, \dots, n$).

The particular switching functions a_{ij} that couple to negative coupling constants k_{ij} are, in a concrete chemical system, caused by the nonlinear coupling elements present. With decreasing concentration of the substrate (x_a) of a catalyst (x_b), the unit value of the switching function a_{ij} ceases to be applicable at a certain (still positive) value of the substrate x_a, so that the linear controlling function of the catalyst, $\dot{x}_a = -kx_b$, gets suspended in favor of a smoothly nonlinear one. Into the latter, the residual concentration of the substrate enters as a factor, so that in the simplest case one has $\dot{x}_a = -kx_b x_a$.

This is usually called the principle of Michaelis and Menten.[42] Hereby the substrate, x_a, can approach the value of zero only asymptotically so that it always remains positive. The larger the linear range of control compared to the nonlinear near-zero range, the less objectionable becomes the functional approximation of the "knee" (at which catalyst saturation occurs) by a switching function. The strictly speaking non-differentiable model of Eq. (3) therefore represents the limiting case of a smooth, continuously integrable (C^∞) nonlinear model, of the following form:

$$\dot{x} = F \cdot x, \tag{5}$$

with matrix elements $f_{ij} = k_{ij} \cdot a_{ij}$. Note that the only difference to Eq. (3) lies in the switching functions a_{ij}. They have now been replaced by smooth functions of the form

$$a_{ij} = \begin{cases} 1 & \\ \dfrac{x_i}{x_i + b_i} & \end{cases} \text{if} \begin{array}{l} k_{ij} \geq 0 \\ k_{ij} < 0, \end{array} \tag{6}$$

that is, the k_{ij} are unchanged while the new small positive constants b_i are given smaller and smaller positive values in the transition towards the limiting Eq. (3).

In realistic reaction systems, of course, many further "constellational conditions" exist under which elements of the system matrix of Eq. (33) become nonlinear. Spatial relations (when eventually included) can also have such effects. Nevertheless, a consideration of the above simplest class of systems can be hoped to facilitate a better understanding of more and more complex chemistry-analogous evolutionary systems.

Now that the system of Eqs. (3) and (5) has been specified, it is about time to list some further properties of this evolutionary model:

(a) The number n of system components (state variables a_i,) is very large:

$$n \rightarrow 10^{2 \text{ million}} \tag{7}$$

This number in the exponent suffices for the realization of a so-called "universal library" in the sense of Kurd Lasswitz (1884), quoted after Flechtner.[43] Such a library contains any possible text of a specified finite length (as is well known).

(b) Strong inhomogeneity of initial conditions:

$$\text{Almost all } x_i(0) = 0, \text{ the rest positive.} \tag{8}$$

(c) As to the frequency H of nonzero matrix elements, the following relation holds true:

$$n \gg H \gg n^2 \tag{9}$$

(d) The numbers of positive and negative matrix elements are approximately equal,

$$n_+ \approx n_- \tag{10}$$

(e) The frequency of nonzero matrix elements depends on their absolute value in a strongly decaying fashion,

$$n_k \approx \mathrm{e}^{-|k|}. \tag{11}$$

(f) The non-zero matrix elements are equidistributed across the matrix,

$$p_k = \text{const.}, \tag{12}$$

here p is the probability. (A modification follows below.)

The preceding constraints determine a large ensemble of systems described by Eqs. (3–12). Every single member of the class is, owing to its size, intractable by either analytic or numerical methods. In the present context, however, it fortunately turns out that only a few questions concerning the behavior of the system class are of primary interest:

(1) Does there exist in the majority of the systems an at least transient *progressive growth* of individual system components?
(2) Do there exist different *types* of progressive growth?
(3) Can the individual growing *components* of a system be associated to specific functional types?
And finally:
(4) Do those components which during a comparable phase in the system's development show the largest values (prevail quantitatively), belong to the same succession of function types in most systems of the class, such that one can speak of an invariant sequence of evolutionary stages or *epochs*?

5. Some Answers to the Questions Posed

The "qualitative" developmental properties sought concern the general evolutionary properties of the system class in question. Leaving aside the complication of the minimal realization of linear systems,[44] we can define the "evolutionary graph" of a dynamical system as that particular weighted directed graph[45–47] which possesses the transpose of the system matrix $F \cdot x$ (that is, the right-hand side of the differential equation) as its adjacency matrix (the latter being sometimes called node matrix). The elements of this matrix $(f_{ij} \cdot x_i)$ represent the weights of the arrows of this graph (while the components of the solution vector x_i give the node values if needed). The totality of possible shapes of this graph is described by the diagram of the transpose of the matrix F. This is the "graph of pre-formed relations." At every moment in time, the latter graph contains the much smaller *graph of momentarily effective relations* (or "evolutionary graph" for short).

The larger one of these graphs (the graph of pre-formed relations) is homeomorphic to the so-called block diagram of the system (up to a minor point again,[44]). The smaller one of these graphs (the evolutionary graph) is isomorphic to the system's signal flow graph.[48] Its development in time gives one information about all functional relations between individual state variables that occur during the development of the system in question. All of this is illustrated, in two steps, in Fig. 5.

In the case of an empty initial vector, the graph of momentarily effective relations is nonexistent. After the "setting" of one or several larger-than-zero initial conditions $x_i(0)$, the graph consists of the corresponding nodes and those edges (arrows) that emanate from these nodes (are negatively incident on them). These arrows carry either a positive or a negative weight. In the former case (increasing-degree of the target node), the graph grows in a centrifugal manner through the sprouting of further arrows. While the path of this growth pattern follows strictly the graph of pre-formed relations, the rates of the individual sproutings depend on the quantitative cooperation between all momentarily involved variables x_i.

In this first stage of the evolution of the graph of momentarily effective relations, only a few state variables act together (corresponding to a small partial system of the whole system being activated). This fact enables one to make certain statistical predictions concerning the initial evolution of the present class of systems. To this end it suffices to "read-off" the number of connections that are active at a so small number of nodes,

$$\begin{bmatrix} 0\,1 \\ 1\,0 \end{bmatrix} \qquad \begin{bmatrix} 1 \\ 0 \end{bmatrix} \qquad \begin{bmatrix} \frac{e^{t}+e^{-t}}{2} \\ \frac{e^{t}-e^{-t}}{2} \end{bmatrix}$$

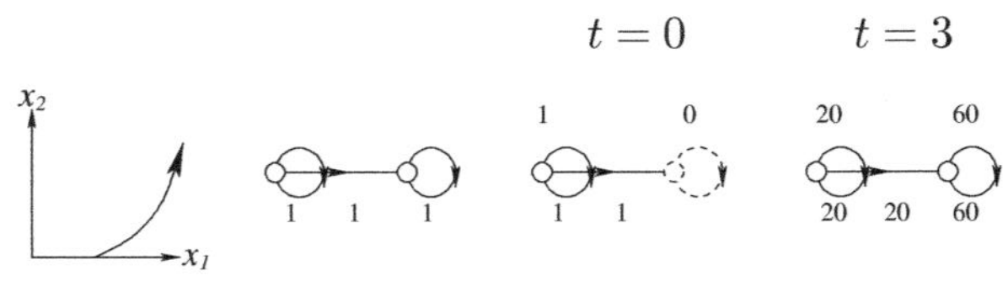

$$\begin{bmatrix} 0\,1 \\ 1\,1 \end{bmatrix} \qquad \begin{bmatrix} 1 \\ 0 \end{bmatrix} \qquad \begin{bmatrix} 0.3\,e^{1.6t} + 0.7\,e^{-0.6t} \\ 0.4\,e^{1.6t} - 0.4\,e^{-0.6t} \end{bmatrix}$$

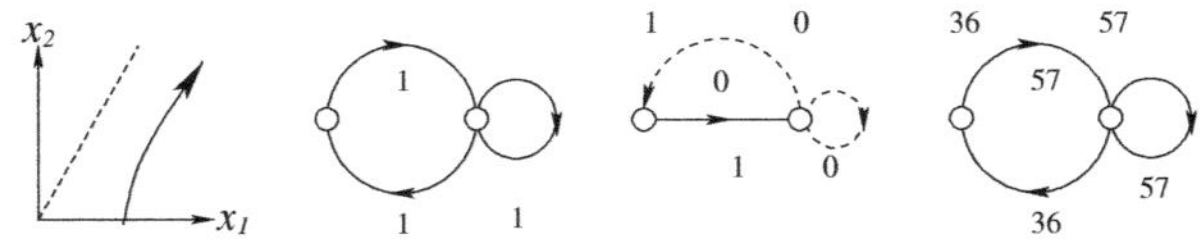

1. 2. 3. 4. 5. 6. 7.

(a) Case of two variables. One sees from left to right: 1. the system matrix, 2. the initial condition, 3. a corresponding particular solution, 4. the trajectory in the nonnegative part of phase space, 5. the graph of pre-formed relations, 6. the graph of momentarily effective relations, and 7. the same graph at a later moment in time.

$$\begin{bmatrix} 0\,0\,0\,1 \\ 1\,0\,0\,0 \\ 0\,1\,0\,0 \\ 0\,0\,1\,0 \end{bmatrix} \qquad \begin{bmatrix} 1 \\ 0 \\ 0 \\ 0 \end{bmatrix} \qquad \begin{bmatrix} \frac{\cosh t+\cos t}{2} \\ \frac{\sinh t+\sin t}{2} \\ \frac{\cosh t-\cos t}{2} \\ \frac{\sinh t-\sin t}{2} \end{bmatrix}$$

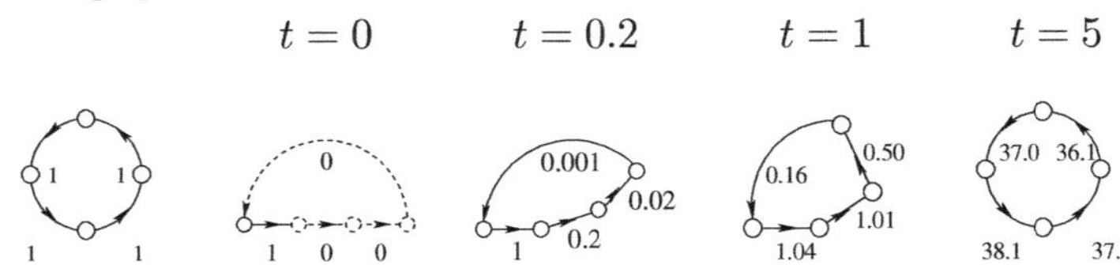

1. 2. 3. 4. 5. 6. 7. 8.

Evolution of the graph of momentarily effective relations of a cyclic system (like that of the last but second line of (a)), only with more variables. The graph is shown at different moments in time. (This is an example of "homeorhesis" in the sense of Waddington.) Cf. text.

Fig. 5. Examples of the "evolution" of a few simple linear systems of differential equations with constant nonnegative coefficients and nonnegative initial conditions.

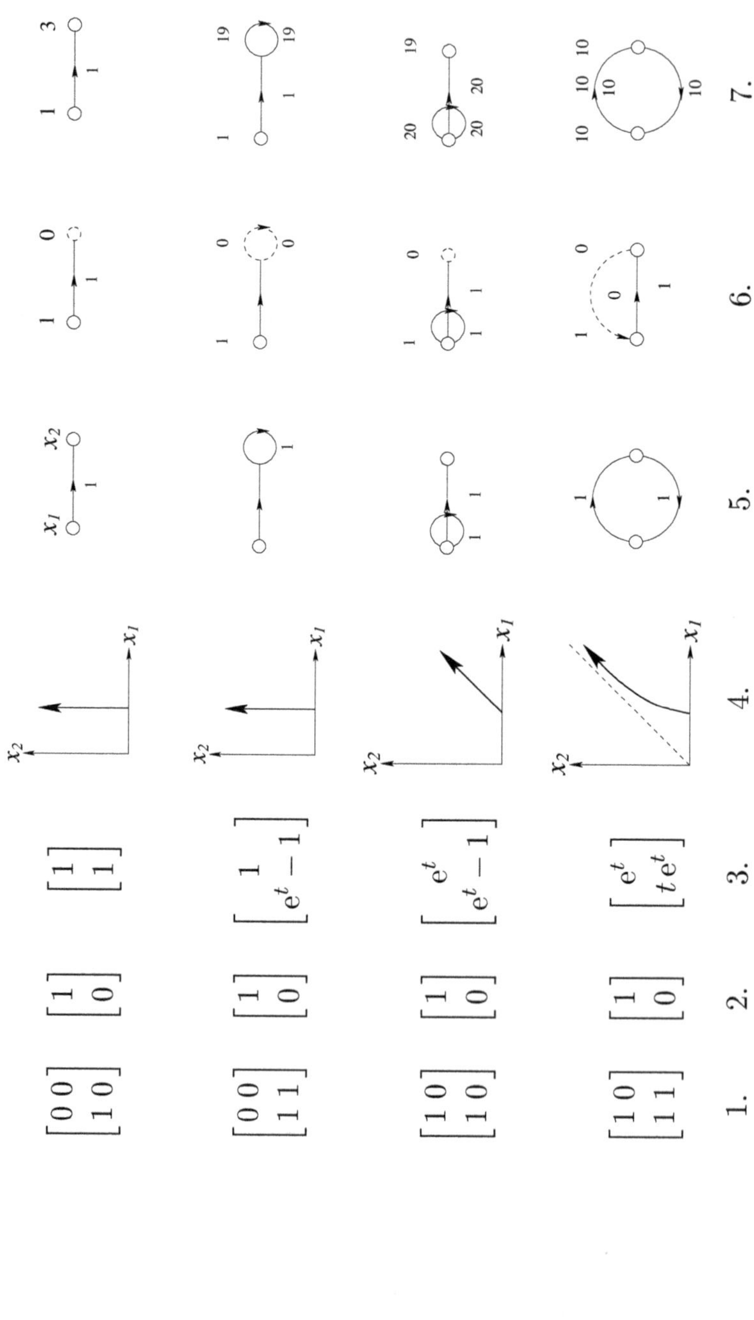

Fig. 5. (Continued)

and to identify their types in order so to arrive at the relative frequencies at which they occur as a function of time in the graph of pre-formed relations. Fortunately, not all possible types of interaction need to be taken into account in the present context: only those which suffer an especially rapid increase are of interest. Later, as the number of interacting nodes (and hence momentarily active connections) increases, rarer and rarer filling mechanisms are included.

When proceeding in this fashion, in the present non-negative system, only the first stage of its *transition-phase* development can be captured. This is no drawback, though. For all nonnegative systems of differential equations of the present type that are meant to describe physico-chemical processes remain in this transitory phase for a very long time — before eventually entering the "second or *stationary phase*.

If this is correct, the famous question as to the *stability* of the behavior of the system's stationary solutions, so often posed in the context of dynamical systems (see the criteria of Lyapunov[49]), is a matter of virtual irrelevance in the present context. For in the class of systems considered here, the relative frequency of systems with *stable* solutions decreases rapidly with increasing n (roughly with 2^n according to Ashby[50]). On the other hand, *linear* homogeneous systems of differential equations with constant coefficients (which enjoy a closed theory and form a class of systems closely related to the one considered here) possess asymptotic solutions that in many cases show characteristic quantitative relations between their components.

During the transition phase of the system's development, every component may successively belong to several different transitory solution types. Therefore this phase of the development of the system is comparable in the complexity of its behavior to an equally large linear automaton in the sense of Gill[51] and Zadeh.[52]

It can be conjectured that the nonlinear system of Eqs. (3) and (5) retains a great similarity to the related fully linear system with unchangeable matrix. On the one hand, the behavior of a subsystems with a strictly positive matrix is by definition identical when looked at in isolation. On the other, the action of the negative matrix elements is in the nonlinear case, strongly restricted. Unlike fully linear systems in which pre-formed negative arrows can participate in growing progressions of arrows, causing a change of sign in every second subsequent arrow, the strictly positive system of Eq. (3) is less flexible since such events cause a total break-off of the chain in question. Since the negative arrows cannot participate in effective cycles, they also cannot cause oscillations about the origin of phase space. This fact greatly reduces the incidence rate of undamped oscillations in the present class of systems.

Pre-formed negative arrows can become effective only if they issue forth from a positive partial graph (like the leaves of a branch). Also, they can only act on already existing arrows — by emptying-up their filling node. Furthermore, those segments of a chain of arrows that are already activated can continue to grow at a reduced rate. The non-permanent effectiveness of negative matrix elements below zero values of their target nodes makes sure that the present nonlinear systems mimic the qualitative behavior of nonnegative linear systems in many respects.

Specifically, the regular segments (without a change of matrix) of the solution curves of Eq. (3), with the constraints of Eqs. (7–12) added, are described by linear combinations of terms of the form $t^K \cdot e^{Ct}$ with K and C nonnegative constants. Hence much as in the purely linear case, there cannot exist any progressive growth proceeding faster than exponentially.

Second, the statistics of partial graphs comprising only a few arrows from the graph of pre-formed relations can be sought, either in a graph-theoretical or in a matrix-theoretical manner (cf. Refs. 47, 53, 54). The (by Eq. (12) equidistributed) matrix F possesses a transpose whose diagram consists of a reduced (i.e. noncyclic) graph R plus many cyclic quasi-components $Q_1, \ldots, Q_n$. The latter are each characterized by a different path length and a different degree of internal connectedness. In particular, the frequency of pre-formed cycles of positive sign increases rapidly with increasing length. So does the frequency of noncyclic preformed chains. In the range of small lengths l, the number of the latter increases almost with the l-th power. Hereby, long cycles tend to be intersected by oriented paths in proportion to l.

The growth rate of nodes that are not drained-out again, in the individual partial graphs, can be classified as follows:

- *Acyclic* progressions of positive-valued effective arrows are characterized by a linear sprouting rate. (Hereby re-attainment of a constant positive value by the first arrow is chosen as the criterion of evaluation presupposing equal preformed cœfficients.

More specifically, the first node of a chain gets filled up at a linear rate ("parabola of the first order"), the second at a quadratic rate (parabola of the second order), and so forth up to the n-th node (parabola of the n-th order). Note that this corresponds to the n-th polynomial member of the series of multiplicative terms accompanying the exponential terms (which in the present case have zero exponents).

- Positive-valued *cycles*, on the other hand, show for their inner nodes, irrespective of the cycle length, an exponential filling law with a nonnegative exponent of the dominating term.

In particular (if the arrows are again assumed to have equal positive values for simplicity), for large t the filling rate of the inner nodes of long cycles is the same as in short cycles. However, the final phase in which the cycle has been "effectively closed" is preceded by a transit phase of parabolic growth that is dependent on the length of the cycle in question.

- *Reduced chains* (with cycles contained in them) and interconnected cycles obey more complicated "fuse-like" effects.

The corresponding unstable partial systems show, after an initial phase of strong input-dependence, a characteristic phase of "input-emancipation",[55,56] see Fig. 6. Note that all results indicated qualitatively in Fig. 6 and its sequel (Fig. 7) can be made quantitative.

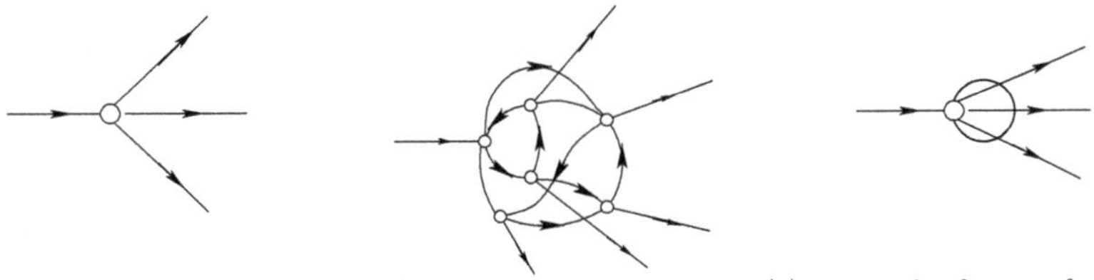

(a) Individual state variables that are connected by positive-valued directed edges

(b) Multiple-connected positive-valued cycles

(c) Network of many functionally connected positive cycles implicit in Eq.(12).

Fig. 6. Description of some functional elements in the graph of pre-formed relations that show "fuse effects".

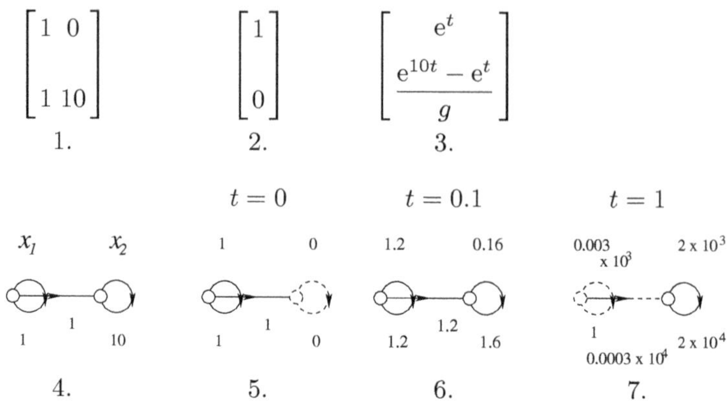

Fig. 7. An example of emancipation-from-input in an unstable partial system (loop). Notation as in Fig. 5.

6. Invariant Evolutionary Properties of the Model Class

The results presented above can be put together as shown in Fig. 8.

The content of Fig. 8 now puts us in the position to answer the four questions posed at the end of Section 4:

(1) In almost all systems of the present ensemble [determined by Eq. (3) or (5) with constraints (7)–(12)], a transitory phase of progressive growth of individual subsystems occurs.

(2) In all these systems, there exist two fundamentally different types of progressive growth: parabolic and exponential growth.

(3) The first type (parabolic growth) occurs in subsystems that lie in a chain-like positive dependence on an in the simplest case constant positive component. Then the matrix of the partial system which determines the momentary behavior of these components is nilpotent. The second type (exponential growth) occurs in components which belong to a directed, positive-valued cycle. In the simplest case this cycle is unbranched and activated by an internal node; the matrix of these partial system is nonsingular and nondecomposable.

(4) In virtually all systems of the ensemble, the great majority of variables that exhibit marked growth in an early phase belong to the chain-like type and hence show parabolic growth. This is **epoch 1**. In virtually every case these subsystems are subsequently overtaken by exponentially growing components. This is **epoch 2**. The latter at first belong in their greatest part to unbranched cycles of medium

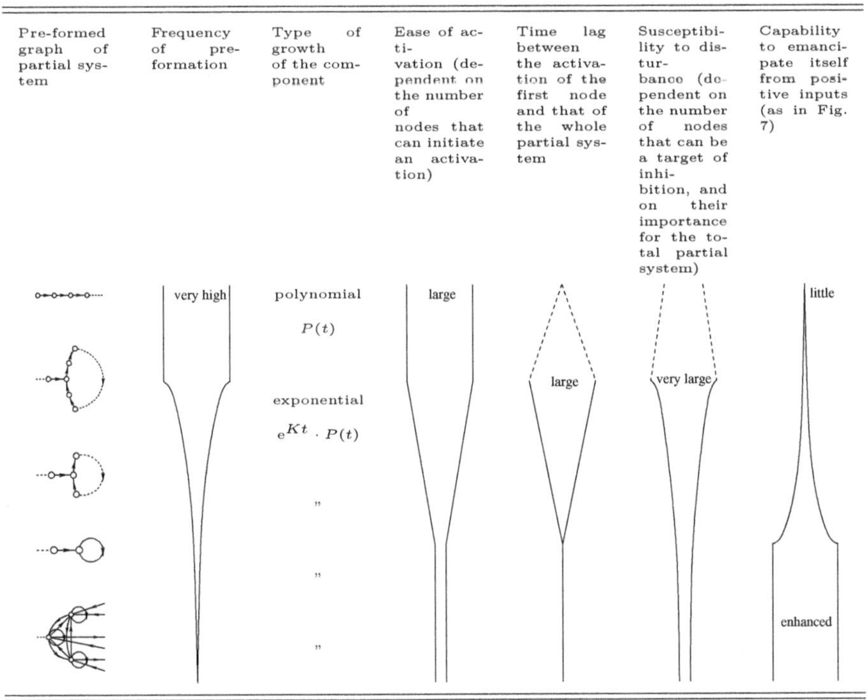

Fig. 8. Synopsis of the results obtained so far. On display are those unstable partial systems of the present class of systems (described by Eq. (3) or (5) with the constraints of Eqs. (7)–(12)) that during the early transitory stage of the systems' evolution possess the greatest chance of being activated individually. The transitory lack of functional integration into a larger partial system is accompanied by a particularly rapid growth of the components.

length (epoch **2A**). Subsequently, shorter cycles of the same type take over. Although pre-formed more rarely, they enjoy, owing to their reduced dependence on negative external influences, a longer period of undisturbed growth (epoch **2B**). Still later, even shorter (single-element) cycles become preponderant due to their possessing the best chance for an undisturbed continued growth (epoch **2C**). Since this is the longest epoch among those still characterized by a preponderance of individual partial systems, it can also be called **epoch 3**. Thereafter, with the activation of more and more subcomponents of the total system, "functionally insulated" subsystems cease to be preponderant in the present class of continuous systems. The initial phase of a "lawful universal evolition" is then over.

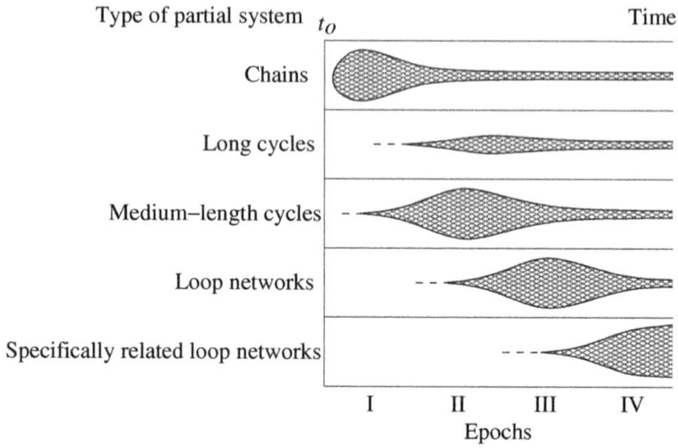

Fig. 9. Overview of the results implicit in the data of Fig. 8. Synoptic representation in the manner of a biopaleontological time scheme. The relative durations of the individual "epochs" are not to scale. See text for epoch 4.

The preceding results can be summed up in the form of the "biopaleontological scheme" shown in Fig. 9.

The results obtained so far were contingent on the assumption of an equidistribution of nonzero matrix elements (the constraint of Eq. (12)). If this constraint is relaxed, a more chemistry-analogous modification of Eq. (3) results. In this case, the derivation of a further invariant epoch can be derived (**epoch 4**). Chemical relatedness expresses itself in a system of catalytic influences of the present type in two ways: Firstly, as active, and secondly as passive chemical relatedness. This fact was already incorporated in Fig. 9.

The graphs that correspond to these mutually related systems as the diagrams of their matrices are called mamillar graphs.[57] They possess shared peripheral nodes and nonshared central nodes. In the case of active relatedness, the nodes in the mamillar graphs have a centrifugal orientation; in the second case of passive relatedness, they have a centripetal orientation. This fact is illustrated in Fig. 10.

When (in the first or second case) the set of central nodes is a genuine subset of the set of peripheral nodes, a special case of combined catalytic relatedness applies: Fig. 10(e). When the distribution of the numerical values characterizing the edges of the graph is taken into explicit account, subforms of catalytic relatedness can be defined. In the subcase of Fig. 10(d), a subgroup with a strong catalytic prevalence is shown.

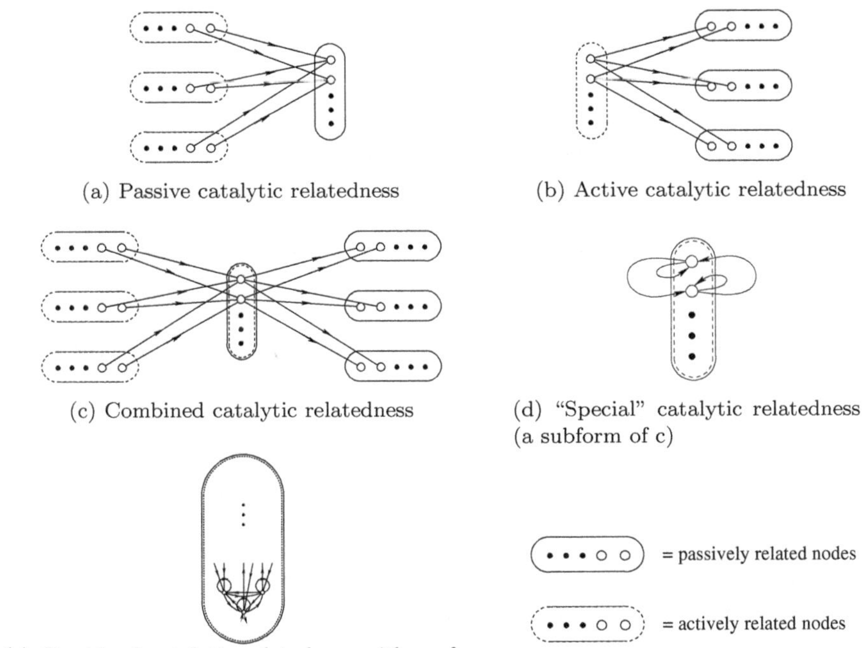

(a) Passive catalytic relatedness

(b) Active catalytic relatedness

(c) Combined catalytic relatedness

(d) "Special" catalytic relatedness (a subform of c)

(e) Combined catalytic relatedness with preferential autocatalysis (a subform of d)

Fig. 10. Catalytic relatedness reflects itself in isomorphic partial graphs in the network of preformed relations.

To focus on large or very large groups of relatedness (in the form of ordered substructures within the system matrix) in accordance with the constraint of Eq. (12), can also be rewarding. This is because the relative frequency of partial graphs of different kinds, in the graph of preformed relations, is unaffected thereby. For the constraint of equidistribution of the matrix elements (Eq. (12)) necessarily remains valid in this case for a certain (larger or smaller) percentage of the matrix elements. Such matrices can be achieved by superimposing an equally large equidistributed set of matrices possessing a much (or very much) larger number of zero elements in accord with the constraint of Eq. (10) above

In a matrix built according to the assumptions just introduced, a principle valid for purely equidistributed matrices is therefore suspended: That the relative frequency of isomorphic partial graphs decreases strongly and nonlinearly with their size (as well as their absolute frequency). Suspension of this principle means that between groups of system components, similar mutual relations become possible as hold true for individual components. That is, within groups, there now occur systematic relations, too. Figure 10(e) shows an example of this type.

The newly introduced modification of Eqs. (9) and (10) leaves the previous results unchanged, but allows a new answer to question 4:

4. Continued: In **epoch 3**, eventually single-element cycles are prevalent (namely, loops). Relative to the previous, superseded, multi-element loops, they stand in the new relation depicted in Fig. 10(e). (Compare in this context also the lowermost lines of Figs. 8 and 9 which had not been discussed before.) The length of sequences of mutually related loops succeeding each other will eventually reach very large numbers. This is **epoch 4**. Within the individual groups, a selection (toward visibility) of the most fertile partial systems occurs as before. That is, the most growth-intensive loops are the ones that eventually become prevalent.

With this last result, the hypothetical model of biogenesis proposed by Darwin in 1851[25] has been re-derived, it appears. It follows that — in contrast to what Darwin had conjectured — autocatalytic systems do not stand at the beginning but are themselves the product of a long evolution already. They mark a stage of development that is already rather close to the biological system. Thus they are analogous to the biological system itself. Nevertheless they already occur, as we saw, in the modelling of a very small subset of chemical functional capabilities. The above derived evolved subsystems lack, in particular, the property of "genome-phenome differentiation" that is so characteristic of the biological system.[20] Moreover, they also appear incapable, under the assumptions made above, of secondarily acquiring this property.

7. Conclusions

The statistical results described above were obtained with the aid of a graph-theoretical procedure applied to the qualitative behavior of very large linear and nonlinear dynamical systems. They represent a methodological contribution to the analysis of systems characterized by a combinatorially high complexity.[58] The price to pay is that one has to confine oneself to a question that is "biologically biased." Compared to the well-known treatment of very large nonlinear ecological systems of the "Volterra type",[59,60] the present approach addresses a much earlier phase of biological development. Recent insights into the logical structure of linear automata[61] and linear systems[44] still leave room for elementary "groping-type" considerations like those presented above. The evolutionary behavior of less idealized systems can be expected to lend themselves to the same treatment. Model investigations of the present type follow in the footsteps of investigations done by Rosen[62] and Rashevksy.[63] The procedure introduced can be hoped to contribute to the goal of an "invariant theoretical biology" in a unified methodology in the footsteps of Bernal[64] and Waddington.[24]

A scientific problem which has so far found little attention has been focused on: The existence of a class of systems that are comparable to a "pot filled with fuses." This class of systems is amenable to a statistical treatment. The system class was introduced in terms of its most idealized representative — nearly linear, deterministic, continuous system. The least natural axiom, used above, was that of unlimited growth of all state variables. When it is dropped, a second, only slightly more complicated system class is obtained which bears a formal resemblance to Ashby's homeostat[50] and includes the model of Cairns-Smith.[23]

Acknowledgments

For helpful discussions I would like to thank Roland Wais, Carl-Friedrich von Weizsäcker, Horst Mittelstaedt, Gretl Lorenz, Konrad Lorenz, Carl-Adam Petri and Friedrich Franz Seelig.

References

1. I. Prigogine and J. M. Wiame, Biologie et thermodynamique des phénomènes irréversibles, *Experientia (Basel).* **2**, 451–453 (1946).
2. L. von Bertalanffy, *Theoretische Biologie*, 1st edn., Vol. 2. Gebrüder Bornträger, Berlin (1932).
3. R. Haase, Der zweite Hauptsatz in der Biologie, *Zeitschrift für Elektrochemie.* **55**, 566–569 (1951).
4. J. S. Kirkaldy, Thermodynamics of terrestrial evolution, *Biophysical Journal.* **5**, 965–979 (1965).
5. L. Onsager, Reciprocal relations in irreversible processes I, *Physical Review.* **37**, 405–426 (1931).
6. L. Onsager, Reciprocal relations in irreversible processes II, *Physical Review.* **38**, 2265–2279 (1931).
7. H. B. Casimir, On Onsager's principle of microscopic reversibility, *Review of Modern Physics.* **17**(2–3), 343–350 (1945).
8. I. Prigogine and R. Balescu, Sur les propriétés différentielles de la production d'entropie II, *Bulletin de l'Académie Royale de Belgique.* **41**, 917–928 (1955).

9. M. Sugita, Kinetic theory and mathematical model of cell metabolism, *Helgoländer wissenschaftliche Meeresuntersuchungen.* **14**, 78–109 (1966). *2nd International Symposium on Quantum Biology and Metabolism.*
10. J. R. Mayer, Die Torricellische Leere und über Auslösung. In J. Weyrauch (ed.), *Die Mechanik der Wärme in gesammelten Schriften von Robert Mayer.* Cotta, Stuttgart (1893), pp. 440–447.
11. A. G. J. MacFarlane, *Engineering Systems Analysis.* Harrap and Co, London (1964).
12. H. Kacser, Some physico-chemical aspects of biological organization. In C. H. Waddington (ed.), *The Strategy of Genes.* Allen and Unwin, London (1957), pp. 191–249.
13. H. Kacser, Kinetic models of development and heredity, *Symposium of the Society for Experimental Biology.* **14**, 13–27 (1960).
14. M. Sugita, Functional analysis of chemical systems *in vivo* using a logical circuit equivalent, *Journal of Theoretical Biology.* **1**, 415–430 (1961).
15. M. Hale, Essays on population. In G. Chalmers (ed.), *An Estimate of the Comparative Strength of Britain During the Present and Four Preceding Reigns; and of the Losses of Her Trade From Every War Since the Revolution.* C. Dilly & J. Bowen, London (1782).
16. T. R. Malthus, *An Essay on the Principle of Population as It Affects the Future Improvement of Society* (1890). 1st edn., 1798.
17. L. von Bertalanffy, *Theoretische Biologie*, 2nd edn., Vol. 2. Francke, Bern (1951).
18. W. R. Ashby, *An Introduction to Cybernetics.* Chapman and Hall, London (1956).
19. G. Schramm, H. Grötsch and W. Pollmann, Non-enzymatic synthesis of polysaccharides, nucleosides and nucleic acids and the origin of self-reproducing systems, *Angewandte Chemie International Edition in English.* **1**(1), 1–7 (1962). doi: 10.1002/anie.196200011.
20. R. W. Kaplan, Molekularbiologische Probleme der Lebensentstehung, *Naturwissenschaften.* **55**, 97–104 (1968).
21. C. H. Waddington, A discussion of some aspects of theoretical biology, in *The Strategy of the Genes.* Allen and Unwin, London (1957), p. 32, 65.
22. H. H. Pattee, The recognition of hereditary order in primitive chemical systems and their molecular matrices. In S. W. Fox (ed.), *Proceedings of the Conference Wakulla Springs Florida 1963* (1965), pp. 385–402.
23. A. G. Cairns-Smith, An approach to a blueprint of a primitive organism. In C. Waddington (ed.), *Towards a Theoretical Biology*, Vol. 1. Aldine and Edinburgh University Press, Chicago (1968), pp. 57–66.
24. C. H. Waddington, The basic ideas of biology. In C. H. Waddington (ed.), *Towards a Theoretical Biology: An IUBS Symposium*, Vol. 1. Aldine and Edinburgh University Press, Chicago (1968), pp. 1–31.
25. C. Darwin, *The Lifetime and Letters of Charles Darwin*, London (1887), p. 202. Including an autobiographical chapter.

26. J. B. S. Haldane, *The Origin of Life* — Rationalist annual reprint 1928. In J. B. S. Haldane (eds.), *The Inequality of Man*. Penguin Books, London (1937).
27. A. I. Oparin, *The Origin of Life*. Macmillan, New York (1938).
28. A. Mittasch, *Berzelius und die Katalyse*. Akademische Verlagsgesellschaft, Leipzig (1935). Quotation of J. J. Berzelius, 1836.
29. B. Chance, The kinetics of the enzyme-subrate compound of peroxidase, *Journal of Biological Chemistry*. **151**(2), 553–577 (1943). doi: 10.1016/S0021-9258(18)44929-0.
30. B. Chance, Analogue and digital representations of enzyme kinetics, *Journal of Biological Chemistry*. **235**, 2440–2443 (1960). doi: 10.1016/S0021-9258(18)64639-3.
31. F. Heinmets, Elucidation of induction and repression mechanisms in enzyme synthesis by analysis of model systems with the analog computer. In B. Pullman (ed.), *Electronic Aspects of Biochemistry*. Academic Press, New York (1964), pp. 415–479.
32. D. Garfinkel, Computer simulation in biochemistry and ecology. In T. Waterman and H. Morowitz (eds.), *Theoretical and Mathematical Biology*. Blaisdell (Ginn), New York (1965), pp. 292–310.
33. B. Hess, Zustände und Kontrollmechanismen metabolischer Systeme, *Studia Biophysica*. **1**, 41–63 (1966).
34. H. Preuss, *Grundriss der Quantenchemie*. Bibliogtaphisches Institut, Mannheim (1962).
35. K. S. Trintscher, *Biology and information: A discussion of problems of biological thermodynamics* (in German) (1966). Original in Russian, Moscow, 1964.
36. G. Pask, A discussion of the cybernetics of learning behaviour, *Progress in Brain Research*. **2**, 177–214 (1963). doi: 10.1016/S0079-6123(08)62069-5.
37. H. Mittelstaedt, Die Regelungstheorie als methodisches Werkzeug der Verhaltensanalyse, *Naturwissenschaften*. **48**, 246–254 (1961). doi: 10.1007/BF00597508.
38. M. Berman and R. Schoenfeld, Invariants in experimental data on linear kinetics and the formulation of models, *Journal of Applied Physics* **27**(11), 1361–1370 (1956). doi: 10.1063/1.1722264.
39. J. Z. Hearon, The kinetics of linear systems with special reference to periodic reactions, *Bulletin in Mathematical Biophysics*. **15**, 121–141 (1953). doi: 10.1007/BF02476377.
40. H. von Foerster, Some aspects in the design of biological computers. In *Proceedings of the 2nd Internation Congress on Cybernetics 1958*. Association Internationale de Cybernétique, Namur, Belgium (1960), pp. 240–255.
41. V. V. Nemytskii, Topological problems of the theory of dynamical systems, *American Mathematical Society Translations*. **1**(5), 414–497 (1962). Original in Russian, 1949.
42. L. Michaelis and M. L. Menten, Die Kinetik der Invertinwirkung, *Biochemische Zeitschrift*. **49**, 333–369 (1913).

43. H.-J. Flechtner, *Basic Notions of Cybernetics: An Introduction* (in German). Wissenschaftliche Verlagsgesellschaft, Stuttgart (1966).
44. R. E. Kalman, New developments in system theory relevant to biology. In M. D. Mesarovic (ed.), *Proceedings of the 3rd Systems Symposium on Case Institute of Technology 1966*. Springer Verlag (1968), pp. 222–232.
45. C. Berge, *The Theory of Graphs and Its Applications*. Wiley (1962). Original in French, Paris, 1958.
46. O. Ore, *Theory of Graphs*, Vol. 38. American Mathematical Society Colloquium Publications (1962).
47. J. Sedlacek, *Einführung in die Graphentheorie*. Harri Deutsch Verlag, Frankfurt (1968). Original in Czech, Prague Academia, 1964.
48. S. J. Mason, Feedback theory — Some properties of signal flow graphs, *Proceedings IRE.* **41**, 1144–1156 (1953).
49. A. M. Lyapunov, Théorie générale de la stabilité du mouvement, *Annals of Mathematical Sudies.* **17** (1949). Original in Russian, Charkov, 1892.
50. W. R. Ashby, *Design for a Brain.* Chapman and Hall, London (1952).
51. A. Gill, State graphs of autonomous linear automata. In E. Caianiello (ed.), *Automata Theory.* Academic Press, New York (1966), pp. 176–180.
52. L. A. Zadeh, The concept of state in system theory. In M. Mesarovic (ed.), *View on General Systems Theory.* Wiley (1964), pp. 39–50.
53. F. Harary, R. Z. Norman and D. Cartwright, *Structural Models — An Introduction to the Theory of Directed Graphs.* Wiley (1965).
54. A. Gill, The reduced form of a linear automaton. In E. Caianiello (ed.), *Automata Theory.* Academic Press, New York (1966), pp. 164–175.
55. W. Kaplan, *Ordinary Differential Equations.* Addison-Wesley, Sao Paulo (1958).
56. R. F. Drennick, System causality and signal predictability. In M. D. Mesarovic (ed.), *Views on General Systems Theory.* Wiley (1964), pp. 157–164.
57. C. W. Sheppard and A. S. Householder, The mathematical basis of the interpretation of tracer experiments in closed steady-state systems, *Journal of Applied Physics.* **22**(4), 510–520 (1951). doi: 10.1063/1.1699992.
58. W. R. Ashby, Views on general systems theory. In M. D. Mesarovic (ed.), *Views on General Systems Theory.* Wiley (1964), pp. 165–169.
59. E. H. Kerner, A statistical mechanics of interacting biological species, *Bulletin of Mathematical Biophysics.* **19**, 121–146 (1957). doi: 10.1007/BF02477883.
60. E. H. Kerner, Gibbs ensemble and biological ensemble, *Annals of New York Academy of Sciences.* **96**, 975–984 (1962). doi: 10.1111/j.1749-6632.1962.tb54115.x.
61. C. A. Petri, Kommunikation mit Automaten. Ph.D. Thesis, Fakultät für Mathematik und Physik der Technischen Hochschule Darmstadt, Schriften des Rheinisch-Westfälischen Institutes für Instrumentelle Mathematik an der Universität Bonn (1962).

62. R. Rosen, A relational theory of biological systems, *Bulletin of Mathematical Biophysics*. **20**, 245–260 (1958).
63. N. Rashevsky, A note on the nature and origin of life, *Bulletin of Mathematical Biophysics*. **21**, 185–193 (1959).
64. J. D. Bernal, Molecular matrices for living systems. In S. Fox (ed.), *The Origin of Prebiological Systems and Their Molecular Matrices, Proceedings of the Conference Wakulla Springs, Florida 1963*. Academic Press (1965), pp. 65–88.

https://doi.org/10.1142/9789819823130_0009

Chapter 9

CHAOS PEOPLE

Franco Vivaldi

University of London, UK
f.vivaldi@qmul.ac.uk

During the early developments of the theory of chaos, I met some remarkable people who shaped my research life and beyond: Joseph Ford, Boris Chirikov, Jeffrey Tennyson. I present some sketches of my experience with them.

1. Joseph Ford

It was the end of May 1977, and Joseph Ford (1927–1995) was waiting for me at Atlanta's airport. He held in his hand a large plastic thermo-mug, filled with Coca-Cola and ice, with his name written on it in large letters; he wore dockside shoes and chino trousers, with his shirt sleeves rolled-up. This was his signature outfit, which he rarely changed, and this is how I remember him.

Joe was welcoming and approachable, and spoke with a charming Southern accent (being my first days in an English-speaking country, I wasn't able to identify it just yet). A pioneer in nonlinear dynamics, he belonged to the generation of physicists that confronted the seemingly paradoxical outcome of the 1955 Fermi–Pasta–Ulam numerical experiment,[1] which contradicted long-held assumptions in statistical mechanics. These three scientists used a computer to investigate the question of equipartition of energy for weakly coupled oscillator systems using Hamiltonians of the form

$$\mathcal{H} = \frac{1}{2}\sum_{k=1}^{N}\left(p_k^2 + \omega_k^2 q_k^2\right) + \alpha \sum_{i,k,l=1}^{N} A_{ikl}\, q_i q_k q_l, \tag{1}$$

where the nonlinear coupling parameter α is chosen sufficiently small that the coupling energy never exceeds some small fraction of the total energy. Joe's early work provided some explanations of the Fermi-Past-Ulam problem[2,3] and this problem remained dear to him, eventually leading to a synthesis of two decades of Hamiltonian chaos written in terms of FPU.[4] With his personal flair, charisma, and limitless energy, Joe had developed a rich network of connections — and indeed of lasting friendships — within a research community that was growing larger and very interdisciplinary. Joseph Ford did physics with the typewriter (Fig. 1), the telephone, and the airplane.

I encountered chaos in the context of statistical mechanics, two years before arriving in Atlanta. For my *Tesi di Laurea* at the University of Milan, supervised by Giulio Casati, I sought to reproduce a normal thermal conductivity (Fourier's law) in models of one-dimensional lattices placed between two thermal reservoirs at different temperatures. It was believed that a combination of instability of motion and scattering by isotopic impurities should cause any coherent (soliton-like) pulse generated by the interaction with the reservoirs to decay, bringing about thermalisation. But my work on the unequal masses hard-point gas — programmed on punch cards for a computer able to simulate just a few hundred collisions per second — was not conclusive. Giulio told me that Joe Ford had funds left

Fig. 1. Joe Ford's Royal typewriter, with which he produced the Nonlinear Sciences Abstracts. He gave it to me when he changed typewriter, and I still have it.

over from a grant, to support a few months' visit to Georgia Tech: I ended up staying seven years. In Atlanta, after many failed attempts, we did observe Fourier's law in a rather special model,[5,6] although it later emerged that in one dimension this property is not typical.[7]

By the late '70s, an uninterrupted flow of preprints on nonlinear systems was reaching Joe's mailbox, and I was in the fortunate position of being up-to-date with the latest developments without having to make an effort (there was no `arXiv` back then). The *Nonlinear Science Abstracts* initiative was born at this point. With commendable dedication, Joe collected, organized, and typed up the abstracts of all those preprints; he then photocopied them, stapled them into thick bundles, and distributed them to an ever-expanding mailing list (Fig. 2). This tiresome process took place every few months, and he never asked for help. Passing by his office, I could hear his Royal typewriter ticking away, and I would often find him during week-ends labouring at the photocopying machine of the School of Physics. This initiative cemented Joe's position as catalyst and focal point of the community of chaos people.

Being around Joe meant meeting very many people. I recall the gentlemanly Yakov Sinaï, for whom I spent a busy day computing on the Lorenz attractor; the personable and mildly eccentric John Greene, with whom I was to collaborate on the period-doubling universal route to chaos[8] in the de Vogelaere map[9]

$$\begin{cases} x_{n+1} = -y_n + p\,x_n - (1-p)\,x_n^2 \\ y_{n+1} = x_n - p\,x_{n+1} + (1-p)\,x_{n+1}^2; \end{cases} \tag{2}$$

the austere Jürgen Moser, who suggested to me the outer billiard of a polygon problem[10,11] (see Fig. 3). And then Michael Barnsley, Michael Berry, Boris Chirikov, Doyne Farmer, Mitch Feigenbaum, Robert MacKay, Benoît Mandelbrot, ...

The founding of *Physica D*, the very first journal devoted to nonlinear phenomena, was the natural next step in Joe's indefatigable effort to support the science of chaos. The first issue appeared in April 1980. A few years later, to honour his 60th birthday, *Physica D* published a special volume. The impressive list of contributors included most big names of Joe's generation: V. Arnol'd, B. Chirikov, J. Greene, M. Hénon, I. Percival, Ya Sinai@Sinaï, G. Zaslavsky. The following year, at the associated birthday conference in San Francisco, Joe was presented with

Bifurcation: analyse.

R. Lozi

NONLINEAR SCIENCE ABSTRACTS

September, 1977

Prepared by: Joseph Ford, School of Physics, Georgia Institute of Technology, Atlanta, Georgia 30332, USA

1. NEW RESULTS IN STOCHASTIC EQUATIONS- THE NONLINEAR CASE, G. Adomian, Center for Applied Mathematics, Tucker Hall, University of Georgia, Athens, Georgia 30602, USA

A method of iteration for determination of the first and second order solution statistics for stochastic operator equations, using the concept of stochastic Green's functions, has recently been extended to nonlinear stochastic differential equations of the form $\mathcal{L}y + \eta(y, \dot{y}, \ldots) = x(t)$, where $x(t)$ is a stochastic process, $\mathcal{L}$ is an nth order linear stochastic operator, η is a nonlinear term involving stochastic process coefficients, and initial conditions may be random as well. A convergent series form of the solution is obtained, A new symmetrized and easily calculable general term is possible and such work is being extended to various classes of nonlinearity. General expressions can be obtained for stochastic Green's functions which determine directly the desired statistics. Related questions and special cases are investigated.

JOURNAL: Proceedings of the Symposium on Nonlinear Analysis and Applications, Ed. by Lakshmikantham, Academic Press, 1977, to appear.

2. SYMMETRIZED SOLUTIONS FOR NONLINEAR STOCHASTIC DIFFERENTIAL EQUATIONS, G. Adomian and L. H. Sibul, See 1. for address

Solution of nonlinear stochastic differential equations in series form can be put into convenient symmetrized forms which are easily calculable. This paper investigates such forms for quadratic nonlinearities. Other classes of nonlinearity are being investigated

JOURNAL: J. Nonlinear Analyses and Applications, to appear

3. STOCHASTIC DIFFERENTIAL OPERATOR EQUATIONS WITH RANDOM INITIAL CONDITIONS, G. Adomian and T. E. Lynch, See 1. for address

The iterative solution of the stochastic differential equation is considered for the case of accompanying random, rather than zero, initial conditions. The mean solution and the correlation are determined

JOURNAL: J. Math. Anal. and Appl. , to appear.

4. NONLINEAR STOCHASTIC DIFFERENTIAL EQUATIONS, G. Adomian, See 1. for address

A convergent series is found by an iterative procedure using the concept of stochastic Green's functions to determine first and second order statistics of the solution process for the stochastic nonlinear differential equation $\mathcal{F}[y(t, \omega)] = x(t, \omega)$ with $x(t)$ a stochastic process, and $\mathcal{F}$ a nonlinear differential operator involving stochastic process coefficients.

JOURNAL: J. Math. Anal. and Appl. 55, 441 (1976).

Fig. 2. First page of the *Nonlinear Science Abstracts* collected and typed by Joe Ford. Here the first bundle sent on September 1977 and received by René Lozi.

an elegant leather-bound copy of this volume. Fittingly, greetings from colleagues were placed in a large Coca-Cola bottle.[a]

Joe wrote very well and cared deeply about it, and it wasn't unusual for him to offer assistance to friends and colleagues — foreigners, particularly — with their writing. In 1978, Boris Chirikov had reached the final stages of preparation of his famous *Physics Reports* review article.[12]

[a]Letter from Hermann Flaschka, February 11, 1989.

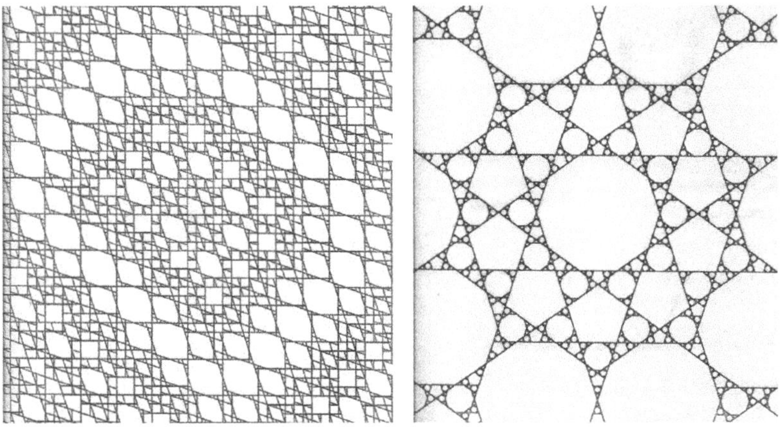

Fig. 3. My first computer plots (1983) of the discontinuity set of the outer billiard of a rational quadrilateral (left) and the regular decagon (right), which provided evidence, respectively, of local finiteness and renormalisability of the tiling.

With characteristic unselfishness, Joe embarked in a thorough revision of the intimidating 120-page manuscript, which became at once a must-read for anyone working in Hamiltonian systems.[b] When I first asked him for comments on my own writing, he sprinkled my manuscript with red marks, while reassuring me that my English was much better than that of other Italians he knew. Many years later, I wrote a book on *Mathematical Writing*,[13] and I have no doubt that it all got started from those red marks.

In the popular article *How random is a coin toss?*,[14] Joe exposed to a broad audience deeper implications of the concept of chaos, and the role played by algorithmic complexity theory in defining and assessing the fundamental limits of our ability to deal with chaotic systems. Joe's stubborn focus on complexity enriched my perspective, and later guided me in the study of dynamical systems with discrete phase space, where the meaning of chaos must be reconsidered from scratch.[15,16]

The theme of complexity was particularly appealing for seminars and colloquia, and he was in high demand — too much by his own admission. This is hardly surprising. He wasn't just a high-profile figure: he also gave memorable talks, peppered with wit and humour, which on occasion would draw a spontaneous applause from the audience.

[b]According to Google Scholar, this article has attracted 4800 citations.

On a human level, you could not fail to admire Joe for his generosity and honesty. But he was also that entertaining guy you'd love to invite to a dinner party.

2. Boris Chirikov and the Cray I

Boris Chirikov (1928–2008) arrived in Atlanta in the summer of 1978, his first and only trip to the US, hosted by Joseph Ford. He was already a legend — his paper on chaos in magnetic traps dated twenty years earlier[17] — but very few people had met him in person, mainly due to travel restrictions in the USSR. In a letter to him some years later, Joe remarked jokingly: "*You have become a legend in your own time by simply not being around!*"[c]

Boris was friendly and ready to dispense a smile. We shared an office, and before long I saw him reading a letter from his beloved wife Olya — a cheerful operetta singer, whom I would meet in Siberia a couple of years later (Fig. 4). Tears came to his eyes, and I concluded that the great scientist was also a really nice fellow.

In conversations with him, you would catch a glimpse of his vast knowledge. When he drew inferences from scarce information — a necessity in physics — he combined a deep physical intuition with an eye for the boundaries of rigour. Thus in his writings, he employed the symbol '$\approx$' as often as '$=$', and he knew exactly what that inexact symbol meant. (A fine example of his style is his masterful model of Halley's comet, extracted from just 46 data-points.[18])

Boris came to America to use the Cray I supercomputers, then the fastest on Earth and immeasurably faster than anything available in Siberia at the time. For years, he had managed to squeeze the impossible out of the spartan facilities of the Institute of Nuclear Physics in Akademgorodok. He told me that sometimes they wrote the inner core of programs in assembler, to improve speed. Some of Boris' clever tricks — such as considering Poincaré recurrence rather than correlations decay in the study of the chaos border[19] — were born out of computational necessity, but ended up affecting theoretical developments. (Shortly afterwards Charles Karney used this technique to carry out an impressive numerical experiment on the chaos border, after securing some 90 h of Cray time.[20])

[c]D. L. Shepelyansky, Boris Chirikov — Sputnik of Chaos, online resources (2008), https://www.quantware.ups-tlse.fr/chirikov/.

Fig. 4. Boris Chirikov and his wife Olya, gardening at their dacia near Akademgorodok in 1982.

By that time, there were already several Cray I at various US research establishments. Joe Ford had planned our Cray-hunting trip across America, and we began by heading West to the Oak Ridge National Laboratory, in Tennessee. From there you could access the Cray I of the Los Alamos National Laboratory, the first such machine on US soil. It was the middle of the Cold War, and it was quite unusual to have a prominent Soviet scientist using state-of-the-art computer facilities at a US research institution. The success of this visit is indeed a tribute to Joe's diplomatic skills.

We arrived at Oak Ridge with a Fortran program that performed long-time evolution of a driven two-dimensional symplectic map. A fast computer was essential to measure reliably the so-called Arnol'd diffusion, which required computing large ensembles of long trajectories. A local expert briefed us on how to approach the machine's novel vector architecture — a form of parallel processing. He showed us graphs and statistics, detailing the machine performance; he then stated the maximum speed attainable in optimal situations, when the so-called full vectorization of a code was implemented. This was some 80 million floating-point operations per second, 80Mflops, in jargon. (To put this in perspective, consider that Apple's iPhone 11 is over one thousand times faster.)

I sat at a terminal, Boris behind me, looking at the screen over my shoulders. Optimizing code had never been a pressing concern of mine. Boris suggested several changes: divisions turned into multiplications, scalar variables became arrays, statistical tools were fine-tuned. Then there were two independent nested loops — the core of the program — which processed two-dimensional arrays. Boris suggested that I swap the order of the loops. Puzzled, I pointed out that mathematically it made no difference, but he explained that, given how array elements are accessed, reversing the order would speed up the execution considerably.

In an hour or so we were ready for a trial run on the Cray I. At the first go, the program ran faster that the declared theoretical limit. The Cray experts were stunned: they downloaded and dissected the compiled code, trying to figure out how on Earth this could have happened. This was vintage Boris Chirikov. Needless to say, it changed my approach to computer codes for good.

The next Cray was at the National Center for Atmospheric Research (NCAR) near Boulder, Colorado, which had been installed two years earlier. NCAR is a beautiful site on the edge of the Rockies, high up, with a commanding view of the great plains. To house the computer, they had to dig a subterranean cave, due to environmental restrictions. You descended with an elevator, and then you saw the awesome machine: an imposing horseshoe-shaped tower, placed at the centre of a large room. The setting was skilfully choreographed: the Cray, shielded behind glass, bathed in suffuse blue light. It felt like the appearance of the monolith in Kubrick's 2001 Space Odyssey, and we resumed our computations with renewed enthusiasm.

Our final destination was Berkeley University. This wasn't merely a Cray stop, as there were many chaos people there: Allan Kaufman, Allan Lichtenberg, Michael Lieberman, and then the Brasilian postdoc Celso Grebogi and the Ph.D. students Steven McDonald, James Meiss, and Jeffrey Tennyson (1950–1992).

Berkeley was a cool place — the buzz on Telegraph Avenue was electrifying — and Jeff, friendly, laid back, and wearing a ponytail, fitted my idea of a cool Californian, although I later discovered that he was born in Vienna and grew up in Washington DC. He was working on theoretical aspects of Arnol'd diffusion,[21] and he was able to make this difficult subject accessible, thanks to his geometrical intuition, carefully crafted mathematics, and strong aesthetic sense. It was Jeff who handled the hurdle of obtaining computer time on the Cray at the Lawrence Livermore Laboratory. In his unique low-rev style, he steered us successfully through the Department of Energy's administrative maze.

Joe Ford had already visited Boris Chirikov in Siberia; for the rest of us who had not, the Siberian connection was now established.

3. Siberia

Akademgorodok is a peaceful town scattered in a forest of birch trees and pines; it is not far from Novosibirsk, near a vast artificial lake fed by the great river Ob. It was home to a large number of research institutes of the Siberian Division of the USSR Academy of Sciences. To Western eyes, the Soviet architecture gave it a distinctive retro appearance, and the scarcity of conventional amenities meant that the pace of life too seemed of yesteryears. It was a closely knit community of scientists, several of whom had arrived twenty years earlier with their young families, when the town was built. They spoke fondly of those early years. If you walked late at night, you could see many windows still lit up, as scientists often worked at odd hours.

Siberia was on the map because of Boris Chirikov, who worked at the Budker Institute of Nuclear Physics. This was a large building complex, prominently placed at the intersection of two oversized avenues, of the kind favoured by the Iron Curtain's urban planners. The institute's rich facilities included a particle accelerator and a plasma laboratory, which meant that research on nonlinear stability problems was relevant. It was a well-funded establishment, thanks to the skills of its founder and long-time director Gersh Budker — a nuclear physicist known for a method of damping particle oscillations in a synchrotron[22] — who made it prosper for a quarter of a century. Two years before my first visit, in 1975, Budker suddenly passed away, and the institute was renamed after him.

Boris kept a photo of Budker in his office, under the glass plate that covered his desk: a patriarchal figure, with a long beard (Fig. 5). Boris had great admiration for him. He told me that to generate extra income for the institute, he had sold to Japan synchrotron radiation sources for medical applications. He also said that one day a week Budker switched off the

Fig. 5. Gersh Budker, founder and long-time director of the Institute of Nuclear Physics in Akedemgorodoc, now named after him.

telephone in his office to do research, because he believed that you cannot direct the research of others if you don't do your own.[d]

The people I met in Siberia were welcoming and warm, and curious about foreigners, which resulted in an intense social life and a lot of food. Much of the action took place inside homes. The combination of intellectual sophistication and a no-frills existence resulted in a quality of human interactions that I had not experienced in previous work-related travel. Then, there was the added bonus of receiving lot of attention, and having privileged access to the store where visitors paid with foreign currency — you could buy vodka for friends.

I soon met Felix Izrailev and Dima Shepelyansky, Boris' local collaborators. The three of them wrote so many papers together, that I came to regard them as the Siberian version of the Grebogi, Ott & Yorke trio in Maryland.

My trips to Siberia were funded by a science exchange programme between Italy and the USSR. Applications for visits to Siberia were viewed

[d]I wish this were true in British Universities.

Fig. 6. Boris Chirikov and I in Moscow, in 1981. Lenin's billboard image was a display for the anniversary of the October revolution.

favourably, and some Italians — Giulio Casati in particular — were regulars there. Eventually, I stumbled into Jeff Tennyson, who had just begun his decade-long meandering as a chaos nomad. During another visit, Michael Lieberman turned up; he was about to finish his book with Allan Lichtenberg, which would become a classic graduate text in Hamiltonian dynamics.[23]

Boris and collaborators were busy on many fronts, in both classical and quantum chaos. The latter was in its infancy (the first conference took place in 1977 in Como), and the early efforts were on the quantum version of the standard map (nicknamed in Italy *il pendolo siberiano* — the Siberian pendulum),[24] and later on the ionisation of Rydberg atoms.[25] Research on classical stability problems, where Jeff and I worked, was connected to accelerator physics: the stability of electron-positron colliding beams,[26] and the weak instability induced by parametric modulation.[27] My acquaintance with Anna Shaidenko, from the Sobolev Institute of

Mathematics, eventually led to a collaboration on the outer billiard problem.[11]

After Siberia, Jeff moved to Texas, working with Dominique Escande and John Cary on the difficult problem of separatrix crossing.[28] Then, after a short visit to Japan, he returned to Siberia for a whole year (1984–85), arriving there from Vladivostok on the mythical trans-Siberian railway. He carried with him a Macintosh, which had just appeared on the market, and his must have been the first Mac to reach Siberian soil. A few months later, while he was away on travel, a mysterious intruder broke into his hotel room and erased selected files from the Mac. Jeff took it phlegmatically, and that became a KGB story to share with friends, speculating on the circumstances of the crime.

Jeff had a fiercely independent mind, and a passion for lengthy discussions of foundational issues, particularly of statistical mechanics. After Siberia, he spent a year at Queen Mary in London, invited by Ian Percival and myself, where he had the time to develop ideas very dear to him, on the flow of information and entropy between microscopic and macroscopic systems.

He could have obtained stable employment in an accelerator laboratory, but he was reluctant to settle down. With a healthy immunity from conventional career worries, he continued his nomadic life, alternating short stays at his home in Berkeley with visits to accelerator laboratories in the US and Europe, where he earned his living, and summer visits to the Aspen Center for Physics in Colorado, where he contributed to the running of the Dynamics workshops, even repairing the center's bicycles.

In 1990, he finally surrendered to a more stable employment at the Stanford Linear Accelerator Center and got married. But stability was not in Jeff's destiny: shortly afterwards he was diagnosed with leukaemia, to which he succumbed within a year. I much admired his courage and serene acceptance during his last, difficult journey. *Physica D* published a special issue to honour his memory.[29]

Acknowledgments

Section 2 is an extended version of an essay I wrote for the web resource, which was subsequently published in;[30] I thank Dima Shepelyansky for supplying the relevant bibliography. I am grateful to Giulia Vivaldi for her careful reading of the manuscript.

References

1. E. Fermi, P. Pasta, S. Ulam, and M. Tsingou, Studies of the nonlinear problems. Technical Report LA-1940, Los Alamos National Laboratory, Los Alamos, NM (1955).
2. J. Ford, Equipartition of energy for nonlinear systems, *Journal of Mathematical Physics.* **2**(3), 387–393 (1961). doi: 10.1063/1.1703724.
3. J. Ford and J. Waters, Computer studies of energy sharing and ergodicity for nonlinear oscillator systems, *Journal of Mathematical Physics.* **4**(10), 1293–1306 (1963). doi: 10.1063/1.1703904.
4. J. Ford, The Fermi-Pasta-Ulam problem: Paradox turns discovery, *Physics Reports.* **213**(5), 271–310 (1992). doi: 10.1016/0370-1573(92)90116-H.
5. G. Casati, J. Ford, F. Vivaldi, and W. M. Visscher, One-dimensional classical many-body system having a normal thermal conductivity, *Physical Review Letters.* **52**, 1861–1864 (1984). doi: 10.1103/PhysRevLett.52.1861.
6. J. Maddox, Fourier's law obeyed — official, *Nature.* **309**, 511 (1984). doi: 10.1038/309511a0.
7. S. Lepri, R. Livi, and A. Politi, On the anomalous thermal conductivity of one-dimensional lattices, *Europhysics Letters.* **43**(3), 271–276 (1998). doi: 10.1209/epl/i1998-00352-3.
8. J. Greene, R. MacKay, F. Vivaldi, and M. Feigenbaum, Universal behaviour in families of area-preserving maps, *Physica D.* **3**(3), 468–486 (1981). doi: 10.1016/0167-2789(81)90034-8.
9. R. de Vogeleare, Study of a conservative algebraic transformation symmetric with respect to an involution. Technical Report AD 292-670, University of California, Berkeley, CA (1962).
10. J. Moser, Is the solar system stable? *The Mathematical Intelligencer.* **1**, 65–71 (1978). doi: 10.1007/BF03023062.
11. F. Vivaldi and A. V. Shaidenko, Global stability of a class of discontinuous dual billiards, *Communications in Mathematical Physics.* **110**, 625–640 (1987). doi: 10.1007/BF01205552.
12. B. V. Chirikov, A universal instability of many-dimensional oscillator systems, *Physics Reports.* **52**(5), 263–379 (1979). doi: 10.1016/0370-1573(79)90023-1.
13. F. Vivaldi, *Mathematical Writing.* Springer-Verlag, London (2014). doi: 10.1007/978-1-4471-6527-9.
14. J. Ford, How random is a coin toss? *Physics Today.* **36**(4), 40–47 (1983). doi: 10.1063/1.2915570.
15. D. Bosio and F. Vivaldi, Round-off errors and p-adic numbers, *Nonlinearity.* **13**(1), 309–322 (1999). doi: 10.1088/0951-7715/13/1/315.
16. J. A. G. Roberts and F. Vivaldi, Arithmetical method to detect integrability in maps, *Physical Review Letters.* **90**, 034102 (2003). doi: 10.1103/PhysRevLett.90.034102.
17. B. V. Chirikov, Resonance processes in magnetic traps, *Atomnaya Energiya.* **6**, 630 (1959) (in Russian).

18. B. V. Chirikov and V. V. Vecheslavov, Chaotic dynamics of comet Halley, *Astronomy & Astrophysics.* **221**(1), 146–154 (1989).
19. B. V. Chirikov and D. L. Shepelyansky, Statistics of Poincaré recurrences and the structure of the stochastic layer of a nonlinear resonance. Technical Report PPPL-TRANS–133, Institute of Nuclear Physics, Novosibirsk (1981). Translated from the Russian, PPPL-TRANS-133, Princeton Plasma Physics Laboratory, Princeton University, 1983.
20. C. F. F. Karney, Long-time correlations in the stochastic regime, *Physica D.* **8**(3), 360–380 (1983). doi: 10.1016/0167-2789(83)90232-4.
21. J. L. Tennyson, M. A. Lieberman, and A. J. Lichtenberg, Diffusion in near-integrable Hamiltonian systems with three degrees of freedom, *AIP Conference Proceedings.* **57**(1), 272–301 (1980). doi: 10.1063/1.32111.
22. G. I. Budker, An effective method of damping particle oscillations in proton and antiproton storage rings, *Atomnaya Énergiya.* **22**, 346–348 (1967) (in Russian).
23. A. Lichtenberg and M. Lieberman, *Regular and Chaotic Dynamics.* Applied Mathematical Sciences, Vol. 38. Springer-Verlag (1992). doi: 10.1007/978-1-4757-2184-3. First edition in 1983.
24. G. Casati, B. V. Chirikov, F. M. Izrailev, and J. Ford, Stochastic behavior of a quantum pendulum under a periodic perturbation, *Lecture Notes in Physics.* **93**, 334–352 (1979). doi: 10.1007/BFb0021757.
25. G. Casati, B. V. Chirikov, and D. L. Shepelyansky, Quantum limitations for chaotic excitation of the hydrogen atom in a monochromatic field, *Physical Review Letters.* **53**(27), 2525–2528 (1984). doi: 10.1103/PhysRevLett.53.2525.
26. A. Gerasimov, F. Izrailev, J. Tennyson, and A. Tyomnikh, The dynamics of the beam-beam interaction, *Lecture Notes in Physics.* **247**, 154–175 (1986). doi: 10.1007/BFb0107350.
27. B. V. Chirikov, M. A. Lieberman, D. L. Shepelyansky, and F. Vivaldi, A theory of modulational diffusion, *Physica D.* **14**, 289–304 (1985).
28. J. L. Tennyson, J. R. Cary, and D. F. Escande, Change of the adiabatic invariant due to separatrix crossing, *Physical Review Letters.* **56**, 2117–2120 (1986). doi: 10.1103/PhysRevLett.56.2117.
29. J. D. Meiss, Selected papers in dynamics. dedicated to the memory of Jeffrey L. Tennyson, *Physica D.* **71** (1994). Special issue.
30. O. V. Zhirov, D. L. Shepelyansky, K. V. Epstein, and A. N. Skrinsky (eds.), *Boris Valerianovich Chirikov — Legislator of Chaos.* Russian Academy of Sciences, Siberian Branch, Budker Institute of Nuclear Physics, Novosibirsk (2014). Partly in Russian. https://www.quantware.ups-tlse.fr/dima/myrefs/chirikovbook2014.pdf.

https://doi.org/10.1142/9789819823130_0010

Chapter 10

CHAOS AND ME

Yves Pomeau

Ecole Normale Supérieure, Paris, France
yves.pomeau@gmail.com

This presents a personal view of the way I got into chaos theory and nonlinear physics in general. The hope is that some readers may benefit of it in one way or another, even though every scientist and even everyone has a different history and so it is hard to draw conclusions useful to others from a personal history, that took place in a given time framework and in a given social, personal and intellectual background. After all, as Newton wrote this is because we are on the shoulder of giants that we can reach just a little higher.

1. Introduction

With Pierre Coullet we wrote a few years back a piece on "*The history of chaos in the French perspective*" which was published in the proceedings of a conference on Chaos held in Paris at Institut Henri Poincaré.[1] When Christophe Letellier asked me to write my part to tell how I got involved in Chaos theory, I thought first I had little to add to my previous contribution.[1] Thinking again to it I changed a bit my mind and believe (perhaps wrongly) now that I could add something of interest, particularly by linking more tightly the teaching years and what I did afterwards. This is perhaps of some interest for scientists (much) younger than me who can feel — wrongly I belive — that everything important is already known and that the only things a scientist can do now are third order improvements.

2. Training at ENS

I had a chance to study at École Normale (ENS below) in the sixties; with many distinguished professors, among them, Yves Rocard, about whom I will talk about soon. The French educational system, from what I know, is quite special because it is organized since a fairly long time. In Science, the topic that were taught at high school (lycée) followed more or less the same pattern as centuries before in Jesuit School. In Science, mathematics were (and still are) prominent with a heavy insistance on geometry, mostly plane geometry based on Euclid. From my own experience, this allows to figure out solution of problems in a non-mechanical way, that is, by combining into one's mind the various possibilities offered by looking at figure in different ways. I think this kind of global reasoning, instead of the step by step algebraic handling of formulae, uses a fundamental strength of human mind making it still superior to machines, computers and even artificial intelligence. The other thing I liked in my high school studies was Latin. I likely belong to one of the last generation where Latin was taught and studied seriously at French high schools. This gave the student access to a langage with a structure totally different of the modern langages, and guessing what is actually said in a text by Cicero or Tacitus is pretty hard, even though the meaning of every word is, most of the time, easy to find. This relationship between grammarand more generally the structure of langages, modern or ancient, and the ability for science of their speakers makes a very fascinating topic anyway, and not much studied as far I know.

As I said I was admitted at ENS in 1961, planning to do Physics but not knowing what kind of Physics. At that time, I had no clear idea of what subfield of Physics I could be working on. This was a time of major changes in France, that turned definitely from an agriculture based society to Industry and what is called "service sector". This change requested a population better educated than before, and so, a massive growth of high schools and Universities. For us, ENS students, this meant that the gates of University got wide open instead of jobs in high school, and that jobs in pure research became available as well. As a high school student and by family connections, I had already heard that Plasma physics was a domain full of promises and in which France did quite well. So, when I had to choose, at the end of the first year at ENS, which field I should begin in, I did choose Plasma physics and thus spent my second ENS-year following special classes in plasma physics with other students who became friends I still have after 60 years. This was my first exposure to modern physics, besides the lectures of Rocard, I will come to. This Plasma physics curriculum had many distinguished lecturers, among them Evry Schatzman who lectured on plasma instabilities, and the most senior of them all, Jacques Yvon, who lectured on statistical physics, an outstanding course I still regret it was never published. We had also courses on quantum mechanics, quite confusing and another one on atomic physics. At that time (1962–1963), there was no computers at all, at least for students, and the calculations had to be made either by hand or with a slide rule or a table of logarithms. Related to that, I remember a visit at an Observatory, where there was still people (mostly women), likely the last ones, specialized in carrying computer work by hand with special log-tables. The single time I ever heard the word "turbulence" was in a private discussion with Pr. Yvon who recommended me to read Batchelor's book on turbulence,[2] actually more a book on Kolmogorov theory than on turbulence in general.

As I alluded to, I was rather unhappy with the course on quantum mechanics, even though this was a fashionable topic for ENS-students in physics. I did not like at all, and stil do not, the presentation we had of this subject, that appeared, so to speak, to imply that quantum mechanics is more than a branch of physics but requires its own logic, different, say, from the Aristotelian logic. When writing this, I am, of course, fully aware, but was not at that time, about the attempt by Garrett Birkhoff and John von Neumann[3] to build a logic specific to quantum mechanics. To my knowledge, this did not lead to any new development in physics *per se*. As far I know too, Kolmogorov did never publish anything on statistical problems posed by quantum mechanics (see however D. A. Slavnov[4]). My understanding is that Everett multiworld idea[5] has settled the "philosophical" questions posed by quantum mechanics and its so-called "interpretation."

I remember discussing my insatisfaction on quantum mechanics as it was taught with a teaching assistant. He gave me one the best advice I ever had: why do not you get the textbooks by Lev Landau and Evgenii Lifschitz on classical mechanics[6] and the one on quantum mechanics?[7] At that time, there was a French translation of Landau–Lifschitz's book sold at bargain price in a communist bookshop near Sorbonne (which does no longer exist). I bought the two volumes and others afterwards. I still have them, in a very bad shape but I stil use them! This was exactly what I was looking for: everything is explained logically and follows from the principles laid down at the beginning in the most elegant way. I hope this monument of the human intellect will be there for a long time and inspire generations of students and scientists.

Reading Landau–Lifschitz'textbook on Classical Mechanics, I had my second encounter with nonlinear phenomena. Landau discusses at length the frequency locking of oscillators driven by an external periodic signal. This gave me a basic understanding of how nonlinearities may deeply change the oscillations of a given dynamical system. Actually, I had already a little knowledge of nonlinear oscillations, thanks to the lectures by Yves Rocard at ENS. Rocard was, at least at the time, mostly interested, as I recall, in designing electronic circuits made of resistors, capacitors, self-inductance and vacuum lamps, like triode, pentode, etc.[8] This was already a doomed field because of the exponential developments of solid state circuitry. To me, the most interesting part of his lecture was on the van der Pol equations and its solutions in the two limits of weak and strong nonlinearity. I read since that Rocard attended a meeting at Institut Poincaré in the early thirties organized by van der Pol himself, where the main topic was the existence of relaxation oscillations in solutions of the van der Pol equation. From what I still remember, the difference between the two limits (weak and strong non linearity) was not that obvious from the teaching we got. To a reader interested in this question, particularly the relaxation limit, I recommend the excellent paper by Anatoly Dorodnitsyn.[9] Rocard did not refer to Poincaré and did not use his deep geometrical methods.

My third year at ENS was devoted top the writing of a "thèse de troisième cycle" (something like a master thesis). I wrote down a rather formal calculation of nonlinear phenomena in plasma oscillations in the framework of kinetic theory. Somehow this extended the classical Landau treatment of plasma oscillations from linear to weakly non linear. The fourth year at ENS, I spent partly completing the master thesis, and also preparing the final exam, the *agrégation de sciences physiques*, in principle, for determining whether I was able to teach students in the upper levels of high school. This exam included also an experimental part and some chemistry, but no mathematics at all. I guess that this has not changed much since. At the end of my ENS period, I was recruited in the laboratory of Plasma physics at Orsay University, an institute head by Jean-Loup Delcroix who was one of the lecturers of my classes in plasma physics[10] in my second year at ENS. I began there to meet regularly Guy Laval and René Pellat, two outstanding plasma physicists in France. I still meet Guy more or less regularly, but René passed away many years ago.

The French organization of Science and Technology was such that plasma physics was cut in two parts: the big plasma machines were built and run at CEA (Atomic energy commission) whereas the teaching and other applications of plasma physics were performed in universities (mostly Orsay and Grenoble). Somehow, this left the most interesting part of plasma physics, the one focused on reaching controlled fusion, far from the place where students were. Therefore, although plasma machines involve a lot of interesting physics, and particularly nonlinear physics and chaos, this was cut out from the Academic world. Nevertheless, I made, I believe, one contribution to nonlinear phenomena in plasma physics. This is related to what is called the inverse Faraday effect.[11] The direct Faraday effect is, as its name indicates, something going back to the nineteenth century. It is the rotary optical power induced in condensed transparent matter by an external constant uniform magnetic field: under this direct Faraday effect, the plane of polarization of a light beam traversing the medium rotates by an amount proportional to the strength of the magnetic field times the length of the beam in the medium. The inverse Faraday effect (called inverse because the coefficients of the direct and inverse effects are related by a relation of symmetry coming from the fact that the two effects are represented by the same contribution to the energy of the medium that is proportional to the square of the electric field of the light beam times the magnetic field). In the inverse Faraday effect, a magnetic polarization is created by a circularly polarized electromagnetic wave. In plasmas, this is explained very simply by noticing that the charged particles (in practice, the electrons in a plasma) rotate along circles under the effect of the electric field of the wave, which creates according to the laws of electrodynamics, a permanent magnetic polarization. This phenomenon was observed[12] in the plasma physics laboratory of Orsay shortly after we predicted it with Daniel Quemada.[11]

3. Ph.D. Thesis

This brings me to the end of the sixties, early seventies. At that time, I had to think to write a Ph.D. thesis, something that was more or less mandatory if I wished, as I did, to continue performing full-time research at CNRS. The French doctoral system was, at the time, totally different from the one in most developed countries. Following, I guess, the old tradition of the medieval Universities, someone wishing to become a Doctor (a Ph.D.) had to find a topic by himself and to work on it almost alone and prove, at the end, that by his/her own capacities, he/she can discover new things, and so, be awarded the title of Doctor of such and such University. So, I had to find something new. I had the opportunity to attend a conference on statistical mechanics in Netherland by the Ocean. This was the first international Summer school, I ever attended. One of the topic taught there was the density expansion in kinetic theory of systems with short range molecular forces: at small densities, one has Boltzmann kinetic theory. At next order, one must take into account the collisions between three particles, etc. It had been discovered that the three body collisions introduces long time correlation due to the possibility that a first collision (1–2) can be followed, for instance, by a collision (2–3). All this correlate the velocities of particles 1, 2 and 3. Hereafter, it is highly possible that particle 2, for instance, comes back to hit again particle 1. This second collision (1–2) cannot be treated as the first one: the first collision has created a correlation between particles 1 and 2, and this correlation therefore forbids to use the Stosszahlansatz of Boltzmann's kinetic theory for this second 1–2 collision. It turns out that, for hard discs, this correlation is so strong that one cannot even find a converging kinetic operator for the three-body collisions (the same divergence exists for collisions between four spheres in 3D). This divergence had been already formally considered in the literature and a theory incorporating the loss of correlations on long trajectories responsible of the divergence by collisions with other particles had been proposed. It was however rather formal and involved the non trivial inversion of linear operators. To the surprise of some, I showed that the "renormalisation" that should have led to the disappearance of the divergence of the kinetic operator in 2D actually does not kill at all the divergence.[13,14] This went against one fundamental tenet of theoretical physics at the time: by summing the most divergent terms at every order in a formal expansion, one gets at the end a result without divergence. At that time, I was unaware of the field where such results had been proved and by whom. Now, I know that the situation was actually a bit more complex than implied by this statement on the sum of the most diverging terms. This renormalization program to eliminate divergences in expansions with respect to a small parameter was started by no less than Heisenberg and Dirac when computing the screening of the electric charge of a proton by particles in the Dirac sea.[15] Thanks to a trick of Heisenberg the summation of the most diverging terms of a perturbation series was avoided and the correct result obtained very elegantly.

4. Lattice Gas Model

In my Ph.D. research, I had to deal with fairy complex linear operator resulting from the linearization of Boltzmann collision operator. Trying to make it simpler, I used an idea that I heard first from Yvon, the one of discrete velocity space with a few possible orientations for the velocities of particles. This made the collision operator a finite matrix that could be handled reasonably. As I had found a divergence of transport coefficients in dense 2D-gases, I tried to find out a possible physical model where this could be put in evidence, including at finite densities. This led me to invent the first lattice gas model where everything is discrete, space, time and velocities. Together with Olivier de Pazzis and Jean Hardy, we did work on this HPP model and made the biggest possible molecular dynamics of this model that showed the expected divergence, including at finite densities.[16] This was the first time, as far I know, when a problem of fluid flow was solved by using a lattice discrete model. This HPP model was however pathological for real fluid mechanics because it had a too strong anisotropy. This was corrected with what is called now te FHP model (Uriel Frisch, Brosl Hasslacher, and Yves Pomeau) model on a triangular lattice,[17] implemented for the first time by Dominique d'Humières and Pierre Lallemand who showed that it could yield a fair representation of the Bénard-von Karman wake behind a flat plate,[18] something that was, at that time, slightly beyond the capabilities of regular DNS methods. My understanding is that the LBM method (Lattice Boltzmann method) is still ahead of all other methods for computing complex flows with boundaries.

In the early seventies, I had many exchanges with Pierre Résibois with whom I spent a very active year (1971) in Prigogine's lab in Brussels. The early death of Pierre has been a major loss for European physics. In Brussels, I made friend with postdoc of about my age, Paul Clavin, Jarek Piasecki and Manuel Velarde, and wee are still friends. Retrospectively, it did not cost much money to run this international Brussels hub of good theoretical physics and it had a long term impact on science. I am unsure that all the European programs for supporting science are working as well as this Brussels school of statistical, physics to foster collaboration between young scientists from all around the world.

5. Intermittencies as Route to Chaos

When I was in Brussels, I heard a few talks on nonlinear science, but at that time, it was not the center of my interests. As I said, we had a very friendly and productive collaboration with Pierre Résibois and some of his postdocs, particularly Jarek Piasecki. In France, I moved to CEA-Saclay, in theoretical Physics. Thanks to my CNRS position, this move was relatively easy and I landed in a lab where pure theory was very well regarded. In Saclay, I had the job of organizing seminars. Knowing a bit David Ruelle (since the Summer school in the Netherland), I invited him to give a talk, as he was already quite famous as a theoretical physicists. He gave, as always, a very clear and interesting talk. His talk was on chaos and this was the first time I heard the word with a scientific meaning. He referred to Edward Lorenz's paper[19] and explained what was in. I thought immediately that this field was full of potentialities. Although Ruelle did not insist on that, chaos theory was in principle applicable to unravel the mysteries of the transition to turbulence. As well known, Ruelle and Takens showed that what is called the Landau scenario for the bifurcation from non-chaotic to chaotic oscillations cannot be correct.[20] Landau had thought that to get chaos in a dynamical system one needs to have infinitely many oscillators at all frequencies in a continuous noise spectrum.[21] Therefore, assuming that each of those frequencies appears at a bifurcation, a real noisy signal requires to appear an infinite number of bifurcations, namely for infinite Reynolds number in a given flow. Using a non trivial topological argument, Ruelle and Takens proved that, after four bifurcations on oscillations, there is a structural instability of the dynamical system to a strange attractor, and to chaotic dynamics. This truly remarkable result made me realize a few useful things in research: first, new things can be found thanks to hard thinking and relying on application of non trivial maths.

At CEA, I had befriended Pierre Bergé and Monique Dubois who I knew a bit from an interest in their work on light scattering by thermal fluctuations, a field where Pierre was a pioneer worldwide. We had collaborated already (mid-seventies) on the transition to instabilities in Rayleigh–Bénard convection.[22] This was a good field for experimental research on nonequilibrium physics because, contrary to most experiments in fluids, the flow is not driven by an external pressure difference fluctuating with time in a more or less controlled way. Their study of the transition to convective pattern showed well a second-order transition, including nice boundary effects known also in other transitions phenomena. As soon, I heard David Ruelle's talk, I suggested Pierre and Monique to look at the transition from steady convection to time dependent and, hopefully turbulent convection. Curiously studies in turbulence had completely neglected this important topic of how a flow becomes turbulent.

Besides Lorenz's paper, there was practically nothing published on the subject. Curiously, the subject of turbulence was seemingly only concerned with the fairly unrealistic situations of homogeneous isotropic turbulence, as imagined by Kolmogorov with little connection with physical situations. I will come to this point later. At that time (end of seventies), I began to work with many bright physicists, among them Bernard Derrida, Paul Mannevile and Annie Gervois. With Annie and Bernard, we began to investigate nonlinear (non invertible) maps of the interval and discovered a very rich mathematical structure showing a kind of self-similartty in the space of parameters.[23] I had also the good luck to present some of the ideas of chaos theory at a talk at the Nice observatory, a talk attended by Michel Hénon who at lunch time managed to find on his pocket calculator what is known now as "the Hénon attractor."[24] This attractor showed obviously the multilayered structure expected for such an attractor but in a system a priori closer to real dynamical systems than our map of the interval because the Hénon map is invertible, a property equivalent to the fact that an ordinary differential equation can be integrated backward in time. With Paul, we did something a bit weird and slightly different from the study of iterations of maps. Using the last existing analog computer (in the nearby center of CEA), we began to study there the Lorenz system for different values of the parameters. This Lorenz system (see Chapter 2) is a set of three coupled nonlinear differential equations, first-order with respect to the derivation and contracting the volume element in phase space. Compared to the computers used at that time, this analog computer had a big advantage and an equally big inconvenience. The advantage, crucial for us, was that we had an instantaneous display of the solution on the screen of a monitor. The inconvenience was that the whole machine was not very reliable and crashed from time to time for a number of reasons, and needed a fairly large support team working full time to fix it. This explains why this kind of device was not developed more, besides of course the exponential development of Californian computer industry that led to the (relatively) cheap and efficient devices we all use today.

At that time, and may be for just a few years, this analog computer was ahead of the pack of all computing machines because it was the one where it was the easiest to watch the effect of a change of parameters in a dynamical system. What is called now the transition by intermittency was observed first on it and was quite obvious: a periodic solution displayed a closed trajectory on the screen. But it happened that for a rather narrow range of parameters this closed trajectory was destroyed by bursts occurring randomly in time filling the screen and followed by a return to closed trajectories.[25] Although it was not so obvious to see, the time lag between two bursts was seemingly growing to infinity as the parameter was increased in one direction and decreased as the parameter changed in the other direction. I thought this should be described in a rational way by using what I knew at the time of dynamical system theory. The only "advanced" theory I knew a bit was the first-return map of Poincaré, and I had the idea, after a lot of thinking, of the tangential bifurcation in the first-return map opening a tiny channel when the representing curve gets ac little above the bissectrix under changes of the system's parameter.[26] This explained two things, the bifurcation to the instability and the diverging duration of the oscillations between two bursts. We discussed that with Paul Manneville whom I saw very often at that time, as he belonged to the same Laboratory as Pierre Bergé. and Paul managed to reproduce the intermittent transition with the digital computer available to us. I had the good luck to see this "scenario" of transition to chaos observed in experiments in small cell of thermal convection[27] and in chemical reactions.[28]

Before to leave my research time on deterministic chaos and transition to chaos, it is perhaps of some interest to digress a bit on the way this domain started scientifically, of course from my own point pf view. During my students time, I attended Yvon's lectures on statistical physics. One point of his teaching was the explanation of the foundations of statistical physics in its relationship to the equations of motion of mechanics for point particles interacting with a two-body potential. Of course this was nothing new and Yvon explained well the relationship between deterministic dynamics and the randomness of the resulting behavior in time. This is called ergodic theory. Yvon said that the proof of the uniqueness of the invariant Lebegue measure made an extremely hard problem. At that time (1962–1963), Sinai had not yet proved the ergodicity of the hard sphere system.[29] After my ENS time, I kept an interest for this problem of ergodicity. I learned that Kolmogorov had found different properties of randomness — Bernoulli, K-mixing, mixing and ergodic —, not equivalent, and each one of this ordered set weaker than the one before. I taught this at a Summer School on Chaos in Sitges. I had read also works by Hadamard on geodesic flows on surfaces of negative curvature[30] that have the strongest randomness in the Kolmogorov list.[31] Therefore, before Ruelle's talk, I had already a fair understanding of what is called sometimes "deterministic chaos." This was also loosely connected with the work of Brussels group on irreversibility of Hamiltonian dynamics and its relation to the growth of entropy. The innovation in Ruelle and Takens was not in the discovery of "deterministic chaos," a long known idea, but in the elucidation of a possible way for a dynamical system to change from non chaotic to chaotic behavior as a parameter changes. This was completely new. There has been before various examples of "deterministic chaos" that is non linear differential equations with random solutions (or iterations, see later). As far I am aware, the earliest example is a very interesting paper by Levinson on a periodically driven van der Pol-like equation,[32] where he showed by a very clever geometrical method the existence of random trajectories, with any random sequence of two kinds of oscillations. Somehow this was reminiscent of Hadamard proof of the randomness of geodesic flows on surfaces of negative (Gaussian) curvature. Of course, after Levinson there has been the famous Lorenz paper[19] in which he showed random oscillations in the generic solution of a set of three nonlinear ordinary differential equations. Being at that time (mid-seventies) a regular visitor of CNLS at Los Alamos National Lab, I had an opportunity to read a report by Metropolis, Stein and Stein on iterations of simple maps of the plane into itself displaying obvious randomness.[33] So, in some way, I was ready to start studying deterministic chaos and transition from laminar to turbulent behavior as soon as I heard Ruelle's talk.

Somehow, the motivation to start this kind of study was not to put in evidence the well-known fact, at that time, that deterministic systems may show chaotic dynamics but to understand how a given system may change from non-chaotic to chaotic behavior as a parameter changes. In this respect, the use of iteration of the line were potentially very useful because they mapped things in a simple way and permitted good guesses. To end this, I should say that this gave me the great opportunity to write with Pierre Bergé and Christian Vidal our book "*Order within Chaos*"[34] which was a bit pioneering the field of books of various levels on Chaos published in the wake of the fashion of chaos theory. The writing of this book has been a chance for the three of us to sharpen our understanding of the subject and, hopefully, to contribute to the spreading of good science.

6. Taylor–Saffman Fingers

I spent an interesting year in the Schlumberger-Doll laboratory in Connecticut, not too far from New-York City. This opened my interest for topics I was unaware of, perhaps less fundamental than what I had done before. One was posed by the science of microscopy: given a cross section of a 3D material what can be said of its structure in 3D. In the geological research, this is important because one would like to know from cuts of stones with a more or less complex structure if this stone has enough empty channels to permit the flow of a pumped liquid like oil and how easily it will flow? This is one of the many interesting issues raised by the science of microscopy. Another, perhaps more technical, topic I looked at when I was at Schlumberger, was the general problem of what determines pattern, some in fluid mechanics, other in other fields. Among those "patterns" two posed manifestly non-trivial problems. The first was the shape of needle crystals, the other was the width of Saffman-Taylor fingers. Those fingers appear when a viscous fluid, like oil, is pushed by a less viscous fluid like air or water in a porous medium. In the simple situation this is done in the so-called Hele-Shaw cell, made of two plates close to each other so that the flow in the gap follows more or less Darcy equations, which are also valid for the flow in a porous rock. As one tries to push the viscous fluid with a less viscous one, this less viscous fluid makes a bubble in the gap that invades the viscous fluid, as a result of an instability. This bubble makes the ultimate stage of the growth of the instability.

Experiments and theory for this bubble had been done by Geoffrey Taylor and Philip Saffman.[35] Using conformal mapping they showed that a continuum of bubbles exists with all possible relative width in the channel. However they noticed that, for some reason they were unable to explain, the bubble of width 1/2 was selected in the experiments and also that the exact solution they had found becomes much simpler for this particular width. This nagging question remained unsolved for almost fifty years, and the solution was found and published in the same issue of *Physical Review Letters* by three independent groups,[36–38] I belonged myself to one of those three groups. Our solution relied on a rather advanced result of Kruskal and Segur on perturbation theory with non converging series.[39] At the same time, using the same mathematical method, Martine Ben Amar and myself solved another problem of shape of structure in the highly nonlinear domain, the one of the needle crystal growing in an undercooled melt that has also escaped for many years the understanding.[40]

7. Closing Remarks

I will close this wandering on chaos theory and more generally non-linear phenomena by explaining another contribution to an important phenomenon, the way shear flows go from laminar to turbulent as the

Reynolds number increases. As I explained above, Ruelle and Takens theory was based on the dynamical system theory, a fancy word for finite set of coupled ordinary differential equations, nonlinear of course, as the Lorenz system. This makes a fair mathematical object for describing the so-called "small boxes," that is, for fluids strained one way or another but kept in a box. Examples are, for instance, the convective cells from Pierre and Monique with an aspect ratio of order one in all directions, like a cube, for instance. If one projects all possible velocity fields on a basis of functions of space, one gets an infinite set of coupled amplitude but the modes representing the fluctuations with the smallest space scale are heavily damped, and so, have their amplitude enslaved to the less damped or even to the unstable modes. This explains why, in small boxes, one can approximate the dynamics by a finite dimensional system, the set of the unstable and less damped modes. Due to that, many thoughts in classical fluid mechanics developed in the nineties with chaos theory are irrelevant for describing the transition to turbulence in "real" flows that matter for fluid mechanics, like an airflow around a wing or in a pipe. In such flows, there are many degrees of freedom involved in the transition: in a long pipe, this number is at least proportional to the length of the pipe.

I discovered this subject when I spent a year at MIT teaching turbulence in 1984. I had several discussions with the head of the math department there, Willem Malkus on precisely this topic: how shear flows do change from laminar to turbulent? Somehow Willem challenged me, a specialist in the transition to chaos in dynamical systems, to explain the observation of this transition in shear flows. This topic had already a long history started by Osborne Reynolds in the nineteenth century. Reynolds observed that in a pipe flow beyond a certain value of what is called now the Reynolds number, turbulence sets in a highly non homogeneous way: it remains inside patches of turbulence separated by laminar fluid. To Willem, this was very hard to understand because, for him, a "system" is either in a turbulent or in a laminar state, as for instance in Rayleigh–Bénard convection: as soon as the temperature difference driving the instability is above the instability threshold, the "system" bifurcates everywhere to the convective state. Being familiar with Landau–Lifschitz's textbook on fluid mechanics, I remind that, there, Landau explains that a bifurcation may be either sub- or super-critical (using different words). To explain Reynolds findings, this was only one part of the puzzle, because in the Landau, theory there was no space dependence at all, and so, no way for explaining the existence of such heterogeneous turbulent state.

Being familiar, thanks to the study of thermal convection, with the amplitude equation for describing bifurcation with space-dependent patterns, I thought I had the good instrument for explaining Reynolds findings by adding a possible space dependence of the turbulent field by adding to Landau's amplitude equation a diffusion term saying that if the amplitude has a certain value at one point, its neighboring points tend to have also the same amplitude. This idea had been used ton derive amplitude equation for the Rayleigh–Bénard convection, where there is also an expansion parameters, the shift from the threshold value of the control parameter. This yields the Newell–Whitehead–Segel amplitude equation. There, the amplitude equation can be derived rigorously from the fluid equations because the bifurcation is super-critical, so that, the nonlinearities remain manageable near the threshold. I made a little jump forward by assuming that the phenomenology by Landau for the sub-critical bifurcations was not so different from the one in the Rayleigh–Bénard convection; thus, one cannot use the same idea to study space dependent situations like the one observed first by Reynolds.

I started thinking to this at a Geophysical fluid dynaics school at Woods Hole in 1984. From there, I moved to the Los Alamos National Lab for a conference I was invited to. I presented there the result of my Summer thoughts on the transition to chaos in shear flows, based on Landau's amplitude equation with space dependence described by a diffusion-like term. In the paper published after the conference,[41] I included some thoughts on the non trivial case where the "upper" branch of the Landau diagram representing the bifurcated state of finite amplitude is actually a dynamically chaotic state. There, a simple reasoning shows that the transition belongs to the class of directed percolation in statistical physics, a transiton very well explained in a paper by Sergei Obukhov[42] I had a chance to read in Los Alamos. When I was back in Saclay, I discussed this interesting question with Pierre and Paul. Pierre did wonderful experiments showing the transition in a very clever realization of the Couette flow, and observed convincingly a transition to turbulence described by the scenario of directed percolation.[43] Paul with his then student, Hugues Chaté, managed to cook up a deterministic model, a partial differential equation, nonlinear, first order in time and fourth-order in space where thee was also a transition to chaotic behavior following the directed percolation scenario, including with the correct values of the various "critical exponents" typical of this transition.[44] Of course, I was very glad to see this idea of directed percolation applicable to "real situations." Amazingly, this was, as far I know, the first time that directed percolation was observed in an experimental set-up.

References

1. P. Coullet and Y. Pomeau, History of chaos from a French perspective. In C. Skiadas (ed.), *The Foundations of Chaos Revisited: From Poincaré to Recent Advancements* (2016), pp. 91–101. doi: 10.1007/978-3-319-29701-9_6.
2. G. K. Batchelor, *The Theory of Homogeneous Turbulence.* Cambridge University Press (1953).
3. G. Birkhoff and J. V. Neumann, The logic of quantum mechanics, *Annals of Mathematics.* **37**(4), 823–843 (1936).
4. D. A. Slavnov, Quantum measurements and kolmogorovian probability theory (2003). https://arxiv.org/abs/quant-ph/0301027.
5. H. Everett, "Relative state" formulation of quantum mechanics, *Review of Modern Physics.* **29**, 454–462 (1957). doi: 10.1103/RevModPhys.29.454.
6. L. D. Landau and E. M. Lifshitz, Mechanics. In *Course of Theoretical Physics*, 3rd edn., Vol. 1. Pergamon Press, Oxford (1977). Translated by J. B. Sykes and J. S. Bell.
7. L. D. Landau and E. M. Lifshitz, Quantum mechanics: Non-relativistic theory. In *Course of Theoretical Physics*, 3rd edn., Vol. 3. Pergamon Press, Oxford (1977). Translated by J. B. Sykes and J. S. Bell.
8. Y. Rocard, *Théorie des oscillateurs.* Livre de science, Éditions de la Revue scientifique (1941).
9. A. A. Dorodnitsyn, Asymptotic laws of distribution of the characteristic values for certain special forms of differential equations of the second order, *Uspekhi Matematicheskikh Nauk.* **7**, 3–96 (1952).
10. J.-L. Delcroix, *Plasma Physics.* John Wiley & Sons, London (1965).
11. Y. Pomeau and D. Quemada, Effet Faraday inverse dans un plasma, *Comptes-Rendus de l'Académie des Sciences B (Paris).* **264**, 517–618 (1967).
12. J. Deschamps, M. Fitaire, and M. Lagoutte, Inverse Faraday effect in a plasma, *Physical Review Letters.* **25**, 1330–1332 (1970). doi: 10.1103/PhysRevLett.25.1330.
13. Y. Pomeau, Transport theory for a two-dimensional dense gas, *Physical Review A.* **3**, 1174–1194 (1971). doi: 10.1103/PhysRevA.3.1174.
14. Y. Pomeau, Correlation length in a nonequilibrium gas, *Journal of Mathematical Physics.* **12**(11), 2286–2295 (1971). doi: 10.1063/1.1665533.
15. W. Heisenberg and H. Euler, Folgerungen aus der Diracschen Theorie des Positrons, *Zeitschrift für Physik.* **98**, 714–732 (1936). doi: 10.1007/BF01343663.
16. J. Hardy, Y. Pomeau, and O. de Pazzis, Time evolution of a two-dimensional model system. I. Invariant states and time correlation functions, *Journal of Mathematical Physics.* **14**(12), 1746–1759 (1973). doi: 10.1063/1.1666248.

17. U. Frisch, B. Hasslacher, and Y. Pomeau, Lattice-gas automata for the Navier-Stokes equation, *Physical Review Letters.* **56**, 1505–1508 (1986). doi: 10.1103/PhysRevLett.56.1505.
18. D. d'Humières, P. Lallemand, and U. Frisch, Lattice gas models for 3D hydrodynamics, *Europhysics Letters.* **2**(4), 291 (1986). doi: 10.1209/0295-5075/2/4/006.
19. E. N. Lorenz, Deterministic nonperiodic flow, *Journal of the Atmospheric Sciences.* **20**, 130–141 (1963). doi: 10.1175/1520-0469(1963)020⟨0130:DNF⟩2.0.CO;2.
20. D. Ruelle and F. Takens, On the nature of turbulence, *Communications in Mathematical Physics.* **20**, 167–192 (1971).
21. L. D. Landau, On the problem of turbulence, *Doklady Akademii Nauk SSSR.* **44**, 339–342 (1944) (in Russian).
22. J. Wesfreid, Y. Pomeau, M. Dubois, C. Normand, and P. Bergé, Critical effects in Rayleigh-Bénard convection, *Journal de Physique.* **39**(7), 725–731 (1978). doi: 10.1051/jphys:01978003907072500.
23. B. Derrida, A. Gervois, and Y. Pomeau, Iteration of endomorphisms on the real axis and representation of numbers, *Annales de l'Insitut Henri Poincaré A.* **29**, 305–356 (1978).
24. M. Hénon, A two-dimensional mapping with a strange attractor, *Communications in Mathematical Physics.* **50**, 69–77 (1976).
25. Y. Pomeau and P. Manneville, Intermittency: A generic phenomenon at the onset of turbulence. In G. Laval and D. Grésillon (eds.), *International Workshop on Intrinsic Stochasticity in Plasmas.* Les éditions de Physique Couraboeuf, Orsay, France (June 17–23, 1979), pp. 329–340.
26. Y. Pomeau and P. Manneville, Intermittent transition to turbulence in dissipative dynamical systems, *Communications in Mathematical Physics.* **74**, 189–197 (1980). doi: 10.1007/BF01197757.
27. P. Bergé, M. Dubois, P. Manneville, and Y. Pomeau, Intermittency in Rayleigh-Bénard convection, *Journal de Physique (Lettres).* **41**, L–341–345 (1980).
28. Y. Pomeau, J.-C. Roux, A. Rossi, S. Bachelart, and C. Vidal, Intermittent behaviour in the Belousov-Zhabotinsky reaction, *Journal de Physique Lettre.* **42**(13), 271–273 (1981). doi: 10.1051/jphyslet:019810042013027100.
29. Y. G. Sinai, Dynamical systems with elastic reflections. Ergodic properties of dispersing billiards, *Russian Mathematical Surveys.* **25**(2), 137–189 (1970).
30. J. Hadamard, Les surfaces à courbures opposées et leurs lignes géodésiques, *Journal de Mathématiques Pures et Appliquées V.* **4**, 27–74 (1898).
31. A. N. Kolmogorov, Three approaches to the quantitative definition of information, *Problemy Peredachi Informatsii.* **1**, 3–11 (1965).
32. N. Levinson, A second-order differential equation with singular solutions, *Annals of Mathematics.* **50**(1), 127–153 (1949). doi: 10.2307/1969357.
33. N. Metropolis, M. L. Stein, and P. R. Stein, On finite limit sets for transformations on the unit interval, *Journal of Combinatorial Theory A.* **15**, 25–44 (1973).

34. P. Bergé, Y. Pomeau, and C. Vidal, *Order within Chaos: Towards a Deterministic Approach to Turbulence*. Wiley-Interscience (1986).
35. G. Taylor and P. G. Saffman, A note on the motion of bubbles in a Hele-Shaw cell and porous medium, *The Quarterly Journal of Mechanics and Applied Mathematics*. **12**(3), 265–279 (1959). doi: 10.1093/qjmam/12.3.265.
36. B. I. Shraiman, Velocity selection and the Saffman-Taylor problem, *Physical Review Letters*. **56**, 2028–2031 (1986). doi: 10.1103/PhysRevLett.56.2028.
37. D. C. Hong and J. S. Langer, Analytic theory of the selection mechanism in the Saffman-Taylor problem, *Physical Review Letters*. **56**, 2032–2035 (1986). doi: 10.1103/PhysRevLett.56.2032.
38. R. Combescot, T. Dombre, V. Hakim, Y. Pomeau, and A. Pumir, Shape selection of Saffman-Taylor fingers, *Physical Review Letters*. **56**, 2036–2039 (1986). doi: 10.1103/PhysRevLett.56.2036.
39. M. D. Kruskal and H. Segur, Asymptotics beyond all orders in a model of crystal growth, *Studies in Applied Mathematics*. **85**(2), 129–181 (1991). doi: 10.1002/sapm1991852129.
40. M. Ben Amar and Y. Pomeau, Needle-crystal growth in an axial flow: Scaling laws, *Comptes-Rendus de l'Académie des Sciences II (Paris)*. **308**, 907–914 (1989).
41. Y. Pomeau, Front motion, metastability and subcritical bifurcations in hydrodynamics, *Physica D*. **23**(1), 3–11 (1986). doi: 10.1016/0167-2789(86)90104-1.
42. S. P. Obukhov, The problem of directed percolation, *Physica A*. **101**(1), 145–155 (1980). doi: 10.1016/0378-4371(80)90105-3.
43. F. Daviaud, J. Hegseth, and P. Bergé, Subcritical transition to turbulence in plane Couette flow, *Physical Review Letters*. **69**, 2511–2514 (1992). doi: 10.1103/PhysRevLett.69.2511.
44. H. Chaté and P. Manneville, Transition to turbulence via spatio-temporal intermittency, *Physical Review Letters*. **58**, 112–115 (1987). doi: 10.1103/PhysRevLett.58.112.

https://doi.org/10.1142/9789819823130_0011

Chapter 11

HOW I BECAME INVOLVED IN CHAOS

Philip Holmes

Princeton University, Princeton, USA
pholmes@math.princeton.edu

1. Starting with the Duffing System

In 1973, I was finishing a Ph.D. thesis on noise transmission in structures at the Institute of Vibration Research at Southampton University (I.S.V.R., U.K.), when I noticed that a course on differential topology and its applications to dynamical systems would be taught by David Chillingworth in the Mathematics Department. This introduced me to René Thom's catastrophe theory and to many analytical tools that I had not known before. While auditing the course, I met David Rand, who was also completing his thesis and who wanted to move toward applied mathematics. We began working together and found an interesting mistake in the interpretation of the amplitude response function for periodic solutions in a preprint of Christopher Zeeman's on Duffing's equation with a stiffening spring ($\alpha > 0$):

$$\ddot{x} + 2\zeta\dot{x} + k(x + \alpha x^3) = f\cos(\omega t). \qquad (1)$$

The mistake involved a misinterpretation of the cusp catastrophe which occurs as a limiting case when two curves of saddle-node bifurcations collide in the (f, ω) parameter space. Zeeman's paper finally appeared in 1976.[1] We corrected the mistake and sent our preprint to Christopher.

He encouraged us to submit and publish, and in due course our paper came out,[2] followed by further papers on van der Pol's equation[3] and on a combined Duffing-van der Pol oscillator.[4] We did not learn of Yoshisuke Ueda's analog simulations of chaos in a similar equation with a cubic damping term until I moved to the Theoretical and Applied Mechanics Department (TAM) at Cornell University in 1977. Shortly thereafter, Ueda visited TAM and described some of his results.

However, while still in Southampton David Rand and I found two preprints on nonlinear oscillations by Floris Takens in the I.S.V.R. library, and things began to become chaotic. Takens' papers appeared in 1974.[5,6] I believe that they established fundamental examples for the classification of bifurcations in dynamical systems.

Cartwright had found a curve of homoclinic bifurcations in a two-dimensional averaged van de Pol equation.[7] David Rand and I were able to relate it to the homoclinic bifurcations studied by Takens, and thus assemble the correct codimension-2 bifurcation set.[3] Motivated by Takens' work, I borrowed an analog computer, the type of which I have forgotten, from the undergraduate laboratory and (physically) programmed it to simulate Duffing's equation in the form

$$\begin{cases} \dot{x} = y, \\ \dot{y} = x - x^3 - \delta y + \gamma \cos(\omega t). \end{cases} \tag{2}$$

This system has a potential energy function with two wells separated by a peak. With damping $\delta > 0$, but without periodic forcing ($\gamma = 0$), almost all solutions approach one or the other of the two stable equilibria $(x, y) = (\pm 1, 0)$, except for the stable manifold of the saddle point $(x, y) = (0, 0)$. More significantly for chaos, when damping $\delta = 0$, there are two homoclinic orbits beginning and ending at the saddle point. I had read and understood relevant parts of Stephen Smale's paper in the Bulletin of the A.M.S.[8] and realized that this implied that, with sufficiently large periodic forcing $\gamma \neq 0$, the Poincaré map of equation (2) has infinitely many transverse homoclinic points, and therefore contains Smale horseshoes.

In recording the resulting chaotic oscillations, I broke the wire controlling the pen on the recording device, but not before recording chaotic time-histories of $x(t)$ and phase portraits as (x, y) traces projected onto the Poincaré cross section (e.g. Fig. 1). Some of this work appeared in my 1979 paper.[9] For further examples of phase portraits, see Figs. 2.2.4–2.2.8 of Section 2 in Chapter 2 of the textbook[10] written with John Guckenheimer.

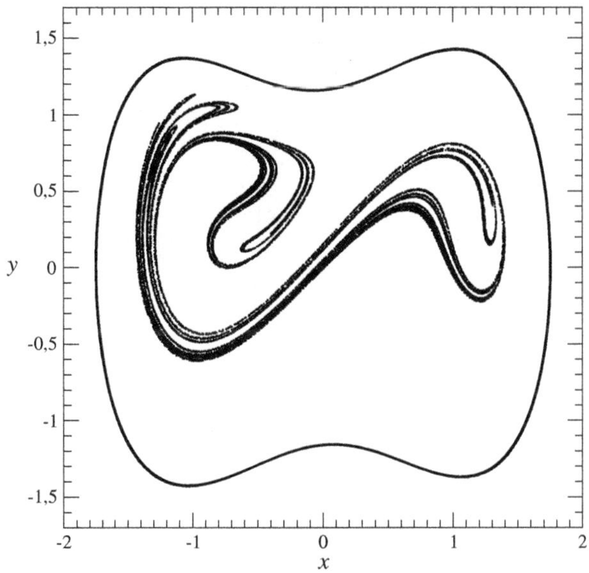

Fig. 1. Coexistence of a chaotic attractor and a large period-1 limit cycle. Parameter values: $\delta = 0.15$, $\gamma = 0.30$, and $\omega = 1$. Initial conditions: $x_0 = 1$ for the chaotic solution and $x_0 = 1.8$ for the period-1 limit cycle, and $y_0 = 0$.

Around this time (I think in 1975), Jerrold Marsden, from U.C. Berkeley, and John Guckenheimer, from U.C. Santa Cruz, came to Southampton for an informal conference on nonlinear dynamics. I have no records of proceedings being published, but their visit started several fruitful collaborations. I had meanwhile become a Research Fellow at I.S.V.R., whose Director, Brian Clarkson, kindly allowed me to choose my own research directions.

2. From Mathematics to Experiments

I interviewed for academic teaching positions in the UK, but failed to find one, and so arranged a trip to the US in 1976, which resulted in the move to Cornell in 1977. Soon after arriving I explained the behavior of equation (2) to a colleague, Francis Moon. He quickly built a mechanical analog: a cantilever beam with two stable and one unstable equilibria, buckled by a pair of magnets. See Fig. 2, and see Moon and Holmes for more details.[11] Subsequently, Marsden and I extended the analysis to a partial differential equation.[12]

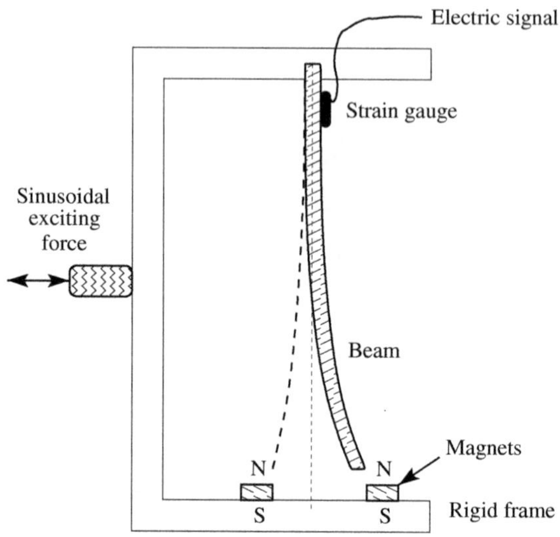

Fig. 2. The magneto-elastic beam.
Source: Redrawn from Ref. 10.

I cannot remember when someone pointed me to the Russian literature and specifically to a paper of Melnikov,[13] which provides a method for proving the existence of transverse homoclinic orbits in oscillators with time-periodic perturbations. This connects directly to classical analyses of nonlinear oscillators and has since been extended to systems with many degrees of freedom.

Two conferences sponsored by the *New York Academy of Sciences* in 1977 and 1979 focused on bifurcation theory and nonlinear dynamics[14,15] and did much to stimulate the field by bringing diverse researchers together. I chaired a related conference in 1979 which both engineers and mathematicians attended.[16] Since I had trained as an engineer and at that time was still trying to become an applied mathematician, this meeting was important for my future.

Throughout my early work and collaborations, I was most excited by the combination of computer simulations, rigorous mathematical theorems, and physical experiments that, together, could create new models of dynamical processes. The classification of bifurcations played a key role in this.

In closing, I would like to mention that David Rand, John Guckenheimer and I began teaching courses on dynamical systems at the University of Warwick (UK), UC Santa Cruz, and Cornell respectively. David and

I started drafting chapters. However, our interests and approaches were diverging and it was finally decided that John and I would write the book.[10] This was done in 1970–1972 with the help of many long-distance telephone calls.

References

1. E. C. Zeeman, Duffing's equation in brain modelling, *Bulletin of the Institute of Mathematics and Applications.* **12**, 207–214 (1976).
2. P. J. Holmes and D. A. Rand, The bifurcations of Duffing's equation: An application of catastrophe theory, *Journal of Sound and Vibration.* **44**(2), 237–253 (1976). doi: 10.1016/0022-460X(76)90771-9.
3. P. J. Holmes and D. A. Rand, Bifurcations of the forced van der Pol oscillator, *Quarterly of Applied Mathematics.* **35**(4), 495–509 (1978). doi: 10.1090/qam/492551.
4. P. Holmes and D. Rand, Phase portraits and bifurcations of the non-linear oscillator: $\ddot{x}+(\alpha+\gamma x^2)\dot{x}+\beta x+\delta x^3=0$, *International Journal of Non-Linear Mechanics.* **15** (6), 449–458 (1980). doi: 10.1016/0020-7462(80)90031-1.
5. F. Takens, Singularities of vector fields, *Publications Mathématiques de l'Institut des Hautes Études Scientifique.* **43**, 47–100 (1974).
6. F. Takens, Forced oscillations and bifurcations, *Communications of the Mathematics Institute.* **3**, 1–59 (1974).
7. M. L. Cartwright, Forced oscillations in nonlinear systems. In *Contributions to the Theory of Nonlinear Oscillations*, Vol. 20. Princeton University Press, Princeton (1950), pp. 149–242. doi: 10.1515/9781400882632-005.
8. S. Smale, Differentiable dynamical systems. I. Diffeormorphisms, *Bulletin of American Mathematical Society.* **73**, 747–817 (1967).
9. P. Holmes, A nonlinear oscillator with a strange attractor, *Philosophical Transactions of the Royal Society of London A.* **292**(1394), 419–448 (1979). doi: 10.1098/rsta.1979.0068.
10. J. Guckenheimer and P. J. Holmes, Nonlinear Oscillations, Dynamical Systems, and Bifurcations of Vector Fields. Applied Mathematical Sciences, Vol. 42. Springer-Verlag, New York (1983).
11. F. Moon and P. Holmes, A magnetoelastic strange attractor, *Journal of Sound and Vibration.* **65**(2), 275–296 (1979). doi: 10.1016/0022-460X(79) 90520-0.
12. P. Holmes and J. Marsden, A partial differential equation with infinitely many periodic orbits: Chaotic oscillations of a forced beam, *Archive for Rational Mechanics and Analysis.* **76**(1), 135–165 (1981).
13. V. K. Mel'nikov, On the stability of a center for time-periodic perturbations, *Trudy Moskovskogo Matematicheskogo Obshchestva.* **12**, 3–52 (1963) (in Russian).
14. O. Gurel and O. E. Rössler (eds.), *Bifurcation Theory and Applications in Scientific Disciplines.* Annals of the New York Academy of Sciences, Vol. 316 (1979), pp. 1–686.

15. R. H. G. Helleman (ed.), *Nonlinear Dynamics.* Annals of the New York Academy of Sciences, vol. 357, pp. 1–505 (1980).
16. P. Holmes (ed.), *New Approaches to Nonlinear Problems in Dynamics.* Proceedings of the Engineering Foundation Conference, SIAM Publications, Philadelphia (1980).

https://doi.org/10.1142/9789819823130_0012

Chapter 12

HOW CHAOS SHAPED MY ACADEMIC LIFE

René Lozi

Université Côte d'Azur, France
Rene.LOZI@univ-cotedazur.fr

René Lozi writes a recollection of his career starting back to 1967, when he entered the university of Nice, France. After five years of classical study in mathematics, lastly focused on numerical analysis and theoretical physics, his advisor introduced him in the study of bifurcation of ordinary differential equations, a very new concept in mathematics at that time in France (although it was already being explored in physics and mathematics in the United States). He had the great fortune to see and participate in the birth of chaos at the end of the 70s. Nonlinear chaotic dynamics became his mantra for the rest of his career, during which he met and collaborated with outstanding scientists from several countries including Otto Rössler who inspired some of his articles.

Fifty years after the beginning of his academic life, he realizes that he was incredibly lucky to begin his career as the same time as the emergence of chaos studies in dynamical systems. He had a great feeling of freedom, being able to discover a new world of extreme richness, as undoubtedly explorers of America had it hundreds of years before him.

1. Introduction

I feel sometimes uneasy to recall what exactly happened nearly fifty years ago in my academic life. In what follows, I hope to reconstitute fairly accurately the beginning of my research in bifurcation and chaos theories at University of Nice (now University Côte d'Azur) where I spent all my life, even under different administrations of research (C.N.R.S.) or education. In fact, my office was always inside or in the same small neighborhood of the Faculté des sciences, where stands the Institute for mathematics Jean Alexandre Dieudonné, in honor to the famous "Bourbakist".

Sometimes people who have lived the same scientific adventure do not share the same memory. Therefore, I warn the reader of this recollection that it only reflects my own point of view. To be mathematician is very special, not only folks thinks that you are different from normal people, not keeping the feet on the ground; but you can develop friendship lasting decades with researcher of other countries, speaking the same common mathematical language. It is why, beyond mathematical considerations, I will take time to speak of friends of mine or outstanding mathematicians, I have encountered during those years. Because the aim of this book is to celebrate Otto Rössler's 80th birthday, I feel the need to link my recollections with his outstanding discovers in chaos theory. Strangely enough, although I was very interested in the first examples of chaotic attractors from the end of the 1970s, I never paid attention too much to the Rössler attractor produced by the system

$$\begin{cases} \dot{x} = -y - z \\ \dot{y} = x + 0.2y \\ \dot{z} = 0.2 + z(x - 5.7) \end{cases} \tag{1}$$

and published in 1976,[1] nor to the equations

$$\begin{cases} \dot{a} = k_1 + k_2 a - \dfrac{a(k_3 b + k_4 c)}{a + K} \\ \dot{b} = k_5 a - k_6 b \\ \mu \dot{c} = k_7 a + k_8 c - k_9 c^2 - k_{10} \dfrac{c}{c + K'}. \end{cases} \tag{2}$$

published as an example of the method explaining the principle for 3-dimensional chaos generation.[2]

I will explain the reasons. However, the qualitative procedure of this method was strongly inspiring me during several years, allowing to propose a geometric model of slow-fast Lorenz-like attractor[3] and the Alpazur oscillator with Hiroshi Kawakami.[4,5] It was only in 2010, following an unusual request from Christophe Letellier, that I investigated the possible origin of equation (2), trying to reconstruct what had been the intellectual journey of Otto Rössler, building his attractor from chemical reaction kinetics.

Moreover, I was interested in his researches on hyperchaos and his prototypic models (in the form of differential equations[6,7] or in the form of mappings[8]), and I published a simple hyperchaotic map (although non-continuous) three years later.[9] But let me start at the beginning, when I finished my studies at the high school in Antibes, in the south of France.

2. Dynamical Systems in France in the 70s

I started my studies at the University of Nice — which is a distance of 20 km from Antibes — in October 1967, in mathematics and physics. It was classical studies in a world that had not known many upheavals since the second world war, except the conquest of space started in 1957 with the launch of the Russian satellite "Sputnik" and the journey of the first astronaut, the Russian Yuri Gagarin in 1961 and the American John Glenn in 1962. I was fascinated by particle physics and I dreamed to make researches in theoretical physics, without knowing which courses to attend to reach this goal. At the university I took my first programming course about FORTRAN IV in 1968, using punched cards. The decade that followed and of which I will try to describe the evolution that I personally perceived in the limited field of the study of bifurcations and chaos, has seen great changes, such as the conquest of space with Neil Armstrong, the first man who walked on the moon on 20 July 1969.

2.1. *From bifurcation to chaos*

During my studies at the University of Nice, in the fields of applied mathematics, I discovered with fascination the methods of numerical integration of ordinary differential equations (ODEs). It was during the academic year 1970–71, I was following my bachelor's degree under the supervision of Professor Martin Zerner (1932–2017). During my previous studies, I had been taught that there was a list of ODEs written by Bernoulli, Lagrange, Clairaut, Riccati, etc. and a list of solving methods. A clever student in mathematics should be able to know which method and which trick was efficient to solve a given equation among a huge list of them available in handbooks. No physical sense, in fact no meaning at all, was attached to these academic exercises. No numerical method was taught. Moreover, between professors, there was a strict separation between "pure mathematician" and the few "applied mathematician" who were able to use a computer. It must be said that only one computer "IBM-1130" was available at the university of Nice (another computer IBM-7040 was installed at the Observatory). Martin was the first guy, considered as weird by the other professors, who was able to use a computer that I ever met (although his Ph.D. superviser was Laurent Schwartz and he was chosen, when working as "attaché de recherches" at C.N.R.S., to help Jacques Hadamard to finalize the last books of his complete works in 1958–59). I knew him fairly well, because during the famous "student revolution" of May 1968, students and professors of the small faculty of science of Nice had exchanged a lot, breaking down the usual hierarchical barriers. With Martin we even guarded together, against hypothetical attacks, the "Château de Valrose", building of the Institute of mathematics, during many nights of this month of May, perched on its roof, like medieval warriors in their dungeon.

During his lectures, my mind knew a breakthrough that changed the paradigm in which I conceived the theory of differential equations: the set of all the equations I was taught, were of zero measure in the set of all ones existing. No closed formula of solution can be found for most of them. Only numerical methods like Runge–Kutta, Adams–Bashforth, etc. were able to provide approximate solution. Of course, in this scope, computer was essential. Moreover, ODEs were useful to model physical, chemical or even biological situations (at the end of the year, to be graduated, I presented a memoir on the Yukawa potential[10]). This new paradigm has guided my research career throughout my whole life. Looking for an superviser for my Ph.D. thesis in 1972, I had the chance to meet Professor Giuseppe ("Pippo") Geymonat who first supervised my "Diplôme d'Études Approfondies" and proposed me to study the particular nonlinear ODE

$$\begin{cases} \ddot{u}(x) - [\dot{u}(x)]^2 + cu(x) = 0, c \in \mathbb{R} \\ u(0) = u(1) = 0 \end{cases} \tag{3}$$

which belongs to the class of two-point boundary value problems.[11] The difficulty in this study comes from the quadratic term of the first derivative. Such nonlinear equations were not in vogue at this time, the focus being more on Floquet theory of linear equations and the study of their stability and Floquet exponents in various spaces.[12,13] A typical example of such problem is the ballistic motion of a projectile, starting from a gun and aiming a target. The solution gives the angle of launch required to hit the target. One often uses numerical methods called shooting methods, for solving such problem. Another example is the deflection of a homogeneous beam of fixed length subject to some transverse load.

The study of (3) led me to defend in 1975, my Ph.D. thesis entitled "*Analyse numérique de certains problèmes de bifurcation*"[14] (distantly supervised by Jean Céa with the invaluable help of Geymonat who was in Turin). I published just after a short article on, perhaps the first example of continuation method, at least in the context of the bifurcation theory.[15] While preparing my Ph.D., the name "bifurcation" was largely unknown in the communities of mathematical and numerical analysis in France. When people in Nice and other universities were asking me on my research topics, they were astonished by this name. Of course, as one can read in Abraham's chapter,[16] that the term bifurcation was introduced 90 years before, by Henri Poincaré[17] who used the term "form of bifurcation" to describe a form of equilibrium belonging to two different linear series of differential equation, which does not correspond exactly to what it is referred nowadays as the solution to nonlinear differential equation. Furthermore, we must consider that the decade 1960–70 was the golden age of the Bourbaki group, whose philosophy was drastically opposed to Poincaré's way of thinking. The bedside books of every young student in maths at university were "*Éléments de mathématiques.*"[18] Moreover, Jean Alexandre Dieudonné, one of the founders of the Bourbaki group, arrived at Nice in 1964, just after the University of Nice was created. He was the most prominent professor from the department of mathematics. During the academic year 1969–70, I attended his lectures about the integration theory, with only a dozen of students. Upon Martin Zerner's request, he spent an entire afternoon in helping me on some problem related to my doctoral work. Poincaré's works were therefore not at all in my mind, although I eventually quoted him[17] in the analytical bibliography of my Ph.D. thesis (Refs. [130, 131] in 14), together with the papers of A. M. Lyapunov (Ref. [107] in 14) (Fig. 1).[a] Abraham[16] referring to 19 mentions also that "*The word bifurcation is indexed by Thom on 18 pages,*" but Thom, Fields medalist in 1958, was doing, in my opinion, theoretic mathematics at a level I never could reach. Moreover, his book in English was published during the middle of my thesis, and at this epoch, it took months, or rather years, for a new published book in English will be delivered to our library. Oddly, few years later, Thom assisted me many times for my "Thèse d'État," later for my Habilitation and to apply for higher position in the university.

[a]I spent a lot of time to recover all these articles, delving in libraries of several universities.

[107] LIAPUNOV A.M. Sur les figures d'équilibre peu différentes des ellipsoïdes d'une masse liquide homogène douée d'un mouvement de rotation. Zap.Akad. Nauk. Vol 1 (1906) pp 1 - 225;puis (1908) pp 1 -175 (1912) pp 1 - 228 ; (1919) pp 1 - 112

[130] POINCARE H. Sur l'équilibre d'une masse fluide animée d'un mouvement de rotation. C.R.A.S. de Paris p346 , 9 fev. 1885 ; p 1068 , 20 avr. 1885 ; p 307 , 27 juil. 1885; p 857 ,12 avr. 1886 ; (note de MATTHIESSEN); p970,27 avr. 1886; p 622,7 mars 1887 (relative à Liapunoff); p 1571,4 juin 1888 ; p 655,16 mars 1903.

[131] ________ Idem. Acta Math. 1 (1885) pp 259 - 380

Fig. 1. The works of Lyapunov and Poincaré in the bibliography of my Ph.D. thesis.

Hence, the first paper in French on this subject (however in the fields of mechanics) was published in 1971 by Gérard Iooss[20] who obtained later, in 1974, a permanent position in the University of Nice (see also Refs. 20–28 quoted in 14). After my Ph.D., I asked him for a new subject for my "Thèse d'État." We hence worked together on the famous dynamo problem explaining the origin of the magnetic earth field.[29] During this period, we both attended the "International Conference on Mathematical Problems in Theoretical Physics," organized by the university of Roma, Italy (June 6–15, 1977), with a team of researchers from the university of Nice (Claude Bardos, Alain Chenciner, Pierre Coullet, ...). The opening talk of this conference was given by David Ruelle,[30] famous for coining the name "strange attractor" with his colleague Floris Takens.[31] In his talk, Ruelle conjectured that, for the Hénon attractor, the theoretical entropy should be equal to the characteristic exponent. It is how I discovered the first example of chaotic and strange attractor [Fig. 2(a)].[32] At that time, the term "strange" was used, referring to Ruelle and Takens' paper. Today, according to the terminology introduced by Grebogi *et al.*,[33] we would use "chaotic" rather than strange which now refers to the fractal properties of the invariant set. Nevertheless, this is rather the fractal properties of this attractor which were highlighted by Hénon and astonished the research community, rather than its chaotic characteristics. Although Michel Hénon was an astronomer at the Nice Observatory, few kilometers abroad from my office, I did not know him at this time. Hénon who explored numerically the Lorenz map using the IBM-7040, found difficult to highlight its inner nature due to its very strong dissipativity. Its rate of volume contraction is given by the Lie derivative of the Lorenz equations which can be solved. For the parameters chosen by Lorenz,[34] $V(t) = V(0)\mathrm{e}^{-\frac{41}{3}}$; thus after 1 time unit,

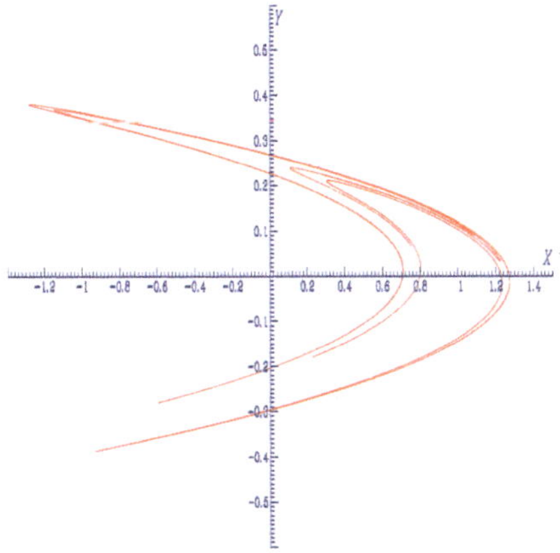

(a) The Hénon attractor

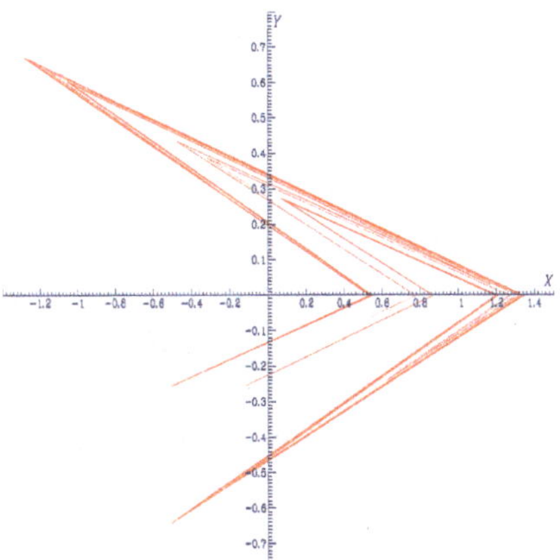

(b) The genuine Lozi attractor

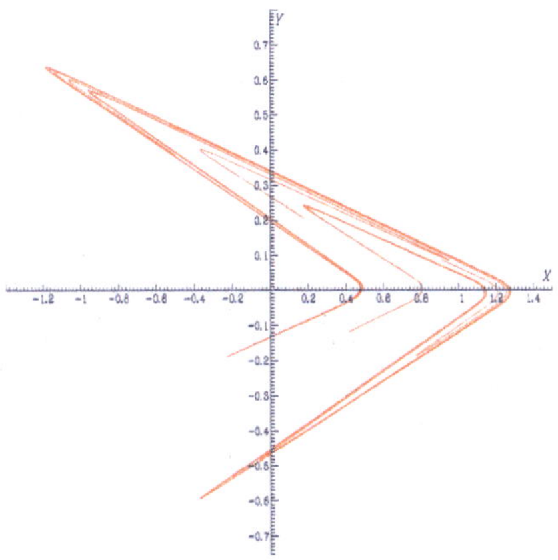

(c) Its smooth version for $\epsilon = 0.1$

Fig. 2. Some simple mappings.

volumes are reduced by a factor 10^{-6}. Hénon built the metaphoric model

$$\begin{cases} x_{n+1} = 1 - ax_n^2 + y_n \\ y_{n+1} = bx_n \end{cases} \tag{4}$$

for such a map whose contracting properties are only determined by the parameter b. With $b = 0.3$, the contraction in one iteration is mild enough that the sheaves of the attractors are visible. Eventually he observed graphically the fractal structure of the attractor.

Beyond bifurcation problem, my main interest was focused to discretization problem and finite element method in which nonlinear functions are approximated by piecewise linear ones. During the Roma conference, I tried to apply the spirit of the method of finite element to the Hénon attractor. Back to Nice on June 15 in the morning, during the defense of a Ph.D. by a colleague of mine, I eventually decided, using paper and pencil, to change the square function of the Hénon attractor, which is U shaped, into the absolute value function, which has a V shape, implying folding property. Few minutes after, in my office, I tested this modification, on my small desktop computer HP 9820 linked to the HP 9862 plotter. I shifted the parameter value a from 1.4 to 1.7 and b from 0.3 to 0.5 (why? I do not remember!) and plotted what is known today as the "Lozi map" [Fig. 2(b)].[35] Iooss and Chenciner encouraged me later to publish the formula

$$\begin{cases} x_{n+1} = 1 - a|x_n| + y_n \\ y_{n+1} = bx_n \end{cases} \tag{5}$$

which appeared in the proceedings of a "*Meeting between physicists and mathematicians about nonlinear problems and their applications*" held in Nice on 26–30 September 1977.[36] This was for me the very beginning of my career in chaotic dynamical systems. I was convinced that few weeks would be enough to explain and give a proof of the structure of a so simple attractor, but I failed. In the next two years I attended two workshops on iteration theory: the first one on 21–23 May, 1979 at La Garde-Freinet (a small town in the south of France), where Alain Arnéodo, Igor Gumowski, Michel Hénon, Christian Mira and Yves Pomeau were also present, and where Michal Misiurewicz, after some questions at the end of my talk, jumped on the stage, and on the blackboard gave some clues of his forthcoming results presented at the famous New-York conference, seven months later.[37]

The second meeting is a summer school in physics on July 1979 in Cargèse (Corsica) in the proceedings of which I eventually published a paper where I used the genuine non differentiable Lozi map

$$T_{z_\epsilon}(x, y) = (y + 1 - az_\epsilon(x), bx)$$

$$\begin{cases} |z_\epsilon(x) - |x|| \leq \epsilon \quad \text{for } |x| < \epsilon \\ z_\epsilon(x) = |x| \quad \text{for } |x| \geq \epsilon \\ z_\epsilon(\cdot) \in C^k, k \geq 2, \epsilon \text{ small enough} \end{cases} \tag{6}$$

with

$$z_c(x) = -\frac{x^4}{8\epsilon^3} + \frac{3x^2}{4\epsilon} + \frac{3\epsilon}{8} \quad \text{for } |x| < \epsilon \tag{7}$$

as an example, in order to prove the existence of one homoclinic point for a smooth version of it [Fig. 2(c)]. Then, applying a theorem proposed by Stephen Smale,[38] I proved the existence of an invariant Cantor set.[39] Before these two workshops, I met Smale in Orsay, and at the end of his talk, I discussed with him about the validity of my proof.

During the workshop in Cargese, I sympathized with Joe Ford from Georgia Institute of Technology who, in the 70s, conceived the idea of disseminating *Nonlinear Science Abstracts*. "*He personally collected, organized, typed, copied and distributed volumes of abstracts of dynamics papers, and thereby forged a global, interdisciplinary community* [...] *Eventually Joe's tireless efforts resulted in the foundation in 1980 of the first journal devoted to nonlinear dynamics, Physica D.*"[40] I will meet him again in New-York (see below) and during the Dynamics Days, Twente University, May 25–26, 1982.

After that, took place the congress organized by the New-York Academy of Science where I am proud to have shook the hand of Edward Lorenz, the father of chaotic attractors and I listen with a mix of anxiety and curiosity the first proof by Misiurewicz for the existence of a chaotic attractor for the map I discovered two and half years before.[37]

This conference entitled *International Conference on Nonlinear Dynamics*, patronized by the New-York Academy of Sciences, held on December 17–21, 1979, in the Barbizon-Plaza Hotel theatre, near the south bound of Central Park. The speakers and the session chairman were the worldwide foremost figures of nonlinear analysis or experimenters in physical sciences or chemistry: M. V. Berry, M. J. Feigenbaum, J. Ford, J. P. Gollub, J. Guckenheimer, D. D. Joseph, A. Katok, J. B. Keller, N. Kopell, O. Lanford III, P. D. Lax, E. N. Lorenz, B. B. Mandelbrot, J. E. Marsden, M. Misiurewicz, J. Moser, S. Newhouse, Y. Pomeau, M. I. Rabinovich, D. Ruelle, M. Shub, Yakov G. Sinaï, S. Smale, E. A. Spiegel, H. L. Swinney, Y. Ueda, C. Vidal, S. M. Ulam, J. Yorke, and many others. From Nice, Uriel Frisch, Claude Froeschlé, Michel Hénon (these three from the Observatory of Nice), Jean Coste (Ph.D. superviser of Coulet), Pierre Coullet, Georges Gonzi, Gérard Iooss, Charles Tresser, as I remember, attended the conference. I knew some speakers (I explain how, in the next section): Ford, Frisch, Joseph, Marsden, Misiurewicz, Moser, Newhouse,

Pomeau, Ruelle, Smale, Spiegel, and Vidal. The topics of the conference was split in six sections: Turbulence, Ergodic and integrable behavior, Physics and chemistry, Chaotic maps and flows, Chemical and fully developed turbulence, Strange attractors.

I was interested in the session devoted to turbulence due to the concept of strange attractor developed by Ruelle and Takens,[41] as a possible route to turbulence, to overcome the classical of theory of Landau[42] promoting an infinite sequence of bifurcations in tori of increasing dimension corresponding to quasiperiodic motion with an increasing number of frequencies which seemed still valid. The talk by Vidal[43] on the Belousov–Zhabotinsky reaction was of a so great interest for me, because I had visited in the early 1979 his Paul Pascal Institute in Bordeaux, during which Claude Lobry who was teaching there, at the mathematics department, introduced me to the team of chemists working on this oscillating reaction.

Of course, the talk by Misiurewicz[37] was a kind of ecstasy for the young researcher that I was. Surprisingly the talk was entitled "*The Lozi mapping has a hyperbolic structure*," differently from the corresponding paper in the *Annals*, and nested in the session "chaotic maps and flows," not in the last session "strange attractors." Even if it was possible to listen every lecture, I did not attend all, nor did Michel Hénon that I met many times visiting computer shops in the streets of New-York city. One has to say that in 1979, there was a huge technologic gap between France and USA. USA was a kind of magic kingdom for me. I saw for the first time microwave oven and cloth dryer, one year before in California (see next section). Therefore, for Hénon who was a geek before the word "geek" was invented, looking for to buy a new pocket calculator, like the HP 25 which was then unavailable in France, was fascinating. Beside the interest of the scientific program, the most memorable memory of my stay in New York is to have crossed the city by night, after dinner, walking during three hours from Battery Park to Central Park, via Greenwich Village, with my colleague Claude Froeschlé! I remember also the first snowfall of the winter with 3.5" of snow fell, on Wednesday 19, on New-York City which forced me to buy warm clothes and gloves![b]

---

[b]https://thestarryeye.typepad.com/weather/2012/02/new-york-weather-highlights-1979.html.

2.2. *Bifurcation, branching, buckling and nonlinear eigenvalues problems*

Let me go back to the early 70s, and, also to fill the gap between July 1977 and May 1979. As I said above, in France, the name bifurcation was largely unknown in the math community of analysts at the beginning of this decade. Referring to,[16] the situation was greatly different in the USA. However, the situation was slowly changing in my country. At the end of March 1973, on the incentive of Geymonat, and with the help of Zerner, I invited to give a seminar in Nice, Paul H. Rabinowich, a former student of Jürgen Moser (the M of the KAM theorem), who travelled from the "Scuola Normale di Pisa" to Paris. He spoke about his last results "Bifurcation from simple eigenvalues."[44] However, the name bifurcation was not yet very popular, it was in competition with "nonlinear eigenvalue problems," "branching" and "buckling" (and "diramazione" in Italian).

On June 16–25, 1974, I attended a C.I.M.E. (Centro Internazionale Matematico Estivo, sponsorized by NATO) seminar in Varenna, Italy, entitled "Eigenvalues of nonlinear problems." The title of the talk given by Rabinowitz was "Variational methods for nonlinear eigenvalue problems."[45] The lecture by R. E. L. Turner was entitled "Positive solutions of nonlinear eigenvalue problems,"[46] Paul Fife, in his talk about "Branching phenomena in fluid dynamics and chemical reaction-diffusion theory,"[47] indicated that

> The study of the bifurcation and branching behavior of problems in partial differential equations is a relatively new endeavor, and is rapidly gaining momentum. [...] Among the other papers which have been most influential, I mention Joseph and Sattinger (1972), [...] Nicolis and Auchmuty (1974), Iooss (1972-73), Kirchgässner and Kielhofer (1973) (a thorough mathematical account of bifurcation and stability of steady-state solutions of Navier-Stokes equations), [...], Hale (1969) (for branching theory of ordinary differential equations). [...] It was my intention also to outline the present state of branching theory of thin elastic plates and shells. [...] Indeed, the "buckling" of plates and rods can be interpreted as the transition from one solution branch of the relevant equation to another.

It is noteworthy that Iooss was the only French quoted by Fife.

The arrival of Iooss in our department of Mathematics in September 1974, was the turning point of the development of the bifurcation theory in Nice.[48] He was in touch with many researchers working on the topic including Klaus Kirchgässner, Daniel D. Joseph, Hans True, and many others. He worked with his Ph.D. student Pascal Chossat.[49] Alain Chenciner arrived the next year and stayed three years in Nice. Some researchers from USA, were visiting our department, like J. F. K. Auchmuty[50] and H. B. Keller.[51] Nevertheless, the study of bifurcation phenomena was often marginally nested in workshop, seminar or conference of nonlinear analysis and application or physics as one can observe in the following events.

From 18 to 24 January 1976, I was invited by Kirchgässner with Iooss and Claude Bardos (visiting professor at Nice in 74–76), at Mathematisches Forschungsinstitut Oberwolfach for a seminar on "Nichtlineare Funktionalanalysis und irhe Anwendungen,"[c] organized by him, H. Amann (Bochum) and N. Bazley (Köln),[52] where I spoke about the numerical computation of branches of bifurcation[15] and Iooss about the "Conjectures on the dynamo problem for the geomagnetic field."[29] Some bifurcation problems were exposed by N. Chafee (Hopf Bifurcation and Arbitrary Perturbations of a Differential Equation), E. N. Dancer (Bifurcation from infinity), R. J. Magnus (The odd multiplicity criterion of bifurcation theory) and Roger Nussbaum (A Hopf Global Bifurcation Theorem for Retarded Functional Differential Equations). There were only six talks on bifurcation among a total of 27.

All the walls of the conference room at Oberwolfach support gigantic blackboards. An amazing time during this seminar was when, at the end of the talk by Bardos (Hölder estimates and time regularity for the Euler equation), all these blackboards were covered with formulas and theorems and that Ernst Hölder (Mainz), the son of Otto Hölder, the mathematician whose name was given to the Hölder spaces, stood up, went to the blackboards and before each word "Hölder" wrote either E. or O., with reference to his father's work, or to his own! Unfortunately, I did not understand the explanations he gave in German.

When as young student, I studied analysis, the name of the mathematicians given to the functional spaces, like Frechet, Banach, Hilbert, Hölder, Sobolev, Besov, had a scent of history, except for Sobolev (1908–1989) who, Céa reminded me from time to time, attended to the first colloquium on

[c]Nonlinear functional analysis and its applications.

numerical analysis he had organized in Paimpol (France) in 1967.[d] However, concerning Hölder, I simply realized that he was not so far in the past as I imagined.

In September 1976, after two years as an assistant (the first grade of permanent position in the French university) I obtained one of the very few positions in France at C.N.R.S. as "Attaché de Recherche" with Céa for superviser. In September 1977, the group around Iooss, Chenciner, Coste, Frish, Coullet, Tresser, was sufficiently strong to organize a "Meeting between physicists and mathematicians about non-linear problems and their applications," gathering around thirty speakers.[53–55]

In June 19–27, 1978, the group from Nice, excepted Iooss who was spending the academic year in Minneapolis, at the University of Minnesota, invited by D. D. Joseph, also attended the C.I.M.E seminar on Dynamical systems in Bressanone (Italia). Chenciner gave one of the five lectures during this event,[56] along with the reputed researchers Guckenheimer, Moser, Misiurewicz and Newhouse. I sympathized with Newhouse and I drove him in my car to the nearby mountains of Trentino.

On July 10–11 of the same year, Marsden[57] organized a "Symposium on nonlinear analysis and mathematical physics" at the university of California, Berkeley, in the Evans building recently achieved. I attended this workshop together with Chenciner, Bardos, Iooss and Jean-Michel Lemaire (geometrician) from Nice. I remember that David Chillingworth who studied also the dynamo reversal of the magnetic Earth field, was present.[58] Other participant were physicists like A. Fisher,[59] I. Singer[60] and P. van Moerbecke,[61] working on general relativity or other theoretic topics. It was my first visit to this university and my first stay in the USA. During a picnic lunch, Marsden tried to initiate us to play frisbee, however the French team was never able to play as a true team, like our American colleagues.

In October 1978, Céa, incited me, from an idea of Lions, to organize a conference on bifurcation in Nice. The Institute of mathematics provided all

[d]Céa, who was very close to Jacques-Louis Lions, his former superviser, was at that time professor at the university of Rennes. He said me recently that Lions phoned to him to invite Sergueï Sobolev, visiting professor at Collège de France, at this colloquium. Sobolev was fluent in French and knew a lot about the history of France. He was willing to meet young students. At first, in Paimpol, all the researchers wanted to stroll with him in the city, but soon they changed their mind because Sobolev was reading the names of the streets and told a lot about the incumbents, while the French mathematicians did not know so much. So, not to show their ignorance, there were fewer candidates to walk with this great man.

the funds. I managed a 4-days conference entitled "*Didactique et recherche en bifurcation*," on December 4–7. Among the participants one can mention Céa (Generalized Lorenz model following J.-H. Curry), Coullet (Successive bifurcations and renormalization group), Geymonat (Example of use of the tools of catastrophe versus functional analysis), Iooss (Tutorial on Hopf bifurcation), Lemaire (A theorem of Alexander–Yorke–Ize on global existence of Hopf bifurcation), Jean-Pierre Puel (Homogenization and bifurcation: multi-perforated plate problem), Roger Temam (Bifurcation in plasmas), Tresser (Local topological entropy for endomorphisms of the interval), Bruno Scheurer (Application of the Malgrange theorem to the problems of multiparameter bifurcation), and Jean-Christophe Yoccoz (Introduction to the work of Yakov Pesin on differentiable manifolds).

It was the first ever invitation of Yoccoz, future Fields medalist (1994), to give a talk in a conference. In fact, I tried to invite his well-known superviser Michael Herman that I visited in Paris; however, Herman was not comfortable in travelling by train (at this time, travelling from Paris to Nice by air was very expensive). He said me "*I know a very gifted student in third year of the École* [Normale Supérieure (Paris)], *he can speak as well as me.*"[e]

Following the advice of some speakers, I tried to publish the proceedings in *Astérisque*. However, after several months of discussion, eventually the editorial board of this journal refused for obscure reasons. I was not enough experimented in research to argue with this board, mainly based in Paris. At this time, it was difficult to challenge the idea that "good mathematics" could be done only in capital of France! Moreover, the city of Nice, was often associated with tourism and idleness, not with mathematics (except for the international Congress of Mathematics in 1970, held in this city, overseen by Dieudonné[f])! Nevertheless, I kept in my records the kind letter from Yoccoz accepting to write his talk (Appendix 1).

[e]In his thesis, Yoccoz improved theorems of his supervisor Herman by giving simpler proofs but also obtaining the same results under weaker hypotheses. After that, he made major progress on an important conjecture about Benoît Mandelbrot's fractal ensemble, leading to a conjecture by A. Douady and J. Hubbard, which earned him the Fields Medal in 1994. In this aim, he introduced the famous "Yoccoz puzzles."[62]

[f]Gérard Tronel[27] in a public report about this Congress indicated that "*J'ai assisté à toutes les conférences plénières et au maximum de sessions parallèles compatibles avec le programme, il fallait faire des choix en fonction de thèmes plutôt en dehors de mes intérêts du moment — analyse numérique et bifurcation — sur ce dernier sujet, aucune communication ne figurait au programme, il n'était pas à la mode à l'Ouest, alors qu'à l'Est une équipe pilotée par M. A. Krasnosel'skii, à Voronej, publiait des articles sur ce thème, mais en russe!*"

After this conference, it had become obvious to me that the works on bifurcation were little known in France. There was not yet any online publication database. The only revues providing a synthetic view of mathematical articles published in the Western world were the *Mathematical Reviews* published by the American Mathematical Society and the *Zentralblatt für Mathematik* founded in 1931, published by Springer, however they did not reference the entry bifurcation. Following the bibliography of 172 titles published in my thesis, I decided to do a review of the articles on bifurcations. In 1982, after several years of hard work spent reading all the volumes of this journal since 1940 for *Mathematical Reviews* and 1971 for *Zentralblatt*, after having written hundreds of letters to the authors and received a huge quantity of reprints, I programmed a text editor in FORTRAN IV and published a "tentative bibliography" including 806 references.[63]

3. In a Search for a Good Realistic Model for Chaos

During the year 1979, I worked on two different directions. One direction was the research of a realistic model for chaos, coming either from physics or chemistry. At this time, I encountered Claude Lobry, at the university of Bordeaux, who was working with the group directed by Adolphe Pacault on the Belousov–Zhabotinsky reaction at the Paul Pascal Institute including Christian Vidal and Jean-Claude Roux. Lobry was also involved in Non-Standard analysis, mainly applied to slow-fast dynamical systems. I detail this field of research in a next subsection.

3.0.1. *Hyperchaos*

The second direction was the research of another very simple map with a chaotic attractor in a dimension greater than 2. As I said above, the same year, Otto Rössler published some examples of hyperchaos, either in EDOs or mapping.[8] One originality of these articles is that Rössler used stereoplot to show 3D images. The sole published 3D map

$$\begin{cases} x_{n+1} = 3.8x_n(1-x_n) - 0.05(y_n+0.35)\,(1-2z_n) \\ y_{n+1} = 0.1\left[(y_n+0.35)\,(1-2z_n)-1\right](1-1.9x_n) \\ z_{n+1} = 3.78z_n(1-z_n) + 0.2y_n \end{cases} \tag{8}$$

therein [Fig. 3(a)] is described by Rössler as a "*generalization of the simple two-dimensional diffeomorphism*"

$$\begin{cases} x_{n+1} = 3.8x_n(1 + x_n) - 0.1y_n \\ y_{n+1} = 0.2(y_n - 1.2)(1 - 1.9x_n) \end{cases} \tag{9}$$

Analyzing this map, Rössler indicates that[g]

> A brick-shaped original domain is, after being compressed and stretched, folded over itself in two directions and then put back into itself, between one-time step and the next ("folded towel"). A non-selected initial point is "plane-filling" with its iterates. More importantly: pairs of adjacent points diverge exponentially in both the x- and the z-direction. Thus, there is "lateral instability in two directions."

To provide a more amenable model, allowing mathematical proof, I tried to build a piecewise linear model, with the constraint of a constant Jacobian on the whole state space, which was a non-trivial generalization of the Lozi map producing the same kind of attractor. Unfortunately, with such tough constraints, I could not find a continuous model. However, I found the map[9]

$$\begin{cases} x_{n+1} = a|x_n| + y_n \operatorname{sgn}(z_n) + 1 \\ y_{n+1} = b(x_n + y_n) \\ z_{n+1} = y_n \operatorname{sgn}(x_n) + c|z_n| + 1 \end{cases} \tag{10}$$

which has the particularity that, for some parameter value, thread and sheet strange attractors coexist [Fig. 3(b)]. Following Rössler presentation, I used stereoplot, thanks to Pierre Magnenat of the Geneva observatory which sent me the subroutine he had written. In August 1983, Ippeï Shimada from the Nihon university, Tokyo specialist of Lyapunov exponents,[65] gave me the results of his computation for this map. For the values of Fig. 3(b), there are two positive exponents, for both initial values.

In 1984, Rössler with J. L. Hudson published a piecewise linear 3D map called "noodle map."[66] Starting from the smooth noodle map

$$\begin{cases} x_{n+1} = \alpha x_n(x_n^2 - 1) - y_n - z_n \\ y_{n+1} = \beta x_n \\ z_{n+1} = \gamma(x_n^2 - 0.33) + \delta z_n \end{cases} \tag{11}$$

[g]Strangely, Rössler did not link its observation to the horseshoe introduced by Smale ten years before[38] as he did two years before for the Lorenz equation,[64] nor used the Lyapunov exponents to quantify the divergence of adjacent points.

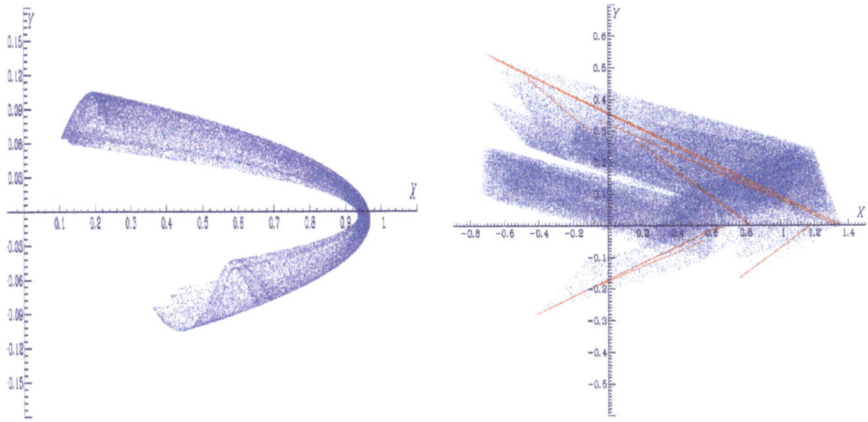

(a) (b)

Fig. 3. (a) Rössler hyperchaotic attractor; initial value $x_0 = 0.085$, $y_0 = -0.121$, $z_0 = 0.075$. (b) Coexistence of thread and sheet piecewise hyperchaotic attractors for $a = -1.25$, $b = 0.2$, $c = -1.25$; initial value threat attractor (red) $x_0 = 0$, $y_0 = 0$, $z_0 = 0$; sheet attractor (blue) $x_0 = 0.37$, $y_0 = 0.8$, $z_0 = 0.075$.

with a constant determinant $\beta\delta$[67] [Fig. 4(a)], they built a piecewise linear analogue [Fig. 4(b)],

$$\begin{cases} x_{n+1} = -y_n - z_n + \left| \begin{array}{ll} 3.3(x_n - 1) - 2 & \qquad x_n > 1 \\ -2x_n & \text{if } \; 1 \geq x_n \geq -1 \\ 3.3(x_n + 1) + 2 & \qquad x_n < -1 \end{array} \right. \\ y_{n+1} = \beta x_n \\ z_{n+1} = \gamma(|x_n| - 1) + \delta z_n \end{cases} \tag{12}$$

arguing

> Piecewise linear maps are of interest because they are more amenable to analysis. After Hénon[32] had indicated his well-known two-dimensional diffeomorphism with constant Jacobian determinant, Lozi[36] gave a variant of this equation that had the same determinant but was piecewise linear. Since then, many analytical studies of Lozi's map have appeared.[37,68,69]

They mentioned that "*So far, no other piecewise linear, invertible map in three dimensions has apparently been described.*" This article is the only one in which Rössler referred to my research, although not for hyperchaos. In the 80s, it was not possible to get access to new results for several years,

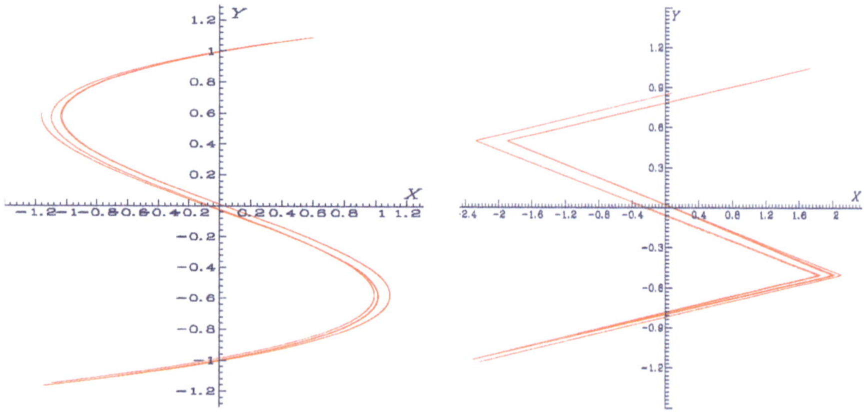

(a) Continuous map
$\alpha = 2.77$, $\beta = 0.01$, $\gamma = 0.1$, and $\delta = 0.05$
$x_0 = 0.0$, $y_0 = 0.0$, and $z_0 = 0.0$

(b) Piecewise map
$\beta = 0.1$, $\gamma = 0.1$, and $\delta = 0.05$
$x_0 = 0.0$, $y_0 = 0.0$, and $z_0 = 0.0$

Fig. 4. Rössler noddle maps. (a) y is magnified 100 times and (b) y is magnified 5 times.

moreover Eq. (10) was not published in a journal, but in a book of limited distribution, this explain the remark of the authors.

3.1. *Slow-fast differential equations*

In the early 1980s, I worked for several years with Claude Lobry who arrived in the department of Mathematics in Nice. We were interested in "slow-fast" mathematical models for the Belousov–Zhabotinsky reaction, like the Brusselator model made of two ODEs developed by Nicolis and Prigogine in 1977,[70] the Oregonator devised by Field and Noyes working at the University of Oregon which is composed of five coupled elementary chemical stoichiometries,[71] and the model with even more ODEs developed at Bordeaux.[43,72] To model the complex bifurcation diagram observed in very precise chemical experiments in Bordeaux, I introduced the idea of bifurcation of "motifs" (patterns) in a note to the *Académie des Sciences de Paris*, presented by René Thom.[73] This was my first indirect exchange with Thom. I developed this model with Lobry,[74] and on June 29, 1983, I defended my "Thèse d'État" that Thom accepted to review.[3] He was very interested by the ontology of the slow-fast metaphoric model I introduced.

I had also a long scientific relationship with José Argémi (1933–1985)[75] who worked at C.N.R.S. in Marseilles and his doctoral student, Bruno Rossetto (now professor emeritus) from the University of Toulon. Argémi was a specialist in slow-fast ODEs systems. He coined co-folded line and pseudo-singular point, very useful mathematical tools for such systems. However, it was very difficult to argue with him because he was locked into his own mathematical universe. During 7–17 September 1982, the three of us attended the colloquium "Logos et théorie des Catastrophes" to celebrate René Thom in Cerisy-la-Salle, organized by Jean Petitot.[76] Numerous renowned mathematicians and physicists like Michael Berry, Chenciner, Chillingworth, Adrien Douady, Georges Lochak, Ruelle, Timothy Poston, Bernard Teissier, and Sir Christopher Zeeman gave a talk. All those, and scientists of other disciplines, philosophes, psychanalysts, linguists, were very impressed by René Thom. Me too much more. It was my second in-person meeting with him. The first one took place in May of the same year, during a meeting in Toulouse.[9]

Conferences in Cerisy-la-Salle are very special. For ten days, all the attendees are living in an old castle near the tiny village of Cerisy-la-Salle which is isolated in Normandy, difficult to access. Everyone is eating at the same long tables in the kitchen, sharing his times between conferences, long discussions, strolls in the park, reading of books of the library, or even playing ping-pong. It is the perfect place to build relationships, exchange ideas. Spending ten days with such outstanding researchers was a singular experience.

An incredible moment of this conference for me, was when Douady[77] who worked on Julia sets, was very anxious to make a presentation in presence of Thom. Therefore, the day before, he asked me to attend the rehearsal of his talk. I have to say that when I was preparing my bachelor's degree, Douady was the second most renowned professor of the Institute of mathematics at the university of Nice after Dieudonné. He was a pure mathematician in geometry at this time. Somewhat eccentric, having a bushy beard, wearing all long the year a "Lederhosen" (kind of knee-length leather breeches that are worn as traditional garments in Bavaria), and sometimes climbing on palm trees of the garden of the university to mull over a math problem. I suddenly found myself, in a very uncomfortable position, all alone in a small room in the castle, listening to him and asked to do some remarks on his talk!

It was precisely during this conference, that Thom accepted to review my "Thèse d'État". I gave a talk entitled "Bifurcation de motifs dans la réaction de Belousov–Zhabotinsky", Argémi speaked about "Autour d'un point singulier sur la fronce" and Rossetto on "Application of the Hill method generalized to the study of stability of periodic solutions of a singular perturbed system." Unfortunately, as Petitot confirmed me in May 2020, the project of publishing the proceedings of all the talks given during the conference in two volumes (one in French, one in English) partially failed and only the volume containing epistemological and philosophical aspects was published,[76] six years after the event.

During the speech of conclusion talk Thom said [76, p. 513]

> [...] So that even where there is apparently chaos, there are always morphological elements to study. The presentations that we have heard subsequently, for example that of Lozi very recently, are in exactly the same direction: there is always interest in trying to understand how things evolve morphologically and, if we cannot, of course, one can always dive into statistics."[h]

Thom was very kind to suggest to Ruelle to invite me to give a talk on the same subject during a meeting on week turbulence held at I.H.E.S. on November 18–19, 1983.

3.2. *Constrained Lorenz model*

During spring 1981, I received the surprise visit of Hiroshi Fujii[78] from Kyoto university, who was spending one year in INRIA near Paris. He was interested by my doctoral dissertation. We discussed about people working in the same area in Japan. He gave me the names and addresses of Shigehiro Ushiki, and his superviser Masaya Yamaguti,[79] both also from Kyoto university. I sent to Ushiki some papers on July 1981 who, in turn, sent me back his own ones on August 81. One year later, during the meeting in Toulouse,[9] I met him. He was fluent in French. Then we worked very closely for twelve years, first on slow-fast systems, and hereafter on the Chua circuit.

In 1983, thanks to Professor Yamaguti, I got three weeks grant from the Japanese Society for the Promotion of Science for an internship at the

[h]De sorte que, même là où apparemment il y a du chaos, il y a toujours des éléments morphologiques à étudier. Les exposés qu'on a pu entendre par la suite, par exemple celui de Lozi tout récemment, vont exactement dans le même sens : il y a toujours intérêt à essayer de comprendre comment les choses se passent morphologiquement et, si on ne peut pas, bien entendu, on peut toujours plonger dans la statistique.

Fig. 5. Masaya Yamaguti, Shigehiro Ushiki and René Lozi during the IUTAM symposium on "Turbulence and chaotic phenomena in fluids" (September 5–10, 1983) in Kyoto.

Faculty of Science of the imperial University of Kyoto. Few weeks after the defense of my Thèse d'État in June, I flied with my family to Japan and worked with both researchers (Fig. 5). I met also Hiroshi Kawakami who I knew since the conference in Toulouse in 1982 and with whom I will discover the Alpazur oscillator in 1991,[4,5] Hiroshi Fujii, Ippei Shimada who was interested in hyperchaos,[65] Ichiro Tsuda then at Bioholonics Research Project in Tokyo,[80] who will develop the concept of nowhere differentiable attractors with Rössler, twenty years later[81] and Masayasu Mimura at Hiroshima university.[82]

Masaya Yamaguti (1925–1998) was also fluent in French. In 1948 his superviser Hiroshi Okamura died. The students of Okamura tried to learn maths by themselves, but they did not understand anything. Then Yamaguti with his close friend Hiroshi Mizohata (1924–2002) decided to go to France by boat, spending 50 days in the travel. They studied French together with literature students. Yamaguti stayed two and half years in France, following the lectures of Laurent Schwartz on topological vector spaces, finding them too much theoretical.

Otto Rössler met Yamaguti eight years earlier than me, as he said[83]

> Fritz Horn in Rochester also became a friend and Okan Gurel, White Plains, and Hans Degn and Lars Folke Olsen in Odense and Dick Noyes, Oregon [...] Okan and Desnet, his chemist wife, brought us to Kyoto in 1975 where we met Anatol Rapoport and Masaya Yamaguti, who had graduated from a Zen master rather than a state school to become the weightless personality to anyone he interacted with — a holy man in mathematics.

May be the weightless personality of Yamaguti came from his father Kayo Yamaguti (1899–1984) who was a famous animal and flower bird painter.[i] Masaya Yamaguti was upset that these painting were too much expensive, he could not buy at least one.

I did not published maths with him; however, he was interested in teaching maths, and we published a paper in pedagogic journal explaining a method to multiply using only fingers.[84] He visited me two years later, gave a talk in French in my Institute on April 1st, 1985 entitled "Ensembles invariants pour un système de contractions." An amazing fact, during the multiple discussions I had with him, he told me part of his life; when I told him that, as a student I worked on the Yukawa potential, Yamaguti explained me that when Hideki Yukawa, who was a former student of the Imperial university of Kyoto, died in 1981, he was Dean of the Faculty of Science, and he had to give a talk on the career of this great savant.

During my stay in Kyoto, on August 22–24, I attended a three-days seminar on "concretization of strange attractors" organized by Yamaguti and Ushiki. Hiroshi Kokubu, a Ph.D. student of Yamaguti and Ushiki, gave a talk on "bifurcations of Rössler attractor." I gave two talks, one on pattern bifurcation in the Belousov–Zhabotinsky reaction, and one on stereoscopic plotting of three-dimensional chaotic attractors. This talk was motivated by the first use in my career of a personal computer with 7 colors screens, a SORD M-343 V [Fig. 6(a)]. I built handicraft anaglyph glasses with green and red transparent papers, and cardboard [Fig. 6(b)] and I showed on a computer the anaglyphic figures of the Rössler and piecewise linear hyperchaotic attractors (8) and (10).

On 31 August–1st September, I traveled to Hiroshima to visit Masayasu Mimura. Discussing with him, I suddenly realized that my vocation

[i] https://www.momak.go.jp/Japanese/exhibitionArchive/2012/395.html.

(a) Desktop computer SORD M-343 V (b) Anaglyph glasses

Fig. 6. Devices used by René Lozi in 1983 for watching the attractors he computed with the help of anaglyphic figures.

in research will be building mathematical models, rather than proving theorems. I turned back to Kyoto on the precise day the Korean Air Lines Flight 007 which was a scheduled flight from New York City to Seoul via Anchorage Alaska,was shot down by a Soviet Su-15 interceptor. All 269 passengers and crew aboard were killed, arousing strong indignation in Japan.

The last week of my stay in Japan was devoted to the IUTAM symposium on "turbulence and chaotic phenomena in fluids" (September 5–10) in Kyoto. Several people from the university of Nice or from France attended too: Arnéodo, Coullet, Frish, Iooss, Albert Libchaber, etc. Yamaguti introduced me to Benoît Mandelbrot. Among one hundred of talks, one can remark those of Arnéodo, Coullet and Spiegel (Chaos and temporal intermittency in mildly unstable fluids); Frisch (Numerical simulation of fully developed turbulence and the appearance of spatial chaos); Iooss (Generation of chaos and transition to turbulence); Libchaber (Rayleigh-Bénard experiments in low Prandtl number fluids); Mandelbrot (Fractal chaos and turbulence: foundations and applications); H. L. Swinney and A. Brandstäter (A strange attractor: chaos in Couette-Taylor system).

After my first stay in Japan, I could find some grants from C.N.R.S. allowing Ushiki to work several months every year in Nice between 1984 and 1992, excepted 1987. During most of these years, I was elected member of the mathematical section of the Comité National de la Recherche of

C.N.R.S. (and secretary of this section between 1984–1987) in Paris, which allowed me to better know how to get such grants!

In my thèse d'État, as I said above, I had introduced the concept of "motif" (pattern) to explain de shape of oscillations in the Belousov–Zhabotinsky reaction, in the frame of "slow-fast" mathematical models. I was strongly inspirated by the work of Rössler as I mentioned in my thesis: *It should be noted that all of this work was strongly inspired by the research of O. E. Rössler on "Chaos by reinjection."*[3] However, the Belousov–Zhabotinsky chemical reaction is very difficult to correctly model, due to the uncertainty of the number of chemical species transitorily involved, which could be up to dozens. I was more attracted to study a classical model of chaotic attractor, using the tools of slow-fast differential equation. The Lorenz model does not belong to this class of ODE. In order to link this model to slow-fast equations, I proposed an ideal geometrical representation of the Lorenz attractor on a cusp [3, Fig. 20] (Fig. 7).

Hiroe Oka and Hiroshi Kokubu, two Ph.D. students of Yamaguti and Ushiki, investigating systematically chaotic attractors modelled by constrained systems, using implicit differential equation (also called generalized

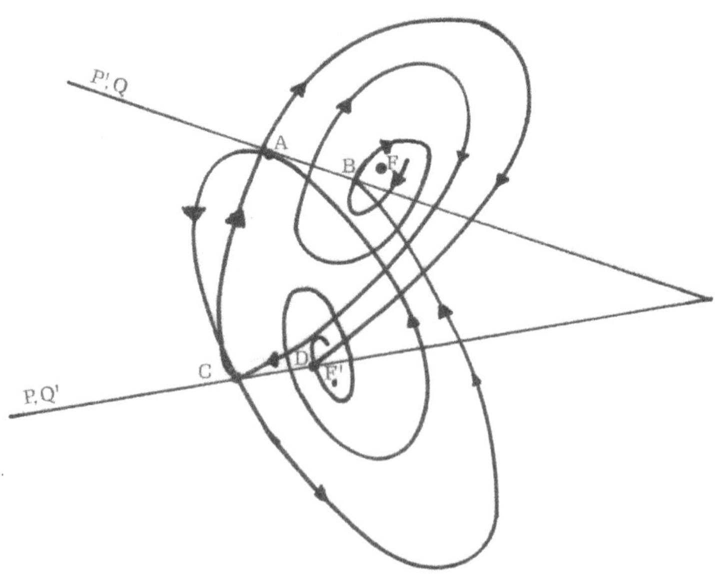

Fig. 7. Sketch of a slow-fast construction for the Lorenz attractor as shown in Fig. 20 in Lozi,[3] and as in Fig. 4 in Oka and Kokubu's paper.[85]

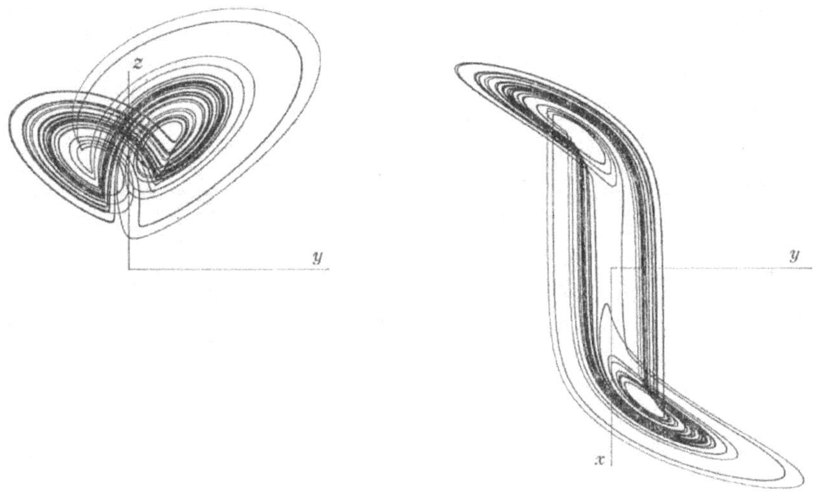

Fig. 8. Chaotic attractor produced by the slow-fast system (13).

Notes: Parameter values: $\epsilon = 0.03$, $A = 0.7$, $B = 0.7$, $C = -1.0$, $\alpha = 0.7$, $\beta = -1.0$ and $\gamma = 1.0$.

Source: From Oka and Kokubu's paper.[85]

vector field), found the set of equations

$$\begin{cases} \epsilon\dot{x} = y + \alpha xz - x^3 \\ \dot{y} = Ax + By + \beta xz \\ \dot{z} = Cz + \gamma x^2 \end{cases} \tag{13}$$

that produces a similar attractor (Fig. 8) to the model I draw by hand.

In their paper,[85] Oka and Kokubu explicitly refer to Rössler and Takens

> Investigations of strange attractors by constrained systems seem to have been begun by Takens and Rössler independently. [...] Takens gave an example of a constrained system.[86] Though he only illustrated a picture, his constrained system seems to model the Lorenz's strange attractor.[34] [...] Rössler[2] also considered such systems on a double fold type surface [Fig. 9], and he proposed four typical constrained sytems as prototype of continuous chaos.[6] One of these systems called "the spiral type chaos" is shown in [Fig. 9(a)]. Among these prototypes, a constrained system modelling the Lorenz attractor is contained, which is essentially the same as the Takens's model, though the latter is constrained on the cusp surface (the Whitney's pleat).

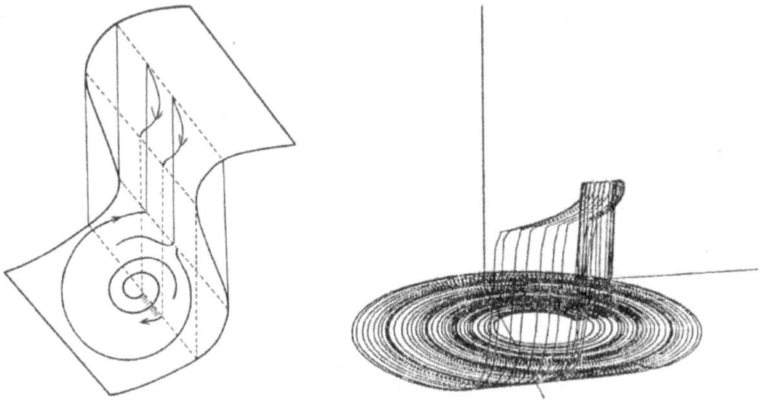

(a) Slow-fast manifold (b) Numerical simulations of (14)

Fig. 9. The spiral type attractor proposed by Rössler.[6]

Sources: Redrawn from Fig. 3 in Oka and Kokubu's paper,[85] largely inspired by Rössler's one.[6]

Moreover, Rössler gave an explicit form of differential equation as follows[6]

$$\begin{cases} \dot{x} = -y + ax - bz \\ \dot{y} = x + 1.1 \\ \epsilon \dot{z} = (1 - z^2)(x + z) - \epsilon z \end{cases} \tag{14}$$

which realizes a spiral type strange attractor by numerical experiment when $\epsilon = 0.03$, $a = 0.1$ and $b = 1$ [Fig. 9(b)].

As for the constrained model of the Lorenz attractor, neither model indicated above reflects the symmetry of the original Lorenz equation, which is invariant under the transformation $(x, y, z) \to (-x, -y, z)$.

Recently, Lozi proposed another constrained model respecting this symmetry.[3] His model (Fig. 7) has also the cusp type constraint surface.

3.3. *Confinor vs. attractor*

At this point of my historical description, I must introduce a philosophical reflection which guided partially my researches during the ten following years. The modern study of dynamical systems is built on the underlying use of infinity. Attractors, first introduced by Smale,[38] play a central role in the description of dynamics. They are Ω-limit sets, that is sets of topological limit points that the considered systems approach as time goes to positive infinity.

This underneath use of infinity faces two problems. The first one is related to numerical computations often essential to study chaotic systems. Infinity cannot be reached even doing very long computations. One can compute only transient states, and many people consider that for mappings, the computation of the firsts thousands of iterates is enough to display an attractor.[j] However, there are many examples in scientific literature, of long chaotic transient regime hiding a periodic attractor with a very short period.[87] The second problem is that one cannot choose the level of analysis of the state space. This level is dictated by the thinnest objects nested inside. Can you imagine the geographical map of the world in which you could not describe states, countries, region, but cities only?

In order to solve both problems, killing two birds with one stone, I considered in my thèse d'État, some particular trapping regions of the phase space, and called them confinors.[3] In fact, at the coarsest level of analysis, a trapping region, smaller than the basin of attraction, is defined and called the main confinor. It can contain one or several attractors. In the special case of slow-fast differential equation, there is a characteristic "motif" (pattern) of the solutions linked to this main confinor. At the second level of analysis, the main confinor is split into confinor(s) and anticonfinor(s), with more complicated "motifs," and so on for the next levels.

With Ushiki, we studied thoroughly equations linked to (13) in this paradigm, using some tools introduced by Argémi: pseudo singular points, fold and co-fold curves, together with others defined by Takens: induced and reduced vector field.[86] Especially, we studied the family of differential equations

$$\begin{cases} \epsilon\dot{X} = Y + \dfrac{3}{4}XZ - X^3 & = f(X,Y,Z) \\ \dot{Y} = \tau^2 AX + \tau BY + \dfrac{3\tau}{4\alpha}\beta XZ & = g(X,Y,Z) \\ \dot{Z} = \tau CZ - \dfrac{4\alpha\tau}{3a}\gamma X^2 & = h(X,Y,Z) \end{cases} \tag{15}$$

[j]In Hudson's work,[66] setting $\alpha = 2.726$, $\beta = 0.1$, the authors follow an initial point through 1,600 iterations only and conclude that it is a noodle chaotic attractor. It seems that they limit their computation to 1,600 iterations because the 1,630th iterates diverge! Instead with $\alpha = 2.77$, $\beta = 0.01$ as in Fig. 4(a), there is no divergence before 10^7 iterations, at least.

constrained on a cusp surface,[88] similar to Eq. (13). The "slow" vector field defined on the stable part

$$S = \left\{ (X, Y, Z) \mid f(X, Y, Z) = 0, \frac{\partial f(X, Y, Z)}{\partial X} = 0 \right\} \tag{16}$$

of the slow surface

$$M = \{(X, Y, Z) \mid f(X, Y, Z) = 0\} \tag{17}$$

is the vector field defined on S, tangent to the cusp surface whose (Y, Z)-components are given by

$$\begin{cases} \dfrac{\mathrm{d}Y}{\mathrm{d}t} = g(X, Y, Z) \\ \dfrac{\mathrm{d}Z}{\mathrm{d}t} = h(X, Y, Z). \end{cases} \tag{18}$$

Under some changes of variable, we were able to simplify the reduced vector field into the form

$$\begin{cases} \dot{x} = -x(z - U - Vx^2) \\ \dot{z} = D\left(z - 4x^2\right)\left(z - Wx^2\right) \end{cases} \tag{19}$$

where

$$\begin{cases} U = \tau^2 A \\ V = \tau\left(B - \dfrac{\alpha\gamma}{a}\right) \\ D = -\dfrac{3C\tau}{4} \\ W = -\dfrac{4\alpha\gamma}{3aC} \end{cases} \tag{20}$$

getting closed formula of the fold and cofold curves (Fig. 10).

A further simplification led to the study of the reduced vector field

$$\begin{cases} \dot{\xi} = -2\xi(z - (W - V) - V\xi) \\ \dot{z} = D(z - 4\xi)(z - W\xi) \end{cases} \tag{21}$$

with fold line and cofold line (Fig. 11).The jump from the fold line to the cofold line being given by $(\xi, z) \mapsto (4\xi, z)$.

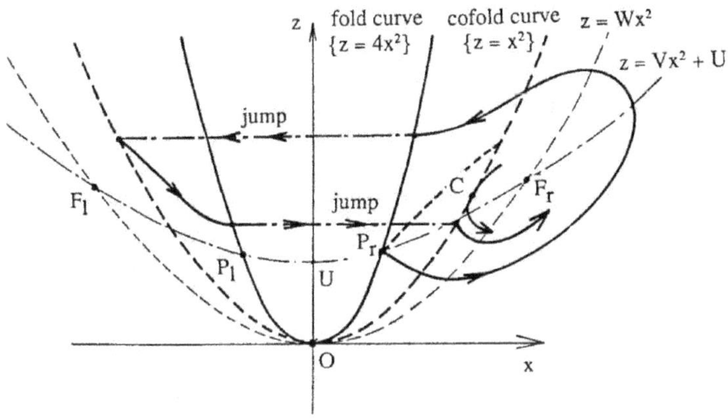

Fig. 10. Phase portrait of the reduced vector field (19).

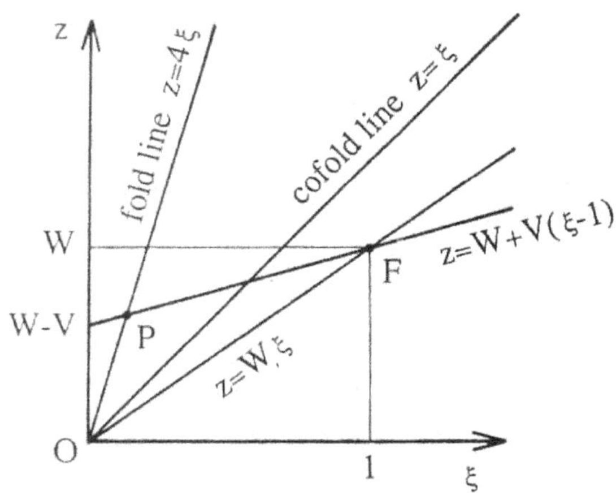

Fig. 11. Phase portrait of the simplified reduced vector field (21).

We considered confinors whose boundary consists of a finite number of arcs of integral curves, arc of fold and cofold curves, and singular points. Figure 12(a) shows an example of nontrivial confinor which has S^1L^3, S^1L^2 and S^1L^1 oscillatory patterns. S being the oscillation linked to the jump from the fold line to the cofold line, and *vice-versa*, and L to the rotation around F [Fig. 12(b)]. We computed bifurcation diagrams of confinors in the space of the parameters (W, V) [Fig. 12(c)] and highlighted different kinds of bifurcation of confinor.[89]

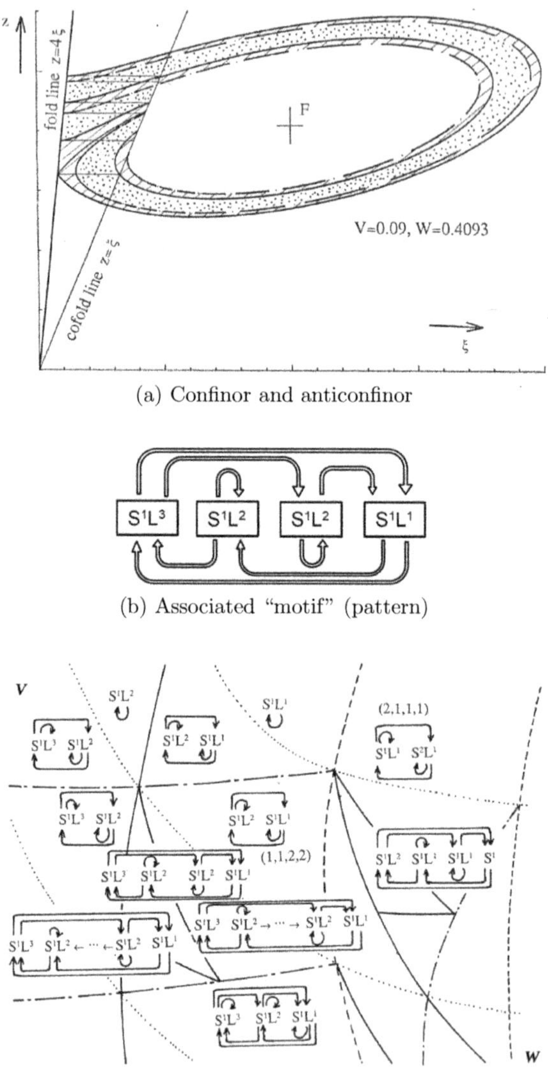

(a) Confinor and anticonfinor

(b) Associated "motif" (pattern)

(c) Bifurcation diagram of confinors

Fig. 12. Example of confinor (hashed part) and anticonfinor (dotted part) of the constrained Lorenz system (a) and its associated "motif" (b). Bifurcation diagram of "motifs" in the parameter space W-V (c).

4. Electric Circuits Model for Chaos

My personal search for a good realistic model for chaos comes to an end with the knowledge of the chaotic electric system by Leon Chua discovered in August 1983 when he was visiting professor in the laboratory of Takashi Matsumoto, at Waseda University, Tokyo.[90]

4.1. *The Chua circuit*

I first presented the results on constrained Lorenz model during the conference on chaotical phenomena at Issy-les-Moulineaux near Paris, on December 3–5, 1986. During this conference, Leon Chua gave a talk on his electric circuit.[91] It was the first time I met him. One year later, on 22 December, traveling in the south of France, he visited me briefly in Nice and invited me to visit him. In turn, from 23 to 26 of February 1988, travelling back from Tahiti, were I was giving one-month lectures in the French university of Pacific recently created, I stopped in California and visited his laboratory in Berkeley. I gave a talk on confinors and worked with one of his student, Orla Feerly (now Vice President for Research, Innovation and Impact and a Professor of Electronic Engineering at University College Dublin). This was the beginning of a long cooperation which lasts more than thirty years with Chua.

With Ushiki, we quickly switched our attention to the Chua circuit, because, from our knowledge, it was the first realistic model for chaos, with only three variables, the minimum number required to produce chaotic solutions. In their historical study of the development of the nonlinear dynamical systems theory from radio engineering to electronics, Letellier and Ginoux have identified chaotic electrical circuits published before that of Chua.[92] However the Pikovsky–Rabinovich circuit was published in 1978 under the title of stochastic instead of chaotic oscillations[93] although that Rabinovich was aware of chaotic attractors, because he gave a talk entitled "*strange attractors in modern physics*" during the New-York Conference of 1979.[94] It was then possible to compare numerical results with signal of this electric circuit on an oscilloscope screen, even if the parameters value of the circuit were not fixed with the same precision. Chua wrote me that the equations governing the Chua circuit could should be considered as a model for a real electronical system in the sense that the numerical simulations match within 5% of error with the actually measured data.

Of course, we used confinor as mathematical tools, even if the equation of the Chua circuit

$$\begin{cases} \dot{x} = \alpha(y - f(x) \\ \dot{y} = x - y + z \\ \dot{z} = -\beta y \end{cases} \tag{22}$$

with

$$f(x) = m_1 x + \frac{1}{2}(m_0 - m_1)\left[|x+1| - |x-1|\right] \tag{23}$$

is not a slow-fast equation.

We published first in 1991 a lengthy paper[95] using several mathematical tools created especially for this equation: half-Poincaré map isochronic lines, selecting and anti-selecting curves. We proved that for given values of the parameters, only patterns in the time waveforms (consisting of m successive peaks belonging to the upper sheet of the attractor, followed by n successive peaks belonging to the lower sheet, per period when the waveform is periodic) which are bounded in time can exist. The idea of isochronic lines was suggested to me by Claus Kahlert that I met in Chua's laboratory during one-month stay in November 1988. He had previously introduced the curves of equal mapping duration.[96] Half-Poincaré map was introduced by Bernhard Uehleke and Rössler,[97] and selecting curves by Kahlert and Rössler.[98] Kahlert from Tübingen was a former student of Rössler at the Institute of Physical and Theoretical Chemistry. He was Humboldt Fellow for two years in Berkeley. We discussed often about Rössler that I did not know at this time.

My first paper on the Chua circuit with Ushiki[95] was published in the first issue of the *International Journal of Bifurcations & Chaos*, created by Chua. Chua was very proud of this journal allowing color picture with no charges, and glossy paper. We discussed many times between 1988 and 1990 about this journal (Chua visited me again in August 1989), and I had an extensive exchange of correspondence with him about this article (Fig. 13).

With Ushiki again, we published two years after a second long paper applying confinors to the Chua circuit, highlighting that "*the main paradox here is that the theory of confinors, which could appear as a theory of rough analysis of the phase portrait of Chua's equation, leads instead to a very accurate analysis of this phase portrait*".[99] In this paper we studied accurately the first example of three coexisting attractors for the same value of the parameters ever known we have published in 1991.[100]

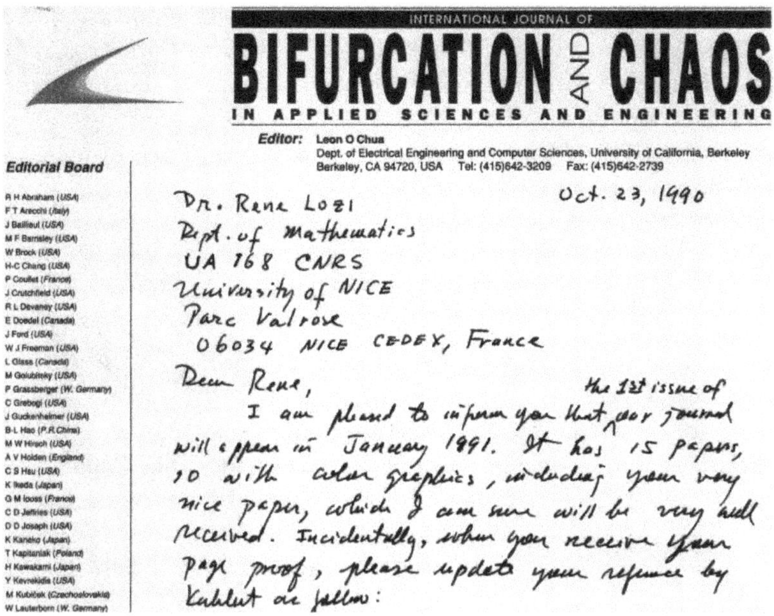

INTERNATIONAL JOURNAL OF
BIFURCATION AND CHAOS
IN APPLIED SCIENCES AND ENGINEERING

Editor: Leon O Chua
Dept. of Electrical Engineering and Computer Sciences, University of California, Berkeley
Berkeley, CA 94720, USA Tel: (415)642-3209 Fax: (415)642-2739

Oct. 23, 1990

Dr. Rene Lozi
Dept. of mathematics
UA 168 CNRS
University of NICE
Parc Valrose
06034 NICE CEDEX, France

Dear Rene,

I am pleased to inform you that the 1st issue of our journal will appear in January 1991. It has 15 papers, 10 with color graphics, including your very nice paper, which I am sure will be very well received. Incidentally, when you receive your page proof, please update your reference by Kahlert as follow:

Fig. 13. Acceptation letter from Chua of the Lozi and Ushiki's paper about the Chua circuit.[95]

During my second stay in Japan, Ushiki arranged me an invitation to present these results in the laboratory of Matsumoto, at Waseda University, Tokyo. The discussion was tense, because Matsumoto, who paradoxically first coined the name Chua to this electrical circuit,[90] was then jealous of the success of Chua's attractor and critized me for using this name, instead of "double-scroll." In 2009, Christophe Letellier, in a careful historical investigation into the authorship of this circuit described in detail the relations between Chua and Matsumoto after the publication of Chua's equation.[101,102]

On September 5–7, 1991, Chua visited me one more time and gave a talk in the Institute of Mathematics in Nice. Then, he invited me in his laboratory in October and November 1991 and I obtained a grant to invite him one month in Nice in May 1993. The focus was then to find some applications of the Chua circuit. During these years, the "control of chaos" was an important topic. Until 1990, the paradigm was that due to the sensibility to initial condition (butterfly effect) researchers thought it is impossible to control chaos. However, in 1990, Edward Ott, Celso Grebogi and James A. Yorke (who coined the name chaos)[k] proposed the OGY method allowing to control chaos.[103] In the same year Pecora and Carroll proposed a method to synchronize two chaotic systems using the master-slave method for the first time.[104] It was then proved that chaos can not only be controlled and synchronized, but also it is the basis of information transmission and processing.

Chua *et al.* applied this method to the Chua circuit.[105] I suggested him to improve the method by cascading two Chua circuits as receivers. The idea was coming from relaxation methods usually used in numerical analysis. The good numerical results were published after the stay of Chua in Nice in 1993.[106] One year later, with Aziz-Alaoui, my former Ph.D. student,[l] we generalized the method, coupling more receivers[107,108] and highlighting a power law in the error of the recovered signal.

Since these years, Chua introduced several new ideas like Cellular Neural Networks (CNN)[109] or memristor and recently was focusing on the "Edge of chaos."[110,111] The history of memristor is amazing. It was described and named in 1971 as the missing fourth element.[112] Then it fell into neglect until 2008, when a team at HP Labs claimed to have found Chua's missing memristor based on an analysis of a thin film of titanium dioxide. They published their result in *Nature*.[113] Since then, the number of publications is still soaring. In August 2016, the Chua Memristor Center (CMC) has been founded in Dresden.[m]

In 2012, with my colleague Mohammed-Salah Abdelouahab from the university of Mila, Algeria and Chua, we developed the general theory of "memfractance" linking properties of first-order, second-order, etc. memristor, meminductor, memcapacitor, using fractional derivative. We spent two entire years to overcome every obstacle and develop a generalized Ohm's law.[114]

Nevertheless, I did not forget completely the Chua circuit. A new mathematical tool, hidden attractor, was introduced in 2011 by Gennady Leonov, Nikolai Kuznetsov and Vladimir Vagaitsev.[115] In 2016, with my other colleague Tidjani Menacer from the university of Biskra (Algeria) and Chua, we extended the method to find hidden attractors, in order to find hidden bifurcations in the multispiral Chua circuit.[116]

[k]I met Grebogi at the university of Le Havre in January 2001, and Yorke in the same city in June 2009.
[l]Now, professor and former Scientific Vice-President of University of Le Havre.
[m]https://cmc-dresden.org/.

4.2. *The Alpazur oscillator*

The Chua's attractor is very interesting and nowadays it is thoroughly studied. However, with Hiroshi Kawakami we were not satisfied, because it is not a slow-fast system, it is therefore not possible to apply all the mathematical tools developed for such category of systems. This is why we searched to build an electric circuit corresponding to these systems.

As already said, I first met Kawakami in 1982, during the meeting in Toulouse,[9] and again in Kyoto during the summer 1983. Like Ushiki, he was fluent in French, mainly because he spent two years of internship at the C.N.R.S. center of physics researches in Marseilles in 1970–1972. Later in 1984 he was invited professor for one year at INSA of Toulouse, to work with Christian Mira. In 1988, during my first long stay in Berkeley, Chua invited him, and we discussed a lot on such slow-fast electric circuit. Two years after, I attended the International Congress of Mathematicians at Kyoto on August 21–29, 1990[n] and spent three weeks in Japan, working with Ushiki, and meeting again Kawakami. Both we did a short visit to Kawakami in his university of Tokushima from August 31 to September 1st. Eventually, I found a grant and invited him in Nice for one month in December 1991. During this month we finalized the Alpazur oscillator. Kawakami gave this weird name to this electric circuit because it was the name of the hotel where he spent one month in Nice. Referring to Rössler, we describe this oscillator as follows.[4,5]

> The Alpazur oscillator is a two-dimensional piecewise circuit with a switch controlled by states variables. Two circuit equations or vector fields which describe the dynamics of the circuit are abruptly interchanged each other by changing the position of the switch. If the characteristics of the switch has hysteresis, then the composite flow generated by the mixed vector fields is not necessarily continuous on initial conditions, which allows chaotic behavior of the state variables. Since an electronic switch acts instantaneously, this model serve almost ideal realization of composite dynamical system proposed by Rössler,[6] as well as constrained systems analyzed by Ushiki and Lozi,[88,89] Lozi and Saidi,[117,118] Saidi.[119]

[n]To attend this conference, I obtained a grant from the National Committee of French Mathematicians which in turn was awaiting a report. The *Gazette des Mathématiciens* rewarded the two most original which were published in the 7th volume, January 1991. Adrien Douady was the first winner; I was the second (see Appendix 2).

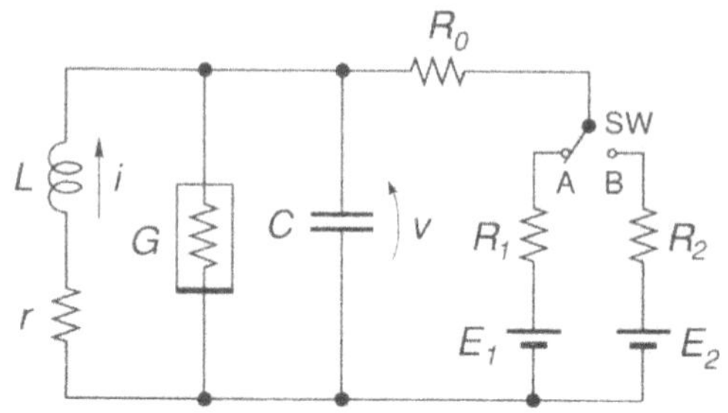

Fig. 14. The Alpazur circuit.

The circuit dynamics of the Alpazur oscillator shown in Fig. 14 is given by

$$\begin{cases} L\dot{I} = -V - rI \\ C\dot{V} = -G(V) + I + \dfrac{E_1 - V}{R_0 + R_1} \end{cases} \tag{24}$$

when the switch SW is thrown into contact with position A and

$$\begin{cases} L\dot{I} = -V - rI \\ C\dot{V} = -G(V) + I + \dfrac{E_2 - V}{R_0 + R_2} \end{cases} \tag{25}$$

when SW is in position B, with

$$G(V) = -a_1 V + a_3 V^3, \quad a_1, a_3 > 0 \tag{26}$$

> The function of the switch SW is controlled by the capacitor's voltage and has hysteresis. This functional characteristics can be realized by electronics switches controlled by comparator and operational amplifiers. Spiral, screw, and composite chaos stated by Rössler[6] may be easily realized.

We studied also a simplified model of this oscillator called planar spiral-linear system[120]

$$\begin{cases} \dot{x} = \sigma x - y \\ \dot{y} = x - \sigma y \end{cases} \tag{27}$$

defined by a standard logarithmic flow on the half-plane

$$H = \left\{(x, y) \in \mathbb{R}^2 \mid y \geq -1\right\}$$

and a linear flow

$$\begin{cases} \dot{x} = 0 \\ \dot{y} = 1 \end{cases} \tag{28}$$

on

$$B = \{(x, y) \in \mathbb{R}^2 \mid y \leq -1 + \alpha, 0 \leq \alpha \leq 1\}.$$

The Alpazur oscillator did not have the expected success, however some papers were published analyzing its properties. Miino and Ueta, 23 years later,[121] showed some results of laboratory experiment for bifurcation phenomena on the forced Alpazur oscillator. Recently, Stefanescu *et al.*[122,123] studied the influence of computer simulation on a chaotic circuit. The Alpazur oscillator being used in their study and the results are interpreted precisionwise and timewise (time needed to run one simulation). Kousaka *et al.*,[124] changed the linear conductor into a nonlinear one.

Kawakami invited me one month in Tokushima in 1994, however due to the fact that I was very involved in administration of my university, I never found the time to publish the results of our research done during this month. Unfortunately, we found neither the opportunity to meet again, nor to work together.

4.3. *Last researches related to confinor, global orbit pattern*

I developed during several years, between 1982 and 2000, the idea of confinors, as already described for the Chua circuit,[125] the constrained Lorenz model and the Alpazur oscillator. In 1989, I had the opportunity to present some results about their use in the Belousov–Zhabotinsky reaction, when I was invited to give a talk in the 48th meeting between theoretical physicists and mathematicians, held in honor of René Thom, in Strasbourg.[126]

During these years, I advised four Ph.D. thesis whose subjects were related to this tool. In 1987, M. Aziz-Alaoui defended his thesis on the analysis of patterns and confinors of a differential system constrained on a slow manifold.[127] The next year, Abdelmajid Saidi studied an algorithm to partition the parameter space for a family of constrained differential equations.[119] I published with him two papers in connection with his doctoral work.[117,118] In 1994, Nasr-eddine Hamri studied the continuity of half-Poincaré maps of the Chua's circuit equations.[128] In 2002, Soraya Boughaba defended her Algerian thesis about the Chua circuit.[129]

The two thesis involved the concept of confinor. We had already published a method based on isochronic lines for fitting trapping regions for Chua's attractor.[130]

However, confinors were not as successful as I expected within the dynamical systems research community. Attractors still have a supreme reign. Nevertheless, the theory of templates, first introduced by Williams,[131] is nowadays improved by a team led by Christophe Letellier[132]

> Describing the topological properties by a template is a powerful technique to classify chaotic attractors. [...] direct templates — in which all mechanisms (torsions, branch permutation, etc.) identified in the attractor are explicitly described without any simplification — are of a great interest to have a better description of the subtleties of the dynamics.

Their use of templates produces successful results to choose the level of analysis of the state space, probably in a more efficient way that the one I tried to do with confinors.

In a very different attempt to get reliable solutions of dynamical systems computed numerically, I tried ten years later with Clarisse Fiol, to consider mappings of finite set.[133, 134] However the results are too much technical to find their place in this recollection.

5. The Russian School

I would like to mention my relationships with three professors (one Ukranian and two Russians) who made major contributions to chaotic dynamical systems during the decades 1960–80: Alexander N. Sharkovskii (Sharkovskii's order), Leonid Shil'nikov (homoclinic bifurcations) and Vladimir Belykh (Belykh's attractor) and my brief encounter with Romen Plykin and Evgenii Sataev. Before the fall of the Berlin Wall on November 9, 1989, it was uneasy to meet researchers from USSR. Even if it was possible for them to travel to West (Paul H. Rabinowitz, Ya. G. Sinaï and few other Russians from the Landau Institute, Moscow, attended the conference organized by the *New-York Academy of Science* twenty years before (see Section 2.1), very often they did not get their visa on time or they had a lack of money to travel, and they missed the conference. As an example, in 1982, Alexander N. Sharkovskii did not attend the meeting organized in Toulouse where his talk was scheduled (Fig. 15). Moreover, most of them

MARDI

9.00 - 9.30	R. VILELA MENDES, Lisbonne (Portugal) «Symmetries and stable periodic orbits for one dimensional maps»
9.30 - 10.00	M.C. ZDUN, Katowice (Poland) «Iteration semigroups with restricted domain»
10.00 - 10.30	A. SMAJDOR, Katowice (Poland) «Measurable semigroups of multivalued functions»
10.45 - 11.15	G. DOVEIL, Palaiseau (France) «Théorie de renormalisation et transition à la stochasticité pour les systèmes hamiltoniens à deux degrés de liberté»
11.15 - 11.45	O.E. RÖSSLER, Marburg (R.F.A.) «Chaos and Tolerance»
11.45 - 12.15	R. LOZI, Nice (France) «Dimensional bifurcations between thread and sheet strange attractors»
	*******
14.00 - 15.00	R. THOM, Paris (France) «Structure des attracteurs»
15.15 - 15.45	M.R. GUEVARA, Montréal (Canada) «One dimensional Poincaré maps for periodically stimulated biological oscillators»
15.45 - 16.15	I. GUMOWSKI, Toulouse (France) «Periodic steady states of dynamic systems and their smooth dependence on parameters»
16.15 - 16.45	A.N. SARKOVSKII, Kiev (U.R.S.S.) «On some properties of discrete dynamical systems»
17.30	Réception offerte à la Mairie de Toulouse

Fig. 15. Program of Toulouse on Tuesday May 18, 1982.[9]

did not speak English. They published in Russian. It took years for few of their results to reach the West community through translation.[o]

[o]See for instance *Doklady Mathematics* is a journal consisting of English translations of articles published in Russian.

5.0.1. *Alexander Nikolaevich Sharkovskii*

A prototypal example of the difficulty to make known their results is the differential treatment of the Li and Yorke theorem "period three implies chaos" published in 1975[135] and the "Sharkovskii's order" published in an Ukrainian journal in 1964.[136,137] In the 70s, as described in Section 2, the studies of chaotic phenomena were soaring. Many amazing results were published like the Feigenbaum's constants for one-dimensional maps,[138] and the Hénon map in dimension 2 or the Lorenz and the Rössler attractors for ODEs. In many seminars, the result of Li and Yorke was hailed as an important mathematical discovery. It is only two years after, in 1977, that the Sharkovskii's order was published in English by Peter Stefan.[139] After that, this order overshadowed the Li–Yorke theorem which was perceived as a peculiar case of this order, published nine years before. However, it is not exactly the case as Yorke himself reminded me during the banquet of the *International Conference on Complex Systems and Applications*, in Le Havre on 28 June 2009. I was sited in front of him and near Leon Chua. With Chua, I was discussing about Sharkovskii whom I met one year before in Yalta. Yorke seemed not very happy about the warm friendship, Chua and me, showed to Sharkovskii. Abruptly he challenged me about the Sharkovskii's order, saying that this order was not giving any result of chaotic properties of continuous function, contrarily to the theorem he proved with Li 34 years before! I was very surprised by his tense language, even if he was not wrong in his argument, their paper analyzing a situation in which the sequence of iterates of a map is non-periodic and might be called "chaotic." The discoveries performed by scientists may have profound repercussions on their feelings and personalities during their lifetimes!

I did not use or work on the Sharkovskii's order, because before 1994, I was not interested by researches on maps of the interval which seemed too much technical, without any application. Hence, I did not try to contact or to meet him. However, one day during my second-long stay in the Chua laboratory in October–November 1991, Chua brought me and my family for a journey to the Tamalpais mountains. This wonderful place, north San Francisco, was his favorite place of promenade for his guests. During the stroll, Chua said me that he brought here, separately, Leonid Shil'nikov and Sharkovskii, and that if he was able to walk quietly on the trail with Shil'nikov who looked healthier than Sharkovskii, he was unable to follow this second who was a terrific hiker! In fact, I verified many times, some years after, that it is impossible to challenge Sharkovskii in mountain!

It is during my third one-month visit at Berkeley from November to December 1993, that I have had the opportunity to meet both Shil'nikov and Sharkovskii. At this time, Chua was planning the special workshop "Chua's circuit: chaotic phenomena and applications" during the *International Symposium on Nonlinear Theory and its Application* (Nolta'93) in Honolulu, December 5–10. As usual, he was a terrific organizer. To get a large audience, he invited in Berkeley many speakers to stay one month in his laboratory and chartered a flight for all of them to Honolulu. Then, early in the morning, Sunday December 5, 1993, with Sharkovskii, Shil'nikov, Vadim Anishenko (Saratov, Russia), Martin Hasler (Lausanne, Switzerland), Maciej Orgozalek (Krakow, Poland), and many others, we started our trip until to be lodged at the Sheraton Waikiki, Honolulu for one week (Fig. 16)!

Diana Chua, Leon's wife organized our dispatching in the rooms. I was not smoking, nor Sharkovskii, contrarily to Anishenko and Shil'nikov who shared the same room. Therefore, she planned that I must share the room with Sharkovskii. Consequently, we both discussed a lot during the week, even if he was not very fluent in English. This was the beginning of a many long years of cooperation!

During the following years, with Rossetto and Aziz-Aloui, we found several grants between 1994 and 2002 to invite Sharkovskii in Toulon, Le Havre and Nice. I published three papers, two with him, Aziz-Alaoui and

Fig. 16. Alexander Nikolaevich Sharkovskii, Leonid Shil'nikov and René Lozi upon their arrival in Honolulu Airport on December 5, 1993.

A. D. Fedorenko from Kiev, and one with him and J. Sousa Ramos from Lisbon. This was initially performed in continuation of some of my previous results.[107, 108] First, we developed a method to recover trajectories of a chaotic dynamical system from its ϵ-perturbed trajectories.[140] Then, we got some results on estimating the parameters of families of piecewise chaotic dynamical systems.[141] Such problems have practical applications in secure communication.

Another work was devoted to the study of the dynamics generated by one-dimensional map linked to certain classes of boundary value problems for one dimensional wave equation

$$w_{tt} - w_{xx} = 0 \tag{29}$$

by the relationship

$$u_{n+1} + au_n = F\left(-bu_{n+1} + u_n\right) \tag{30}$$

where the parameters a, b, and the nonlinearity F are determined by the boundary conditions.[142]

Sharkovskii invited me to the Summer school on dynamical system, Katsiveli, Crimea, in August 2000, and to the European Conference on Iteration Theory (ECIT) held in Yalta, Crimea, in September 2008. I worked with him and Sousa Ramos in Lisbon in May 2004. I last met him during the ECIT 2012 in Ponta del Gada, Azores in September 2012.

5.0.2. *Leonid Pavlovich Shil'nikov*

The year after the Honolulu workshop, Shil'nikov visited Christian Mira in Toulouse in 1994. From there, he phoned me to visit my Institute. I arranged a stay for him and his assistant Dmitry Turaev from Nizhny Novgorod, on February 18–23, 1994. Shil'nikov gave a talk on Friday 18th. The next day, I drove them to visit Monte-Carlo. Shil'nikov absolutely wanted to have a picture of the two of us in front of the famous Casino where Anton Tchekov and his friend Ignaty Potapenko lost all their money in 1898. Next, visiting the Oceanographic Museum of Monaco, he told me about his fishing trips in Siberia north of the polar circle.

In order to work together we applied twice unsuccessfully for INTAS programs (Fig. 17). I never met him again. I attended the conference *Dynamics, Bifurcations and Strange Attractors* dedicated to his memory, held at the Lobachevsky State University of Nizhny Novgorod, Russia, on July 1–5, 2013.

```
From adm@inforis.nnov.su Thu Nov 30 14:53:30 1995
Return-Path: <adm@inforis.nnov.su>
Received: from inforis.nnov.su by math.unice.fr.noname (4.1/SMI-4.1)
        id AA16079; Thu, 30 Nov 95 14:52:38 +0100
Received: from focus.UUCP by inforis.nnov.su with UUCP id AA08020
  (5.65.kiae-1  for lozi@math.unice.fr); Thu, 30 Nov 1995 16:55:51 +0300
Received: from sabena (sabena.focus.nnov.su [155.0.0.6]) by focus.nnov.su (8.6.9
/8.6.9) with SMTP id QAA00320 for <lozi@math.unice.fr>; Thu, 30 Nov 1995 16:37:2
8 +0300
Received: by sabena; (5.65/1.1.8.2/27Nov95-0119PM)
        id AA04060; Thu, 30 Nov 1995 16:47:29 +0300
Date: Thu, 30 Nov 1995 16:47:29 +0300
From: Andrey Shilnikov <andrey@focus.nnov.su>
Message-Id: <9511301347.AA04060@sabena>
Apparently-To: lozi@math.unice.fr
Status: RO

Dear Rene,

I hope you are doing well. Perhaps, I will be able to see you early 1996
while visiting France.
I am now writing to you concerning one more INCAS project; this tome in
collaboration with Russian Foundation of Basic Research which has been
recently announced in local scientific echelons. As previous grant the new
one aims: supporting fundamental research in Russia, travel to and from Russia (
in
our case) of participants of both teams: INTAS and Russia, no real gain for
INSTAS teams excluding 20% and travels to here.
The rules are more strictly the following:

1. Total amount should not exceed 60 000 ECU for 2 years;

2. 2 teams (at least) are to belong to INTAS countries and 2 teams to Russia.

3. <= 20% of a grant is allocated to INTAS participants for covering only travel
expenses as well as overhead and other indirect expenses necessary for
governing the project. An INTAS researcher  may apply for salary  provided
he is engaged temporary in the project and some other variations.

Visiting in Poland few days ago, I had a talk to G. Nocolis about this
matter. He agreed his group and himself to take part. Our discussion
resulted that it would be great if you might agree to participate and even
more  would not mind to become a coordinator of the project entirely.
What is your opinion?

In case of a positive answer, let us know of your preferancies which  another Ru
ssian group
could become a second participant.

You can find further information on http:/www.ib.be  at /intas. There is the
only Russian text here. We don't have plenty time because the application
deadline is December 15, at 12 a.m. CET. By the date, all proposals have to
come to INTAS headquarter in Brussels.

I hope to hearing from you soon. Many regards,

Yours Leonid
```

Fig. 17. Letter from Leonid Shil'nikov in November 1995, suggesting applying for another INTAS program.

5.0.3. *Vladimir Belykh, Romen Plykin and Evgenii Sataev*

Aside the Hénon and Lozi maps, an important chaotic map was published (in Russian) in 1976 by Vladimir Belykh.[143] This two-dimensional map f is defined on the square $S = \{(x, y) \mid 0 \leq x \leq 1, 0 \leq y \leq 1\}$ by

$$\begin{cases} x_{n+1} = \lambda x_n \\ y_{n+1} = \gamma y_n \end{cases} \quad \text{if } L(x_n, y_n) \leq 0 \tag{31}$$

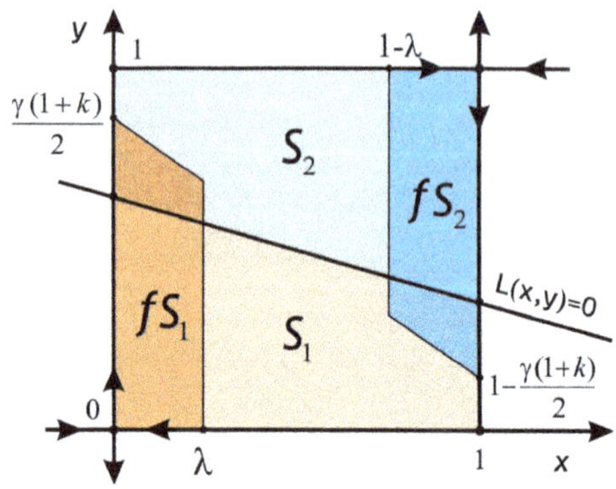

Fig. 18. First iterate of the unit square $S = S_1 \cup S_2$ under the map f, where the discontinuity line $L(x, y) = 0$ splits it into S_1 and S_2. One iterate of f transforms the trapezoids S_1 (light brown) and S_2 (light blue) into fS_1 (dark brown) and fS_2 (dark blue), respectively.

and

$$\begin{cases} x_{n+1} = \lambda\,(x_n - 1) + 1 \\ y_{n+1} = \gamma\,(y_n - 1) + 1 \end{cases} \quad \text{if } L(x_n, y_n) > 0 \tag{32}$$

with

$$L(x, y) = k(2x - 1) + 2y - 1.$$

The square S is cut into two parts S_1 and S_2 by a line $L(x, y) = 0$ (Fig. 18). The dynamics of the map f is governed by (31) on the lower part S_1, where $L(x, y) \leq 0$, and by (32) on the upper part S_2.

The Belykh map which is invertible if $0 < \lambda < \frac{1}{2}$ is a special case of a hyperbolic map with singularities like the Lozi map and a dispersing dynamical billiard.[144] The ergodic properties of this map in the parameter range of invertibility were studied by many authors including Belykh himself, Yakov Pesin[145] and Evgueni Sataev.[146]

It was a great pleasure to meet Vladimir Belykh for the first time in 2013, during the conference dedicated to the memory of Shil'nikov, in Nizhny Novgorod. Four years after, on June 28–30, 2017, I was invited again in the Lobachevsky State University in order to give the inaugural talk of the Conference NOMA-2017. Belykh was invited speaker too. We were happy to meet again,and at the end of the Banquet, I suggested him to sing with me some popular songs of my favorite Russian singer Bulat Okudzhava (Fig. 19).

Fig. 19. Vladimir Belykh and René Lozi, singing popular songs of Bulat Okudzhava[147] during the banquet of the Conference NOMA-2017 on June 2017, in Nizhny Novgorod.

Fig. 20. From left to right, René Lozi, Romen Vasilievich Plykin and Evgueni Sataev on August 2000, during the Summer school on dynamical system, Katsiveli, Crimea.

I had the pleasure, during the ten days spent in Katsiveli, Crimea, for the Summer school on dynamical system organized by Sharkovskii, to meet two other important figures of the dynamical systems in USSR: Evgueni Sataev and Romen Vasilievich Plykin (1935–2010),[148] both from Obninsk State Technical University (Fig. 20).

The venue of the Conference, the building of the Ukrainian academy of sciences, was standing near the beach, on the Black Sea. The attendants were spending the free long hours in the afternoon, swimming during this hot summer, before the scheduled evening sessions. The banquet took place on a pontoon during the night of August 28th. I strongly disappointed

Plykin by not accepting his insistent invitation to dive down with him from the pontoon into the Black Sea, around midnight, after the traditional speeches washed down with vodka!

5.1. *The Rössler attractor revisited*

The first meeting with Otto Rössler that I missed should have taken place during the *workshop on iteration theory* at La Garde-Freinet in May 1979 (Fig. 21). He never attended this workshop. The second one was during the meeting in Toulouse in 1982, where his talk was scheduled just before mine (Fig. 15). Again, he did not attend this meeting.

During these years I was inspired by Otto's papers for my Thèse d'État and my researches on hyperchaos. However, I did not try to exchange with him. As I said above, it is only in 1988, when I was visiting the laboratory of Leon Chua in Berkeley, that I had some discussion about him with Claus Khalert.

Eventually, the first time I had the honor and the great pleasure to meet Otto, was in Rouen at CORIA Institute. It was a very brief encounter, during the *4th Conference on temporal Chaos and Spatio-temporal chaos* organized by Letellier and Aziz-Alaoui, on December 15–16, 2003.[149] None of us had enough time to discuss in particular. In fact, I was overwhelmed by my position of Director of the *Institut Universitaire de Formation*

COLLOQUE SUR LES COMPORTEMENTS GENERAUX DES PROCESSUS ITERATIFS

21-22-23 Mai 1979

La Garde Freinet

PROGRAMME SCIENTIFIQUE

ARNEODO Alain, COULLET Pierre et TRESSER Charles	*Physique Théorique* - NICE	Un groupe de renormalisation avec comportement périodique.
LOZI René	*Mathématiques* - NICE	Attracteurs étranges : Une classe d'applications du plan dans lui-même laissant invariant des ensembles de Cantor.
LÜCK Herman et LÜCK Jacqueline	*Botanique* - MARSEILLE	Titre non parvenu.
MIRA Christian	I.N.S.A. - TOULOUSE	Récurrences et bifurcations d'un domaine d'attraction avec modifications qualitatives importantes.
POMEAU Yves et MANNEVILLE Paul	C.E.A. - GIF	L'intermittence : Un phénomène générique dans l'apparition de la stochasticité et de la turbulence.
RICHELLE Jean	*Génétique* - BRUXELLES	Titre non parvenu.
RIGUET Jacques	*Math. Logique* - PARIS V	Titre non parvenu.
RÖSSLER Otto	*Chimie Théorique* - TÜBINGEN	Titre non parvenu.

Fig. 21. Excerpt of the program of the *Workshop on iteration theory* at La Garde-Freinet in May 1979.

des Maîtres in Nice and spent most of the time during these two days to discuss by phone with my administrative staff. The second time was again in Rouen, in June 2011, during the conference *From Laser Dynamics to Topology of Chaos* (Rouen), for celebrating the 70th birthday of Robert Gilmore, organized again by Christophe Letellier. We had more time to discuss many philosophic topics.

In between these two encounters, Christophe during the Colloque DYCOEC, December 14–16, 2009 asked some people including myself, to do a review of Otto's original paper,[150] as if we were in 1976 and Otto was not reputed. It was an original and unusual demand. I was mainly interested by the link between the chemical scheme proposed by Otto and the resulting system of ODE. The background I had acquired studying the different chemical models of Belousov–Zhabotinsky reaction many years ago, encouraged me in this task. I include below some excerpts of the report I sent to Letellier, who used it in his paper.[151]

> In this paper, the author proposes a 3-component abstract chemical reaction showing the same qualitative behaviour of the deterministic nonperiodic flow (or "chaotic" flow) previously observed by E. N. Lorenz, in 1963, in a 3-component differential system. The corresponding system consists of a 2-dimensional differential equation coupled with an ordinary single-variable chemical hysteresis system. Under numerical simulation, the proposed chemical system provides an almost ideal illustration to the underlying dynamical prototype,the "3-dimensional blender".
>
> This chaotic system is based on the reaction scheme displayed on Fig. 22 which is a combination of an Edelstein switch with a Turing oscillator.
>
> It combines a 2-variable chemical oscillator (variables a and b) with a single-variable chemical hysteresis system (c) as prescribed by the recipe. The author affirms that the system

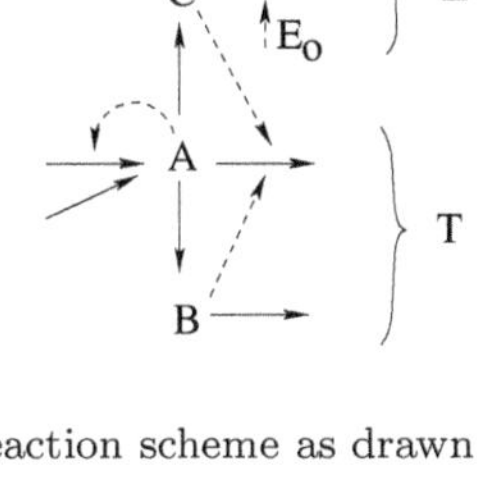

Fig. 22. Reaction scheme as drawn by Rössler.[2]

obeys, under the usual assumptions of wellstirredness and isothermy as well as an appropriate concentration range, the following set of rate equations

$$\begin{cases} \dot{x} = k_1 + k_2 x - \dfrac{x\,(k_3 y + k_4 z)}{x + K} \\ \dot{y} = k_5 x - k_6 y \\ \dot{z} = \dfrac{1}{\mu}\left(k_7 x + k_8 z - k_9 z^2 - \dfrac{k_{10}\, z}{z + K'}\right) \end{cases} \tag{33}$$

(for the sake of mathematical convenience, we use x, y, and z as variables instead of a, b, and c).

Describing a model of chaotic attractor coming from chemical reaction this paper seems interesting as it gives some substance to this notion conversely to the Lorenz's equation which is derived mathematically from a more complicated model of turbulence, with no direct link with physics.

However, there are three main issues:

(1) The exactness of the relationship between the reaction scheme depicted in Fig. 22 and the system of equations (33),
(2) The nature of the slow manifold of equation (33),
(3) The number of parameters in equation (33).

Then analyzing the chemical scheme, I concluded that

Finally the system (33) should be

$$\begin{cases} \dot{x} = k_1 + k_2 x - \dfrac{x\,(k_3 y + k_4 z)}{x + K} \\ \dot{y} = k_5 x - k_6 y - \dfrac{k_3 x y}{x + k} \\ \dot{z} = \dfrac{1}{\mu}\left(k_7 x + k_8 z - k_9 z^2 - \dfrac{k_{10}\, z}{z + K'} - \mu \dfrac{k_4 x z}{x + K}\right) \end{cases} \tag{34}$$

Analyzing the slow manifold, I wrote

Then the Michaelian kinetics used in order to fold the manifold, which actually appears to be the central of the idea of this paper, should not be the main mechanism which produces the chaos for this system of equations. Instead the nonlinearity in z should be the genuine underneath mechanism, as in the case of the Lorenz attractor.

[...] In conclusion I do not recommend the publication of this paper, until some clarifications are given by the author concerning the three issues I have developed here.

Fortunately, the paper was already published. Letellier and Messager[151] highlight some underneath insight behind the proposed equation (33).

6. To Sum Up

Forty-five years after the beginning of my academic life, I realize that I was incredibly lucky to begin my career as the same time as the emergence of chaos studies in dynamical systems. I had a great feeling of freedom, being able to discover a new world of extreme richness, as undoubtedly explorers of America had it hundreds of years before me.

I had also the privilege to discuss with outstanding researchers, among them five Fields medalists: Thom, Yoccoz, Smale, Alain Connes (in the national committee of C.N.R.S.) and Pierre-Louis Lions (during seminars in Paris), one Abel prize laureate: Yves Meyer (in the same committee), and many renowned scientists: Belykh, Chua, Hénon, Lorenz, Rössler, Sharkovsky, Shil'nikov, Yorke, etc.

In this contribution, I have described roughly half of my research life, between 1972 and the beginning of the third millemium. Since the 2000s, I focused my research on concrete applications of chaos theory like pseudorandom number generators,[152] cryptography,[153] global optimization,[154] etc. I explored the new ideas of fractional mappings[155] and fractional derivative.[114, 156]

Nowadays, the Hénon, Lorenz, Lozi maps, the Chua, Lorenz, Rössler systems and many others are routinely used for many applied purposes. They are living their own life, without the help of their "fathers", like emancipated children. The most amazing fact for me is that, in 1975 and during many years, I tried unsuccessfully to find an efficient algorithm for global optimization. Now, the map I introduced in an entirely different aim is often used within evolutionary algorithms to find global solution of optimization problems involving hundreds of variables. This emphasizes the intimate unity of mathematics.

Another strange thing is when I attend a conference in which one lecture uses explicitly this map, and that I speak with young researchers who read my name which is quite uncommon on my badge, they never make the connection between the map and me. I did the same when I was in Oberwolfach with Ernst Hölder in the room!

I met Rössler only two times, however our attractors continue to be used together, as for instance,[157]

> In particular we study the Hénon map, which is an example of a nonhyperbolic dynamical system, like the Rössler attractor, and the Lozi map as an example of a hyperbolic system.

or in Changaival *et al.*[158]

> Chaotic Traversal (CHAT) which integrates chaotic dynamics to traverse large unknown graphs via the Lozi map and the Rössler system.

In conclusion to this long story of the beginning of my researches and their interaction with the work of Rössler, although being eight years younger and far less known than him, I would like to make mine his words in his last paper.[159]

> Turning 80 this month, I am somehow relieved from the burden of having to be totally up-to-date and having to respond to specifics, being rather allowed to summarize what I consider the most important issue from my momentary point of view which automatically is very narrow and totally outdated.
>
> You see, turning 80 has its advantages: you can dream freely about winning the interest of an ingenious writer. And whether or not a single young scientist will decide to embark on the golden road you have laid out for her or him is a matter of both irrelevancy and happy daydreaming.

Acknowledgments

René Lozi is indebted to the following friends and colleagues who helped him to refresh his memory: Ahmed Aziz-Allaoui, Lenore Blum (Carnegie Mellon University, Pittsburgh), Jean Céa, David Chillingworth, Brian Quigley (Head, Engineering & Physical Sciences Division/Mathematics, Statistics & Computer Science Librarian, University of California, Berkeley), Jean Petitot, Paul H. Rabinowitz, Shigehiro Ushiki, Monique and Jeanne Zerner, wife and daughter of Martin Zerner, to Stéphane Junca for shrewd remarks on continuation methods, to Leon Chua for stimuling discussions on the history of nonlinear circuit theory and to Christophe Letellier for helpful discussions and formatting the text.

References

1. O. E. Rössler, An equation for continuous chaos, *Physics Letters A*. **57**(5), 397–398 (1976). doi: 10.1016/0375-9601(76)90101-8.
2. O. E. Rössler, Chaotic behavior in simple reaction system, *Zeitschrift für Naturforschung A*. **31**, 259–264 (1976).

3. R. Lozi, Modèles mathématiques simples et consistants pour l'étude de quelques systèmes dynamiques expérimentaux. Thèse d'État, University of Nice, France (June, 1983).
4. H. Kawakami and R. Lozi, Switched dynamical systems — Dynamics of a class of circuits with switch. In S. Ushiki (ed.), *Avanced Series in Dynamical Systems*, Vol. 11. World Scientific Publishing, Singapore (1992), pp. 39–58.
5. H. Kawakami and R. Lozi, Switched dynamical systems — A switch defines semi-flows on a pair of half planes. In *Proceedings of the RIMS Conference*, Vol. 804. RIMS Kokyuroku, Japan (1992), pp. 38–54 (in Japanese).
6. O. E. Rössler, Continuous chaos: Four prototype equations, *Annals of the New York Academy of Sciences.* **316**, 376–392 (1979).
7. O. E. Rössler, Chaotic oscillations: An example of hyperchaos, *Lectures in Applied Mathematics.* **17**, 141–156 (1979).
8. O. E. Rössler, An equation for hyperchaos, *Physics Letters A.* **71**, 155–157 (1979).
9. R. Lozi, Dimensional bifurcations between thread and sheet strange attractor. In R. Thibault (ed.), *Théorie de l'itération et ses applications*, Vol. 322. Colloques Internationaux du CNRS (1982), pp. 145–152.
10. J. C. del Valle and D. J. Nader, Toward the theory of the Yukawa potential, *Journal of Mathematical Physics.* **59**(10), 102103 (2018). doi: 10.1063/1.5050621.
11. H. B. Keller, *Numerical Methods for Two-Point Boundary-Value Problems.* Dover, Mineola, New York (1992).
12. J. L. Daleckii and M. G. Kreĭn, *Stability of Solutions of Differential Equations in Banach Space.* Translations of Mathematical Monographs, Vol. 43. American Mathematical Society, Providence, Rhode Island (1974).
13. V. A. Yakubovich and V. M. Starzhinskii, *Linear Differential Equations With Periodic Coefficients*, Vols. 1 & 2. John Wiley & Sons, New York (1975).
14. R. Lozi, Analyse numérique de certains problèmes de bifurcation. Ph.D. Thesis, University of Nice, France (April, 1975).
15. R. Lozi, A computing method for bifurcation boughs of nonlinear eigenvalue problems, *Bulletin of the American Mathematical Society.* **81**(6), 1127–1129 (1975).
16. R. H. Abraham, The Smale program. In C. Letellier (ed.), *Some Elements for an History of the Dynamical Systems Theory* (2020).
17. H. Poincaré, Sur l'équilibre d'une masse fluide animée d'un mouvement de rotation, *Comptes Rendus de l'Académie des Sciences.* **100**, 346–348 (1885).
18. N. Bourbaki, Éléments de mathématiques. Masson (1959–2019). Collective Work.
19. R. Thom, *Structural Stability and Morphogenesis: An Outline of a General Theory of Models.* W. A. Benjamin, New York (1973). Translated from the French by David Fowler.
20. G. Iooss, Bifurcation des solutions périodiques de certains problèmes d'évolution, *Comptes Rendus de l'Académie des Sciences I.* **273**, 624–627 (1971).

21. G. Iooss, Théorie non linéaire de la stabilité des écoulements laminaires. Ph.D. Thesis, Paris (1971).
22. G. Iooss, Sur la stabilité de la solution périodique secondaire intervenant dans certains problèmes d'évolution, *Comptes Rendus de l'Académie des Sciences A.* **273**, 912–915 (1971).
23. G. Iooss, Nonlinear stability theory of laminar flows in the case of exchange of stabilities, *Fluid Dynamics Transactions.* **6**(2), 307–316 (1971).
24. G. Iooss, Théorie nonlinéaire de la stabilité des écoulements laminaires dans le cas de l'échange des stabilités, *Archive for Rational Mechanics and Analysis.* **40**, 166–208 (1971).
25. G. Iooss, Stabilité de la solution périodique secondaire intervenant dans certains problèmes d'évolution, *Comptes Rendus de l'Académie des Sciences A.* **274**, 108–111 (1972).
26. G. Iooss, Existence et stabilité de la solution périodique secondaire intervenant les problèmes problèmes d'évolution du type Navier-Stokes, *Archive for Rational Mechanics and Analysis.* **47**(4), 301–329 (1972).
27. G. Iooss and G. Tronel, Exemples de bifurcation, *Comptes Rendus de l'Académie des Sciences.* **276**, A1353–A1356 (1973).
28. G. Iooss, Bifurcation of a t-periodic flow towards and nt-periodic flow and their non linear stabilities, *Archives of Mechanics.* **26**(5), 795–804 (1974).
29. G. Iooss and R. Lozi, Convection entre deux plaques planes en rotation et effet dynamo résultant d'une bifurcation secondaire, *Journal de Physique — Colloques.* **39**, C5-15–C5-16 (1977). doi: 10.1051/jphyscol:1978508.
30. D. Ruelle, Dynamical systems with turbulent behavior, *Lecture Notes in Physics.* **80**, 341–360 (1978). *International Mathematics Physics Conference*, Roma, 1977.
31. D. Ruelle and F. Takens, On the nature of turbulence, *Communications in Mathematical Physics.* **20**, 167–192 (1971).
32. M. Hénon, A two-dimensional mapping with a strange attractor, *Communications in Mathematical Physics.* **50**, 69–77 (1976).
33. C. Grebogi, E. Ott, S. Pelikan, and J. A. Yorke, Strange attractors that are not chaotic, *Physica D.* **13**(1-2), 261–268 (1984).
34. E. N. Lorenz, Deterministic nonperiodic flow, *Journal of the Atmospheric Sciences.* **20**, 130–141 (1963). doi: 10.1175/1520-0469(1963)020⟨0130:DNF⟩2.0.CO;2.
35. E. Zeraoulia, *Lozi Mappings — Theory and Applications.* CRC Press (2013).
36. R. Lozi, Un attracteur étrange (?) du type attracteur de Hénon, *Journal de Physique.* **39**(8), C5-9–C5-10 (1978). doi: 10.1051/jphyscol:1978505.
37. M. Misiurewicz, Strange attractors for the Lozi mappings, *Annals of the New York Academy of Sciences.* **357**(1), 348–358 (1980). doi: 10.1111/j.1749-6632.1980.tb29702.x.
38. S. Smale, Differentiable dynamical systems. I Diffeormorphisms, *Bulletin of American Mathematical Society.* **73**, 747–817 (1967).

39. R. Lozi, Strange attractors: A class of mappings of $\mathbb{R}^2$ which leaves some cantor set invariant. In G. Laval and D. Grésillon (eds.), *International Workshop on Intrinsic Stochasticity in Plasmas.* Editions de Physique, Orsay, France (1979), pp. 373–381.
40. T. Uzer, B. Chirikov, F. Vivaldi, and G. Casati, Joseph Ford, *Physics Today.* **48**(10), 88 (1995).
41. D. Ruelle, Measure describing a turbulent flow, *Annals of the New-York Academy of Sciences.* **357**, 1–9 (1980).
42. L. D. Landau, On the problem of turbulence, *Doklady Akademii Nauk SSSR.* **44**, 339–342 (1944) (in Russian).
43. C. Vidal, J.-C. Roux, S. Bachelart, and A. Rossi, Experimental study of the transition to turbulence in the Belousov-Zhabotinsky reaction, *Annals of the New-York Academy of Sciences.* **357**, 377–396 (1980).
44. M. G. Crandall and P. H. Rabinowitz, Bifurcation from simple eigenvalues, *Journal of Functional Analysis.* **8**(2), 321–340 (1971). doi: https://doi.org/10.1016/0022-1236(71)90015-2.
45. P. Rabinowitz, Variational methods for nonlinear eigenvalue problems. In G. Prodi (ed.), *Eigenvalues of Non-Linear Problems*, Vol. 67. Springer, Berlin (1974), pp. 139–195.
46. R. Turner, Positive solutions of nonlinear eigenvalue problems. In G. Prodi (ed.), *Eigenvalues of Non-Linear Problems*, Vol. 67. Springer, Berlin (1974), pp. 213–239.
47. P. C. Fife, Branching phenomena in fluid dynamics and chemical reaction-diffusion theory. In G. Prodi (ed.), *Eigenvalues of Non-Linear Problems*, Vol. 67. Springer, Berlin (1974), pp. 23–83.
48. G. Iooss (ed.), *Bifurcation of Maps and Applications*, Vol. 36. North-Holland Mathematics Studies (1979).
49. P. Chossat, Étude, par la théorie des bifurcations, de la convection dans un domaine sphérique en rotation. Ph.D. Thesis, University of Nice, France (1977).
50. J. F. G. Auchmuty and G. Nicolis, Bifurcation analysis of nonlinear reaction-diffusion equations — I. Evolution equations and the steady state solutions, *Bulletin of Mathematical Biology.* **37**, 323–365 (1975). doi: https://doi.org/10.1016/S0092-8240(75)80036-X.
51. H. B. Keller, Numerical solution of bifurcation and nonlinear eigenvalue problems. In P. Rabinowitz (ed.), *Applications of Bifurcation Theory.* Academic Press, New York (1977), pp. 359–384.
52. H. Amann, N. W. Bazley, and K. Kirchgässner (eds.), Nichtlineare Funktionalanalysis und ihre Anwendungen. In H. Amann, N. W. Bazley, and K. Kirchgässner (eds.), *Workshop Report*, Vol. 5. Mathematisches Forschunginstitut Oberwolfach (1976). https://oda.mfo.de/handle/mfo/1405.
53. M. Hénon, Une transformation à deux dimensions avec un "attracteur étrange", *Journal de Physique Colloques.* **39**(C5-5), 5 (1978).
54. P. Coullet and C. Tresser, Itération d'endomorphismes et groupe de renormalization, *Journal de Physique Colloques.* **39**, C5–25–C5–28 (1978).

55. A. Chenciner, Stabilité structurelle et ergodicité (ou catégorie et mesure), *Journal de Physique Colloques.* **39**, C5–69–C5–77 (1978).
56. A. Chenciner, Hopf bifurcation for invariant tori. In J. Guckenheimer (ed.), *Dynamical Systems: Lectures Given at a Summer School of the Centro Internazionale Matematico Estivo.* Springer-Verlag (1980), pp. 197–207.
57. J. E. Marsden, Qualitative methods in bifurcation theory, *Bulletin of the American Mathematical Society.* **84**(6), 1125–1148 (1978).
58. D. R. J. Chillingworth and P. J. Holmes, Dynamical systems and models for reversals of the earth's magnetic field, *Journal of the International Association for Mathematical Geology.* **12**, 41–59 (1980).
59. Y. Choquet-Bruhat, A. E. Fischer, and J. E. Marsden, Maximal hypersurfaces and positivity of mass. In J. Ehlers (ed.), *Isolated Gravitating Systems in General Relativity.* Italian Physical Society (1979), pp. 396–456.
60. O. Alvarez, I. M. Singer, and B. Zumino, Gravitational anomalies and the family's index theorem, *Communications in Mathematical Physics.* **96**(3), 409–417 (1984).
61. H. P. McKean and P. van Moerbeke, Hill and Toda curves, *Communications on Pure and Applied Mathematics.* **33**(1), 23–42 (1980). doi: 10.1002/cpa.3160330103.
62. M. Lyubich, Dynamics of quadratic polynomials, I-II, *Acta Mathematica.* **178**, 185–297 (1997).
63. R. Lozi, Bifurcation and nonlinear eigenvalue problems: A tentative bibliography. Technical Report 18, INRIA (1982).
64. O. Rössler, Horseshoe-map chaos in the Lorenz equation, *Physics Letters A.* **60**(5), 392–394 (1977). doi: 10.1016/0375-9601(77)90029-9.
65. I. Shimada and T. Nagashima, A numerical approach to ergodic problem of dissipative dynamical systems, *Progress of Theoretical Physics.* **61**(6), 1605–1616 (1979).
66. J. L. Hudson and O. E. Rössler, A piecewise-linear invertible noodle map, *Physica D.* **11**(1), 239–242 (1984). doi: 10.1016/0167-2789(84)90446-9.
67. O. E. Rössler, Classical quantization: Two possible approaches. In G. Casati (ed.), *Chaotic Behavior in Quantum Systems.* NATO ASI Series B, Vol. 120. Springer, Boston, MA (1985), pp. 345–351. doi: 10.1007/978-1-4613-2443-0_23.
68. T. Tél, Invariant curves, attractors, and phase diagram of a piecewise linear map with chaos, *Journal of Statistical Physics.* **33**, 195–221 (1983).
69. R. L. Devaney, A piecewise linear model for the zones of instability of an area-preserving map, *Physica D.* **10**(3), 387–393 (1984). doi: 10.1016/0167-2789(84)90187-8.
70. G. Nicolis and I. Prigogine, *Self-Organization in Nonequilibrium Systems — From Dissipative Structures to Order through Fluctuations.* John Wiley & Sons, New York (1977).
71. R. J. Field and R. M. Noyes, Oscillations in chemical systems. IV. Limit cycle behavior in a model of a real chemical reaction, *The Journal of Chemical Physics.* **60**(5), 1877–1884 (1974). doi: 10.1063/1.1681288.

72. C. Vidal, S. Bachelart, and A. Rossi, Bifurcations en cascade conduisant à la turbulence dans la reaction de Belousov-Zhabotinsky, *Journal de Physique.* **43**(1), 7–14 (1982).
73. R. Lozi, Sur un modèle mathématique de suite de bifurcations de motifs dans la réaction de Belousov-Zhabotinsky, *Comptes Rendus de l'Académie des Sciences I.* **294**, 21–26 (1982).
74. C. Lobry and R. Lozi, Bifurcations of motifs in families of mixed two-vectors fields. In Springer Series in Synergetics, Vol. 12. Springer, Berlin (1981), pp. 67–70.
75. J. Argémi, Points singuliers non élémentaires de quelques classes de systèmes dynamiques plans, *Annali di Matematica Pura ed Applicata.* **89**, 321–351 (1971).
76. Logos et théorie des catastrophes à partir de l'œuvre de René Thom. In J. Petitot (ed.), *Actes du colloque international de 1982.* Patino (1988).
77. A. Douady, Julia sets and the Mandelbrot set. In *The Beauty of Fractals.* Springer, Berlin (1986), pp. 161–174.
78. H. Fujii, M. Mimura, and Y. Nishiura, A picture of the global bifurcation diagram in ecological interacting and diffusing systems, *Physica D.* **5**(1), 1–42 (1982). doi: 10.1016/0167-2789(82)90048-3.
79. M. Yamaguti and S. Ushiki, Chaos in numerical analysis of ordinary differential equations, *Physica D.* **3**(3), 618–626 (1981). doi: 10.1016/0167-2789(81)90044-0.
80. I. Tsuda, Self-similarity in the Belousov-Zhabotinsky reaction, *Physics Letters A.* **85**(1), 4–8 (1981). doi: 10.1016/0375-9601(81)90624-1.
81. O. E. Rössler, C. Knudsen, J. L. Hudson, and I. Tsuda, Nowhere-differentiable attractors, *International Journal of Intelligent Systems.* **10**(1), 15–23 (1995). doi: 10.1002/int.4550100104.
82. M. Mimura, M. Tabata, and Y. Hosono, Multiple solutions of two-point boundary value problems of Neumann type with a small parameter, *SIAM Journal on Mathematical Analysis.* **11**(4), 613–631 (1980). doi: 10.1137/0511057.
83. O. E. Rössler, My life in chaos, short autobiography (September, 2004). http://www.atomosyd.net/spip.php?article6.
84. M. Yamaguti and R. Lozi, Multiplication using fingers. In *Sugaku Seminar* (July, 1984), pp. 30–31 (in Japanese).
85. H. Oka and H. Kokubu, An approach to constrained equations and strange attractors. In T. Nishida, M. Mimura, and H. Fujii (eds.), *Patterns and Waves.* Studies in Mathematics and Its Applications, Vol. 18. Elsevier (1986), pp. 607–630. doi: 10.1016/S0168-2024(08)70149-8.
86. F. Takens, Implicit differential equations: Some open problems, *Lecture Notes in Mathematics.* **535**, 237–253 (1976).
87. R. Lozi, Can we trust in numerical computations of chaotic solutions of dynamical systems? In C. Letellier and R. Gilmore (eds.), *Topology and Dynamics of Chaos in Celebration of Robert Gilmore's 70th Birthday.* World Scientific Series on Nonlinear Science, Vol. 84. World Scientific Publishing (2013), pp. 63–98.

88. S. Ushiki and R. Lozi, Confinor and anticonfinor in constrained "Lorenz" system, *Japan Journal of Applied Mathematics.* **4**(3), 433–434 (1987).
89. R. Lozi and S. Ushiki, Organized confinors and anticonfinors and their bifurcations in constrained "Lorenz" system, *Annales des Télécommunications.* **43**(3-4), 187–208 (1988).
90. T. Matsumoto, A chaotic attractor from Chua's circuit, *IEEE Transactions on Circuits & Systems.* **CAS-31**(12), 1055–1058 (1984).
91. L. Chua, A rigorous proof of the existence of chaos in double scroll, *Annales des Télécommunications.* **42**(5-6), 239–246 (1987).
92. C. Letellier and J.-M. Ginoux, Development of the nonlinear dynamical systems theory from radio-engineering to electronics, *International Journal of Bifurcation & Chaos.* **19**(7), 2131–2163 (2009). doi: 10.1142/S0218127409023986.
93. A. S. Pikovski, M. I. Rabinovich, and V. Y. Trakhtengerts, Onset of stochasticity in decay confinement of parametric instability, *Zhurnal Eksperimentalnoi i Teoreticheskoi Fiziki.* **74**, 1366–1374 (1978).
94. M. I. Rabinovich, Strange attractors in modern physics, *Annals of the New York Academy of Sciences.* **357**(1), 435–452 (1980). doi: 10.1111/j.1749-6632.1980.tb29709.x.
95. R. Lozi and S. Ushiki, Confinors and bounded time — Patterns in Chua's circuit and the double scroll family, *International Journal of Bifurcation & Chaos.* **1**(1), 119–138 (1991).
96. C. Kahlert and L. O. Chua, Transfer maps and return maps for piecewise-linear three-region dynamical systems, *International Journal of Circuit Theory and Applications.* **15**(1), 23–49 (1987). doi: 10.1002/cta.4490150103.
97. B. Uehleke and O. E. Rössler, Complicated Poincaré half-maps in a linear system, *Zeitschrift für Naturforschung A.* **38**, 1107–1113 (1983).
98. C. Kahlert and O. E. Rössler, Analytical properties of poincaré halfmaps in a class of piecewise-linear dynamical systems, *Zeitschrif für Naturforschung A.* **40**, 1011–1025 (1985).
99. R. Lozi and S. Ushiki, The theory of confinors in Chua's circuit: Accurate analysis of bifurcations and attractors, *International Journal of Bifurcation & Chaos.* **3**(2), 333–361 (1993).
100. R. Lozi and S. Ushiki, Coexisting chaotic attractor from Chua's circuit, *International Journal of Bifurcation & Chaos.* **1**(4), 923–926 (1991).
101. L. Chua, The genesis of Chua's circuit, *Archiv für Elektronik and Übertragnugstechnik.* **46**(4), 250–257 (1992).
102. C. Letellier, *Chaos in Nature*, 2nd edn. Series on Nonlinear Science Series A, Vol. 94. World Scientific Publishing (2019). doi: 10.1142/11305.
103. E. Ott, C. Grebogi, and J. A. Yorke, Controlling chaos, *Physical Review Letters.* **64**(11), 1196–1199 (1990). doi: 10.1103/PhysRevLett.64.1196.
104. L. M. Pecora and T. L. Carroll, Synchronization in chaotic systems, *Physical Review Letters.* **64**(8), 821–824 (1990).
105. L. J. Kocarev, K. S. Halle, K. Eckert, and L. O. Chua, Experimental demonstration of secure communication via chaotic synchronization, *International Journal of Bifurcation & Chaos.* **2**(3), 709–713 (1992).

106. R. Lozi and L. O. Chua, Secure communications via chaotic synchronization. II Noise reduction by cascading two identical receivers, *International Journal of Bifurcation & Chaos.* **3**(5), 1319–1325 (1993). doi: 10.1142/S0218127493001070.
107. R. Lozi and A. Aziz-Alaoui, Secure communications via chaotic synchronization in Chua's circuit: Numerical analysis of the errors of the recovered signal. In *Proceedings of the 1994 International Symposium on Nonlinear Theory and Its Applications* (October, 1994), pp. 145–148.
108. R. Lozi and A. Aziz-Alaoui, Secure communications via chaotic synchronization in Chua's circuit and Bonhöeffer-van der Pol equation. Numerical analysis of the errors of the recovered signal. In *Proceedings of 1995 IEEE International Conference on Circuits and Systems* (May, 1995), pp. 684–687.
109. L. O. Chua and L. Yang, Cellular neural networks: Theory, *IEEE Transactions on Circuits and Systems.* **35**(10), 1257–1272 (1988).
110. M. Itoh and L. O. Chua, Chaotic oscillation via edge of chaos criteria, *International Journal of Bifurcation & Chaos.* **27**(11), 1730035 (2017).
111. X. Zhang, Z. Wu, and L. O. Chua, Hearts are poised near the edge of chaos, *International Journal of Bifurcation & Chaos.* **30**(9), 2030023 (2020).
112. L. Chua, Memristor — The missing circuit element, *IEEE Transactions on Circuit Theory.* **18**(5), 507–519 (1971).
113. D. Strukov, G. Snider, D. Stewart, and R. S. Williams, The missing memristor found, *Nature.* **453**, 80–83 (2008).
114. M.-S. Abdelouahab, R. Lozi, and L. O. Chua, Memfractance: A mathematical paradigm for circuit elements with memory, *International Journal of Bifurcation & Chaos.* **24**(9), 1430023 (2014).
115. G. A. Leonov, N. V. Kuznetsov, and V. I. Vagaitsevbh, Localization of hidden Chua's attractors, *Physics Letters A.* **375**(23), 2230–2233 (2011). doi: 10.1016/j.physleta.2011.04.037.
116. M. Tidjani, R. Lozi, and L. O. Chua, Hidden bifurcations in the multispiral Chua attractor, *International Journal of Bifurcation & Chaos.* **26**(14), 1630039 (2016).
117. R. Lozi and A. Saidi, Étude des motifs dans les oscillations de la réaction de Belousov-Zhabotinsky I Existence d'un confineur principal pour un modèle contraint particulier, *Annales de l'ENIT.* **3**(2), 69–98 (1989).
118. R. Lozi and A. Saidi, Étude des motifs dans les oscillations de la réaction de Belousov-Zhabotinsky II un algorithme de partition de l'espace des paramètres pour une famille d'équations différentielles contraintes, *Annales de l'ENIT.* **4**(2), 91–124 (1990).
119. A. Saidi, Un algorithme de partition de l'espace des paramètres pour une famille d'équations différentielles contraintes. Ph.D. Thesis, University de Nice, France (May, 1988).
120. H. Kawakami and R. Lozi, Switched dynamical systems — simple example of spiral linear system. In *Regional Conference on Electrical and Electronical Engineering* (1992), pp. 21–22.

121. Y. Miino and T. Ueta, Laboratory experiment for bifurcation phenomena of the forced Alpazur oscillator. In *International Symposium on Nonlinear Theory and Its Applications NOLTA2015*, Kowloon, Hong-Kong, China (2015), pp. 169–175.
122. D. Chirita, V. Stefanescu, B. C. Florea, and D. A. Stoichescu, An analogical control method of the Alpazur oscillator. In *9th International Symposium on Advanced Topics in Electrical Engineering (ATEE)* (2015), pp. 854–857.
123. V. Stefanescu, D. Stoichescu, M. Frunzete, and B. Florea, Influence of computer computation precision in chaos analysis, *Scientific Bulletin–University Politehnica of Bucharest A.* **75**(1), 151–162 (2013).
124. T. Kousaka, T. Ueta, and H. Kawakami, Bifurcation of switched nonlinear dynamical systems, *IEEE Transactions on Circuits and Systems II: Analog and Digital Signal Processing.* **46**(7), 878–885 (1999).
125. S. Ushiki and R. Lozi, Piecewise-linear vector fields and 3-dimensional confinors, *Transactions of the Japan Society for Industrial and Applied Mathematics.* **2**(4), 219–248 (1992) (in Japanese).
126. R. Lozi, Motifs et confineurs organisés dans la réaction de Belousov-Zhabotinskii et le "double scroll". In D. Bennequin (ed.), *Comptes Rendus de la 48ème Rencontre entre Physiciens et Mathématiciens*, Vol. 41. Publications de l'I.R.M.A. (1990), pp. 1–62.
127. M. A. Aziz-Alaoui, Analyse des motifs et confineurs d'un système différentiel contraint sur une variété lente. Ph.D. Thesis, Université de Nice, France (1987).
128. N.-E. Hamri, Continuité de la demi-application de Poincaré dans les équations du circuit de Chua. Ph.D. Thesis, Université de Nice, France (December, 1994).
129. S. Boughaba, Phénomènes de chaos déterministe et de turbulence — L'attracteur étrange de Chua: coexistence d'attracteurs, dynamique symbolique et confineurs. Ph.D. Thesis, Université Mentouri de Constantine, Algérie (March, 2002).
130. S. Boughaba and R. Lozi, Fitting trapping regions for Chua's attractor — A novel method based on isochronic lines, *International Journal of Bifurcation & Chaos.* **10**(1), 205–225 (2000).
131. R. F. Williams, The structure of Lorenz attractors, *Publications Mathématiques de l'I.H.E.S.* **50**, 73–99 (1979).
132. M. Rosalie and C. Letellier, Systematic template extraction from chaotic attractors: I. Genus-one attractors with an inversion symmetry, *Journal of Physics A.* **46**(37), 375101 (2013). doi: 10.1088/1751-8113/46/37/375101.
133. C. Fiol and R. Lozi, Global orbit patterns for discrete maps. In A. N. Sharkovsky and I. M. Sushko (eds.), *European Conference on Iteration Theory*, Vol. 354. *Grazer Mathematische Berichte*, 112–144 (2009).
134. C. Fiol and R. Lozi, Global orbit patterns for dynamical systems on finite sets. In *9th National Conference of Indian Society of Industrial and Applied Mathematics.* AIP Conference Proceedings (2009), pp. 303–331.

135. T.-Y. Li and J. A. Yorke, Period three implies chaos, *The American Mathematical Monthly.* **82**(10), 985–992 (1975).
136. O. M. Sarkovskii, Coexistence of cycles of a continuous map of a line into itself, *Ukrains'kyi Matematychnyi Zhurnal.* **16**, 61–71 (1964). doi: 10.3842/umzh.v76i1.8026.
137. A. N. Sarkovskii, Coexistence of cycles of a continuous map of a line into itself, *International Journal of Bifurcation & Chaos.* **5**, 1263–1273 (1995).
138. M. J. Feigenbaum, Universality in complex discrete dynamics, *Los Alamos Theoretical Division Annual Report.* **LA-6816-PR**, 98–102 (1976).
139. P. Stefan, A theorem of Sarkovskii on the existence of periodic orbits of continuous endomorphisms of the real line, *Communications in Mathematical Physics.* **54**, 237–248 (1977).
140. M. A. Aziz-Alaoui, A. D. Fedorenko, R. Lozi, and A. N. Sharkovsky. Recovering trajectories of chaotic piecewise linear dynamical systems. In F. L. Chernousko and A. L. Fradkov (eds.), *Proceedings of the International Conference in Control of Oscillations and Chaos*, Vol. 2 (1997), pp. 230–233. doi: 10.1109/COC.1997.631332.
141. M. A. Aziz-Alaoui, R. Lozi, A. D. Fedorenko, and A. N. Sharkovsky, Recovering parameters of chaotic piecewise linear dynamical systems. In *Proceedings of the International Conference in Control of Oscillations and Chaos*, Vol. 2 (2000), pp. 380–383.
142. R. Lozi, J. S. Ramos, and A. N. Sharkovsky, One-dimensional dynamics generated by boundary value problems for the wave equation. In J. S. Ramos, D. Gronau, C. Mira, L. Reich, and A. Sharkovsky (eds.), *Proceedings of the 14th European Conference on Iteration Theory.* Grazer Mathematische Berichte, Vol. 346 (2004), pp. 255–270.
143. V. N. Belykh, On models of phase synchronization systems and their study, *Dinamika System.* **11**, 23–32 (1976) (in Russian).
144. Y. G. Sinai, Dynamical systems with elastic reflections. Ergodic properties of dispersing billiards, *Russian Mathematical Surveys.* **25**(2), 137–189 (1970).
145. Y. B. Pesin, Dynamical systems with generalized hyperbolic attractors: Hyperbolic, ergodic and topological properties, *Ergodic Theory and Dynamical Systems.* **12**(1), 123–151 (1992). doi: 10.1017/S0143385700006635.
146. E. A. Sataev, Ergodic properties of the Belykh map, *Journal of Mathematical Sciences.* **95**, 2564–2575 (1999).
147. Bulat okudzhava — His circle and his times: A discussion, *Russian Studies in Literature.* **41**(1), 5–59 (2004). doi: 10.1080/10611975.2004.11062149.
148. N. E. Klinshpont, E. A. Sataev, and R. V. Plykin, Geometrical and dynamical properties of Lorenz type system, *Journal of Physics: Conference Series.* **23**, 96–104 (2005). doi: 10.1088/1742-6596/23/1/011.
149. O. E. Rössler, Fermi deceleration reencountered: recurrent grazing-type collisions with supermassive slow particule imply distance-proportionnal cooling. In C. Letellier, A. Aziz-Alaoui, J. Maquet, and I. Mutabazi (eds.), *Proceedings of the 4ème Colloque Chaos temporal et chaos spatio-temporel.* Université de Rouen et du Havre (2003), p. 3.

150. O. E. Rössler, Different types of chaos in two simple differential equations, *Zeitschrift für Naturforschung A*. **31**, 1664–1670 (1976).
151. C. Letellier and V. Messager, Influences on Otto E. Rössler's earliest paper on chaos, *International Journal of Bifurcation & Chaos*. **20**(11), 3585–3616 (2010).
152. J.-P. Lozi, O. Garasym, and R. Lozi, The challenging problem of industrial applications of multicore-generated iterates of nonlinear mappings, In P. Manchanda, R. Lozi, and A. Siddiqi (eds.), *Industrial Mathematics and Complex Systems. Industrial and Applied Mathematics*. Springer, Singapore (2017), pp. 43–67. doi: 10.1007/978-981-10-3758-0_4.
153. O. Jallouli, S. E. Assad, M. Chetto, and R. Lozi, Design and analysis of two stream ciphers based on chaotic coupling and multiplexing techniques, *Multimedia Tools and Applications*. **77**, 13391–13417 (2018).
154. T. Hamaizia, R. Lozi, and N.-E. Hamri, Fast chaotic optimization algorithm based on locally averaged strategy and multifold chaotic attractor, *Applied Mathematics and Computation*. **219**(1), 188–196 (2012).
155. A.-A. Khennaoui, A. Ouannas, S. Bendoukha, G. Grassi, R. P. Lozi, and V.-T. Pham, On fractional-order discrete-time systems: Chaos, stabilization and synchronization, *Chaos, Solitons & Fractals*. **119**, 150–162 (2019). doi: 10.1016/j.chaos.2018.12.019.
156. M.-S. Abdelouahab, R. Lozi, and G. Chen, Complex canard explosion in a fractional-order Fitzhugh-Nagumo model, *International Journal of Bifurcation & Chaos*. **29**(8), 1950111 (2019).
157. N. Semenova, A. Zakharova, E. Schöll, and V. Anishchenko, Does hyperbolicity impede emergence of chimera states in networks of nonlocally coupled chaotic oscillators?, *Europhysics Letters*. **112**(4), 40002 (2015). doi: 10.1209/0295-5075/112/40002.
158. B. Changaival, M. Rosalie, G. Danoy, K. Lavangnananda, and P. Bouvry, Chaotic traversal (CHAT): Very large graphs traversal using chaotic dynamics, *International Journal of Bifurcation & Chaos*. **27**(14), 1750215 (2017). doi: 10.1142/S0218127417502157.
159. O. E. Rössler, On the Rössler attractor, *Chaos Theory and Applications*. **2**(1), 1–2 (2020).

https://doi.org/10.1142/9789819823130_0013

Chapter 13

MY ROAD TO CHAOS

Leon Glass

McGill University, Canada
glass@cnd.mcgill.ca

This essay was written in honor of Otto Rossler's 80th birthday. I recount early phases in my career, starting from my time as a graduate student at the University of Chicago, continuing as a Postdoctoral Fellow at the University of Edinburgh, the University of Chicago, and Rochester University, and then assuming a faculty position at McGill University. During this time, I developed nonlinear dynamical models of gene expression with Stuart Kauffman, worked with Michael Mackey on time delay equations for control of breathing and blood cell production, and discovered a visual effect found when two identical random dot patters are superimposed on each other with a small rotation.

1. Introduction

As an undergraduate student at Brooklyn College I tried several majors before settling on Chemistry. First I was a pre-Med student (much to my mother's delight), but I disliked intensely Comparative Vertebrate Anatomy. We were told that liking and excelling in Comparative Vertebrate Anatomy was the *sine qua non* for a medical career. Inability to remember the gestation period of an aardvark was one of the barriers in that course. Then I was a psychology major and greatly enjoyed the Introductory Psychology course that dealt with experimental psychology. But I fell down in my Normal Psychology not being able to regurgitate the three characteristics of a normal personality. So I shifted to Chemistry, which I enjoyed.

For my Ph.D., I followed two former Brooklyn College students, Bruce Berne, who was a Professor at Columbia University, and the late Robert Silbey, who was a Professor and Dean at MIT, to the Department of Chemistry at the University of Chicago where they were both working with another Brooklyn College alumnus, Stuart Rice. I studied dynamics of molecules in liquid argon. In particular, I was interested in a time dependent correlation function. Given a particle at time 0 moves to position r at time t, what is the probability that another particle is at position r' and time t'?[1] There was little work on that problem at the time, and whenever I check, it seems that there is still little work on that problem.

2. Pattern in Visual System

For postdoctoral studies, I was interested in going back to my original fascination with medicine and psychology. I received a postdoctoral fellowship from the National Institutes of Health (NIH) to study the brain at the newly formed Department of Machine Intelligence and Perception at the University of Edinburgh. I worked with Christopher Longuet-Higgins, a brilliant Cambridge chemist who had decided to set aside Chemistry to work on the brain along with the psychologist Richard Gregory and computer scientist, Donald Michie.

At Edinburgh, I was fascinated by an experiment that Longuet-Higgins had learned about when he visited the physicist Eric Harth in Syracuse. Take a Xerox copy of a blank piece of paper, and then take a copy of the copy. Keep iterating always taking a copy of the last copy. After about 10–15 cycles you get a stable pattern of dots (Fig. 1). To compute the spatial autocorrelation function of the pattern (this was the same sort of thing I was thinking about for liquid argon), I made a transparency of the dots and projected it on a bull's eye pattern — I could put one dot in the center and count the number in concentric rings. But when I did this, I observed circular patterns from the superimposition of the transparency on the original dot pattern with a slight rotation. With Longuet-Higgins' encouragement and help, I wrote up the work which appeared in *Nature*, pointing out that the visual system must compute the localized autocorrelation function of the dot patterns to enable perception.[2]

During the year in Edinburgh, Jack Cowan invited me to return to Chicago to continue Postdoctoral studies on vision. I left Edinburgh in May

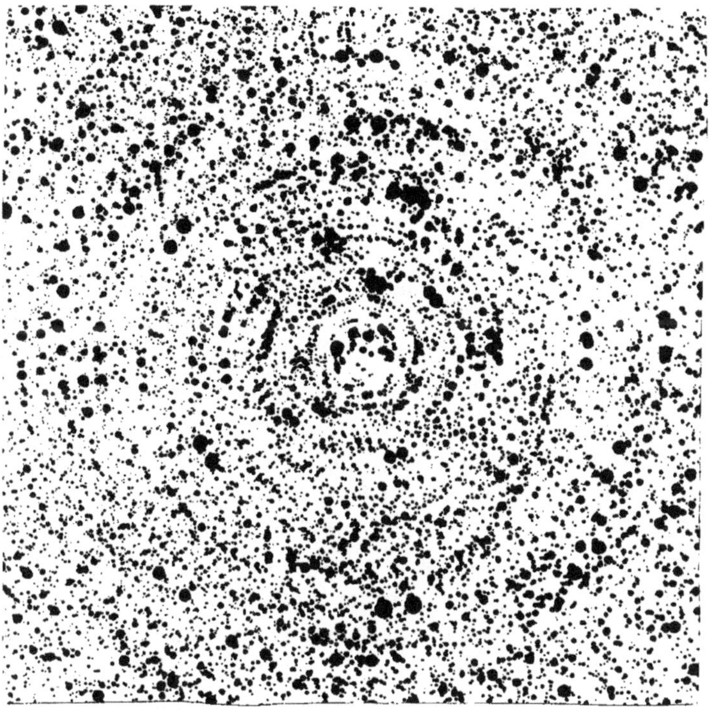

Fig. 1. Random pattern of dots rotated a few degrees and superimposed on itself. *Source*: Adapted from Glass.[2]

and took an extended trip around Europe, including spending a month in Brussels learning about the work that Ilya Prigogine and his students were doing on the generation of structures in chemical and biological systems.

3. Genetic Network

Back in Chicago, Jack Cowan was in the process of setting up the Department of Theoretical Biology. Cowan had started seminars that I had attended during my final year of Ph.D. studies. He had also hired two remarkable young scientists, Arthur Winfree and Stuart Kauffman for their first faculty positions. Although my initial plan was to continue working on vision, I became intrigued by Kauffman's studies of randomly constructed genetic networks. Kauffman had constructed random Boolean switching networks and found that for networks in which each element only had a couple of inputs, the dynamics was amazingly orderly.[3] Depending on the initial state, the dynamics evolved to one of a small number of comparatively short cycles. For example, in a network of 100 Boolean switches modeling genes, a typical cycle length would be of the order of 10 or 20 states out of the possible 2^{100} states. As someone, who had learned about the ergodic trajectories in statistical mechanics, the result was compelling. There was not a visit of all states in the system, but rather the dynamics was focused into a small number of basins of attraction. Equally compelling were Kauffman's supreme enthusiasm and confidence, and I started working with him. I rejected the notion of discrete states and discrete times, but embedded the switching network logic in differential equations.[4,5] This was really my first research that involved nonlinear dynamics. I learned about some of the basic notions including bifurcation and stability theory — topics that were not considered appropriate to include in graduate physical science programs at the time. As I learned later, at about the same time (1972–1975) Otto Rössler was also studying simple mathematical models of chemical systems that could lead to bistability and oscillation.[6] This was immediately before the explosion of interest in chaos.

My attempts to find a faculty position were not successful, but I did find another Postdoctoral position. This time in the Institute of Fundamental Studies that had just been set up by physicist Elliott Montroll in the Department of Physics and Astronomy at the University of Rochester. Montroll who had been selected as an Albert Einstein Professor in New York State, was renowned for his abilities in statistical mechanics. He thought that people who knew some statistical mechanics could tackle any problem. I continued working on problems I had started previously in vision, pattern formation, and dynamics in models of genetic networks. This work was done independently of Montroll, whose style tended to be more technically oriented than mine. One significant project during those years were studies done with graduate student Ronald Shymko and fellow Postdoc Rafael Perez of genetic networks with feedback inhibition. One gene would inhibit a second gene, whereas the second gene would activate the first gene. In the simplified model, if the two genes were in a homogeneous system then there would be a steady state, but if the two genes were located in different spatial locations, then there could be oscillations.[7,8] I realized that the time delays induced by the diffusion led to the oscillation and carried out some simulations (never published) of negative feedback systems with time delays. Although the position at Rochester morphed into a part-time Assistant Professor position, it became clear that the Physics and

Astronomy Department was not a good fit for me. I was also interested in collaborating with experimentalists.

4. Physiology

After a few years in Rochester, I learned of a job opening in Montreal. Michael Mackey, who had training in Biophysics and Mathematics, was a young faculty member in the Department of Physiology at McGill University in Montreal. Mackey initially came to McGill to form a quantitative team with David Firth, a Biophysicist. But when Firth decided to leave McGill to farm in Nova Scotia, there was an opening at McGill. I had met Mackey at Gordon Conferences in Theoretical Biology in the early 1970s, and I was delighted when the opportunity came to apply to McGill. I moved to Montreal in March 1975 and a few months later went out west to spend a month at the Aspen Center for Theoretical Physics. A talk by Stephen Smale, about the period-doubling route to chaos was intriguing. Mitchell Feigenbaum was also there and he attributed that meeting also to piquing his interest in chaos.[9]

When I got back to Montreal, I was excited to discuss chaos with Mackey. I asked him if physiological systems could display chaos. He said he did not know. We decided to write a team grant application to the Natural Sciences and Engineering Council of Canada on the theme "Oscillation and Chaos in Physiological Systems." Much to my dismay, the grant application is now lost, so it is hard to be sure exactly what it said. But we certainly proposed to study both difference and delay differential equation models for two physiological systems: Mackey would look at hematopoiesis and I would look at respiration. In one of the models for hematopoiesis there were several novel features. First there was just one variable — the blood cell concentration. Instead of just using negative feedback, the equation

$$\dot{x} = \beta \frac{x_\tau}{1 + x_\tau^n} - \gamma x \quad \gamma, \beta, n > 0 \tag{1}$$

had a non-monotonic feedback term. Blood cells would only be strongly produced if the number blood cells was neither too high nor too low. And since it took some time to produce red blood cells once the signal was received, the production incorporated a time delay term — the current rate of blood cell production was a function of the concentration of blood cells in the past. We searched for chaos in the model. Most exciting was the day when Mackey and I both sat in front of a primitive computer screen

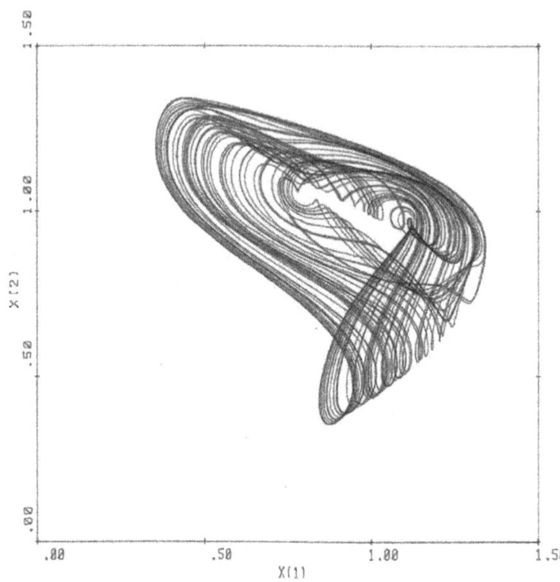

Fig. 2. Time delay embedding of a chaotic time series from the model for blood cell production using the time delay equation (1) that we proposed in 1977.[10] The trace is similar to the time delay embeddings presented at the New York Academy of Sciences November 1977 conference organized by Okan Gurel and Otto Rössler.

Source: Image from the original tracings made in 1976 or 1977 where $x(2)$ is the delayed variable and $x(1)$ is the current variable.[11]

and watched the trajectory. Since there was only one variable, we plotted two coordinates, the current value and the value in the past. To the best of my knowledge, this was the first use of time delay embedding to examine complex dynamics and bifurcations. We tweaked parameters and eventually found what we were seeking. We observed as a simple trajectory consisting of a closed loop. This developed a kink, then a second loop, then doubled, and finally appeared to be irregular.[10,11] It was chaotic (Fig. 2), and the route to chaos seemed similar to what had been observed in simple quadratic maps that had been studied in the past few years by May, Oster, Li, and Yorke.[12,12,13]

General interest in chaos had been piqued by Robert May's 1976 review published in *Nature* on "Simple Mathematical Models with Very Complicated Dynamics."[14] We submitted our findings on chaos in simple mathematical models of physiological systems to *Science*, and were delighted when it appeared.[10] Given the requirements of *Science* for short papers (people did not include lengthy supplemental material those days) we decided to omit the time delay embedding and just had time series from the models that we compared with published traces of hematological and respiratory diseases that were characterized by complex dynamics. We emphasized the concept that diseases could be characterized by abnormal dynamics that might be associated with bifurcations in nonlinear equations.

5. The 1977 NYAS Meeting and Otto E. Rössler

In November 1977, Okan Gurel and Otto Rössler organized a meeting on Bifurcation Theory and Applications in Scientific Disciplines in the New York Academy of Sciences.[15] I was invited to speak, and had the opportunity to develop the ideas on disease dynamics further in a presentation on "Pathological Conditions Resulting from Instabilities in Physiological Control Systems." We used slides in those days, and I ran through a sequence showing the various dynamics in the chaotic time delay equation as a parameter changes using the time delay embedding. The paper eventually appeared in 1979.[11]

There were many people at the 1977 NY Academy of Sciences meeting who went on to have major impact on the development of nonlinear dynamics. One was Robert Shaw, who at the time was a graduate student in Physics at the University of Santa Cruz spending significant time with his fellow graduate students (the "Dynamical Systems Collective" also included Norman Packard, James Crutchfield and Doyne Farmer) to develop a way to beat roulette by entering data from the roulette wheel into a computer program in a shoe.[16] The Dynamical Systems Collective wrote an influential paper in *Physical Review Letters* in 1980 showing how you could get a two dimensional portrait of a time series by plotting the value of one variable on one axis and its derivative, or as suggested in a footnote its value at an earlier time, on the other axis.[17] One member of the group, Farmer, went on to study the time delay equation modeling blood cell dynamics for his doctorate, referring to it as the Mackey–Glass equation![18] Another person at the meeting was David Ruelle, who was already well known for his expertise in statistical mechanics. On a cafeteria food line after my presentation, Ruelle suggested that I could look at the return map to a cross section on the time delay embedding. He correctly thought that it would be parabolic. As far as I know the only published return plot for this equation appeared in a *Scholarpedia* review article that Mackey and I presented many years later, when we finally took Ruelle's suggestion.[19]

Another person at the meeting was Otto Rössler. I believe that meeting was the first time I met Rössler. I have a clear memory of Rössler giving a talk in a small crowded room. He had developed a way to sonify a chaotic time series. He translated the value of one of the coordinates of his equation into a pitch and played the value of the pitch in real time as the integration proceeded. It was intriguing to perceive the chaotic dynamics by listening. An early demonstration is available on YouTube as of the current writing.[a] In addition to the three dimensional system for which Rössler is best known, he discovered several other simple low dimensional systems that displayed chaos. To help develop insight into the geometry of these attractors Rössler drew some simple pictures showing main aspects of the flows deduced from observation of the trajectories. These were very inventive and have had a deep impact in nonlinear dynamics.[20]

Back in the early days of chaos, I attended several meetings with Rössler, and always enjoyed hearing about his work, but I never used any of his equations for my own research. I was more focused on trying to develop collaborations with experimentalists that would lead to complex dynamics that we could model directly. But Rössler's idea to have sonification of complex time series, did influence a project carried out by Mark Ballora, who was a graduate student in music technology at McGill who carried out sonification of heart rhythm records of healthy people and of patients with medical problems resulting in abnormal heart rhythms.[21] indexheart rhythm

Early studies of chaos were motivated not only by the unexpected complexity in simple mathematical equations, but also by the notion that a lot of the irregular, aperiodic dynamics observed in the natural world could be modeled and explained by comparatively simple low dimensional chaotic systems. Although many well controlled experimental models are well described by nonlinear equations displaying low dimensional chaotic dynamics, there is still a big gap in understanding complex dynamics arising in the natural world. And it is still a challenge to translate mathematical insights concerning bifurcations and chaos to practical applications.

---

[a] https://www.youtube.com/watch?v=Tmmdg2P1RIM.

Acknowledgments

I thank the Natural Sciences and Engineering Research Council (Canada) for its continuous support of curiosity-driven research for over 40 years starting with the events recounted here. I also thank colleagues and collaborators including Stuart Kauffman, Rafael Perez, Ronald Shymko, Michael Mackey for their wonderful insights, friendship, and collaboration.

References

1. L. Glass and S. A. Rice, Unified approximation for the velocity autocorrelation function and the structure function of a simple liquid, *Physical Review.* **176**(1), 239 (1968).
2. L. Glass, Moiré effect from random dots, *Nature.* **223** (5206), 578–580 (1969).
3. S. A. Kauffman, Metabolic stability and epigenesis in randomly constructed genetic nets, *Journal of Theoretical Biology.* **22**(3), 437–467 (1969). doi: 10.1016/0022-5193(69)90015-0.
4. L. Glass and S. A. Kauffman, Co-operative components, spatial localization and oscillatory cellular dynamics, *Journal of Theoretical Biology.* **34**(2), 219–237 (1972).
5. L. Glass and S. A. Kauffman, The logical analysis of continuous, non-linear biochemical control networks, *Journal of theoretical Biology.* **39**(1), 103–129 (1973).
6. F. F. Seelig and O. E. Rössler, Model of a chemical reaction flip-flop with one unique switching input, *Zeitschrift für Naturforschung B.* **27**(12), 1441–1444 (1972). doi: 10.1515/znb-1972-1201.
7. L. Glass and R. Pérez, Limit cycle oscillations in compartmental chemical systems, *The Journal of Chemical Physics.* **61**(12), 5242–5249 (1974).
8. R. Shymko and L. Glass, Spatial switching in chemical reactions with heterogeneous catalysis, *Journal of Chemical Physics.* **60**(3), 835–841 (1974).
9. M. J. Feigenbaum, Universal behavior in nonlinear systems, *Los Alamos Science.* 4–27 (1980).
10. M. C. Mackey and L. Glass, Oscillation and chaos in physiological control systems, *Science.* **197**, 287–289 (1977). doi: 10.1126/science.267326.
11. L. Glass and M. C. Mackey, Pathological conditions resulting from instabilities in physiological control system, *Annals of the New York Academy of Sciences.* **316**(1), 214–235 (1979). doi: 10.1111/j.1749-6632.1979.tb29471.x.
12. R. M. May, Simple mathematical models with very complicated dynamics, *Nature.* **261**(5560), 459–467 (1976). doi: 10.1038/261459a0.
13. T.-Y. Li and J. A. Yorke, Period three implies chaos, *The American Mathematical Monthly.* **82**(10), 985–992 (1975).
14. R. M. May and G. F. Oster, Bifurcations and dynamic complexity in simple ecological models, *The American Naturalist.* **110**(974), 573–599 (1976).

15. O. Gurel and O. E. Rössler, *Bifurcation Theory and Appliations in Scientific Disciplines.* Annals of the New York Academy of Sciences, Vol. 316. New York Academy of Sciences (1979). Papers from the Conference Held on October 31–November 4, 1977.
16. T. A. Bass, *The Eudaemonic Pie.* Houghton Mifflin Harcourt (1985).
17. N. H. Packard, J. P. Crutchfield, J. D. Farmer, and R. S. Shaw, Geometry from a time series, *Physical Review Letters.* **45**(9), 712–716 (1980). doi: 10.1103/PhysRevLett.45.712.
18. J. D. Farmer, Chaotic attractors of an infinite-dimensional dynamical system, *Physica D.* **4**(3), 366–393 (1982). doi: 10.1016/0167-2789(82)90042-2.
19. L. Glass and M. Mackey, Mackey-Glass equation, *Scholarpedia.* **5**(3), 6908 (2010).
20. O. E. Rössler, Continuous chaos: Four prototype equations, *Annals of the New York Academy of Sciences.* **316**, 376–392 (1979).
21. M. Ballora, B. Pennycook, P. C. Ivanov, L. Glass, and A. L. Goldberger, Heart rate sonification: A new approach to medical diagnosis, *Leonardo.* **37**(1), 41–46 (2004).

https://doi.org/10.1142/9789819823130_0014

Chapter 14

ENCOUNTERING CHAOS THROUGH LIGHT, AND BEYOND

Kensuke Ikeda

Ritsumeikan University, Japan
ahoo@amber.plala.or.jp

My research life happened to start from research of the dynamics of nonlinear optical systems. As if by fate, I encountered chaos there. I was caught by the stable manifold of the chaos boom in nonlinear optics and for a while I was entangled in that area. After some time, I met an unstable manifold and was thrown out in a different direction. I become aware that some basic problems of quantum dynamics, which I was originally interested in, appear anew in the context of chaos. And I found myself wandering around them. I wish to write about how that happened, and about what I know, saw and felt during the dynamical evolution that I experienced, as I experienced it.

1. Quite Personally

First, let me begin with my education environment at the time that I first encountered physics in my highschool days. It may sound extreme, but many Japanese high schools, especially selective high schools that collected the best students who scored high on the high school entrance examination, were then (and still are) like preparatory schools for university entrance examamination. National universities, including the top universities, Tokyo University and Kyoto University, collected the best students with high examination scores. If one could enter one of the top universities, one could get a job at a big company, employment would be guaranteed by an age-based seniority system, and a happy life would be promised. Aiming for such a guaranteed life, high school students worked hard to enter top universities with good entrance examination scores.

Since the entrance examinations were given in the form of paper tests, training to get high scores on paper tests was a major part of high school education. Complex combinations of physical situations were presented as problems to solve in the paper tests of physics. In my view, winning the game of solving such complex problems on paper was a common view of the pleasure to learn physics.

In the high school physics classes, there were almost no physics experiments in which students could actively participate. "Experiment demonstrations" demonstrated by the teacher on the teacher's desk were the only experiment experiences provided to students.

However, basic laws of physics can be examined using only simple tools, such as a ruler, a watch, and a scale in the case of mechanics. I loved handcraft, so I made crude devices to observe and measure physical phenomena: such as the time for a body to fall, the period of a gravity pendulum, the period of vibration of rubber string with suspended weight... In particular, measuring the static and dynamic friction coefficients was an impressive but unsuccessful experience.

The experiments were quite rough, the measurements were not very accurate, and the experiments often ended in failure. Nevertheless, there were rare cases that "theory" and "experiment" coincided beautifully. Such experiences left me with a vague but convincing feeling that physics was surely related to the real world.

One of the high school experiences that inspired me to pursue physics, was reading an introductary book on astro-mechanics, where I learnt that Newton's law of mechanics is "initial state determinism." Initial conditions determine the evolution of everything in the world, even seemingly uncertain evolution!! This had an intense impact on me.

Then, I heard rumors that quantum theory overturned classical determinism. As is the customary course for high school students who become addicted to reading popular commentaries on quantum theory and are deceived by such matters, I entered the faculty of science at a university where there was a department of physics.

At that time, elementary particle physics was the most popular subject among the students of physics departments. In particular, at Kyoto University which I went to, the Yukawa Institute (formally the Research Institute for Fundamental Physics, RIFP) founded by Hideki Yukawa who received the Nobel prize for the theory of meson[1,2] was an aspiration of the students. The nuclear astrophysics laboratory which was headed by Chushiro Hayashi, and researched a wide range of astrophysics,[3] from the origin of the universe to the evolution of the solar system, was also very popular for students. But I felt little attraction to those kinds of issues far from everyday life, and I was interested in how the fundamental laws of classical and quantum physics are relevant to phenomena of ordinary space-time scales.

In graduate school, I requested to be assigned to the laboratory of Kazuhisa Tomita which studied nonequilibrium statistical mechanics[4] and I was soon accepted as a member in 1972. In those days, the main subjects of nonequilibrium-nonlinear statistical physics were spatiotemporal pattern formation, nonequilibrium ordered states and the transition phenomena between them, and these were explored from the viewpoint of stochastic processes. At one end of the spectrum were quite general approaches, like Ilya Prigonine's non-equilibrium thermodynamics[5,6] and Hermann Haken's synergetics,[7] that view everything from the top side, and at the other end were quite phenomenological approaches such as Belousov–Zhabotinsky chemical reaction, propagation of excitations along nerve cells, and laser oscillations, etc.

The first research paper that I was impressed was on the relaxation process of spin system, which was written by Ryogo Kubo and Kazuhisa Tomita in 1954.[8] I was much fascinated with it, by its harmonic logic unifying mathematical description, and intuitive physical assumptions. At that time, I was misleading that the irreversible relaxation process near equilibrium could be solved by linear response theory, typified by Kubo's theory. I studied Kubo's linear response theory and related theories. The theory is certainly very beautiful, but I could not fully understand how a reversible dynamical process in quantum mechanics or classical mechanics could realize irreversible relaxation behavior.

On the other hand, I could not follow the mainstream of Tomita's laboratory. I did not know what to do. It was at this time that Hiroshi Hasegawa and Tomohei Sasada invited me to study superradiance — an invitation that led me to the field of nonlinear optics which I had not considered before, and to my encounter with chaos.

2. To the World of Nonlinear Optics: Superradiance

Superradiance — which was theoretically predicted by Robert H. Dicke in 1954, see[9] — is an irreversible radiation process through which an assembly of excited atoms, which interact cooperatively via the emitted radiation, spontaneously form a macroscopic classical polarization, and emit radiation at a rate proportional to the square of the number of atoms. Many theoretical approaches were presented at this time. I was intensely interested in this phenomenon because it involves two factors; one is the irreversible dynamical process and another one is the spontaneous macroscopic symmetry breaking. The mainstream theories were based on damping theory with adiabatic elimination of the radiation field as a heat reservoir of temperature 0.

Sasada suggested that we treat the main radiation field not as a reservoir but as active dynamical variables exchanging energy with the atoms. I was immediately attracted by his idea and started theoretical research. I did not achieve significant results in that direction. However, Sasada's way of thinking — to regard the external world (the reservoir) as active elements that interact with the interior system, and remove the boundary between the interior and exterior, had a profound influence on me. Sasada's idea was observed later in experiments as 'ringing' radiation, which suggests active exchange of energy between radiation field and excited atoms.

I turned in a slightly different direction and started research on the superradiance from three-level systems, which had been experimentally observed by Masahiro Matsuoka's laboratory located next to my room. Ordinarily, in a two-level system, the decoherence of atomic polarization prevents superradiant emission. Contrary to this, we theoretically found that the three-level superradiance can not occur without decoherence. Through this study, I learned about the theoretical treatment of light propagation in a decohering atomic system, which became the technical basis of my research on optical chaos.

3. Optical Bistability and Chaos

In the late 1970s optical communication using optical fiber as the medium was close to commercial deployment. The idea of directly processing the information carried by the light, without converting it into an electrical signal, was being advocated. As the source of coherent light signal the laser system working as an active optical resonator filled with excited atomic medium, plays the central role. But optical devices playing the roles of transistors working as memory elements, logical gates, switches, and other basic functional elements for communications were needed. Optical bistability exhibited by nonlinear optical resonators was a promising phenomenon applicable to such functions and was being extensively investigated experimentally by Samuel L. McCall, Hyatt M. Gibbs, and others.[10]

The standard model for optical bistability is two-level atoms confined in an optical resonator and interacting with a coherent electromagnetic field that propagates through the resonator according to the Maxwell equations. This model shares much with the model for superradiance. The essentially different point is that the optical bistability is a stationary phenomenon under the steady excitation by a laser field and is not a transient behavior like superradiance. Optical bistability exhibits typical nonlinear and nonequilibrium responses for strong external excitation by coherent laser light. It had been predicted that the stationary bistable states may become unstable under proper conditions leading to a new spatio-temporal nonequilibrium state.

Such researches were led by European groups in Italy, Germany, England, and Belgium under the strong influence of the general concept of "dissipative structure" proposed by Prigogine. A dissipative structure is a new state due to the nonlinearity of the system under nonequibrium conditions. The study of dissipative structures was a global trend of the research in nonequilibrium statistical mechanics, and optical bistability provides a good example for the study of dissipative structure. At that time, in Japan, a workshop on nonequilibrium and nonlinear statistical mechanics was held every year as a long-term project by the Yukawa Institute of Kyoto University, which was a center of theoretical physics research in Japan.

As shown in Fig. 1, a simple model of a passive resonator is a nonlinear ring resonator filled with two-level atoms. Laser light is inserted from the entrance and makes a round-trip inside the resonator interacting with atoms. The round-trip light reaching the entrance mirror is partially reflected and reinjected into the ring resonator, but some part of the light is transmitted and output through the entrance mirror. The output light exhibits bistable (sometimes multistable) response to the input laser light.

I was initially only watching the thriving optical bistability community as an outsider, but I became aware that the method I used in my superradiance research to treat the light propagation process can be applied to an optical bistability resonator. Applying the methods under the boundary conditions of the ring resonator, I was able to derive a novel scheme of equations. The simpler form

$$\begin{aligned} E(t) &= A + Be^{i(\phi(t)-\phi_0)}E(t-t_{\mathrm{R}}) \\ \dot{\phi} &= -\phi + |E(t-t_{\mathrm{R}})|^2 \end{aligned} \tag{1}$$

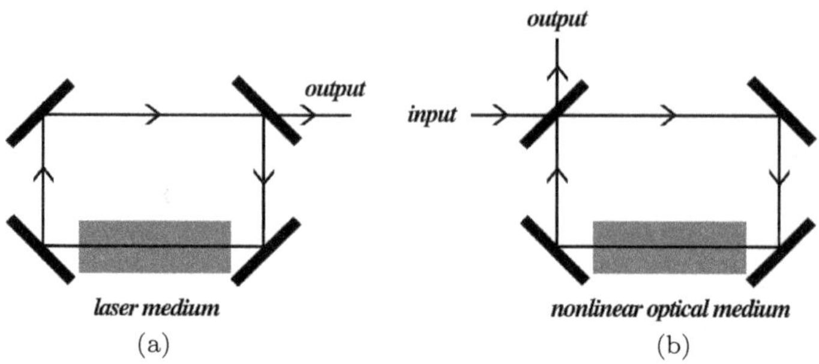

Fig. 1. Optical resonators: (a) active resonator (laser) and (b) passive resonator (optical bistable system). The light goes around along a ringed path.

of the rather complicated original equations is obtained by keeping only the essential parts, the electric field E at the entrance, and the spatially integrated medium ϕ, as two dynamical variables.[11] In Eq. (1), t_{R} is the round-trip time of the light, A is the incident field, $B \leqslant 1$ is the disipation parameter characterizing the loss of the resonator, and the relaxation time of the two level medium is normalized to unity. The equations contain the time delay t_{R} and form a coupled pair of difference and differential equations, which was recently called a delay-differential equation.

At the time, I didn't even know whether such a strange class of equation has a unique solution for a properly given initial condition. How can these equations be solved? I remember having a heated discussion about this issue at a mountain lodge in the Japan Alps with my mountain friend Okikazu Akimoto, with whom I collaborated on another topic.

Not having a way to solve the equations, I tried to reduce Eq. (1) into two classes of equations in the two opposite limits, that is, into

$$E(t) = A + BE(t - t_{\mathrm{R}})e^{i|E(t-t_{\mathrm{R}})|^2 - \phi_0} \tag{2}$$

and

$$\dot{\phi} = -\phi(t) + 2A^2 B \cos\left[\phi(t - t_{\mathrm{R}}) - \phi_0\right]. \tag{3}$$

Equation (2) is for the fast medium relaxation limit and Eq. (3) is for the large resonator loss limit. Equation (2) is a map with the time-unit t_{R}, whereas Eq. (3) is a differential equation with a time delay. In both cases, the existence of the time-delay seemed curious to me. Equation (2) is now called the Ikeda map.

At Kyoto University in the late 1970s, the postgraduate students of theoretical physics shared common rooms, and the interaction among the students of different laboratories was quite active. There were no mind "walls" between the students of different laboratories. Not a few students pursued research and wrote papers independently of the boss of their laboratory. I was one such student.

I heard about chaos from students of the Tomita laboratory, in particular, Hiroaki Daido and Ichiro Tsuda. They encountered chaos through their research of dissipative structures such as limit cycles and spatial patterns of chemical reactions. Through them, I learnt about the papers by Lorenz,[12] Rössler,[13] Ruelle and Takens,[14] and became acquainted with Yoshisuke Ueda who had a laboratory in the same University. Ueda allowed me to examine a simplified version of Eq. (1) with his big analogue computer system that occupied a few rooms.

Daido showed me the paper by Robert May[15] and suggested that my difference model Eq. (2) might have a chaotic solution. He recommended me to plot the trajectory of the numerically solved solution. Equation (2) is a simple map which can nowadays be solved numerically even by undergraduate students, but I believed that the evolution rules of physics should be expressed as (partial-)differential equations, and both Eqs. (2) and (3) seemed to me to be improper and dubious evolution rules. If model (2) has a chaotic solution, its solution has quickly many singularities evolving in every time interval $[nt_{\mathrm{R}}, (n+1)t_{\mathrm{R}}]$ $(n \in \mathbb{N})$.

I had my first experience in plotting numerical solutions in the plotter room of the Kyoto University computer center with the help of Daido. I was very surprised that the pen of the xy-plotter moved irregularly on a wide paper sheet and plotted a quite strange set of points (each point was represented by a "cross"). The set of points formed a spiral shape shown in Fig. 2. On each whorl of the spiral, there is an unstable fixed point corresponding to a destabilized "multistable" state.[16]

I was not convinced that such an evolution rule describes the dynamics of real optical bistability, but I published my first paper about optical chaos in 1979, reporting my formulation of the issue and numerical evidence of chaotic behavior. This was my first paper related to optical bistability. Subsequently, I published, together with Akimoto and Daido, the next report in which the model Eqs. (1) is proposed as well as the analyses of the reduced equations Eqs. (2) and (3) including the instability conditions, and the bifurcation diagram to the chaotic solutions.[16] At that time, it was not known in the nonlinear optics community and the laser community, that

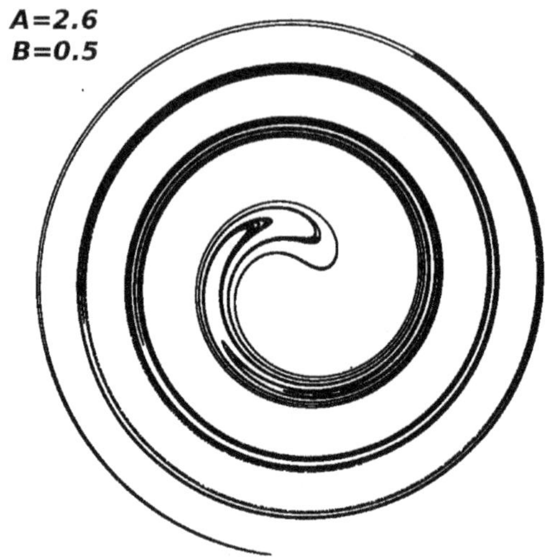

Fig. 2. The strange attractor produced by the mapping (2).

the time-delayed feedback due to the round trip of light in the optical resonator leads to instabilities and chaos to the stationary response of optical bistability.

4. Outbreak of Chaos in Nonlinear Optics

The laser oscillator is an active optical resonator, such as shown in Fig. 1, containing an unstable lasing medium continuously excited and emitting coherent light. Hence, laser oscillations are often accompanied by unstable oscillations, and the instability of laser oscillation was a crucial practical problem that had to be overcome in order to develop stable continuous-wave laser light sources. So, it is ironic that the study of unstable oscillation phenomena as a topic in itself only attracted attention after techniques for achieving stable laser oscillation were well established in the 1970s.

Theoreticians became aware that unstable oscillation phenomena lead to new dynamical states peculiar to nonequilibrium systems. In particular, Hermann Haken and his group studied the laser as a typical example of a nonequiliblium system far from equiliblium in the context of synergetics, and laser instabilities were studied as realistic examples of mechanisms yielding dissipative structure.[7]

However, it seems to me that the interest in laser instability was limited to theoreticians, and there was not clear connexion between theory and experiment until Lee W. Casperson developed a detailed theoretical analysis of the experimentally observed oscillatory instability in gas lasers in 1980s.[17]

On the other hand, the optically bistable system is a passive optical resonator containing nonlinear absorptive and/or dispersive medium modelled by two-level atoms. So, such basic model shares much with a laser resonator, but it contains unexcited atoms as the optical medium and it only operates under irradiation by laser light. For studying optically bistable systems, there was the theoretical group supervised by Rodolfo Bonifacio in Italy, and some others in Europe.[18, 19] Self-pulsations produced by a more simple mechanism was proposed and observed by McCall.[20] They used the unstable mode analysis and Landau expansion with respect to the small amplitude of unstable modes. The model is essentially the same as mine, but the approach is quite different. In particular, the round-trip time does not occur explicitly in their approach. Some other theoreticians in England, France, and Belgium also contributed to the instability analyses of optically bistable systems.

It is well known that more than one optical mode resonating with the resonator may contribute to laser oscillation. Indeed, in a pioneering work by Hannes Risken and Kjell Nummedal, the single-mode CW oscillaton of a multi-mode ring laser system may become unstable and multiple modes participate in self-oscillation.[21] However, if it is allowed to suppose that only a single mode exactly resonates with the two quantum levels of the lasing medium, then the laser equation of motion is described by a set of differential equations

$$\begin{cases} \dot{E} = -\kappa E + \eta\rho \\ \dot{\rho} = -\gamma\rho + \mu E w \\ \dot{w} = -\gamma_{\|}(w - w_{\infty}) - \mu E\rho \end{cases} \tag{4}$$

only involving three dynamical variables, namely the electric field E, the medium polarizaton ρ, and the difference between the populations of the two levels w. γ and $\gamma_{\|}$ are the relaxation rates of the lasing medium, κ is the loss rate of the resonator, and μ and η are the field-medium coupling strengths. Haken realized that the above set of equations is equivalent to the Lorenz equations.[22] He was the first to point out that a laser may produce deterministic chaos under the condition of Lorentz instability. The self-induced oscillation causing Lorenz chaos is due to the Rabi-precession $\rho + iw \propto \mathrm{e}^{-i\mu E t}$, which can readily be confirmed by Eq. (4) in the dissipationless limit $\gamma = \gamma_{\|} = 0$. However, in ordinary lasers, the relaxation rate γ of polarization is so large that the condition for Lorentz instability is not satisfied.

It seems to me that the theoretical researchers in Europe were the first ones to be interested in the instabilities of nonlinear optical systems, and that they were under the influence of the large paradigms of "nonlinear-nonequiliblium thermodymamics" (statistical mechanics), and of "dissipative structures,"[5,6] and "synergetics."[7]

When my 1979 and 1980 papers were published, there were some proposals for research on chaos in the field of nonlinear optics, but there did not seem to be momentum to start experimental research on chaos. However, there was one group of experimentalists who did take notice of our papers. Interestingly, it was not a "philosophy-oriented" Europe group, but an "implementation-oriented" US group, the group of Gibbs and Fred A. Hopf at Arizona University that lead optical bistability research in a wide range of applied and fundamental aspects.[23–26]

My boss Hiroshi Hasegawa gave me a chance to attend the Japan-US symposium on laser physics held in 1980 in Hawaii where Gibbs attended. Gibbs requested a meeting with me, and we meet at a restaurant. We had not had any previous personal contact. I was very impressed that Gibbs, a leader of the optical bistability community, wanted to personally meet me, a novice researcher who was unknown even in Japan at that time. He asked me many questions about my paper of 1979, while correcting many English typos in my preprint. I emphasized that the presence of delay time is essential for the instability leading to chaos. He said that they would like to try observe the onset of chaos with their experimental system, but it did not meet the onset condition. So, I suggested to insert an electric circuit introducing the delay, as proposed in our 1980 paper.[16]

After a few months I was informed that the period-doubling bifurcation and the onset of chaos were successfully observed in their experimental setup, as shown in Fig. 3.

Their paper about the experiment, published in the early 1981,[27] was the first to report the observation of bifurcation leading to deterministic chaos in nonlinear optics. For purely all-optical devise, the first observation of bifurcation to chaos was published by the group from the Kyoto University in 1983,[28] which was followed by the group from the Heriot-Watt University in 1984.[29] These papers triggered a rapid increase in the interest for experimental and theoretical research on chaos in nonlinear optics.

In the following years, many research results related to instabilities and chaos in lasers and optical bistability began to be presented: many sessions

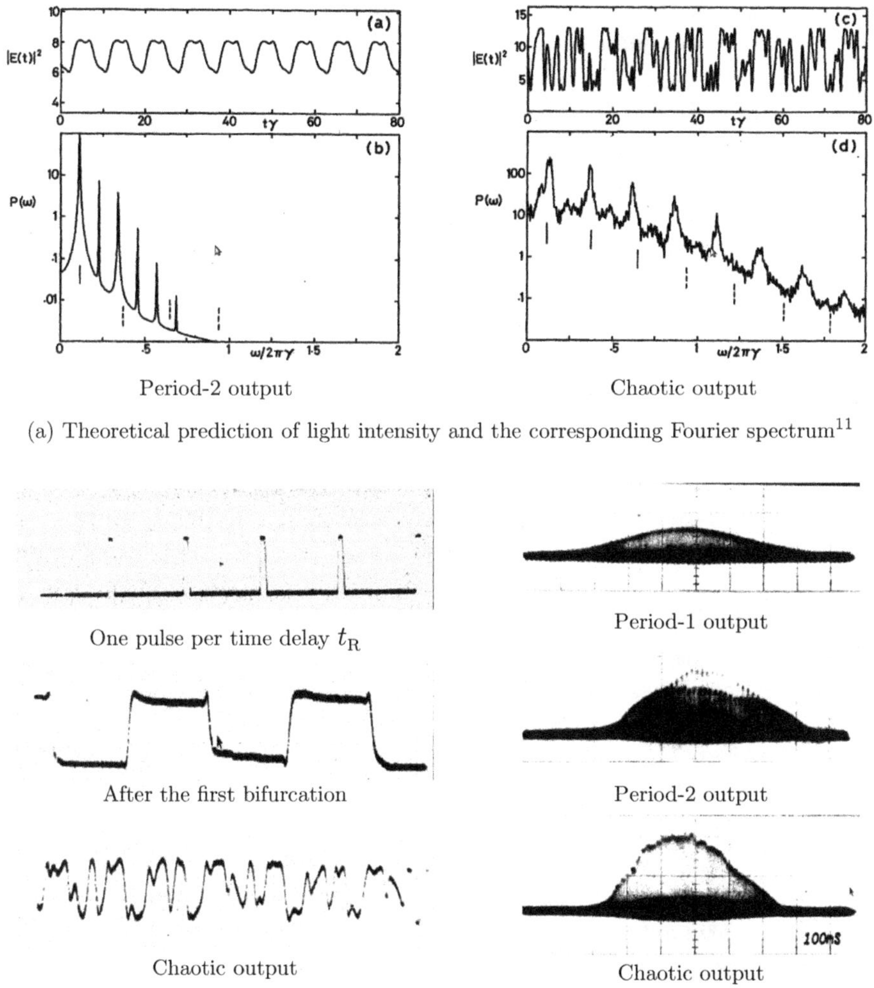

Period-2 output

Chaotic output

(a) Theoretical prediction of light intensity and the corresponding Fourier spectrum[11]

One pulse per time delay t_R

After the first bifurcation

Chaotic output

(b) Arizona group's experiments[27]

Period-1 output

Period-2 output

Chaotic output

(c) Kyoto group's experiments[28]

Fig. 3. Theoretical prediction (a) and experimental intensity [(b) and (c)] of optical chaos in optically bistable system. (a) Time series of the light intensity and the corresponding Fourier spectrum from the delay differential equation (1).[11] (b) Experimental results from the Arizona group. (c) Experimental results from the Kyoto group using all-optical fiber resonator. [(b) and (c) adapted with permission from Refs. 27, 28, respectively. Copyrighted by the American Physical Society]. A long laser light pulse is the input.

with "chaos" in the title were held at international conferences such as the *International Quantum Electronics Conference* (IQEC), *Conference on Lasers and Electro-Optics* (CLEO), and *Rochester Conference on Coherence and Quantum Optics*. From the late 1970s to the 1990s, the physics world was abuzz with the impact of rising chaos research. In Japan, an innovative conference named *Chaos and Related Topics* was held annually from 1984 to 1990 at the Yukawa Institute of Theoretical Physics (YITP) from the Kyoto University. The conference was held over 4 days at the end of each year, and researchers from various different fields all over Japan gave presentations in sessions on bifurcation phenomena, dissipative systems, conservative systems, many dimensional systems, biological systems, chemical reactions, quantum chaos, and so on.[a]

5. Chaos in Laser Oscillations

Chaos had a large impact on nonlinear optics. In the 1960s, during the development of the laser, it was quite usual for laser oscillation to be unstable. Incoherent oscillation far from single-mode oscillation, such as irregular pulsation and multi-mode oscillation, was common. However, at that time, instability was a troublesome phenomenon, and achieving steady oscillation was the most important task. Ironically, the interest in troublesome instability phenomena only became widely popular in the chaos boom after stable operation of lasers had already been achieved. However, while nonlinear dynamics and chaos were booming in the nonlinear optics world, there was one researcher, Lee W. Casperson, pursuing a study of laser instability that seemed to be less related to the chaos boom and more deeply motivated by his earlier original experimental experiences. He developed a brilliant theoretical consideration to explain complex periodic pulsation experimentally observed for the He-Xe gas laser.[17]

The frequency of the gas laser medium is widely broadened by the Doppler effect (Doppler broadening). The atoms with frequencies on both sides of the Doppler broadened frequency band couple resonantly with the oscillating laser light whose frequency is shifted by the anomalous dispersion of the laser medium, giving rise to a limit cycle oscillation at

[a]YITP is a center jointly used by theoretical physicists in Japan. In its early years, all plans related to research projects and personnel affairs were decided by a joint-use management committee with delegates elected from theoretical physicists throughout Japan. It was then a symbolic example of a democratically managed institute established in post World War II.

the difference frequency. This is a brilliant example of a complex instability phenomenon whose experimental results are naturally explained by the theory. His followers found that the instabilities that he discovered (often called Casperson's instability) lead to chaos.

Lorenz instability and chaos proposed for the single-mode laser model by Haken cannot be observed in ordinary visible-light lasers. But it was observed for infrared lasers which oscillates with the rotational levels of CO_2 and NH_3. As it was stated, Lorenz instability is the self oscillation of the Rabi-precession which oscillates at the frequency f_E in the small limit of γ. For usual lasers, γ is so large that the polarization ρ follows the light field E as set $\dot{\rho} = 0$ in the second equation of Eq. (4), leading to

$$\rho(t) = \frac{\mu}{\gamma} E(t) w(t) \tag{5}$$

Under this condition, the laser equations reduce to a system with two degrees of freedom, E and w, for which chaotic instability does not occur. This set of equations is called the laser rate equation. However, if a subsystem controlling laser action, such as a saturable absorber, is introduced, the additional degree of freedom often leads to instabilities and chaos.

The simplest description of a semiconductor laser is basically a single-mode model, or more simply the laser rate equation for E and w, which is obtained by adiabatically eliminating ρ with Eq. (5), and so the operation is quite stable if it is isolated. But a practical problem is that the laser may be destabilized by light reflected by an object, such as a laser disk, at a distance away from the laser. The re-injected light is delayed along the return path, and the laser rate equation contains a delayed-feedback field $E(t - t_R)$ as

$$\begin{cases} \dot{E} = -\kappa E + \eta\mu \dfrac{wE}{\gamma} + \epsilon E(t - t_{\mathrm{R}}) \\ \dot{w} = -\gamma_{\|}(w - w_{\infty}) - \mu^2 \dfrac{wE^2}{\gamma}. \end{cases} \tag{6}$$

It contains the delay time t_{R} like Eq. (3), and its solution may be chaotic. This type of delayed-feedback equation was proposed by Roy Lang and Kohroh Kobayashi in 1980.[31] Strictly speaking, in the case of semiconductor laser, Eq. (5) for ρ should be slightly changed from to include the saturation factor.[31,32] Semiconductor lasers with feedback-induced chaos have been studied extensively for various applications. By this way, research of chaos in optical systems, which was initiated by the instabilities of passive resonators

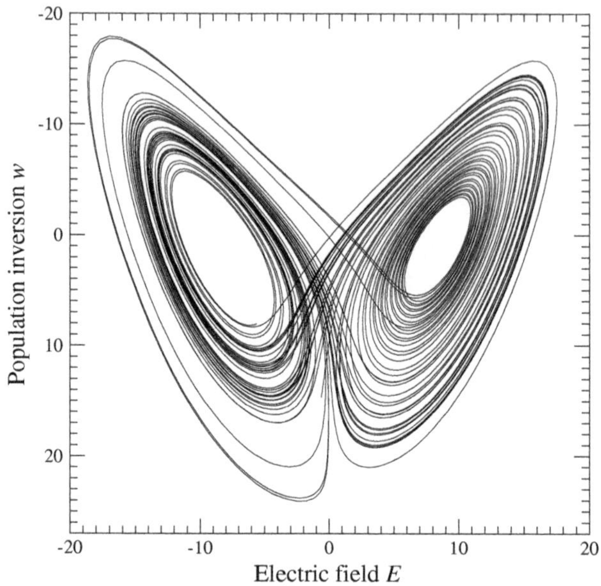

Fig. 4. Chaotic trajectory of Lorenz chaos (a) in the (E, w) plane produced by the Haken equations (4), and corresponding to Lorenz-chaos-like outputs observed in a NH_3 ring laser.[30]

irradiated by lasers, spread to the study of instabilities of lasers themselves. Modeling nonlinear optical oscillators with delayed feedback have been extensively studied in many forms and contexts. For example, models with coupled networks of such oscillators have aimed to simulate some aspects of brain dynamics.

6. Optical Chaos as High-Dimensional Chaos

Equation (3) is a relatively simple class of nonlinear delay differential equation. However, this seemingly simple system is a system with an infinite number of degrees of freedom. Indeed, the return time t_R represents the fundamental period to make a round trip in the resonator, and so the system is effectively composed of an infinite number of modes with the frequency of $2\frac{\pi}{t_R}\times$ [integer] in resonance with the resonator. Therefore, the dimension of the attractor in the infinite-dimensional phase space can in principle be arbitrarily large and a very complicated high-dimensional chaos that may rival fluid turbulence may be generated on the attractor. To emphasize this, I used the term "optical turbulence" in our paper of 1980.[16]

Such is the case also for laser dynamics: there are many light modes resonating with the resonator, and once an instability occurs, complex turbulent output that can not be described by the single mode model (4) is generated. If many modes, which may result in turbulent motion, are modulated by a periodic forcing with the period t_R, they can be phase-locked to form an intense sharp pulse train of period t_R with the width inversely proportional to the number of modes in the gain band. However, if the locking fails, the regular waveform turns into a very complicated high-dimensional chaotic pattern.

I was thinking that a significant aspect of studying optical chaos is to investigate chaotic motion associated with a high-dimensional attractor. When multiple modes taht are resonant with the resonator become unstable, it often occurs that multiple limit cycle attractors coexist. Indeed, Eq. (3) has such multi-stable limit cycle solutions and so does the multi-mode laser equation, which is the many-modes version of Eq. (4), if it exhibits the resonant Rabi instability.[33] In the case of the multi-mode laser equation, each of the limit cycle oscillators is contributed to by the resonant modes in the laser gain band. In many cases, these limit cycles are unified to form a single chaotic attractor, and under some appropriate conditions, the unified attractor enables a chaotic motion that appears to be an itinerant journey among the "ruins" of the limit-cycle attractor which are now destabilized. I noticed the existence of such chaotic motions in numerical simulations of multi-mode laser equation.[33,b]

At the same time, Ichiro Tsuda also noticed a quite similar phenomenon in his network model of brain recognition.[35] Kunihiko Kaneko also noticed it in simulations of coupled map lattices,[36] Both agreed to call this type of behavior "chaotic itinerancy." Experimental observations of chaotic itinerancy have also been reported for several laser systems. We present here an interesting example of chaotic itinerancy inherent to semiconductor laser with feedback by external reflection, which is modelled by the Lang–Kobayashi equation described above. Another example of chaotic itinerancy will be presented later in connection with spatial pattern formation. The external feedback described by the delayed feedback like Eq. (6) is due to the reflection of an external object, which plays the role of a mirror to form

[b]I would like to comment that Eq. (3) also has many higher-harmonic solutions that coexistent with the fundamental one shown in Figs. 3(a) right and 3(b) bottom, which are rich enough to allow binary coding. Their destabilization results in the merging of their basins and chaotic itinerancy is observed at the onset of this merging.[34]

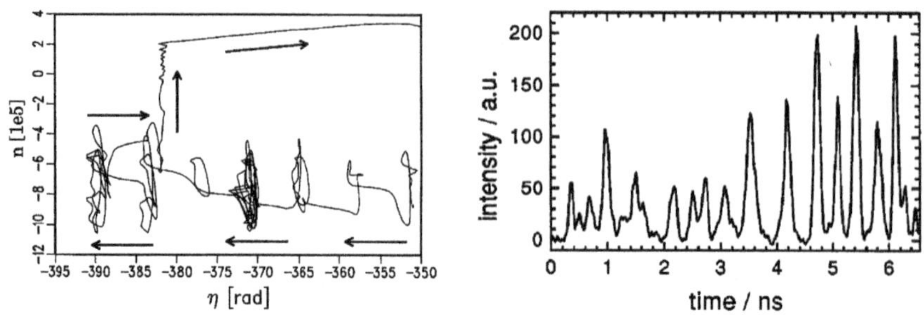

(a) Trajectory in the w-η plane
(b) Experimental output intensity

Fig. 5. (a) Trajectory in the w-η plane exhibiting Chaotic Itinerance among mode oscillations characterized phase lag $\eta_n = 2\pi n$ for n around 60. Numerical result for the delayed feedback model of semiconductor laser with external feedback. (b) Experimental result of output intensity vs time.

Source: With permission from Ref. 32. Copyrighted by the American Physical Society.

an external resonator for the semiconductor laser, and so the whole system is composed of many resonant modes of the external resonator. Each of the modes oscillate with the phase lag $\eta := \psi(t + t_{\mathrm{R}}) - \psi(t)$ being kept at a constant value $\eta_n = 2\pi n$ $(n \in \mathbb{N})$ to form a limit cycle attractor, but the oscillations become unstable under certain conditions and chaotic itinerancy itinerance among the destabilized attractors characterized by η_n with different n is induced as is shown in Fig. 5(a). The trajectory stays for a while on a ruin exhibiting oscillatory chaotic motion at some n, but eventually makes a transition to another ruin with different value n'. At the transition, the laser oscillation stops and so the output signal is fragmented to form randomized pulse series as shown in Fig. 5(b). It is thought that such chaotic itinerancy is the origin of the low frequency noise often observed in semiconductor lasers with external feedback.

The instability properties of lasers and optical instabilities related to the multi-degree-of-freedom property of the laser system began to be discussed in the mid-1980s. Many experimental and theoretical works treating instabilities and chaos related to not only longitudinal resonator modes but also transversal resonator modes have been presented. Figures 6 and 7 show such an example. Arecchi and his coworkers observed a 2-dimensional pattern formation in the 2D-direction transversal to the propagating direction of the emitted laser beam as shown in Fig. 6.[37] Moreover the emitted beam pattern exhibits itenerance among these 2D mode patterns as shown in Fig. 7.

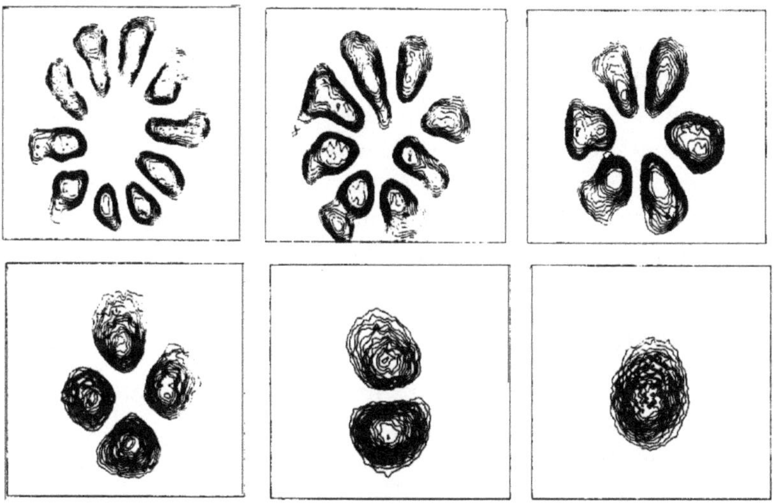

Fig. 6. Typical transverse mode patterns observed for a photorefractive active resonator when the azimuthal quantum number changes from $q = 10$ to around 0.[37]

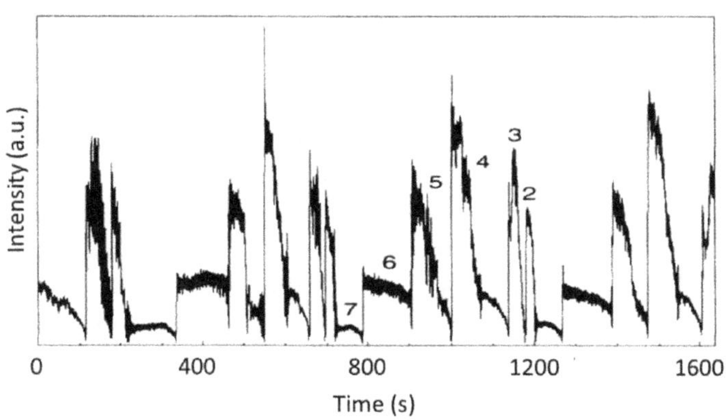

Fig. 7. Output intensity during itinerancy among various transverse modes tagged by angular frequency of transverse mode pattern angular quantum number.[37]

I had hoped that research in this direction would lead to more basic investigations of multi-degree-of-freedom chaos, which may be somewhat similar but less complex than fluid turbulence. But after about 1990, the interest in chaos rapidly faded away, except for its technological applications such as chaotic communication. My interest in optical chaos was still in high-dimensional chaotic motion which may be described with relatively simple equations, such as Eq. (3). I published some papers trying to develop an information theoretical method to elucidate a part of the high-dimensional phase space structure in which the chaotic variety is generated in turbulent condition.[38,39] However, in my own opinion, those papers were not successful. While I kept my interest in the problem of high-dimensional chaos and turbulence, I shifted my attention to more fundamental problems of chaos, and I left the research area of nonlinear optics.

7. Returning to the Beginning

Around 1990, I began to view chaos from a viewpoint that went back to my original interest in physics. As stated before, my interest was in the question of how the irreversible transportation and dissipation of energy is generated in reversible quantum dynamics. The main reason why I was attracted by chaos is that the loss of memory necessary for irreversibility may be self-generated in a system with a small numer of degrees of freedom, without coupling with an infinite number of degrees of freedom.

Let us consider a system governed by a Hamiltonian. A (completely) integrable system in which all degrees of freedom perform periodic motion is the most regular system in Hamiltonian mechanics. The fate of periodic systems is that resonances always occur between degrees of freedom, and when a perturbation is introduced into a completely integrable system, localized chaos always occurs near the resonances. No matter how localized and weakly unstable the chaos may be, when the chaotic system comes into contact with other degrees of freedom, no matter how simple those degrees of freedom may be, it can potentially transform their motion in a complex way. Since chaos does not have a specific frequency, the degrees of freedom that come into contact with chaos is driven by the mixed chaotic force that loses the time correlation and can diffuse globally. In fact, as Arnold showed, even very weak chaos can induce very slow but global diffusion. This is called "Arnold diffusion."[40,41]

Following convention, we will refer to a quantum system whose corresponding classical system exhibits chaos as a quantum chaos system. Furthermore, a quantum chaos system that is classically confined to a narrow phase space, which I call "small quantum chaos," is a collection of a finite number of linear oscillations and cannot be the essential cause of mixing and loss of time correlation. Nevertheless, could large-scale transport occur when small quantum chaos comes into contact with other degrees of freedom, as occurs in corresponding classical systems? The reason I am interested in this problem is that, as I will explain later, it is closely related to the question of whether small quantum chaos systems can have irreversible dissipative properties.

Recently, lead by the "eigenstate thermalization hypothesis" of Michael Berry, the irreversible time evolution processes in isolated quantum systems composed of a large number of degrees of freedom have been very actively studied.[42] There it is supposed that it does not matter whether or not the system has a classical limit.

In contrast to this approach, I have been most interested in whether small isolated quantum systems, such as isolated molecules and small particles, could exhibit irreversible energy transport, in particular, small quantum systems that have chaos in the classical limit. On the other hand, triggered by the transport problem of quantum chaos systems, I also become very interested in how the tunneling processes, which are unique quantum transport (transmission) processes not seen in classical systems, is influenced if the classical limit of the system has chaos.

In the following, I will discuss the above two topics that have been my main interest in the past 20 years; small-quantum chaos and chaos tunneling.

7.1. *Small quantum chaos induces diffusion and dissipation*

As the perturbation strength becomes sufficiently large, small chaos which was localized on a resonant surface that traverses the entire phase space begins to expand and merge, transforming into globalized chaos and spreading throughout the entire phase space. Globalized chaos would realize the ergodic world envisioned by Boltzmann. However, in quantum mechanics, such globalization encounters unexpected obstacles.

In classical systems, the intertwining of chaotic orbits, which leads to the merging of localized chaos, occurs quite easily. However, the corresponding intertwining of quantum wave functions is extremely unlikely. Boris Chirikov and Giulio Casati were the first to recognize this fact in a typical situation of a quantum map system (time-discrete system).[43] On the other hand, it has been discovered that delocalization of chaos resulting into the formation of globalized chaos can also be restored in quantum systems when the chaotic degrees of freedom are coupled to other degrees of freedom.[44] This has been understood by analogy with the Anderson transition[45] and experimentally observed as a phenomenon of global quantum diffusion in optical lattices.[46]

However, if chaos is classically small in the sense that it is confined to a finite region of phase space, its quantum-classical correspondence is still unknown. In particular, only negative results are known for the existence of the quantum counterpart of the very slow Arnold diffusion.

Around 1990, I was immersed in the question of whether irreversible processes like energy relaxation could be observed in small quantum chaos systems. My question is wether even small quantum chaos systems could lead to irreversible energy transfer. As mentioned above, since there are only a finite number (say, N) of quantum levels in a small quantum chaos system, the motion is nothing more than a superposition of linear periodic motions. However, what happens if other degrees of freedom which are dynamically very simple are very weakly coupled additively to the main system as illustrated in Fig. 8(a)? To make the situation more explicit, we suppose additive degrees of freedom are completely integrable if isolated, as was assumed in Arnold's example.[40]

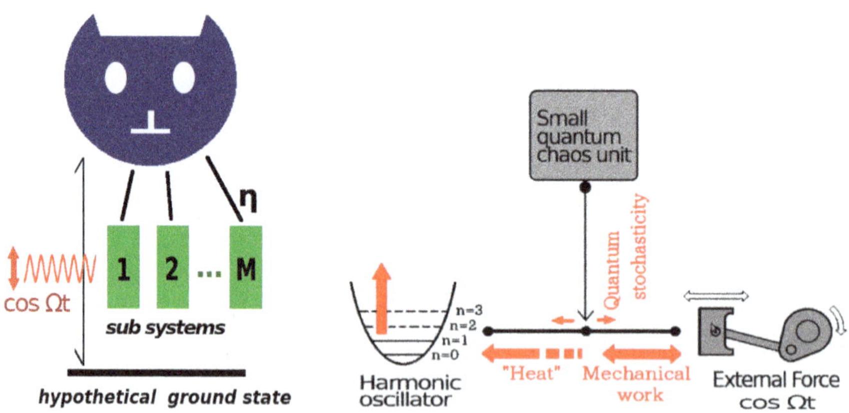

(a) Small quantum chaos (b) A quantum dumper

Fig. 8. (a) A schematic illustration of the system we discuss here. A small quantum chaos system (whose classical version is typically the hyperbolic Arnold cat) is interacting with M sub-systems which is integrable if isolated. A hypothetical ground state is very weakly coupled with the chaos system via an oscillating force with frequency Ω to examine the irreversible characteristics of the system. (b) A possible situation in which a small quatum chaos works as a "quantum dumper". It converts irreversibly the mechanical work into the "heat energy" stored in a harmonic oscillator reservoir.[47]

To simplify the situation further, let's approximate the additional integrable systems by linear oscillators by applying a local linear approximation around the initial states to the action-variable representation. If the number of additional degrees of freedom (say, M) becomes infinite, the additional system becomes the traditional model of a heat reservoir. On the other hand, the effect of chaos on the additional degrees of freedom is related to the "transport by small chaos" problem mentioned above. In classical mechanics, weak coupling with additional degrees of freedom does not qualitatively change the chaotic behavior of the main system. However, in small quantum chaos systems, additional degrees of freedom have a tremendous impact.

Let's attach a hypothetical ground state to the main system that has much lower energy than the main system. Suppose the system is in the ground state. Let's consider a virtual experiment in which light with a frequency Ω approximately equal to the energy difference between the ground state and the main system is irradiated.[48] As shown in Fig. 9(a) if the main system-additional system coupling constant η is small, the energy $I(\Omega, t)$ absorbed by the system just oscillates, which is the normal behavior for small quantum systems. However, if $M \geqslant 3$ and η becomes larger than a certain small threshold value determined by Planck constant, irreversible

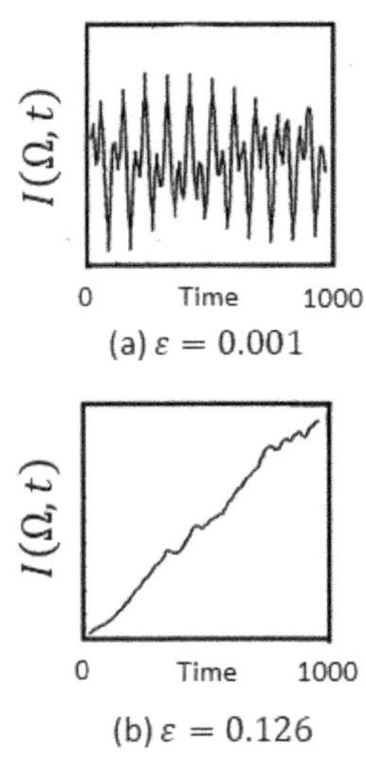

Fig. 9. Time series of the absorbed energy $I(\Omega, t)$ at a fixed Ω and for two different values of the coupling ϵ.

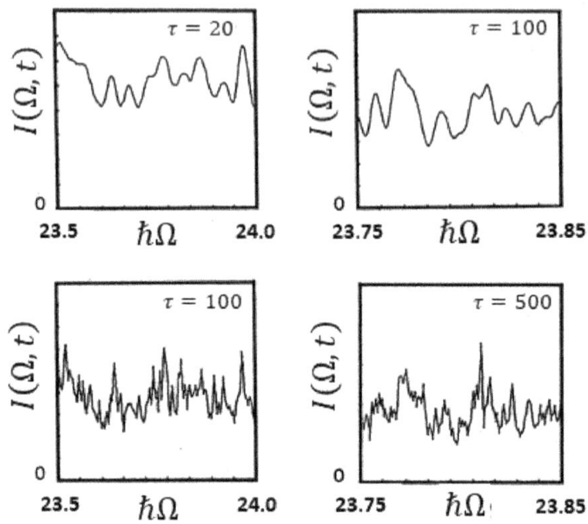

Fig. 10. The absorption spectrum for different durations τ and for different ranges in $\hbar\Omega$ exhibiting the self-similarity.[48] See the main text.

steady-state energy absorption occurs, as in Fig. 9(b). At that time, the absorption spectrum $I(\Omega, t)$ becomes a peculiar continuous spectrum, and as time progresses, peaks within peaks are formed successively, as in Fig. 10 and the structure becomes complex and fractal-like.

Next we discuss what happens to the additional degrees of freedom when such a transition occurs. It requires a large computational cost to see the behavior of the entire additional degrees of freedom, but recently we have developed an efficient method to calculate this in collaboration with Hiroaki Yamada.[49]

The results are demonstrated in Fig. 11. We show the time evolution of the mean-square-displacement, which represents the spread of the wave function in the action space spanned by the additional degrees of freedom. As η increases and the irreversible energy absorption described above occurs, global diffusion occurs in the additional system. The details of the transition from non-diffusion to diffusion, as described in the figure caption, can be seen from the change in the characteristic exponent $\alpha(t) := \frac{\mathrm{d}}{\mathrm{d}t} \log \mathrm{MSD}(t)$ curve accompanying the mean-square-displacement curve shown in the inset of Fig. 11. This is a quantum transition phenomenon that passes through a critical state that exhibits anomalous diffusion.

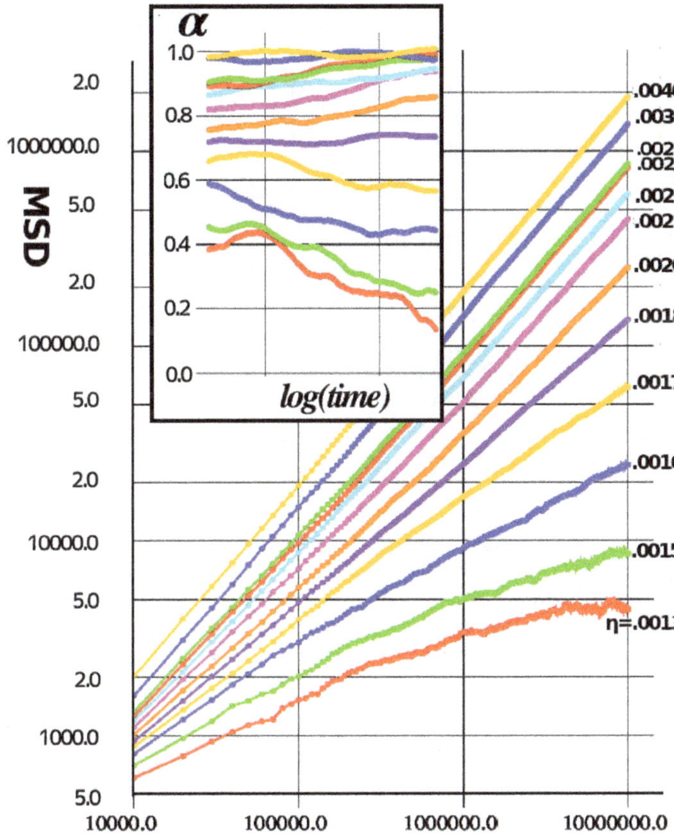

Fig. 11. Diffusion transition that occurs in the additive modes. The mean-square-displacement in the action space of the additional modes is shown as a function of time. The time series of the power exponent $\alpha(t)$ (see the main text) of all the mean-square-displacement curves are shown in the inset. As t increases, $\alpha(t) \to 0$ for small η, but $\alpha(t) \to 1$ for large enough η, which implies the normal diffusion mean-square-displacement $\sim t$. At an intermediate $\eta = \eta_c$, $\alpha(t)$ keeps a constant value $\alpha_c \sim 0.7$ implying an anomalous diffusion MSD$\sim t^{\alpha_c}$, which is just the critical state of the diffusion transition.

Occurence of diffusion in the additional modes followed by the irreversible energy absorption of the main quantum chaos system is observed even when the main system is spanned by only four quantum states. So far, we have considered the case where the classical version of the main system exhibits sufficiently unstable chaos. At the opposite extreme, we can consider the case that the main system is completely integrable.

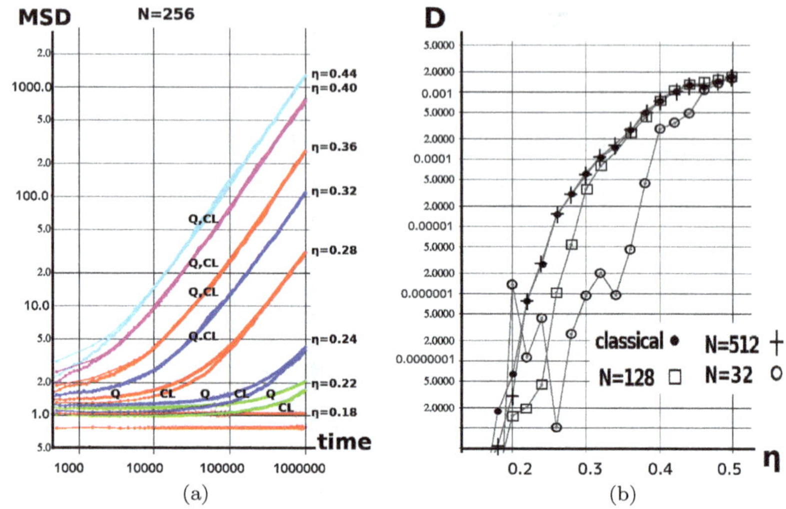

Fig. 12. (a) Very slow evolution of MSD in the additional modes when the main system is completely integrable, and with $M = 3$. The quantal evolution indicated by "Q" follows the extremelly slow classical diffusion whose diffusion rate increases slowly with η. (b) Comparison of the "exponentially small" diffusion constants D of the classical system and its quantum counterpart. N is the number of quantum states spanning the main system, which is related to the Planck constant by $\hbar = \frac{2\pi}{N}$.

Even in such a case, a very weak chaos begins to grow in the classical counterpart, as the coupling constant η with the additional degrees of freedom exceeds a threshold value η_0. In this case, as shown in Fig. 12(a), an exponentially slow but global diffusion with the diffusion constant $D \propto e^{-\mathrm{const.}/(\mu-\mu_0)}$ is induced in the additional system.

Even in the case that the main system is fully chaotic and the induced diffusion in the additional system has a large enough diffusion constant of O(1), its quantum version can not mimic classical diffusion. But Fig. 12(a) manifests that quantum diffusion that roughly follows such an anomalously slow classical diffusion appears as η increases beyond the classical threshold η_0. Figure 12(b) shows that even if Planck constant is not very small, a very slow classical diffusion coefficient is reproduced by the quantum system.[49]

The classical Arnold diffusion also exibits similar exponentially small diffusion rate when the main system is weakly chaotic.[41] The relationship between our observations and the quantum Arnol'd diffusion, whose existence has not been confirmed, seems to be very interesting.

7.2. *Chaos draws imaginary tunneling trajectories toward the real world*

In classical dynamics, chaos causes perforation of the robust action space of fully integrable degrees of freedom and allows transport by diffusion, as exemplified by Arnol'd diffusion. However, quantum effects inhibit such a transport process. On the other hand, quantum dynamics allows for transmission between areas of phase space that are inaccessible in classical dynamics. This is the tunneling process. So how does chaos affect tunneling processes?

In classical dynamics the dynamical variables are restricted to real values. However, what happens, if we discard this restriction and release the classical dynamics into the world of complex numbers by continuing to complex time? Consider two sets of points in phase space, say, $\mathcal{A}_\alpha$ and $\mathcal{B}_\beta$, which are not connected in real number classical dynamics. Here α and β are dynamical variables such as position, monentum, action and angle and so on, characterizing the sets. As the simplest example, let us consider a one-dimensional hamiltonian system If there is a potential barrier, we can take $\mathcal{A}_\alpha$ and $\mathcal{B}_\beta$ as disjoint classical orbits separated by it. However there always exists classical complex orbits connecting $\mathcal{A}_\alpha$ and $\mathcal{B}_\beta$, going over the potential barrier, if their energies are equal. Such a complexified classical orbit is called an *instanton* and is known to describe the quantum tunneling between $\mathcal{A}_\alpha$ and $\mathcal{B}_\beta$.

In nonintegrable systems, the phase space is a mixture of chaos and torus. Here, we consider a typical case that $\mathcal{A}_\alpha$ is in a torus region and $\mathcal{B}_\beta$ is located in a chaotic region. Thus, they are not connected in the real phase space. Even in such a case, $\mathcal{A}_\alpha$ and $\mathcal{B}_\beta$ may be connected by complexified classical orbits. Can complex orbits play a role for tunneling as in 1D system?

The semi-classical dynamics is the only theory that can describe quantum dynamics in terms of classical dynamical concepts. Suppose the wavefunction starts initially from $\mathcal{A}_\alpha$, then its component at the arrival set $\mathcal{B}_\beta$ and at the time t is expressed by the *path-integral*. Among all the possible paths connecting $\mathcal{A}_\alpha$ and $\mathcal{B}_\beta$, the principle paths significantly contributing to the integral is the paths making the action functional connecting $\mathcal{A}_\alpha$ and $\mathcal{B}_\beta$ for time t extreme. It is nothing more than the classical orbit, which we denote by γ, connecting $\mathcal{A}_\alpha$ and $\mathcal{B}_\beta$ for t according to Lagrangian variational principle. Let the classical orbit be γ and let $S_\gamma(\alpha,\beta)$ be the extremum value of the action functional. Then, the wavefunction at $\mathcal{B}_\beta$ and at the time t is approximated by

$$\Psi_t(\alpha,\beta) \propto \sum_{\gamma} \sqrt{\frac{\partial^2 S_\gamma}{\partial\beta\partial\alpha}} \exp\left(i\frac{S_\gamma(\alpha,\beta)}{\hbar}\right) \tag{7}$$

This is a time-domain semiclassical wavefunction. Even though $\mathcal{A}_\alpha$ and $\mathcal{B}_\beta$ are not connected by a real classical orbit, they are connected by a complexifed orbit γ which contributes the weight $\propto \exp\{-\mathrm{Im}S_\gamma(\alpha,\beta)/\hbar\}$ to Eq. (7), which is just the tunneling effect. Satoshi Adachi was the first to apply the time-domain semiclassical wavefunction to the numerical evaluation of a coherent-state representation of a wavefunction. Stimulated by his work, Akira Shudo and I applied the time-domain semiclassical wavefunction to investigate the effects of tunneling in the presence of chaos by using descrete-time quantum map model.[50,51] For continuous-time nonintegrable models, Kin'ya Takahashi and I have developed complex semiclassical theory of tunneling.[52,53] Some results obtained by using quantum maps are shown below.

Figure 13 shows an example that demonstrates how complex orbits in the presence of chaos have remarkable features. In the first example, the starting set $\mathcal{A}_\alpha$ is a torus in a torus region surrounded by chaos. Figure 13(a) shows the initial torus state $\mathcal{A}_\alpha$, marked as a blue circle. The set of complexified initial values of the orbit γ on the torus $\mathcal{A}_\alpha$, which we call the $\mathcal{M}_t$-set, is computed by scanning the chaotic region for the arrival set $\mathcal{B}_\beta$ with given position value $q = \beta$. Since the initial torus $\mathcal{A}_\alpha$ is parameterized by the angle variable θ, the $\mathcal{M}_t$-set is in the complexified angle space.

The sets of points in Fig. 14(a) exhibit an extremely complicated structure. The enlargement Fig. 14(b) shows that the structure consists of collection of separate curved strings arranged in a chain. We called such a chain a Laputa chain. As β is fixed, a single initial θ from which a trajectory reaches β at a fixed t is decided on each string. The number of trajectories reaching $\mathcal{B}_\beta$ is equal to the number of strings. The strings form a hierarchical structure, and the number of levels in the hierarchy increases with t, which is a reflection of the positive topological entropy of chaos. Each orbits has non-null imaginary $S_\gamma(\alpha,\beta)$, which contributes to the tunneling probability with the weight $\exp\left(-\mathrm{Im}\frac{S_\gamma(\alpha,\beta)}{\hbar}\right)$. A troublesome fact is that almost half of the string has negative $\mathrm{Im}S_\gamma(\alpha,\beta)$, which makes the weight diverge in the semiclassical limit and is thus physically meaningless. However, such a meaningless part can be removed according to a certain rule. Summing up all the meaningful contributions from

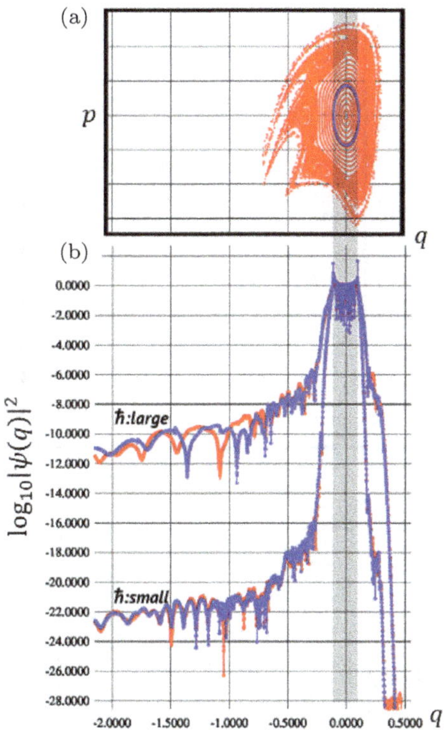

Fig. 13. (a1) The Poincaré section of the phase space (q, p) showing the presence of a torus region (covered with ellipses) and a chaos region. The blue circle is the initial torus state $\mathcal{A}_\alpha$. (a2) The wavefunction started from $\mathcal{A}_\alpha$ extends over the chaotic region ($q \leq -1$) with complicated tunneling tails. The blue curves are the semiclassical wavefunction constructed by using the classical orbits started from the Laputa-chains in (b2). The red curve is the fully quantum results. Two cases of different values of $\hbar$ are shown. The thin shaded region indicates the range of q of $\mathcal{A}_\alpha$.

complex classical orbits, the obtained semiclassical tunneling wavefunction reproduces a very complicated tunneling tail obtained by a fully quantum mechanical treatment, as is shown in Fig. 13(b).

When the chaotic set is on the real plane, the complexified stable manifold W_s of the unstable periodic orbits forming the skeleton of the chaotic sets plays a crucial role: the points of a Laputa chain are close to $W_s \cap \mathcal{A}_\alpha$ and so the tunneling orbit starting trom a Laputa chain follows W_s and approaches exponentially toward the periodic orbit on the real plane, before being scattered far away along the real unstable manifold W_u. Figure 15 illustrates this scene. Such a motion makes the imaginary action

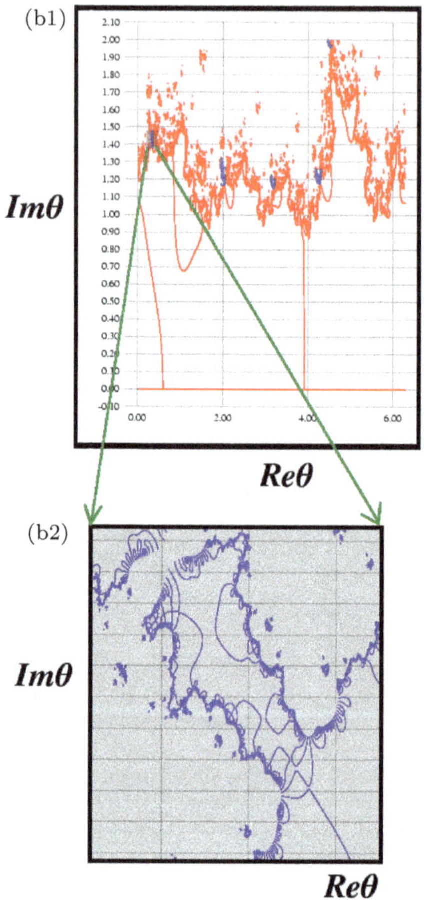

Fig. 14. (b1) The initial set $\mathcal{M}_t$ shown in the complexified angle variable plane. All the complexified classical orbits starting at $\mathcal{M}_t$ arrive at $\mathcal{B}_\beta$ at t. (b2) The enlargement of a tiny region in (a1) indicates that the set consists of chains of curved strings, which are called Laputa chains.

$\mathrm{Im}S_\gamma(\alpha,\beta)$ extremelly small, and the contributed tunneling probability $\exp\left(-\mathrm{Im}\frac{S_\gamma(\alpha,\beta)}{\hbar}\right)$ is exhanced remarkably. It has been said that chaos "assists" tunneling.[54] The facts presented above show clearly that tunneling closely mimics chaos: led by W_s, the complex tunneling trajectory quickly traces close to a real-space chaotic orbit, which reduces the imaginary component and enhances the tunneling probability remarkably. We call such a phenomenon "chaotic tunneling".

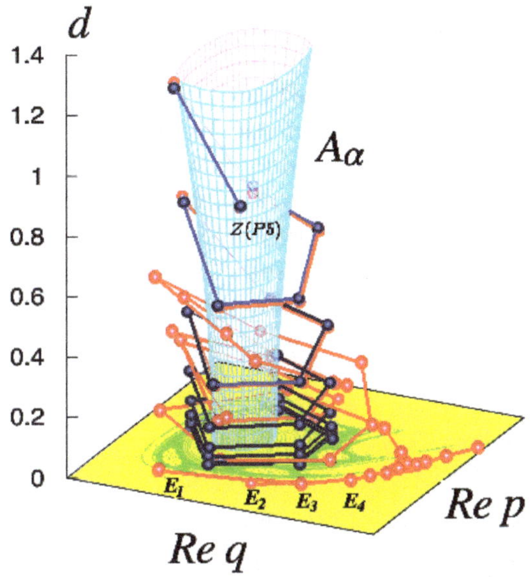

Fig. 15. Typical behavior of a complex classical trajectory starting from a Laputa chain (red line), where d is the distance from the real (q, p) plane. The trajectory approaches exponentially close to the real phase space following the stable manifold (blue line). Such orbits significantly contribute to Eq. (7).

The above complexified semiclassical analysis requires handcraft-like skill and labor, together with considerable computational cost. However, it reveals clearly the role of chaos in tunneling phenomena which can never be seen by a purely quantum approach. With the cooperation of mathematician Yutaka Ishii, it has become clear that the characteristic orbits originating from Laputa have a close relationship with the *Julia set*, which is well-known as a fractal set in the field of complex dynamical systems theory.[55] Roughly speaking, the result is that the Laputa set is included in the Julia set.[c] The dendritic and complex structure of the Laputa chain is a reflection of the fractal structure and chaotic nature of the Julia set. I still do not know how to understand and face the fact that the Julia set extracted out of purely mathematical interests is closely related to realistic dynamics of quantum theory.

[c]The set of points which do not go to infinity for the forward time evolution is called the filled-in forward Julia set K^+. The Laputa set is included in K^+.

8. Finally

My interest in quantum dynamical processes led me to step into the world of nonlinear optics, where I encountered chaos. With chaos, I returned again to my original interests of quantum dynamics including quantum irreversibility and further quantum tunneling. I have seen remarkable effects, in particular those described above, that are manifest in quantum systems that have chaos in classical limit. I don't know where this will take me in the rest of my life, but there is no other way for me to go.

Acknowledgement

First of all, I would like to express my deepest gratitude to all of my co-researchers. My works were supported by Japanese people's tax via MEXT/JSPS KAKENHI Grants, and I would like to acknowledge them. I am also grateful to Doctors Shoji Tsuji and Takuji Tsuji for allowing me to use the facility, "Konton-tei," during my research. Finally, I would like to thank Miwako Ikeda, my okan, who has supported me spiritually for more than 30 years.

References

1. H. Yukawa, On the interaction of elementary particles. I, *Proceedings of the Physico-Mathematical Society of Japan.* **17**, 48–57 (1935).
2. H. Yukawa, On the interaction of elementary particles. I, *Progress of Theoretical Physics Supplement.* **1**, 1–10 (1955). doi: 10.1143/PTPS.1.1.
3. C. Hayashi, Proton-neutron concentration ratio in the expanding universe at the stages preceding the formation of the elements, *Progress of Theoretical Physics.* **5**(2), 224–235 (1950). doi: 10.1143/ptp/5.2.224.
4. K. Tomita and H. Tomita, Irreversible circulation of fluctuation, *Progress of Theoretical Physics.* **51**(6), 1731–1749 (1974). doi: 10.1143/PTP.51.1731.
5. I. Prigogine, *Introduction to Thermodynamics of Irreversible Processes.* Charles C. Thomas Publisher, Springfield, Illinois (1955).
6. G. Nicolis and I. Prigogine, *Self-Organization in Nonequilibrium Systems — From Dissipative Structures to Order through Fluctuations.* John Wiley & Sons, New York (1977).
7. H. Haken, *Synergetics: An Introduction Non-equilibrium Phase Transitions and Self-Organization in Physics, Chemistry and Biology.* Springer-Verlag, Berlin (1977).
8. R. Kubo and K. Tomita, A general theory of magnetic resonance absorption, *Journal of the Physical Society of Japan.* **9**, 888 (1954).
9. M. Gross and S. Haroche, Superradiance: An essay on the theory of collective spontaneous emission, *Physics Reports.* **93**, 301–396 (1982).

10. H. M. Gibbs, *Optical Bistability: Controlling Light with Light.* Elsevier (1985).
11. K. Ikeda, Multiple-valued stationary state and its instability of the transmitted light by a ring cavity system, *Optics Communications.* **30**(2), 257–261 (1979). doi: 10.1016/0030-4018(79)90090-7.
12. E. N. Lorenz, Deterministic nonperiodic flow, *Journal of the Atmospheric Sciences.* **20**, 130–141 (1963). doi: 10.1175/1520-0469(1963)020⟨0130:DNF⟩2.0.CO;2.
13. O. E. Rössler, An equation for continuous chaos, *Physics Letters A.* **57**(5), 397–398 (1976). doi: 10.1016/0375-9601(76)90101-8.
14. D. Ruelle and F. Takens, On the nature of turbulence, *Communications in Mathematical Physics.* **20**, 167–192 (1971).
15. R. M. May, Simple mathematical models with very complicated dynamics, *Nature.* **261**(5560), 459–467 (1976). doi: 10.1038/261459a0.
16. K. Ikeda, H. Daido, and O. Akimoto, Optical turbulence: Chaotic behavior of transmitted light from a ring cavity, *Physical Review Letters.* **45**(9), 709–712 (1980). doi: 10.1103/PhysRevLett.45.709.
17. L. Casperson, Spontaneous coherent pulsations in laser oscillators, *IEEE Journal of Quantum Electronics.* **14**, 756–761 (1978).
18. R. Bonifacio and L. A. Ligiato, Cooperative effects and bistability for resonance fluorescence, *Optics Communications.* **19**, 172–176 (1976).
19. R. Bonifacio, M. Gronchi, and L. A. Ligiato, Self-pulsing in bistable absorption, *Optics Communications.* **30**, 129–133 (1979).
20. S. L. McCall, Instability and regenerative pulsation phenomena in Fabry-Perot nonlinear optic media devices, *Applied Physics Letters.* **32**(5), 284–286 (1978). doi: 10.1063/1.90021.
21. H. Risken and K. Nummedal, Self-pulsing in lasers, *Journal of Applied Physics.* **10**, 4662–4672 (1968).
22. H. Haken, Analogy between higher instabilities in fluids and lasers, *Physics Letters A.* **53**(1), 77–78 (1975). doi: 10.1016/0375-9601(75)90353-9.
23. H. M. Gibbs, S. L. McCall, T. N. C. Venkatesan, A. C. Gossard, A. Passner, and W. Wiegmann, Optical bistability in semiconductors, *Applied Physics Letters.* **35**(6), 451–453 (1979). doi: 10.1063/1.91157.
24. M. W. Derstine, H. M. Gibbs, F. A. Hopf, and D. L. Kaplan, Bifurcation gap in a hybrid optically bistable system, *Physical Review A.* **26**, 3720–3722 (1982). doi: 10.1103/PhysRevA.26.3720.
25. M. W. Derstine, H. M. Gibbs, F. A. Hopf, and D. L. Kaplan, Alternate paths to chaos in optical bistability, *Physical Review A.* **27**, 3200–3208 (1983). doi: 10.1103/PhysRevA.27.3200.
26. F. A. Hopf, D. L. Kaplan, M. H. Rose, L. D. Sanders, and M. W. Derstine, Characterization of chaos in a hybrid optically bistable device, *Physical Review Letters.* **57**, 1394–1397 (1986). doi: 10.1103/PhysRevLett.57.1394.
27. H. M. Gibbs, F. A. Hopf, D. L. Kaplan, and R. L. Shoemaker, Observation of chaos in optical bistability, *Physical Review Letters.* **46**, 474–477 (1981). doi: 10.1103/PhysRevLett.46.474.

28. H. Nakatsuka, S. Asaka, H. Itoh, K. Ikeda, and M. Matsuoka, Observation of bifurcation to chaos in an all-optical bistable system, *Physical Review Letters.* **50**, 109–112 (1983). doi: 10.1103/PhysRevLett.50.109.
29. R. G. Harrison, I. A. Al-Saidi, and W. J. Firth, Observation of bifurcation to chaos in a passive all-optical Fabry-Perot resonator: A high-frequency optical modulator, *Philosophical Transactions of the Royal Society of London A.* **313**, 405–409 (1984).
30. C. O. Weiss, R. Vilaseca, N. B. Abraham, R. Corbalán, E. Roldán, G. J. de Valcárcel, J. Pujol, U. Hübner, and D. Y. Tang, Models, predictions, and experimental measurements of far-infrared NH3-laser dynamics and comparisons with the Lorenz-Haken model, *Applied Physics B.* **61**, 223–242 (1995). doi: 10.1007/BF01082041.
31. R. Lang and K. Kobayashi, External optical feedback effects on semiconductor injection laser properties, *IEEE Journal of Quantum Electronics.* **16**(3), 347–355 (1980). doi: 10.1109/JQE.1980.1070479.
32. I. Fischer, G. H. M. van Tartwijk, A. M. Levine, W. Elsässer, E. Göbel, and D. Lenstra, Fast pulsing and chaotic itinerancy with a drift in the coherence collapse of semiconductor lasers, *Physical Review Letters.* **76**, 220–223 (1996). doi: 10.1103/PhysRevLett.76.220.
33. K. Ikeda, K. Otsuka, and K. Matsumoto, Maxwell-Bloch turbulence, *Progress of Theoretical Physics Supplement.* **99**, 295–324 (1989). doi: 10.1143/PTPS.99.295.
34. T. Aida and P. Davis, Oscillation mode selection using bifurcation of chaotic mode transitions in a nonlinear ring resonator, *IEEE Journal of Quantum Electronics.* **30**(12), 2986–2997 (1994). doi: 10.1109/3.362706.
35. I. Tsuda, Chaotic itinerancy as a dynamical basis of hermeneutics in brain and mind, *World Futures.* **32**(2–3), 167–184 (1991). doi: 10.1080/02604027.1991.9972257.
36. K. Kaneko and I. Tsuda, Chaotic itinerancy, *Chaos.* **13** (3), 926–936 (2003). doi: 10.1063/1.1607783.
37. F. T. Arecchi, G. Giacomelli, P. L. Ramazza, and S. Residori, Experimental evidence of chaotic itinerancy and spatiotemporal chaos in optics, *Physical Review Letters.* **65**, 2531–2534 (1990). doi: 10.1103/PhysRevLett.65.2531.
38. K. Ikeda and K. Matsumoto, Study of a high-dimensional chaotic attractor, *Journal of Statistical Physics.* **44**, 955–983 (1986). doi: 10.1007/BF01011917.
39. K. Ikeda and K. Matsumoto, High-dimensional chaotic behavior in systems with time-delayed feedback, *Physica D.* **29**(1), 223–235 (1987). doi: 10.1016/0167-2789(87)90058-3.
40. V. I. Arnol'd, Instability of dynamical systems with many degrees of freedom, *Doklady Akademii Nauk SSSR.* **156**, 9–12 (1964).
41. B. V. Chirikov, A universal instability of many-dimensional oscillator systems, *Physics Reports.* **52**(5), 263–379 (1979). doi: 10.1016/0370-1573(79)90023-1.

42. A. P. Luca D'Alessio, Y. Kafri and M. Rigol, From quantum chaos and eigenstate thermalization to statistical mechanics and thermodynamics, *Advances in Physics.* **65**(3), 239–362 (2016). doi: 10.1080/00018732.2016.1198134.
43. G. Casati, B. V. Chirikov, F. M. Izrailev, and J. Ford, Stochastic behavior of a quantum pendulum under a periodic perturbation, *Lecture Notes in Physics.* **93**, 334–352 (1979). doi: 10.1007/BFb0021757.
44. M. S. Santhanam, S. Paul, and J. B. Kannan, Quantum kicked rotor and its variants: Chaos, localization and beyond, *Physics Reports.* **956**, 1–87 (2022). doi: 10.1016/j.physrep.2022.01.002.
45. S. Fishman, D. R. Grempel, and R. E. Prange, Chaos, quantum recurrences, and Anderson localization, *Physical Review Letters.* **49** (8), 509–512 (1982). doi: 10.1103/PhysRevLett.49.509.
46. M. Lopez, J.-F. Clément, P. Szriftgiser, J. C. Garreau, and D. Delande, Experimental test of universality of the anderson transition, *Physical Review Letters.* **108**, 095701 (2012). doi: 10.1103/PhysRevLett.108.095701.
47. F. Matsui, H. S. Yamada, and K. S. Ikeda, A quantum damper, *Physica A.* **523**, 631–641 (2019). doi: 10.1016/j.physa.2019.02.052.
48. K. Ikeda, Self-organization of dissipation in classically chaotic quantum systems, *Annals of Physics.* **227**(1), 1–75 (1993). doi: 10.1006/aphy.1993.1075.
49. H. S. Yamada and K. S. Ikeda, Quantum diffusion induced by small quantum chaos, *Physical Review E.* **109**, L062201 (2024). doi: 10.1103/PhysRevE.109.L062201.
50. A. Shudo and K. S. Ikeda, Complex classical trajectories and chaotic tunneling, *Physical Review Letters.* **74**, 682–685 (1995). doi: 10.1103/PhysRevLett.74.682.
51. A. Shudo and K. S. Ikeda, Chaotic tunneling: A remarkable manifestation of complex classical dynamics in non-integrable quantum phenomena, *Physica D.* **115**(3), 234–292 (1998). doi: 10.1016/S0167-2789(97)00239-X.
52. K. Takahashi and K. S. Ikeda, Complex-classical mechanism of the tunnelling process in strongly coupled 1.5-dimensional barrier systems, *Journal of Physics A.* **36**(29), 7953 (2003). doi: 10.1088/0305-4470/36/29/305.
53. K. Takahashi and K. S. Ikeda, Spectroscopic signature of the transition in a tunneling mechanism from the instanton path to a complexified stable-unstable manifold, *Physical Review A.* **79**, 052114 (2009). doi: 10.1103/PhysRevA.79.052114.
54. S. Tomsovic and D. Ullmo, Chaos-assisted tunneling, *Physical Review E.* **50**, 145–162 (1994). doi: 10.1103/PhysRevE.50.145.
55. A. Shudo, Y. Ishii, and K. S. Ikeda, Julia sets and chaotic tunneling: II, *Journal of Physics A.* **42**(26), 265102 (2009). doi: 10.1088/1751-8113/42/26/265102.

https://doi.org/10.1142/9789819823130_0015

Chapter 15

FIRST CHAOTIC STEPS

Arkady Pikovsky

University of Potsdam, Germany
pikovsky@uni-potsdam.de

Here I describe how studies of chaotic dynamical systems began in the mid-1970s at the Chair of Oscillation Theory of the Radiophysics Department of Gorky University. Under the supervision of M. Rabinovich, I participated in the exploration of the Rabinovich system and of an electronic circuit with slow-fast chaotic dynamics. Numerical simulations were particularly challenging at that time.

1. From Convection to Strange Attractors

In the winter of 1974, as a second-year physics student at the Department of Radiophysics at the University of Gorky, my hometown, I had to decide on the direction of my studies. At that time it was not a bachelor-master program; it was a five-year course that ended with a diploma. The first two years were general courses, but from the third year on there was a specialization: you had to choose one of the chairs with the corresponding specialization courses. I had no idea what to choose, so I asked my cousin who had recently graduated from the same department. He explained to me that it was not crucial which subject to choose, but who to choose as a teacher, and gave me some names of mostly bright and active researchers. So in early 1974, I approached Mikhail Rabinovich, then reader at the Theory of Oscillations chair and asked if I could specialize under his supervision, and he agreed. He gave me some review articles[1,2] to read. [*Please note when looking at these early Russian papers that there was no concept of the first/the last author, so all the authors were placed alphabetically according to the Russian Cyrillic alphabet.*] I understood very little of them [*I was even more confused when M. Rabinovich asked me not what I understood, but just what surprised me at most in these papers.*] Nevertheless, when he formulated a first research project — the derivation of a kinetic equation for modes for the Rayleigh-Bénard convection problem, I started reading books and articles, and struggled for most of my third year of university with nonlinear equations for convection. So I read what was relevant in the literature, and in the early 1975, I read a paper by J. McLaughlin and P. Martin.[3] They wrote about a strange attractor in convection, and I understood nothing.

As I later realized, this was the publication that triggered the physical studies of of chaotic dynamics. According to Web of Science, this paper contained only the 13th citation of the famous 1963 Lorenz paper[4] (now more than 10^4 citations), and only the 11th citation of the equally important 1971 paper by Ruelle and Takens[5] (now about 1800 citations), and it was the first publication to cite both. McLaughlin and Martin matched Lorenz's nonperiodic flow with the strange attractor concept. In fact, stochastic dynamics (that is how *deterministic chaos* was called in the Russian literature at that time) was a well-known concept due to the works of Boris Chirikov and his group[6] (see also a contribution of D. Shepelyansky, Chapter 5, p. 123, in this volume), but a general belief was that conservative Hamiltonian chaos studied by Chirikov does not survive dissipation, and that in dissipative systems only steady states and limit cycles could be stable (robust) attractors. Lorenz's model and the theoretical concept of Ruelle and Takens showed that dissipative chaos can be perpetual, and more and more examples of it appeared in 1975. For us, these novel ideas, as I mentioned above, did not come in a direct way by reading the McLaughlin-Martin paper, but through the mathematical group on dynamical system theory around Yakov Sinai in Moscow. M. Rabinovich, together with Svetlana Vyshkind (a lecturer at the Theory of Oscillations Chair), studied low-dimensional models of nonlinearly coupled modes and observed irregular dynamics there.[7] [*As far as I know, this paper sparked the interest of E. Ott in chaotic dynamics;*[8] *somehow Web of Science does not attribute Ott's et al. reference to the Vyshkind-Rabinovich paper correctly*]. He discussed these results with Yakov Sinai and learned from him about the concept of strange attractors. M. Rabinovich came back from Moscow very enthusiastic about it, and I was eager to learn more in this direction (at this point I also realized that I had already read about strange attractors in the McLaughlin-Martin paper). M. Rabinovich brought from Moscow some preprints and lecture notes that Sinai had given him (I remember it was a text by O. Lanford III — just notes of a lecture, probably never published, which I tried to read). [*On this occasion I visited the website at the University of California homepage devoted to O. Lanford,*[9] *there F. Alberto Grünbaum recalls that "His style consisted in writing up extremely polished "unpublished notes" or "drafts", which in many cases were only seen by his closest friends. He would lecture to students at all levels using these notes and improve them every time he was in front of the chalkboard." Maybe I had one of those notes.*] But this was rather difficult for a non-mathematician, with some objects like "*Axiom A*" that I could not identify. So I took a step back and started from more basic texts like S. Smale's review on differentiable dynamical systems[10] and lectures by A. Katok and others at Russian mathematical schools. I presented what I learned at seminars of Rabinovich's group, and after some time we began to understand at least the basic concepts of chaos (such as hyperbolicity and the organization of the Lorenz attractor). Later I translated the 1963 Lorenz paper into Russian for the collection *Strange attractors*, published in 1981.[11]

2. In the Footsteps of Lorenz: Rabinovich System

Around the middle of 1976, M. Rabinovich formulated a couple of problems that I pursued in my diploma thesis. First, he realized that there is a close analog of the Lorenz system in the realm of coupled oscillators (or oscillatory modes). The linear terms in the Lorenz model can be interpreted as a combination of dissipation and parametric excitation, while the nonlinear terms correspond to a "classical" three-mode resonant interaction. This model is known as the Rabinovich system:[12]

$$\begin{cases} \dot{a}_1 = -\nu_1 a_1 + h a_2^* - a_2 a_3 \\ \dot{a}_2 = -\nu_2 a_2 + h a_1^* + a_1 a_3^* \\ \dot{a}_3 = -a_3 \qquad\qquad\;\; + a_1 a_2^*, \end{cases} \tag{1}$$

where the system is written in three columns, with dissipation, excitation, and nonlinear coupling terms, correspondingly. The variables in the Rabinovich system are complex amplitudes $a_{1,2,3}$ of three interacting modes, frequencies of which fulfill the resonance condition $\omega_1 = \omega_2 + \omega_3$. Additionally, modes 1 and 2 are pumped parametrically by an external mode having fixed amplitude $\sim h$ and frequency $\omega_1 + \omega_2$. Remarkably, the complex solutions of the Rabinovich system appeared to lie on the real-valued three-dimensional manifold [Fig. 1(a)], what made the analogy with the Lorenz system ($x \leftrightarrow a_2$, $y \leftrightarrow a_1$, $z \leftrightarrow a_3$) nearly perfect, leading to

$$\begin{cases} \dot{x} = -\nu_2 x + hy + yz \\ \dot{y} = -\nu_1 y + hx - xz \\ \dot{z} = -z + xy, \end{cases} \tag{2}$$

to compare with the Lorenz system[4]

$$\begin{cases} \dot{x} = -\sigma x + \sigma y \\ \dot{y} = -y + rx - xz \\ \dot{z} = -bz + xy. \end{cases} \tag{3}$$

3. In the Footsteps of Rössler: Electronic Generator of Chaos

The second idea came after reading a paper by O. E. Rössler.[13] There he argued that in a three-dimensional slow-fast system, with a two-dimensional S-shaped slow manifold, one can reduce the dynamics to a one-dimensional non-invertible map and thus get chaos. Slow-fast systems were a popular object at the Theory of Oscillations chair, in the context of electronic circuit dynamics. M. Rabinovich proposed to construct a chaotic electronic generator with slow-fast dynamics.

4. Computation and Visualization of Chaos

Working on these two problems was an exciting time for me. By computations one understood mainly "analytical calculations," there was almost no know-how in computer simulations; moreover, there were very few computers available. The only option for a diploma (master) student was an outdated Soviet computer M-220, accessible at the university's computer center. It had no compiler, and one had to program in machine codes. I had a book with a detailed description of these three-address-instructions. The computer just stood in a room at the computer center, I came there, had to start it up, enter some codes by pressing buttons on the board (not a *keyboard*), and the output was a narrow paper ribbon (like one in cashier machines) with numbers only. Fortunately, I did not have to program a Runge–Kutta routine, S. Vyshkind gave me some punched cards with this code that I could use (but I could not make my own punched cards;

all the parameters where inserted from the board). The computer's time was allocated according to a schedule, and for a diploma student with the lowest priority only slots (about 1–1.5 h) late in the evening were available; I had to mind not to miss the last bus to return home. The computations' results had to be put to the graphs (on a graph paper) by hand. This handmade plotting was possible for parts of a trajectory only, or for a Poincaré map. To get beautiful graphs of trajectories (like in the papers from the West), we used an analog computer with a plotter, which was available at another department of the University. One could easily arrange a simple set of equations (like the Lorenz system) on this analog computer, but the accuracy was miserable. So one just adjusted parameters (by turning a potentiometer) to get a beautiful plot. Moreover, while plotting a long trajectory, the parameters could deviate, and the solution suddenly exploded. In our early papers, we used these plots (choosing the nicest ones). I do not have original graphs, so in Fig. 1 I present analog computer phase portraits (with an additional loss of quality due to scanning) of the Rabinovich system and of the slow-fast dynamics in an electronic circuit from papers.[12,14] Still, it was a nightmare for the journals: we provided photos of the plots, where some pieces of lines were nearly broken (the plotter used a pen with red dye — faster movements were thin and slower movements were bold), and the retouchers at the journal production did not know how to make these images clear.

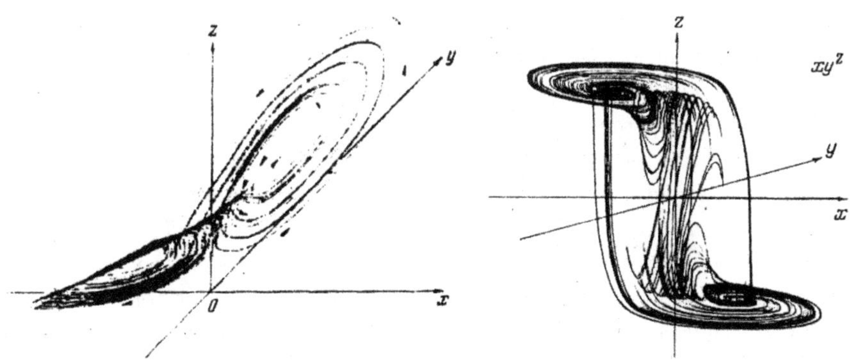

(a) Rabinovich system (b) Electronic chaotic generator

Fig. 1. Phase portraits of two strange attractors created on an analog computer produced by (a) the Rabinovich system (1) and (b) the electronic chaotic generator (4). Although in the original papers we gave the values of the parameters, here I do not reproduce them, because of the mentioned desastrous accuracy of the analog computer.

5. Analytical Approach to Chaos in a Slow-Fast System

For the Rabinovich system, the only analytical part was a derivation of the equations for plasma waves, all the rest was numerics. For the chaotic slow-fast electronic generator, there was an analytical technique to obtain the Poincaré map: one had to approximate the S-shaped slow manifold by a piecewise-linear one, solve linear equations on different branches, and match them. This technique was well-documented in the second edition of the Andronov, Witt, Khaikin book on Theory of Oscillations. *[Chapter VIII with detailed calculations was added to the second edition by Nikola Zhelestzov; it is also available in the English translation.[15]]* The problem, however, was that we started with a picture of the dynamics, not with a circuit and equations. Our first attempt was to combine an oscillator with a negative resistance with a relay element. A relay element is well-described by a two-state model, so I solved equations for each state, matched them, and got an expanding one-dimensional map demonstrating chaos. We wrote a draft of the paper, and Rabinovich showed it to his teacher Andreï Gaponov-Grekhov. [*Here it is appropriate to make some remarks about the Gorky school on nonlinear oscillations. In the Soviet Union and to a certain extent now in Russia as well, it has been crucial to attribute researchers to schools, founded by prominent scientists (in physics, the schools of Landau and Bogoljubov are famous). The Gorky school on nonlinear oscillations was founded by Andronov, who came to Gorky in the mid-1930s. This school split into physical and mathematical branches, the former represented by A. Gaponov-Grekhov, a Ph.D. student of Andronov. He was mainly active in electronics and plasma physics, but was also very interested in general nonlinear phenomena. He organized and led spring scientific schools (not meaning "a group of scholars" as above, but in the sense of "an event") "Nonlinear waves", which since 1972 served as one of the central meetings in the field of nonlinear science in Russia; the 19th school was held shortly before the corona outbreak in February–March 2020. Another mathematical branch, to which in particular Leonid Shil'nikov, Valentin Afraimovich and Dmitry Turaev belong, can be traced back to another doctoral student of Andronov, Yu Neimark, and to Andronov's collaborator and wife E. Leontovich-Andronova. The son of Andronov, also Alexander Alexandrovich, a physicist in the field of plasma physics and solid-state electronics, became in 1976–77 interested in strange attractors; he was the referee of my Diploma thesis.*]

Gaponov-Grekhov liked the idea but did not like the implementation: a relay element is not really a genuine electronic one. Thus we took a real electronic element with an N-shaped current-voltage characteristic, a tunnel diode, and tried to combine it with an unstable (due to negative resistance) linear oscillator. One of my attempts, placing the tunnel diode in series with the inductance, was successful: the equations[14]

$$\begin{cases} \dot{x} = y - \delta z \\ \dot{y} = -x + 2\gamma y + \alpha z + \beta \\ \mu \dot{z} = x - f(z) \end{cases} \tag{4}$$

(using the idealized characteristics of the tunnel diode $f(z) = z^3 - z$ in the analog computations, and as a piecewise-linear function in the analytical calculations) produced irregular dynamics. After constructing the Poincaré map, we realized that although the phase portrait was as assumed by O. E. Rössler, the Poincaré map, contrary to his hypothesis, was not continuous but discontinuous. Therefore, one could not derive chaos from the period-3 argument, which is valid only for continuous maps, but we had to check that the Poincaré map is expanding.

6. In the Footsteps of Li and Yorke

Together with these two problems, in my diploma thesis,[16] with which I graduated in June 1977, I also had a third chapter. The role of period-3 in continuous maps appeared to be very important in those days after the seminal paper by Li and Yorke.[17] I realized that the existence of period-3 in continuous maps means that there exists a subset of dynamical trajectories, which admits a simple symbolic description with two symbols. This symbolic representation immediately gives a lower bound for the topological entropy. I formulated this observation as a "theorem", although I do not think I had a real proof. At least, when I showed this "proof" to Yakov Sinai, he was somewhat skeptical and asked: *"Arkasha, do you need to publish this?"* I understood that I must remain on the physical ground and not enter the mathematical realm. Anyway, I was pleased to read a paper by Smale and Williams[18] with practically the same observation.

References

1. A. Gaponov, L. Ostrovskii, and M. I. Rabinovich, One-dimensional waves in disperse nonlinear systems, *Radiophysics and Quantum Electronics*. **13**, 121–161 (1970). doi: 10.1007/BF01031524.
2. B. B. Kadomtsev and V. I. Karpman, Nonlinear waves, *Soviet Physics Uspekhi*. **14**(1), 40–60 (1971). doi: 10.1070/pu1971v014n01abeh004441.

3. J. B. McLaughlin and P. C. Martin, Transition to turbulence of a statically stressed fluid, *Physical Review Letters.* **33**(20), 1189–1192 (1974). doi: 10.1103/PhysRevLett.33.1189.
4. E. N. Lorenz, Deterministic nonperiodic flow, *Journal of the Atmospheric Sciences.* **20**, 130–141 (1963). doi: 10.1175/1520-0469(1963)020⟨0130:DNF⟩2.0.CO;2.
5. D. Ruelle and F. Takens, On the nature of turbulence, *Communications in Mathematical Physics.* **20**, 167–192 (1971).
6. G. M. Zaslavskiĭ and B. V. Chirikov, Stochastic instability of non-linear oscillations, *Soviet Physics Uspekhi.* **14**(5), 549–568 (1972). doi: 10.1070/pu1972v014n05abeh004669.
7. S. I. Vyshkind and M. I. Rabinovich, The mechanism for phase stochastization and the structure of wave turbulence in dissipative media, *Soviet Journal of Experimental and Theoretical Physics.* **44**(2), 292–299 (1976).
8. J.-M. Wersinger, J. M. Finn, and E. Ott, Bifurcations and strange behavior in instability saturation by nonlinear mode coupling, *Physical Review Letters.* **44**(7), 453–456 (1980). doi: 10.1103/PhysRevLett.44.453.
9. F. Alberto Grünbaum, In memoriam: Oscar E. Lanford III, Professor of Mathematics UC Berkeley 1940-2013 (2018). https://senate.universityofcalifornia.edu/in-memoriam/files/oscar-e-lanford-iii.html. Visited July 24, 2020.
10. S. Smale, Differentiable dynamical systems. I Diffeormorphisms, *Bulletin of American Mathematical Society.* **73**, 747–817 (1967).
11. Y. G. Sinai and L. P. Shilnikov (eds.), *Strange Attractors.* MIR Publishing, Moscow (1981).
12. A. S. Pikovski, M. I. Rabinovich, and V. Y. Trakhtengerts, Onset of stochasticity in decay confinement of parametric instability, *Zhurnal Eksperimentalnoi i Teoreticheskoi Fiziki.* **74**, 1366–1374 (1978).
13. O. E. Rössler, Chaotic behavior in simple reaction system, *Zeitschrift für Naturforschung A.* **31**, 259–264 (1976).
14. A. S. Pikovski and M. I. Rabinovich, A simple autogenerator with stochastic behavior, *Dokladi Akademii Nauk SSSR.* **239**, 301–304 (1978).
15. A. A. Andronov, A. A. Vitt, and S. E. Khaikin, *Teoriya kolebanii.* Fizmatgiz, Moscow (1959). *Theory of Oscillators*, Pergamon Press, 1966.
16. A. Pikovsky, On stochastic self-oscillations in radiophysical systems. Master's Thesis, Department of Radiophysics, Gorky State University, Gorky (June 1977) (in Russian).
17. T.-Y. Li and J. A. Yorke, Period three implies chaos, *The American Mathematical Monthly.* **82**(10), 985–992 (1975).
18. S. Smale and R. F. WIlliams, The qualitative analysis of a difference equation of population growth, *Journal of Mathematical Biology.* **3**, 1–4 (1976). doi: 10.1007/BF00307853.

https://doi.org/10.1142/9789819823130_0016

Chapter 16

TÜBINGEN BLUES

Lars F. Olsen

University of Southern Denmark, Denmark
lfo@bmb.sdu.dk

The peroxidase-oxidase reaction was the first (bio)chemical reaction system to show chaotic dynamics. The finding of chaos in this iconic reaction system was very much inspired by the early publications of Otto Rössler on chaos in simple chemical reaction systems and by a four week visit to Otto's lab in Tübingen in early 1976.

1. Introduction

Biological systems are inherently non-linear systems and even the simplest biological system may outcompete many physical systems when it comes to complexity. The history of chaos in biological systems takes its beginning with the study of rhythms. Rhythmic phenomena in biology have been known for centuries and in the last part of the 20th century it fascinated many biomathematicians. This led to the establishment of a small but very dedicated community studying non-linear dynamics in biology.

My own introduction into the field came when I first went to the department of biochemistry, Odense University (now University of Southern Denmark) as a graduate student in 1975 to study bistability and oscillations in a single enzyme reaction known as the peroxidase-oxidase reaction. My supervisor was professor Hans Degn, who had studied this and other oscillating chemical reactions since the early 1960ies and published some seminal papers on the subject.[1,2] Officially I was a student at the University of Copenhagen and in those days the typical project for a biochemistry student was to purify a new enzyme (or a known enzyme in a new organism), establish an assay to measure its activity and finally determine its K_M and turnover. However, since my high-school days I had a crush for mathematics and physics and the *sine* function was my absolute favourite. Since my teenage years I have also been a keen amateur blues guitarist. Therefore, studying a "sinusoidal" biochemical reaction had the scent of both mathematics and music!

In the fall of 1975 Hans Degn urged me to attend a meeting on "*Rhythmic Functions in Biological Systems*" in Vienna (September 8–12). The meeting was mostly on circadian rhythms and I did not know any of the participants and also had nothing to contribute. However, I had the pleasure of meeting two scientists who have had a great influence on my later career. One was Arthur Winfree and the other was Otto Rössler. I do not remember so much of Otto's talk on "*Deductive Chronobiology.*"[3] It was not so much the content of the talk that caught my attention, but more the enthusiasm by which it was delivered. Back in Odense Hans Degn informed me that the department had some unspent money which I could use for a month visit to a lab of my own choice. Hans suggested that I went to the Max Planck Institute of Nutrition Physiology (Ernährungsphysiologie) in Dortmund headed by Benno Hess. Professor Hess, who visited our department that fall had agreed to host me, but I had a different plan, namely to ask Otto, if he would be willing to have me around for a few weeks. Hans wrote a letter to Otto and Otto replied back that I was welcome there.

2. My First Visit to Tübingen in 1976

Therefore, in early January 1976 I took a train from Odense to Tübingen and Otto picked me up on the station. When I arrived Otto had just submitted his first paper on chaos in a (bio)chemical system,[4] but I knew nothing of this at the time. My plan was to make a model of the PO reaction that could unify its ability to show both bistability and oscillatory behavior based on some enzyme-kinetic measurements done in the lab. Otto helped me with the model and in fact we did get it to work. The equations were as follows:

$$\begin{cases} \dot{a} = K(a_0 - a) - \dfrac{V\left(a + \kappa a^2\right) b}{\lambda b + a + \mu a^2} \\ \dot{b} = \sigma - 2\dfrac{V\left(a + \kappa a^2\right) b}{\lambda b + a + \mu a^2} \end{cases} \tag{1}$$

where a represents O_2 and b represents NADH. K is a constant that determines the rate of diffusion into the reaction mixture and a_0 represents the O_2 concentration at equilibrium. V, κ, λ and μ are enzyme kinetic paramaters and σ is the inflow rate of NADH (see below and Fig. 1 for further explanation of the experimental setup). The model showed coexistence of steady state and limit cycle oscillations with an unstable limit cycle separating the steady state and the stable limit cycle. It was never published in full, but the experimental data underlying the model appeared in a later publication.[5] However, when discussing this model Otto was also telling me about an interesting new kinetic model he had made which could show non-periodic oscillations, sensitive to initial conditions, which he referred to as chaos.

When Otto was not around in the Institute of Physical Chemistry, I was discussing his ideas with his colleagues and students, and although they were all very friendly and forthcoming, some were also skeptical about this new "chaos" concept. Some even suggested that the phenomenon was nothing but a numerical artefact, and they were not alone. In the next 5–6 years many physical chemists were trying to kill the idea of chaotic behavior in chemical and biochemical systems.[6] After my visit Otto wrote a kind letter to Hans Degn saying that he thought I had performed rather well and my visit there was to everyone's satisfaction.

3. Finding Chaos in the Peroxidase-Oxidase Reaction

The peroxidase-oxidase (PO) reaction is the oxidation of an organic electron donor (YH_2) by molecular oxygen catalyzed by the enzyme peroxidase:

$$2YH_2 + O_2 \rightarrow 2Y + 2H_2O \tag{2}$$

Several organic molecules can take the role as YH_2, but the most commonly used in the laboratory is reduced nicotinamide adenine dinucleotide (NADH) which is abundant in all cell types. The PO reaction takes place *in vivo* in many organisms and cells. Following my short stay in Tübingen I started a new series of experiments with an open system where both substrates NADH and O_2 were supplied continuously to the reaction mixture containing the enzyme. This had only been done once before by a Japanese group headed by prof. Isao Yamazaki, a well-known peroxidase researcher.[7] But instead of regenerating the NADH using a second enzyme as Yamazaki's group had done, we decided to use a syringe pump for the supply of NADH. The idea with this setup (see Fig. 1) was to test the model that I had developed in Tübingen. Much to our surprise,

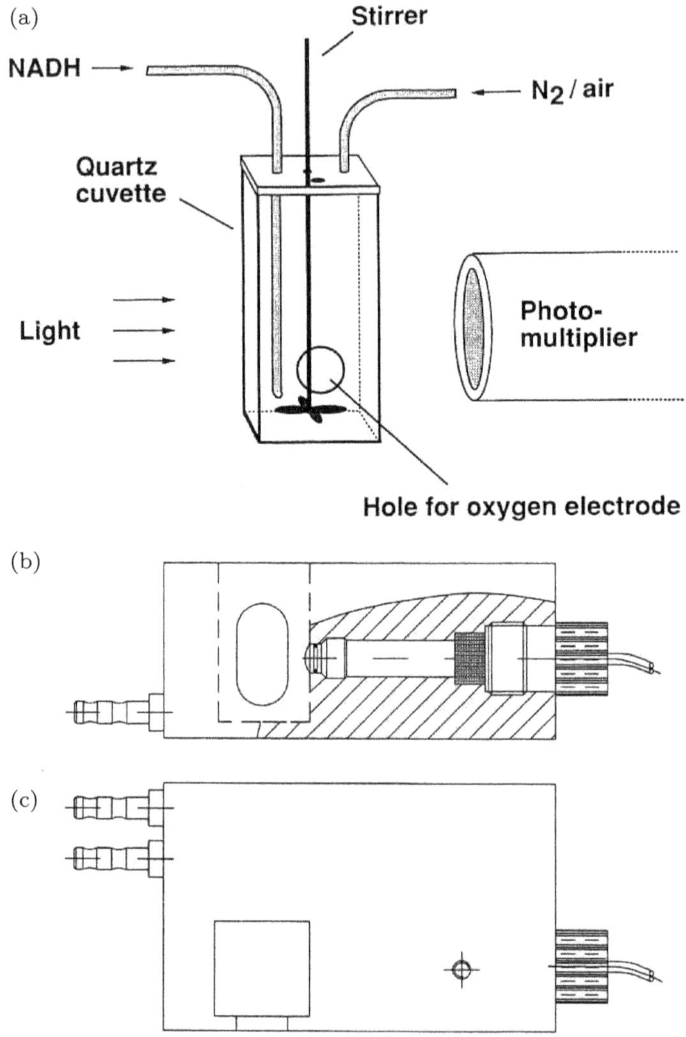

Fig. 1. Schematic illustration of the experimental setup for measurements of absorption and O_2 concentration in the peroxidase-oxidase reaction. (a) The setup consists of a 4 cm × 4 cm square (or hexagonal) quartz cuvette mounted in a thermostatting jacket; The cuvette is covered with a lid with holes for the stirring rod, inflow of a 5% O_2/N_2 mixture and NADH inflow. Light at 360 nm (for measuring NADH) and 418 nm (for measuring an enzyme intermediate) is passing through the cuvette and is collected by a photomultiplier tube. (b) Setup seen from the side. (c) Setup seen from the top.

the resulting oscillations were different and far more complex than we had anticipated. One observation that was at variance with the model developed in Tübingen was that O_2 and NADH oscillated in complete phase and not phase-shifted as predicted by Eqs. (1). Furthermore, in addition to regular periodic oscillations we observed mixed-mode oscillations and bursting oscillations, none of which could be explained by my simple two-variable model. Sometimes we also observed non-periodic oscillations as mixtures of small and large-amplitude oscillations in a seemingly random order. Initially, Hans Degn dismissed these oscillations as artefacts generated by small random fluctuations in the pumping rate of NADH inflow, but they appeared consistently when repeating the experiments with the same experimental settings. I had an idea that these non-periodic oscillations could be chaos, but I was not sure how to analyze them and therefore I wrote a letter to Otto with an extensive description of what I had done with free-hand drawings of the data. Otto's reply was: "*Read Lorenz!*"[8]

I read Edward Lorenz's fascinating paper[8] and more recent papers by Robert May[9] and Li and Yorke[10] and suddenly I understood. Following Lorenz's instructions I constructed a return map by plotting each amplitude from our irregular PO oscillations against the preceding amplitude (I would probably have drawn the map a little differently today, but that would not have changed anything) and applied the Li and Yorke theorem.[10] I showed the plot together with the papers by Lorenz and Li and Yorke to Hans Degn, who immediately changed his opinion on the results. Within a week we had written the manuscript and submitted it as a letter to *Nature* by the end of October 1976. We also sent copies of the manuscript to Otto and to Art Winfree, from whom we received very enthusiastic responses. It took *Nature* nearly 4 months to find a reviewer that would evaluate the paper, but finally in March 1977 we received a positive answer and the paper was published two months later.[11] The time series of periodic and chaotic oscillations are shown in Fig. 2 and the return map in Fig. 3. A few months later the paper by Schmitz, Graziani and Hudson on chaos in the Belousov–Zhabotinskii (BZ) reaction appeared.[12] In 1978 Otto and Klaus Wegmann also published a paper on chaos in the BZ reaction where they compared the electrochemical measurements to a simple model exhibitng screw-type chaotic behavior.[13] It is important to note that in those days Takens' embedding method[14] had not yet been published.

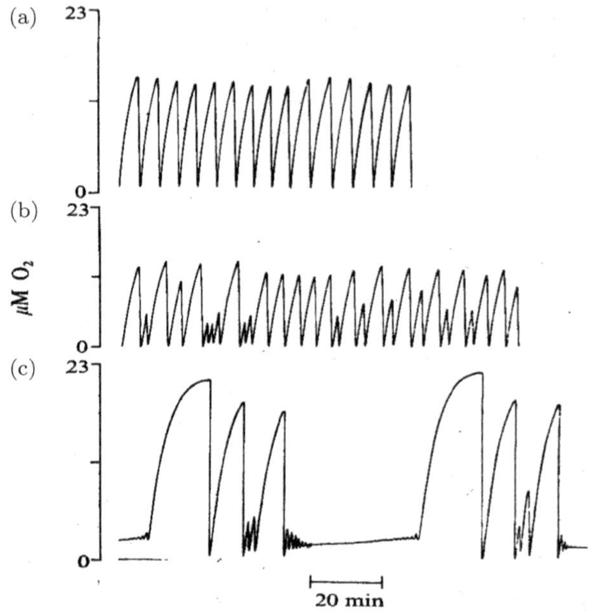

Fig. 2. Simple periodic (a), chaotic (b) and bursting (c) oscillations of O_2 in the peroxidase-oxidase reaction.

Source: Reproduced with permission from Ref. 11.

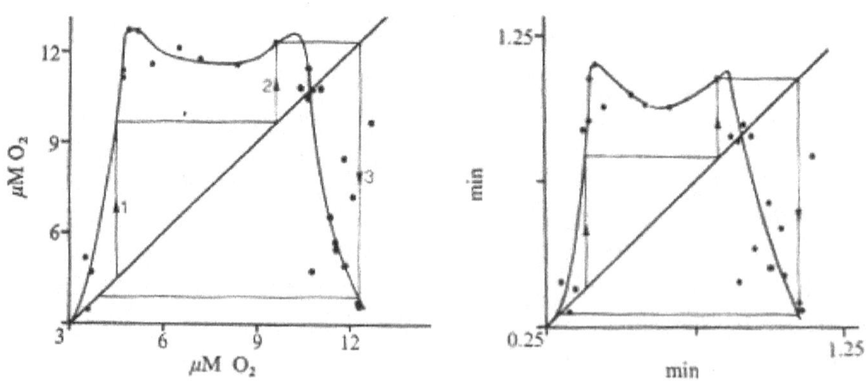

(a) On the amplitude

(b) On the time-period

Fig. 3. (a) Next-amplitude map of oscillations of O_2 from Fig. 2(b); (b) same type of plot for the oscillation periods. Trajectories with arrows were drawn to show that the transition function allow the period 3.

Source: Reproduced with permission from Ref. 11.

4. First Biology Conference on Chaos in Wales

The visit to Tübingen in January 1976 completely changed my career. Non-linear dynamics was not highly prioritised in Danish science in the late 1970's, so I had to take a small divergence into photosynthetic electron transport and proton pumping. However, while I found many of the topics there interesting my heart was still with chaos research and blues music so I spent almost all my spare time on these two "hobbies". One achievement made in this period was the formulation of a simple model based on autocatalysis that reproduced the experimental next-amplitude maps rather well.[15] In 1984 I finally obtained a permanent position as assistant professor, so I could return to my non-linear dynamics research on full time. One of my aims was to write a review on "*Chaos in Biological Systems*" and another one to organize a conference on the subject. Both aims were completed in cooperation with Hans Degn. The review, which took us almost a year to write, was published in *Quarterly Review of Biophysics*[16] and introduced the theory of chaos in simple biological models and discussed the few discoveries of chaos in experimental systems or in the field. However, writing this review was a good exercise to find out what had been done on chaos research in biology and who was doing it, that is, who should we invite to participate in the planned conference. This conference was held in Dyffryn House near Cardiff, Wales on December 8–12, 1986 and was organized by Hans Degn, myself and profs. David Lloyd and Arun V. Holden. It was sponsered by grants from NATO and the EU. Many prominent researchers were participating, e.g. Robert May, Michael Guevara, William Schaffer, Albert Goldbeter, K. Aihara, Paul Glendinning to mention a few and of course also Otto.[17] A group picture of the participants can be seen in Fig. 4 with Otto in the last row to the left. At this meeting I met Bill Schaffer, who was also a strong admirer of Otto's work and he and I started a long term collaboration on the outbreaks of childhood diseases (measles, chicken pox, mumps and rubella) in large cities in Europe and the US. Our analyses of the monthly notifications of these diseases suggested that at least measles outbreaks followed a chaotic pattern.[18] On two occasions my family and I spent longer periods (one semester in 1987 and two semesters in 1992–1993) at the University of Arizona, Tucson. These were commemorable times and besides getting a lot of research done I had the pleasure of running into Art Winfree, who was then a professor at the University of Arizona, from time to time.

Fig. 4. Group picture of participants at the NATO ASI conference on "*Chaos in Biological Systems*" at Dyffryn House near Cardiff, Wales December 8–12, 1986. Otto can be spotted in the last row on the left near the pillar (no 31). Hans Degn is in the last row to the very right (no 26). List of participants on the photo: 1, D. Lloyd; 2, S. Ishizuka; 3, K. Aihara; 4, H. Hayashi; 5, I. Labouriau; 6, M. D. R. A. Ribeiro; 7, S. Sinha; 8, N. Pagett (conference secretary); 9, J. Friend (NATO representative); 10, A. V. Holden; 11, A. J. Mandell; 12, G.P. Drago; 13, L. Cruzeiro; 14, S. Ridella; 15, O. Decroly; 16, C. Kaas-Petersen; 17, I. Gumowski; 18, A. Schierwagen; 19, G. Matsumoto; 20, J. Bélair; 21, Representative of Manchester University Press; 22, L. F. Olsen; 23, R.M. May; 24, A. Goldbeter; 25, F. Moran; 26, H. Degn; 27, J. Parisi; 28, R.M. Everson; 29, J. Gorski; 30, E. Labos; 31, O. E. Rössler; 32, P. E. Rapp; 33, W. M. Schaffer; 34, P. Glendinning; 35, M. Guevara; 36, M. Markus; 37, K. S. Jensen; 38, J. S. Nicolis; 39, J. L. Martiel; 40, B. J. West; 41, J. Demongeot.

5. Later Studies

Around 1991 we started a new series of experiments with the PO reaction in collaboration with professor Raima Larter at Indiana University-Purdue University at Indianapolis and her Ph.D. student Curt Steinmetz. This resulted in the establishment of a period-doubling route to chaos in the reaction.[19] This later led to development of new and detailed models that could reproduce the dynamics of the experimental system.[20] This work was done in collaboration with Drs. Valentina Fed'kina and Tatiana Bronnikova from Pushchino, Russia and Bill Schaffer, who all spent a couple of months at Odense University in the summer of 1994. The following year Dr. Marcus Hauser came from Würzburg to Odense as a postdoctoral fellow to work on the PO reaction. In an impressive series of meticulous measurements

Marcus established a period-adding route to chaos and provided evidence for homoclinic chaos.[21] We were thrilled by these experiments, especially since the dynamics seemed to be so similar to the behavior of Otto's first model.[4] Interestingly, a few years later Marcus coauthored a paper in which the 10-variable Bronnikova–Fed'kina–Schaffer–Olsen (BFSO) model[20] of the PO reaction could be reduced to a 6-variable model and three 4-variable subnetworks and homoclinic chaos could be verified in both the original model and the subnetworks![22]

Over the years I have visited Tübingen a few times and it is always a pleasure to meet with Otto and discuss science and other matters. Strangely enough, given that Otto has had a profound influence on my scientific life, we have not written a single joint paper. However, my research has been directed by his many advices and I have had the pleasure and privilige of collaborating with his students Gerold Baier and Sven Sahle and in particular Gerold's Ph.D. student Ursula Kummer, who is now a professor in Heidelberg. Since the early 1990s Ursula has visited my lab in Odense a number of times, and I spent a semester with her in Heidelberg in 2001. We have published a number of papers together on oscillations in the PO reaction[23] and on calcium dynamics.[24] Interestingly, Ursula, Marcus and I also found experimental evidence for coexistence of steady-state and oscillation in the PO reaction,[25] as predicted by the model I developed in Tübingen in early 1976. Thus, to me Otto has not only left a lasting scientific imprint through his many seminal papers but also through the work of his scientific "children" and "grandchildren", and I humbly count myself as one.

References

1. H. Degn, Effect of bromine derivatives of malonic acid on the oscillating reaction of malonic acid, cerium ions and bromate, *Nature.* **213**, 589–590 (1967). doi: 10.1038/213589a0.
2. H. Degn, Bistability caused by substrate inhibition of peroxidase in an open reaction system, *Nature.* **217**, 1047–1050 (1968). doi: 10.1038/2171047b0.
3. O. E. Rössler, Deductive biology — Some cautious steps, *Bulletin of Mathematical Biology.* **40**(1), 45–58 (1978). doi: 10.1007/bf02463129.
4. O. E. Rössler, Chaotic behavior in simple reaction system, *Zeitschrift für Naturforschung A.* **31**, 259–264 (1976).
5. H. Degn, L. F. Olsen, and J. W. Perram, Bistability, oscillation, and chaos in an enzyme reaction, *Annals of the New York Academy of Sciences.* **316**(1), 623–637 (1979). doi: 10.1111/j.1749-6632.1979.tb29503.x.

6. N. Ganapathisubramanian and R. M. Noyes, A discrepancy between experimental and computational evidence for chemical chaos, *Journal of Chemical Physics*. **76**(4), 1770–1774 (1982). doi: 10.1063/1.443216.
7. S. Nakamura, K. Yokota, and I. Yamazaki, Sustained oscillations in a lactoperoxidase, NADPH and O_2 system, *Nature*. **222** (1969). doi: 10.1038/222794a0.
8. E. N. Lorenz, Deterministic nonperiodic flow, *Journal of the Atmospheric Sciences*. **20**, 130–141 (1963). doi: 10.1175/1520-0469(1963)020⟨0130:DNF⟩2.0.CO;2.
9. R. M. May, Simple mathematical models with very complicated dynamics, *Nature*. **261**(5560), 459–467 (1976). doi: 10.1038/261459a0.
10. T.-Y. Li and J. A. Yorke, Period three implies chaos, *The American Mathematical Monthly*. **82**(10), 985–992 (1975).
11. L. F. Olsen and H. Degn, Chaos in an enzyme reaction, *Nature*. **267**, 177–178 (1977). doi: 10.1038/267177a0.
12. R. A. Schmitz, K. R. Graziani, and J. L. Hudson, Experimental evidence of chaotic states in the Belousov–Zhabotinskii reaction, *The Journal of Chemical Physics*. **67**(7), 3040–3044 (1977). doi: 10.1063/1.435267.
13. O. E. Rössler and K. Wegmann, Chaos in the Zhabotinskii reaction, *Nature*. **271**, 89–90 (1978). doi: 10.1038/271089a0.
14. F. Takens, Detecting strange attractors in turbulence, *Lectures Notes in Mathematics*. **898**, 366–381 (1981). doi: 10.1007/BFb0091924.
15. L. F. Olsen, An enzyme reaction with a strange attractor, *Physics Letters A*. **94**(9), 454–457 (1983). doi: 10.1016/0375-9601(83)90853-8.
16. L. F. Olsen and H. Degn, Chaos in biological systems, *Quarterly Reviews of Biophysics*. **18**(2), 165–225 (1985). doi: 10.1017/S0033583500005175.
17. L. F. Olsen, A. V. Holden, and H. Degn (eds.), *Chaos in Biological Systems*, vol. 138, *NATO ASI Series A: Life Sciences*. Plenum Press, New York (1987).
18. L. F. Olsen and W. M. Schaffer, Chaos versus noisy periodicity: Alternative hypotheses for childhood epidemics, *Science*. **249**(4968), 499–504 (1990). doi: 10.1126/science.2382131.
19. T. Geest, C. G. Steinmetz, R. Larter, and L. F. Olsen, Period-doubling bifurcations and chaos in an enzyme reaction, *The Journal of Physical Chemistry*. **96**(14), 5678–5680 (1992). doi: 10.1021/j100193a004.
20. T. V. Bronnikova, V. R. Fed'kina, W. M. Schaffer, and L. F. Olsen, Period-doubling bifurcations and chaos in a detailed model of the peroxidase-oxidase reaction, *The Journal of Physical Chemistry*. **99**(23), 9309–9312 (1995). doi: 10.1021/j100023a001.
21. M. J. B. Hauser and L. F. Olsen, Mixed-mode oscillations and homoclinic chaos in an enzyme reaction, *Journal of the Chemical Society, Faraday Transactions*. **92**, 2857–2863 (1996). doi: 10.1039/FT9969202857.
22. A. Sensse, M. J. B. Hauser, and M. Eiswirth, Feedback loops for Shil'nikov chaos: The peroxidase-oxidase reaction, *The Journal of Chemical Physics*. **125**(1), 014901 (2006). doi: 10.1063/1.2207140.

23. U. Kummer, M. J. B. Hauser, K. Wegmann, L. F. Olsen, and G. Baier, Oscillations and complex dynamics in the peroxidase-oxidase reaction induced by naturally occurring aromatic substrates, *Journal of the American Chemical Society.* **119**(9), 2084–2087 (1997). doi: 10.1021/ja9636855.
24. U. Kummer, L. F. Olsen, C. J. Dixon, A. K. Green, E. Bornberg-Bauer, and G. Baier, Switching from simple to complex oscillations in calcium signaling, *Biophysical Journal.* **79**(3), 1188–1195 (2000). doi: 10.1016/S0006-3495(00)76373-9.
25. L. F. Olsen, M. J. B. Hauser, and U. Kummer, Mechanism of protection of peroxidase activity by oscillatory dynamics, *European Journal of Biochemistry.* **270**(13), 2796–2804 (2003). doi: 10.1046/j.1432-1033.2003.03655.x.

https://doi.org/10.1142/9789819823130_0017

Chapter 17

FROM CHEMICAL CHAOS TO CHAOTIC BRAIN

Ichiro Tsuda

Sapporo City University, Japan
tsuda@isc.chubu.ac.jp

We deal with the chaos study in the Belousov–Zhabotinsky reaction system conducted at Kyoto university in the 1970s and the early 1980s, and its related works at some other places. The interplay of chaos and noise was particularly studied, whereby a dynamical mechanism for complex behaviors observed in laboratories was clarified. This small element of findings led to a new concept in the information processing of brain, called Cantor coding.

1. Motivation for Chaos Study

As everyone does, I enthusiastically studied Otto E. Rössler's pioneering works on chaos[1–4] in my graduate student days, in the late 1970s. I tried to understand the mathematical structure of Lorenz chaos[5] in relation to Smale's horseshoe map,[6] the relationship between Lorenz chaos and a strange attractor[7] in a sense of mathematical representation of hydrodynamic turbulence, and also the relationship between such chaos and chaos in a sense of Li–Yorke,[8] while I kept thinking of "real" chaos observed in the Belousov–Zhabotinsky (BZ) reaction system.[9–13] Otto's contribution[14] to real chaos in that system with Klaus Wegmann encouraged me to pursue this direction of research.

Here, let me add some more comments about my early research with the late Kazuhisa Tomita, concerning chaos in the Belousov–Zhabotinsky reaction. We noticed early reports of chaotic behaviors in this chemical reaction system. One report was published by Wegmann and Rössler[14] mentioned above, while the other was by Schmitz, Graziani, and Hudson.[9] Although they reported "chaotic" behaviors in laboratory experiments, it was not clarified that those behaviors can be characterized by mathematical structure recognized as deterministic chaos or strange attractors. On the other hand, an authority in the world of chemistry such as Richard Noyes strongly insisted the impossibility of the appearance of chaos in any chemical reaction system. Actually, his group published a paper showing the presence of complicated periodic states in their Belousov–Zhabotinsky reaction model.[15] With this computation result, Noyes *et al.* emphasized that chaotic behaviors that look like deterministic chaos must be complicated limit cycles. We wanted to show definite evidence for the presence of chaos in the Belousov–Zhabotinsky reaction. We thought that finding evidence was easy. The reason was that Otto already showed the presence of chaos in three-dimensional continuous chemical reaction systems even with one quadratic nonlinear term, and further the Belousov–Zhabotinsky reaction system should include more than one quadratic nonlinear terms due to molecular collisions of two different chemical substances.

2. Interplay of Chaos and Noise

For the first time, we made a three-dimensional continuous model for the Belousov–Zhabotinsky reaction, based on an original Oregonator proposed by Field and Noyes.[16] In 1978, Kazuhisa Tomita asked Yoshisuke Ueda to allow us to use analog computers in his laboratory since Kazuhisa had already recognized Yoshisuke's outstanding research on chaotic behaviors in electric circuits. Yoshisuke kindly allowed us to use analog computers in his laboratory. I was told that those analog computers were the same type of analog computers that Yoshisuke had used when he found the "Japanese attractor" in 1961. He kindly accepted our presence despite our different scientific disciplines, that is, his way of thinking was based on electric engineering, while ours were based on physics and mathematics. At that time in Japan, both disciplines had not communicated with each other, except a few special meetings on the nonlinear problems research. Kazuhisa Tomita, a physicist and Masaya Yamaguti, a mathematician had already known the concept of deterministic chaos by the theorem of

Li–Yorke published in 1975.[8] In contrast, Yoshisuke reported his findings as irregular transitions in noisy environments, because it was believed, in the world of electric engineering in Japan, especially in Kyoto university, that deterministic chaos was unable to exist in real electric circuits because of the occurrence of heat in the circuits, thus any chaotic behavior in electric circuits should appear as irregular transitions by noise. We physicists and mathematicians were rather interested in the interplay of deterministic chaos and noise.

I finally found "chaotic" behaviors in our model

$$\begin{cases} \dot{\xi} = (1-\phi)\xi + \eta - \xi\eta - \xi\zeta \\ \dot{\eta} = -(1+\phi)\eta + \zeta - \xi\eta + m \\ p\dot{\zeta} = \xi - (1+p\phi)\zeta - \xi\zeta \end{cases} \tag{1}$$

using analog computers and published a paper in 1979 (see also Fig. 1).[12] Christian Vidal's Bordeaux group found very similar chaotic behaviors to our findings in their laboratory experiment.[11] We were excited by this experimental finding. Unfortunately, however, I could not find any chaotic

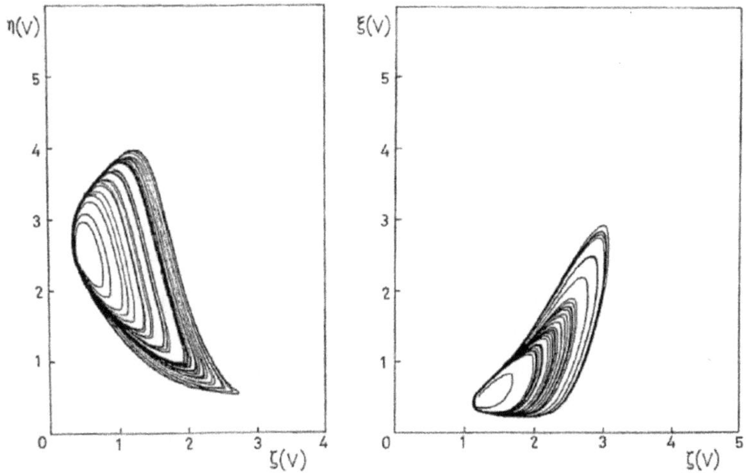

Fig. 1. "Chaotic" orbits yielded by the analog computer at Yoshisuke Ueda's laboratory in 1978. Two sides views of the orbit are shown. The BZ reaction model with flow terms were computed. The parameter values used for this behavior were $p \approx 90$, $\phi \approx 0.048$, and $m \approx 3.44$ in Eqs. (1).

Source: Figures reproduced from Ichiro Tsuda's Master thesis, Kyoto University, 1979.

behaviors by digital computations with our model. The equations used in analog and digital computations were the same, but the computation results were different: one showed chaotic behaviors and the other showed simply periodic ones. I guessed that the chaotic behaviors found in the analog computer could be a kind of noise-induced chaos caused by the weak stability of limit cycles. The reason is as follows. In our continuous model for the Belousov–Zhabotinsky reaction, the domain of the parameter space for which there is a coexistence of two unstable foci and a saddle appears when the bifurcation parameter stems from the flow of chemical substances, which is necessary for maintaining the system in non-equilibrium steady states. The dynamic behaviors of the model were characterized by the interplay between the following two limit cycles. One limit cycle can be induced by the introduction of flow term, which was called a flow-induced limit cycle (FLC). The other limit cycle appears to surround the FLC, two unstable foci, and a saddle, which was called a comprehensive limit cycle (CLC). The switching between FLC and CLC can be induced by a change in the bifurcation parameter. "Chaotic behaviors" that we observed in the analog computers looked chaotic switching between these two limit cycles. However, our model does not have either an additional geometric structure producing chaos as in the Rössler's system or an additional dynamical rule for changing the bifurcation parameter. One more variable was necessary for yielding deterministic Belousov–Zhabotinsky chaos. Therefore, I guessed the findings to be noise-induced chaos.

Christophe Letellier kindly pointed out that there are no descriptions about parameter values in our 1979 *Physics Letters A* paper. Honestly, I had not noticed this before he pointed it out. It is not good not to describe the parameter values used in computations. If I may remember our aim at that time, we had concentrated only on finding chaotic behaviors, which looked like deterministic chaos even in mathematical models. This may be a reason why we were careless about the parameter values. Finally, I gave up making a continuous model for deterministic chaos of the Belousov–Zhabotinsky reaction. Instead, I concentrated on finding mathematical structures showing the existence of deterministic chaos in the experimental data.

I then tried to draw one-dimensional plots from the figures in the paper published by Jack Hudson.[10] This was a very "analog" style of data analysis, but I obtained very nice one-dimensional maps. Just looking at the empirically obtained one-dimensional maps for around a week, I could clearly imagine a mathematical function describing those empirical maps, that should be a class of gamma functions. By introducing a bifurcation parameter, which plays a role in simply shifting a map, I obtained a one-parameter family of one-dimensional discrete maps for the Belousov–Zhabotinsky reaction. I performed computer simulations of my map model with a digital computer (a FACOM-V) and confirmed all dynamic states observed in Hudson's experiment. Furthermore, I could predict numerically the presence of other periodic and chaotic states in addition to the experimentally observed ones. I also investigated the generic changes of the structures of dynamical orbits, and found a similar structure to the Arnold tongue in the form of Farey series, whereby I succeeded in describing bifurcation structures in the Belousov–Zhabotinsky reaction.[13]

We wrote a letter to Jack Hudson to inform him of these findings and to obtain his permission to use his figures in our paper. Jack kindly agreed to all these requests. We then published a paper.[13] Within half a year after that, Jack confirmed my findings and predictions in his further experiments.[17] The correspondence between computed chaos and experimental chaotic behaviors became consistent. I was convinced that deterministic chaos actually exists in the Belousov–Zhabotinsky reaction system. However, the nonlinear response of chaos to external noise was still questionable. I invited Kenji Matsumoto, a graduate student in Tomita's laboratory at that time who was three years younger than me, to perform digital computer simulations of my discrete Belousov–Zhabotinsky map with noise, as we shared our interests in chaotic dynamical systems from an information theoretical point of view and he was talented in computer programming and digital computations. Finally, we confirmed the robustness of our Belousov–Zhabotinsky chaos for external noise in the case of a small amplitude of noise. On the other hand, in the case of a large amplitude of noise, chaos disappears and periodicity became dominant, which we called noise-induced order.[18] The appearance of noise-induced order was confirmed by the appearance of a negative largest Lyapunov exponent, drastic reduction of KS entropy, the appearance of a steep peak in the power spectrum, and a split in dynamical orbits. Thirty-four years after this finding, Galatolo *et al.* succeeded in rigorously proving the existence of noise-induced order in my Belousov–Zhabotinsky map.[19]

3. Cantor Coding and Chaotic Memory in Brain

Otto accepted even my provocative paper about cerebral hermeneutics[20] with his insightful comments. I was also stimulated by his thoughts on endophysics[21] founded by David Finkelstein, as I considered the framework to capture both observers and the rest of the world in a sense of cerebral hermeneutics. When the brain observes environments that may be chaotic, the brain dynamics representing an optimized recognizer could be contracting, thereby adapting to a chaotically changing environment. I wanted to know what the product of such an interplay should be. Just then, Otto invited me to join his paper on continuous but nowhere-differentiable attractors.[22] We extended such attractors to singular-continuous but nowhere-differentiable attractors, which imply continuous but nowhere-differentiable attractors on Cantor sets. Based on this joint paper, I applied this idea to neural networks. During research on this subject, I was able to describe the memory of sequential events such as episodic memory as an interplay between chaotic dynamics showing chaotic itinerancy[23,24] and hyperchaos,[3] and contracting dynamics, in the hippocampus, where Cantor sets may encode sequential memories.[25] This computational result, which has been observed in the rat experiment,[26] implies that the hippocampus plays a role in an internal observer, where we again meet Otto Rössler's deep thoughts.

References

1. O. E. Rössler, An equation for continuous chaos, *Physics Letters A*. **57**(5), 397–398 (1976). doi: 10.1016/0375-9601(76)90101-8.
2. O. E. Rössler, Chemical turbulence: Chaos in a simple reaction-diffusion system, *Zeitschrift für Naturforschung A*. **31**(10), 1168–1172 (1976). doi: 10.1515/zna-1976-1006.
3. O. E. Rössler, An equation for hyperchaos, *Physics Letters A*. **71**, 155–157 (1979).
4. O. E. Rössler, Continuous chaos: Four prototype equations, *Annals of the New York Academy of Sciences*. **316**, 376–392 (1979).
5. E. N. Lorenz, Deterministic nonperiodic flow, *Journal of the Atmospheric Sciences*. **20**, 130–141 (1963). doi: 10.1175/1520-0469(1963)020⟨0130:DNF⟩2.0.CO;2.
6. S. Smale, Differentiable dynamical systems. I diffeormorphisms, *Bulletin of American Mathematical Society*. **73**, 747–817 (1967).
7. D. Ruelle and F. Takens, On the nature of turbulence, *Communications in Mathematical Physics*. **20**, 167–192 (1971).
8. T.-Y. Li and J. A. Yorke, Period three implies chaos, *The American Mathematical Monthly*. **82**(10), 985–992 (1975).

9. R. A. Schmitz, K. R. Graziani, and J. L. Hudson, Experimental evidence of chaotic states in the Belousov–Zhabotinskii reaction, *The Journal of Chemical Physics.* **67**(7), 3040–3044 (1977). doi: 10.1063/1.435267.
10. J. L. Hudson, M. Hart, and D. Marinko, An experimental study of multiple peak periodic and nonperiodic oscillations in the Belousov-Zhabotinskii reaction, *The Journal of Chemical Physics.* **71**(4), 1601–1606 (1979). doi: 10.1063/1.438487.
11. J. Roux, A. Rossi, S. Bachelart, and C. Vidal, Experimental observations of complex dynamical behavior during a chemical reaction, *Physica D.* **2**(2), 395–403 (1981). doi: 10.1016/0167-2789(81)90018-X.
12. K. Tomita and I. Tsuda, Chaos in the Belousov-Zhabotinsky reaction in a flow system, *Physics Letters A.* **71**(5), 489–492 (1979). doi: 10.1016/0375-9601(79)90644-3.
13. K. Tomita and I. Tsuda, Towards the interpretation of Hudson's experiment on the Belousov-Zhabotinsky reaction: Chaos due to delocalization, *Progress of Theoretical Physics.* **64**(4), 1138–1160 (1980). doi: 10.1143/PTP.64.1138.
14. K. Wegmann and O. E. Rössler, Different kinds of chaotic oscillations in the Belousov-Zhabotinskii reaction, *Zeitschrift für Naturforschung A.* **33**(10), 1179–1183 (1978). doi: 10.1515/zna-1978-1010.
15. K. Showalter, R. M. Noyes, and K. Bar-Eli, A modified Oregonator model exhibiting complicated limit cycle behavior in a flow system, *The Journal of Chemical Physics.* **69**(6), 2514–2524 (1978). doi: 10.1063/1.436894.
16. R. J. Field and R. M. Noyes, Oscillations in chemical systems. IV. Limit cycle behavior in a model of a real chemical reaction, *The Journal of Chemical Physics.* **60**(5), 1877–1884 (1974). doi: 10.1063/1.1681288.
17. J. L. Hudson and J. C. Mankin, Chaos in the Belousov–Zhabotinskii reaction, *The Journal of Chemical Physics.* **74**(11), 6171–6177 (1981). doi: 10.1063/1.441007.
18. K. Matsumoto and I. Tsuda, Noise-induced order, *Journal of Statistical Physics.* **31**, 87–106 (1983). doi: 10.1007/BF01010923.
19. S. Galatolo, M. Monge, and I. Nisoli, Existence of noise-induced order, a computer aided proof. arXiv:1702.07024 (2017).
20. I. Tsuda, A hermeneutic process of the brain, *Progress of Theoretical Physics Supplement.* **79**, 241–259 (1984). doi: 10.1143/PTPS.79.241.
21. O. E. Rössler, *Endophysics.* World Scientific (1998). doi: 10.1142/3183.
22. O. E. Rössler, C. Knudsen, J. L. Hudson, and I. Tsuda, Nowhere-differentiable attractors, *International Journal of Intelligent Systems.* **10**(1), 15–23 (1995). doi: 10.1002/int.4550100104.
23. I. Tsuda, Chaotic itinerancy as a dynamical basis of hermeneutics in brain and mind, *World Futures.* **32**(2–3), 167–184 (1991). doi: 10.1080/02604027.1991.9972257.
24. K. Kaneko and I. Tsuda, Chaotic itinerancy, *Chaos.* **13**(3), 926–936 (2003). doi: 10.1063/1.1607783.

25. I. Tsuda, Toward an interpretation of dynamic neural activity in terms of chaotic dynamical systems, *Behavioral and Brain Sciences.* **24**(5), 793–848 (2001). doi: 10.1017/s0140525x01000097.
26. Y. Fukushima, M. Tsukada, I. Tsuda, Y. Yumaguti, and S. Kuroda, Spatial clustering property and its self-similarity in membrane potentials of hippocampal CA1 pyramidal neurons for a spatio-temporal input sequence, *Cognitive Neurodynamics.* **1**, 305–316 (2007). doi: 10.1007/s11571-007-9026-9.

https://doi.org/10.1142/9789819823130_0018

Chapter 18

IN RESONANCE WITH CHAOS

Ulrich Parlitz

University of Göttingen, Germany
ulrich.parlitz@ds.mpg.de

In this article, I describe my personal *route to chaos*, including brief sketches of the scientific topics I worked on during my early career as a "nonlinear dynamist," as well as important encounters and interactions with various teachers, colleagues, and friends.[a]

1. Introduction

In 1978 I started studying physics at the University of Göttingen and in 1982 I was looking for an interesting topic for my diploma thesis. In those days I read popular science books on synergetics, self-organization and evolution theory written by Hermann Haken,[2] Ilya Prigogine & Isabelle Stengers[3] and Manfred Eigen & Ruthild Winkler[4,5] and I was fascinated by these new emerging fields, because they addressed very fundamental questions of human live and existence, on how structure comes into being, how units of increasing complexity and functionality arise as a consequence of natural dynamical laws, and in which sense all these processes can be predicted (or not). When talking to fellow students about these topics and the question whether one could work on them in Göttingen, I got the hint that Werner Lauterborn works on nonlinear dynamics and chaos and I should talk to him about writing a diploma thesis in his group. I did so,

[a]This chapter is an extended version of a similar contribution to an article on the history of dynamical systems theory by Letellier *et al.*[1]

and soon after, he suggested me to investigate the dynamics of a periodically driven Duffing oscillator

$$\ddot{x} + d\dot{x} + x + x^3 = f\cos(\omega t). \quad (1)$$

Werner was primarily interested in nonlinear resonances he found in his pioneering work on acoustically driven gas (cavitation) bubbles in a liquid[6] and one of my first tasks was to look for such phenomena in the parameter space of the Duffing oscillator. Experimentally, it was shown by him and Eckehart Cramer in 1981[7] that bubble dynamics can exhibit period-doubling cascades to chaos, in this context also called *acoustic cavitation noise*. Therefore, searching for chaos was also on my to-do list. It was known[8] that chaotic attractors for the Duffing equation

$$\ddot{x} + d\dot{x} - x + x^3 = f\cos(\omega t) \quad (2)$$

with a double-well potential exist, because for $f = 0$ there is a saddle point in the state space and it had been proved that with driving ($f > 0$) a Smale horseshoe occurs in that system.[8] For the single-well oscillator (1) such theoretical results did (to our knowledge) not exist and therefore it was not clear, whether this system could generate a chaotic attractor or not. This was the main motivation to investigate the Duffing oscillator (1) with a single symmetric potential well in more detail.

2. The Beauty and Richness of Bifurcation Sets

To study the dynamics of the Duffing oscillator (1) I ran countless numerical simulations, most of the time in the university computer center filling the queue of their main computer, a Sperry UNIVAC 1100/82, and using a VAX-11/780 for interactive exploration of the Duffing oscillator to learn more about its dynamics and how it changes when varying the driving frequency ω or the driving amplitude f. The simulation results were stunning. A surprisingly rich, complex but ordered structure of resonances, bifurcations and coexisting periodic and chaotic attractors emerged which fascinated both of us, Werner and me. Just by adding an innocent looking cubic term the periodically driven, damped harmonical oscillator I learned about already at school turned into a system exhibiting a plethora of beautiful complex but ordered self-similar structures in parameter space[9,10] as illustrated in Fig. 1. If this nonlinear system, which looks so simple,

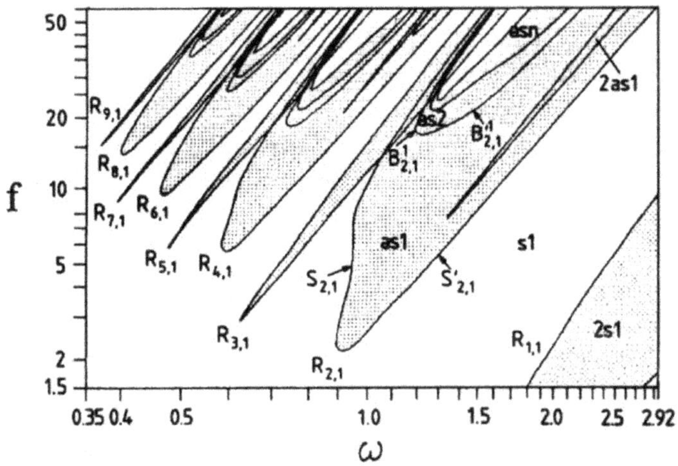

Fig. 1. Parameter space of the Duffing system (1). In the differently dotted domains there are symmetrical period-1 limit cycles (s1), two coexisting symmetry-related period-1 limit cycle (as1), two pairs of asymmetry related period-1 limit cycles (2as1), a pair of period-2 symmetry-related limit cycles (as2) and pairs of period-n symmetry-related limit cycles (as$n$) with $n = 4, 8, 16, \ldots$.

Source: Reproduced with permission from *Physics Letters A*. **107**, 351 (1985). Copyright 1985, Elsevier.

already produces such a wide variety of dynamic behavior, what would happen with more complex, higher dimensional systems?

With these results, I had the feeling after a short time that I had made the right decision, especially since there was a very stimulating, liberal and creative atmosphere in Werner Lauterborn's group.

This was also the time when I listened for the first time to a seminar talk given by Otto Rössler in the Institute for Theoretical Physics at the University of Göttingen. It was so impressive that I still remember the situation in the seminar room when Otto Rössler showed his slides with the taffy puller to explain the mechanism of stretching and folding underlying chaotic dynamics in state space. Werner also took us with him to attend the *Dynamics Days* first organized by Robert H. G. Helleman in Twente (1980–1986) and later by Gert Eilenberger in Düsseldorf where we became immediate whitnesses of the mood of departure in Nonlinear Dynamics at that time. Inspired by Feigenbaum's work on iterated maps there was a huge interest to find generic phenomena and universal laws in nonlinear systems. These were new concepts, because for a long time nonlinear systems were considered to be that much different from each other that there was no hope to make any general statements (in contrast to linear systems where a closed theory exists). The *Dynamics Days* were also a place where you could see the broad range of nonlinear phenomena and applications in diverse scientific fields including not only physics, but also chemistry, biology, geosciences and more.

3. Linking Resonances to Torsion

After I had finished the diploma thesis on superstructure in the bifurcation set of the Duffing oscillator[9] I continued with my Ph.D. studies in Werner's group and delved deeper into the dynamics of periodically driven oscillators. How could one characterize and label the nonlinear resonances and bifurcation curves seen in the Duffing equation (1), the periodically driven bubbles,[6] and other nonlinear oscillators? We knew the dynamics of the circle map and the theory of Arnold' tongues but this was applicable only to self-sustained (or self-excited) systems like the van der Pol oscillator, where the (asymptotic) dynamics takes place on an invariant torus in state space. Such a torus, however, does not exist in the case of the Duffing equation and similar "passive" oscillators where Hopf bifucations can be excluded and where any trajectory asymptotically reaches a steady (resting) state if the oscillator is not externally driven. Therefore, the question was, where can we find a second "rotational motion" in addition to the periodic driving in order to compute and analyse their frequency ratio? The solution was to consider the winding of neighbouring trajectories around the periodic orbits and to quantify this motion by torsion numbers.[11–14] Studies of several other periodically driven nonlinear oscillators similar to the Duffing equation (but with different nonlinear restoring forces) revealed a general characteristic pattern of resonances and bifurcation curves in parameter space that can be comprehensively labeled by torsion numbers.[15] The (average) local torsion frequency Ω (computed, like Lyapunov exponents, with linearized equations of motion) can also be used to define a winding number $w = \Omega/\omega$ that fulfills (like torsion numbers[16,17]) particular recursive schemes in period-doubling cascades and is also defined for chaotic attractors of this class of systems.[10] This approach was not only applied to the Duffing oscillator, a model for a periodically driven gas bubble[18] and many other passive nonlinear oscillators but also to the periodically forced van der Pol oscillator[19]

$$\ddot{x} + d(x^2 - 1)\dot{x} + x = f\cos(\omega t). \quad (3)$$

The van der Pol oscillator is (in contrast to the Duffing equation) a self-sustained oscillator generating for $f = 0$ a stable limit cycle.[20] With external forcing ($f > 0$) an invariant torus in state space may occur,[21] where tangent space based computation of winding numbers was shown to coincide with the direct approach referring to orbits on the torus.[19] The main motivation for our study of the van der Pol oscillator was, however, the fact that it was known since the seminal analytical work of Mary-Lucy Cartwright and John Edensor Littlewood[22] in 1945 that this system may exhibit aperiodic oscillations,[23] but we could not find in the literature any numerically computed example of a chaotic attractor for Eq. (3). In fact, it took some detailed numerical simulations until we found a complete period-doubling cascade to chaos.[19]

In summer 1986 I had the opportunity to participate in the *International Conference on Nonlinear Acoustics* in Toronto and after that I visited in a round trip Albert Libchaber (Chicago), Harry Swinney (Austin), and Gottfried Mayer-Kress at the Center for Nonlinear Studies in Los Alamos. These researchers and their research groups were leaders in chaos research, and I was very grateful for the hospitality and the very interesting and stimulating conversations we had.

4. From Resonances to Synchronization

After receiving my Ph.D. in 1987, I did my civilian service and moved in 1989 to Darmstadt (Germany), where a new DFG Collective Research Center on "Nonlinear Dynamics" was founded. There, I started to work on nonlinear time series analysis, e.g. for estimating Lyapunov exponents from time series. In 1992 I spent four months at the Santa Fe Institute (New Mexico, USA), where I worked with Gottfried Mayer-Kress and enjoyed the very inspiring atmosphere of multidisciplinary complexity research. At that time new results on discrete wavelet transforms due to I. Daubechies and others stimulated also research in nonlinear dynamics and we aimed at using them for analyzing and predicting spatio-temporal time series.[24] Later, in August 1992 I participated in the second Bryn Mawr *Workshop on Measures of Complexity and Chaos*, where I learned about the latest results of chaos research in the US. Immediately after the conference I spent one month in Leon Chua's lab at UC Berkeley, a very lively spot with many visitors from abroad giving lectures or even spending some time in the lab working on nonlinear dynamics in electronic circuits. This was the time after the iron curtain was lifted and therefore scientist from Eastern Europe and the former Soviet Union used the opportunity to visit and work in Leon Chua's lab and to present there their research. One of them was Ljupco Kocarev, a visiting scholar from St. Cyril and Methodius University (Skopje, North Macedonia, former Yugoslavia). We immediately identified common scientific interests and since I heard in Bryn Mawr about recent results on chaos synchronization by Lou Pecora and Tom Carroll[25] we decided to jointly explore synchronization and its potentional applications in electronic circuits. This work with Ljupco turned out to become a very fruitful collaboration. First, in Leon Chua's lab we tried to exploit the idea of using chaotic dynamics for encrypting messages and synchronization of chaos for decrypting them,[25–28] an idea which became very popular in the 1990s. Many schemes for "secure communication" were devised but it was also shown how to attack them using methods from nonlinear time series analysis like delay embedding, for example. In retrospective one might say that this work was full of enthusiasm but also a little bit naive and partly ignoring a lot of work done on cryptography before (also exploiting essential features of chaos, but with different names like *confusion* and *diffusion*). After our common work in Berkeley Ljupco and I continued and further intensified our cooperation (sending emails and fax back and forth between Skopje and Göttingen). One of the most exciting topics we worked on was *generalized synchronization*[29–33] which may occur with *different* dynamical systems (for example, a Rössler system driving a Lorenz system[32,34]) as well as identical systems[35] that are unidirectionally coupled, and plays an important role, for example, in reservoir computing.[36–38] We also studied chaotic phase synchronization, synchronization due to sporadic coupling,[33,39] synchronization of spatially extended systems,[40,41] and many other aspects of synchronization. Here very often the Rössler model turned out to be the ideal example, because of its simple structure and since in allows to introduce a phase variable required for studying chaotic phase synchronization on digital[42] and analog[43] computers. In 1997 this research culminated in my habilitation thesis on *Synchronization of uni-directionally coupled chaotic dynamical systems*. After receiving the Venia Legendi at the University of Göttingen I continued my research there and in 2002 and 2003 I was invited by Henry D.I. Abarbanel to spend a few months at the Institute for Nonlinear Science at UC San Diego. There I had the opportunity to work with outstanding colleagues in a very lively research atmosphere on problems of parameter estimation from time series.

In 2010 Stefan Luther offered me a position in the Research Group Biomedical Physics at the Max Planck Institute for Dynamics and Self-Organization, Göttingen. There I was (and am) involved in fascinating research on cardiac dynamics. The scientific questions range from understanding the complex spatio-temporal dynamics of electromechanical excitation waves in the heart during cardiac arrhythmias to new concepts for controlling and terminating such potentially lethal conditions.[44] The main subject of this research are various types of stable or unstable spiral waves — a topic addressed by O. E. Rössler and C. Kahlert[45] on the basis of the pioneering work V.I. Krinsky[46] and A. T. Winfree,[47] as early as 1979!

Acknowledgments

I thank all fellow students, teachers, colleagues, and friends for many interesting discussions, support, and for making research in nonlinear science a real adventure that I enjoy(ed) very much.

References

1. C. Letellier, R. Abraham, D. L. Shepelyansky, O. E. össler, P. Holmes, R. Lozi, L. Glass, A. Pikovsky, L. F. Olsen, I. Tsuda, C. Grebogi, U. Parlitz, R. Gilmore, L. M. Pecora, and T. L. Carroll, Some elements for a history of the dynamical systems theory, *Chaos.* **31**, 053110 (2021).
2. H. Haken, *Erfolgsgeheimnisse der Natur — Synergetik: Die Lehre vom Zusammenwirken.* Deutsche Verlags-Anstalt, Stuttgart (1981).
3. I. Prigogine and I. Stengers, *Dialog mit der Natur — Neue Wege naturwissenschaftlichen Denkens.* R. Piper & Co. Verlag, München Zürich (1981).
4. M. Eigen and R. Winkler, *Das Spiel — Naturgesetze steuern den Zufall.* R. Piper & Co. Verlag, München Zürich (1975).
5. M. Eigen and R. Winkler, *The Laws of the Game: How The Principles of Nature Govern Chance.* Princeton University Press (1975).
6. W. Lauterborn, Numerical investigation of nonlinear oscillations of gas bubbles in liquids, *The Journal of the Acoustical Society of America.* **59**(2), 283–293 (1976). doi: 10.1121/1.380884.
7. W. Lauterborn and E. Cramer, Subharmonic route to chaos observed in acoustics, *Physical Review Letters.* **47**(20), 1445–1448 (1981). doi: 10.1103/PhysRevLett.47.1445.
8. J. Guckenheimer and P. J. Holmes, *Nonlinear Oscillations, Dynamical Systems, and Bifurcations of Vector Fields.* Applied Mathematical Sciences, Vol. 42. Springer-Verlag, New York (1983).
9. U. Parlitz and W. Lauterborn, Superstructure in the bifurcation set of the Duffing equation $\ddot{x} + d\dot{x} + x + x^3 = f\cos(\omega t)$, *Physics Letters A.* **107**(8), 351–355 (1985). doi: https://doi.org/10.1016/0375-9601(85)90687-5.
10. U. Parlitz, Common dynamical features of periodically driven strictly dissipative oscillators, *International Journal of Bifurcation & Chaos.* **03**(03), 703–715 (1993). doi: 10.1142/S0218127493000611.

11. Y. Aizawa and T. Uezu, Topological aspects in chaos and in 2^k period-doubling cascade, *Progress in Theoretical Physics.* **67**(3), 982–985 (1982).
12. Y. Aizawa and T. Uezu, Topological aspects in chaos and in 2k-period doubling cascade, *Progress of Theoretical Physics.* **67**(3), 982–985 (1982). doi: 10.1143/PTP.67.982.
13. T. Uezu, Topology in dynamical systems, *Physics Letters A.* **93**(4), 161–166 (1983). doi: 10.1016/0375-9601(83)90038-5.
14. U. Parlitz and W. Lauterborn, Resonances and torsion numbers of driven dissipative nonlinear oscillators, *Zeitschrift für Naturforschung A.* **41**(4), 605–614 (1986). doi: 10.1515/zna-1986-0404.
15. C. Scheffczyk, U. Parlitz, T. Kurz, W. Knop, and W. Lauterborn, Comparison of bifurcation structures of driven dissipative nonlinear oscillators, *Physical Review A.* **43**(12), 6495–6502 (1991). doi: 10.1103/PhysRevA.43.6495.
16. P. Beiersdorfer, J.-M. Wersinger, and Y. Treve, Topology of the invariant manifolds of period-doubling attractors for some forced nonlinear oscillators, *Physics Letters A.* **96**(6), 269–272 (1983). doi: 10.1016/0375-9601(83)90177-9.
17. P. Beiersdorfer, Universality of the topology of period doubling dynamical systems, *Physics Letters A.* **100**(8), 379–382 (1984). doi: 10.1016/0375-9601(84)90624-8.
18. U. Parlitz, V. Englisch, C. Scheffczyk, and W. Lauterborn, Bifurcation structure of bubble oscillators, *The Journal of the Acoustical Society of America.* **88**(2), 1061–1077 (1990). doi: 10.1121/1.399855.
19. U. Parlitz and W. Lauterborn, Period-doubling cascades and devil's staircases of the driven van der Pol oscillator, *Physical Review A.* **36**(3), 1428–1434 (1987). doi: 10.1103/PhysRevA.36.1428.
20. G. Datseris and U. Parlitz, *Nonlinear Dynamics.* Undergraduate Lecture Notes in Physics, Springer Nature, Cham, Switzerland (2022).
21. R. Mettin, U. Parlitz, and W. Lauterborn, Bifurcation structure of the driven van der pol oscillator, *International Journal of Bifurcation & Chaos.* **03**(06), 1529–1555 (1993). doi: 10.1142/S0218127493001203.
22. M. L. Cartwright and J. E. Littlewood, On non-linear differential equations of the second order, I, *Journal of the London Mathematical Society.* **20**(3), 180 (1945).
23. J.-M. Ginoux and C. Letellier, Van der Pol and the history of relaxation oscillations: Toward the emergence of a concept, *Chaos.* **22**(2), 023120 (2012). doi: 10.1063/1.3670008.
24. U. Parlitz and G. Mayer-Kress, Predicting low-dimensional spatiotemporal dynamics using discrete wavelet transforms, *Physical Review E.* **51**(4), R2709–R2711 (1995). doi: 10.1103/PhysRevE.51.R2709.
25. L. M. Pecora and T. L. Carroll, Synchronization in chaotic systems, *Physical Review Letters.* **64**(8), 821–824 (1990).

26. L. J. Kocarev, K. S. Halle, K. Eckert, and L. O. Chua, Experimental demonstration of secure communication via chaotic synchronization, *International Journal of Bifurcation & Chaos.* **2**(3), 709–713 (1992).
27. U. Parlitz, L. Chua, L. Kocarev, K. Halle, and A. Sshang, Transmission of digital signals by chaotic synchronization, *International Journal of Bifurcation & Chaos.* **02**(04), 973–977 (1992). doi: 10.1142/S0218127492000562.
28. K. M. Cuomo and A. V. Oppenheim, Circuit implementation of synchronized chaos with applications to communications, *Physical Review Letters.* **71**(1), 65–68 (1993). doi: 10.1103/PhysRevLett.71.65.
29. V. S. Afraimovich, N. N. Verichev, and M. I. Rabinovich, Stochastic synchronization of oscillation in dissipative systems, *Radiophysics and Quantum Electronics.* **29**, 795–803 (1986). doi: 10.1007/BF01034476.
30. N. F. Rulkov, M. M. Sushchik, L. S. Tsimring, and H. D. I. Abarbanel, Generalized synchronization of chaos in directionally coupled chaotic systems, *Physical Review E.* **51**(2), 980–994 (1995). doi: 10.1103/PhysRevE.51.980.
31. H. D. I. Abardanel, N. F. Rulkov, and M. M. Suschik, Generalized synchronization of chaos: The auxiliary system approach, *Physical Review E.* **53**, 4528–4535 (1996). doi: 10.1103/PhysRevE.53.4528.
32. L. Kocarev and U. Parlitz, Generalized synchronization, predictability, and equivalence of unidirectionally coupled dynamical systems, *Physical Review Letters.* **76**(11), 1816–1819 (1996). doi: 10.1103/PhysRevLett.76.1816.
33. U. Parlitz, L. Junge, and L. Kocarev, Subharmonic entrainment of unstable period orbits and generalized synchronization, *Physical Review Letters.* **79**(17), 3158–3161 (1997). doi: 10.1103/PhysRevLett.79.3158.
34. U. Parlitz, Detecting generalized synchronization, *Nonlinear Theory and Its Applications, IEICE.* **3**(2), 113–127 (2012). doi: 10.1587/nolta.3.113.
35. U. Parlitz, L. Junge, and L. Kocarev, Nonidentical synchronization of identical systems, *International Journal of Bifurcation & Chaos.* **09**(12), 2305–2309 (1999). doi: 10.1142/S0218127499001796.
36. H. Jaeger, A tutorial on training recurrent neural networks, covering bppt, rtrl, ekf and the "echo state network" approach, *GMD Report.* **159**, 1–48 (2002).
37. H. Jaeger and H. Haas, Harnessing nonlinearity: Predicting chaotic systems and saving energy in wireless communication, *Science.* **304**(5667), 78–80 (2004). doi: 10.1126/science.1091277.
38. U. Parlitz and A. Hornstein, Dynamical prediction of chaotic time series, *Chaos and Complexity Letters.* **1**(2), 135–144 (2005).
39. T. Stojanovski, L. Kocarev, and R. Harris, Applications of symbolic dynamics in chaos synchronization, *IEEE Transactions on Circuits and Systems I.* **44**(10), 1014–1018 (1997). doi: 10.1109/81.633892.
40. L. Kocarev, Z. Tasev, T. Stojanovski, and U. Parlitz, Synchronizing spatiotemporal chaos, *Chaos.* **7**(4), 635–643 (1997). doi: 10.1063/1.166263.

41. L. Junge, U. Parlitz, Z. Tasev, and L. Kocarev, Synchronization and control of spatially extended systems using sensor coupling, *International Journal of Bifurcation & Chaos.* **09**(12), 2265–2270 (1999). doi: 10.1142/S0218127499001735.
42. M. G. Rosenblum, A. S. Pikovsky, and J. Kurths, Phase synchronization of chaotic oscillators, *Physical Review Letters.* **76**(11), 1804–1807 (1996). doi: 10.1103/PhysRevLett.76.1804.
43. U. Parlitz, L. Junge, W. Lauterborn, and L. Kocarev, Experimental observation of phase synchronization, *Physical Review E.* **54**(2), 2115–2117 (1996). doi: 10.1103/PhysRevE.54.2115.
44. T. Lilienkamp, U. Parlitz, and S. Luther, Taming cardiac arrhythmias: Terminating spiral wave chaos by adaptive deceleration pacing, *Chaos.* **32**(12) (2022). doi: 10.1063/5.0126682.
45. O. E. Rössler and C. Kahlert, Winfree meandering in a 2-dimensional 2-variable excitable medium, *Zeitschrift für Naturforschung A.* **34**(5), 565–570 (1979). doi: 10.1515/zna-1979-0507.
46. V. I. Krinskii, Spread of excitation in an inhomogeneous medium (state similar to cardiac fibrillation), *Biophysics.* **11**, 776–784 (1966).
47. A. T. Winfree, Scroll-shaped waves of chemical activity in three dimensions, *Science.* **181**(4103), 937–939 (1973). doi: 10.1126/science.181.4103.937.

https://doi.org/10.1142/9789819823130_0019

Chapter 19

REMINISCIENCE OF CHAOS

Celso Grebogi

University of Aberdeen, Scotland, UK
grebogi@abdn.ac.uk

It is not a turning point, but a look back merrily to see how much of my path I have already trodden, asking myself if it still leads upwards. I think of the time I lived through, since my young days, working on my father's corn mill and attending rudimentary schools, except for the better gymnasium (5–8th grades). I look back at my 77 years as I look now from my house in Aberdeen with the river gently flowing away. I must admit that it would be perverse of me not to feel grateful. After all, I had been given, far, far more than I hoped or expected. My wish to work on research and to express myself through writing works of some scientific merit had been granted beyond my wildest childhood dreams. My works and ideas have reached perhaps one hundred thousand readers, fifty thousand works by others refer to my work. My existence had unfathomable influence beyond the confines of my own life. I had made friends with many the finest people, mathematicians and scientists of my time. I had seen and enjoyed wonderful artistic performances, immortal cities and pictures, and the most beautiful landscapes in the world. I have remained free, independent of official positions, my work has been a pleasure to me, and I trust, it has given pleasure to others. By the by, CHAOS has amazingly enlightened me that the sequence of "roads not taken" has steered me to where I am now.

1. Beginning: From General Relativity to Plasma Physics

The awakening to physics emerges for most youngsters when still young, fascinated by the cosmos while looking at the night sky, by learning about the solar system, by imagining both the immensity of the universe and the tininess of molecules that fills a room. That was not different with me. However, no university physics degree course was offered in Curitiba, the Brazil city in which I grew up. I chose then to study chemical engineering at the Federal University of Paraná because that faculty had the best physics and mathematics professors. Their particular research interest was in relativity, a subject which, thank the stars, addressed questions about the cosmos and the origins of the universe. Those professors got mostly their doctor degrees, at the time called Doctorat d'État, at the Institute Henri Poincaré in Paris under Marie-Antoinette Tonnelat. As a result of this special liaison, not only Tonnelat would come to the university to deliver seminars and short courses in relativity, but also others like Achilles Papapetrou from the same Institute and André Lichnerowicz from the École Normale Supérieure. The initiation in relativity from those renown relativists led me to the Pontifical Catholic University in Rio de Janeiro as a postgraduate student in 1971. Besides the standard physics course, I continued my studies in relativity.[1] In 1973, John Wheeler came to deliver a stimulating seminar based on his just published *Gravitation* book with Charles Misner and Kip Thorne.[2]

Right after, motivated by that seminar, I applied to the University of Maryland for a Ph.D. degree in theoretical physics, having Misner as my supervisor. I was in his group, being even his teaching assistant, until 1976, when Misner invited me to his office, enquiring about my plans after my Ph.D. degree. He put forward the non-importance of relativity as a field for the development of a country, suggesting that he was not going to supervise Americans any longer, but foreigners with the understanding that they should return to their countries after their Ph.D. degrees. I thought, then, to be the time to bail out of relativity. I joined the recently arrived and most influential plasma physics professor from Princeton, Chuan-Sheng Liu, working on magnetic confinement and laser induced thermonuclear fusion, the area in which I got my Ph.D. degree. That career choice change became pivotal for me to eventually get into chaos.

In mid-seventies, while working on my Ph.D. thesis in thermonuclear fusion, I took a course in the qualitative theory of differential equations with a visiting mathematician. In it, we discussed solutions of a few classical equations. For a given equation with a set of functional terms and a fixed set of parameters, a solution was presented, arguing about its existence and uniqueness. The functional terms and/or parameters were changed in the same equation, the whole procedure was repeated, and that went on for many weeks about that equation. Then, another equation was considered, and everything was repeated all over again. It was a cautious, abstract four-month long course on specific differential equations. As I was about to discover soon after, in that course there was none of the bold, intuitive philosophical generalisations that James Clerk Maxwell,[3] a physicist, and Henri Poincaré,[4,5] a mathematician, felt to be justified. Both understood the importance of systems having sensitive dependence on initial data, the kind of dynamics that is vibrant, compelling and exciting.

2. The Berkeley Years: The Theory of Dynamical Systems

In 1978–1981, I became a postdoc in Berkeley under Allan Kaufman. He typically had one postdoc at a time and very few Ph.D. students. I was the first one that he kept for three years. During that time, a few markedly important events occurred related to chaos. Still during the Soviet times, Boris Chirikov came from Novosibirsk to visit Kaufman in the Autumn of 1978. He brought and left with us a preprint copy of his seminal work, "*A universal instability of many-dimensional oscillator systems*," published in 1979 as a *Physics Reports*.[6] With Chirikov preprint on hand, Kaufman organised a discussion group, three hours every Thursday afternoon, initially to study Chirikov's paper, later to go over V. I. Arnol'd's recently published book[7] on symplectic dynamics. In the discussion group, there were we — Kaufman's group, his former students working then at the Lawrence Livermore Laboratory on the mirror machine confinement, and some mathematicians from the Berkeley mathematics department. The latter ones were really important because we learned from them the fundaments of the theory of dynamical systems and ergodic theory, necessary to embark in this new science. They included Oscar Lanford, Jerrold Marsden and Allan Weinstein. They not only helped to clarify abstract concepts, but they offered to give semester long courses to us. The learning of a new science, chaotic dynamics, supported by both the ergodic theory and the theory of dynamical systems, was the most exciting aspect of the multiple-year discussions. Stephen Smale, the author of the horseshoe map, typically sitting in his office by himself, did not participate in our discussions, though I spoke with him a few times about his rare crystal collection, which included rare Brazilian pieces. Similarly, both Allan Lichtenberg and Michael Lieberman at the Berkeley electrical engineering department did not participate in our group discussions, though I interacted with them rather a lot at the time when they were drafting their book on

Regular and Chaotic Dynamics.[8] Since they both were engineers, I learned from those interactions about the applicability and interdisciplinarity of the research in chaotic dynamics.

Allan Kaufman was, and still is, a scholar in the full sense of its meaning. He would spend hours in the library, consulting all kinds of books. He especially enjoyed having blackboard discussions with us about dynamical systems, statistical mechanics, and plasma physics. He would bring to discussions the newly learned chaotic dynamics and ergodic theory, then use them in his masterful courses in statistical mechanics and plasma theory, including his symplectic phase space wave theory formalism.[9] Regarding research work and, say, student's projects, he would have some gut feeling idea, often vague, about an important problem, and that would become a student Ph.D. project, allowing subsequently the student to work on the project by himself until its completion. At that time, he asked Robert Littlejohn, to look at the guiding centre problem, a charged particle moving in a magnetic field. It is a very hard problem. Littlejohn developed a geometric noncanonical Hamiltonian formalism to tackle the problem. His work[10] was of such originality and depth that it led to his hiring as a Berkeley physics professor. Motivated by Littlejohn's work on the guiding centre theory, I embarked to investigate the classical Lev Landau problem,[11] the ponderomotive effect of an electromagnetic wave,[12] but using the new noncanonical Hamiltonian formulation combined with Lie transforms[13] and with the help of the newly rediscovered Darboux Transformation.[14] This led to a subsequent work on the relativistic formulation with Littlejohn.[15] The generality and the beauty of the symplectic phase space wave theory formalism allowed me to apply it to many practical problems such as beam propagation and free-electron lasers.[16]

Motivated by the studying of the Chirikov preprint, Kaufman asked the student Steven McDonald to solve the Helmholtz equation in the chaotic Bunimovich stadium. Their seminal work, "*Spectrum and Eigenfunctions for a Hamiltonian with stochastic trajectories*" published in *Physical Review Letters* in 1979,[17] opened up a new interdisciplinary research field, quantum chaos. Kaufman–McDonald uncovered in their work that, based on a semi-classical quantum approximation, the energy levels in a chaotic system repel each other, contrary to what occurs in integrable systems. About the same time, Sir Michael Berry, came to Berkeley to deliver the physics colloquium, a colloquium typically attended by five Nobel prize winners. He spoke about his work on the swimming pool hot spots and on the twinkling of the stars,[18] perhaps the two most important examples that can be understood by employing the catastrophe theory of René Thom.[19] It was a fascinating talk that stimulated me to read Thom's catastrophe work. We tried to apply the theory to particle and wave propagation but without much success. While in Berkeley, Berry took advantage to visit Kaufman's group, being told then about Kaufman–McDonald's work on the stadium. Berry promptly realised that the distribution of energy levels followed a Wigner distribution instead of the typical Lorentzian of integrable systems. Another important consequence from Kaufman–McDonald's work on the stadium was that, if one takes the figures from the simulations and look at them side-ways, we immediately realised that the wavefunctions concentrate on the classical periodic orbits embedded in the chaotic dynamics. This phenomenon was named as quantum scars by Eric Heller[20] many years later. To grasp the many "first" in physics by Kaufman, it is a delight to read his autobiographical memoir as a plasma theorist, in response to the 2007 Kaufmanfest to honour his eightieth birthday.[9]

3. Maryland: Chaos Theory

In the Autumn of 1981, I moved back to the University of Maryland, where another chapter in chaotic dynamics was about to take place. Upon arriving in Maryland, I delivered a course on symplectic dynamics and Lie transforms at a Navy lab. That invitation came from Robert Cawley, who felt that the theory of dynamical systems was the way of the future. There I met Louis Pecora,. who, with the support of Michael Shlesinger from the Office of Naval Research, who founded the journal *Fractals* in 1993, built the major Navy dynamics group at the Naval Research Laboratory. As part of that course, I invited James Yorke to deliver a seminar as a guest. I have never met him before, though I saw him walking on campus around 1977. I was slightly aware of his work while at Kaufman's group. That encounter in 1981 was actually the first time I met him. After his talk, we sat on the stairs in front of the building, chatting about his mathematical work on dynamical systems, and about the naming of "chaos" for the phenomenon that I knew, that in his 1975 paper, "*Period three implies chaos.*"[21] I became aware then the importance of giving catchy and proper names to describe new phenomena in science; of course, we know that some of the names will last while others will fade away. The name chaos sounded unbelievably stunning and remarkable to me. I should mention that the science and technology communities used other names for the same phenomenon at the time. They included (internal) stochasticity, irregular or aperiodic motion.

Chaos eventually became the official name in mathematics, sciences and technology.

That initial conversation Yorke was the beginning of a two-decade long collaboration, involving the renowned physicist, Edward Ott, resulting in well over one hundred papers on the fundaments of chaotic dynamics in such a collaboration. We identified, conjectured and showed the occurrence of many basic phenomena in chaotic dynamics, such as crisis,[22] fractal basin boundaries,[23] final state sensitivity,[24] strange nonchaotic attractors,[25] fat fractals,[26] metamorphosis,[27] multifractality,[28] shadowing,[29] chaotic scattering,[30] controlling chaos,[31] antimonotonicity,[32] communicating with chaos,[33] controlling complexity,[34] riddling,[35] multi-stability,[36] chaotic advection,[37] robust chaos,[38] and modelling of complex systems in presence of unstable-dimension variability.[39] Those are some of the fundamental work on chaotic dynamics that I was directly involved in the previous century. Our work, grounded on the theory of dynamical systems and ergodic theory, and often argued in terms of point set topology, were developed with the use of mathematical maps and differential equations. The latter, typically the pendulum equation[27]

$$\ddot{x} + \nu\dot{x} + \omega^2 \sin x = f \cos t \tag{1}$$

(a Josephson junction model) was usually employed to argue that the phenomenon we were addressing was not particular to a mathematical framework but it was pervasive in science and technology. In fact, the objective of the research was to establish basic mathematical principles so that researchers could then apply those principles to understand and analyse the systems they were investigating in their own fields. There, we also contributed towards the formation of major researchers in chaotic dynamics, who became influential group leaders on their own right. My collaboration with Yorke and Ott was, perhaps, the longest-lasting close collaboration of three people working on dynamical systems. We would often have lunch together at a university cafeteria and, after the lunch, we would continue the informal and relaxed conversations in Yorke's office for another couple of hours. The success of the collaboration was grounded on keeping personalities aside, and in having, to a certain extent, complementary expertise that led to the advancing research work. Because of this successfully collaboration, our group was ranked number 1 in the US by the National Research Council.

Among others, the Shadowing proof was very important. There was a lot of pressure by people from all sides, especially mathematicians, claiming that the numerical work done on chaotic systems was nonsense because the numerical trajectories did not have anything to do with the true dynamical trajectories, so the arguments and validations were based on non-valid trajectories. Before our proofs on Shadowing in non-hyperbolic dynamics, people would claim the *Shadowing Lemma*[40–42] to justify their work, but clearly the Lemma was not valid for the systems that people were working on. The important push in chaotic dynamics was to understand non-hyperbolic systems, the systems that occur in nature and man-made applications. Rössler system is such a one, though it is a differential one, it has the advantage of being algebraic and of easier computation. We developed a computer-aided rigorous proof of the shadowing property for non-hyperbolic systems. I developed this work with James Yorke and postdocs. Preliminary results were published in 1987[43] and then the mathematical proof in the highest raking *Bulletin of the American Mathematical Society* in 1988.[44] This result was reported in *Science News*.[45]

Visualisation was a major component in the early scientific developments of chaotic dynamics. It was essential to be able to draw pictures of attractors, basins of attraction, and other invariant sets on a sheet of paper or project them on a screen, both two-dimensional media, so that one could visualise and analyse its topological structures and other features. In the late seventies and early eighties, computer capabilities in terms of capacity, computation speeds and graphics were at their infancy. It was difficult, or often not doable, to carry out more intricate calculations in order to help visualisation and understanding, and to validate the theories and predictions, especially if it involved differential equations, such as the Rössler system or the pendulum equation.[46] When President Reagan came to the US Presidency in 1981, he launched the *Star Wars Programme* with the objective to create a defence shield to protect the country from incoming enemy missiles. It was an expensive Programme, but we decided that our mathematics could also benefit from it.

The justification that we gave was that we had an unmatched expertise in nonlinearity, especially chaotic dynamics, and that the *Star Wars Programme* would benefit tremendously from it. Furthermore, we argued that to investigate nonlinear chaotic systems, we needed to be able to draw pictures to visualise the topology of dynamical sets and their intricate fractal geometry. Happily, we were awarded close to US$ 1.5 million (equivalent to US$ 4.6 million today) from that Programme through the *Defense Advanced Research Projects Agency* (DARPA). We got the permission to buy a unique set of floating-point array processors (US$ 900K), market by Floating Point Systems, and a photographic camera (US$ 60K), as we required. We hired a technician from NASA to programme and to deal with the idiosyncrasies of that computer array. Therefrom, we were able to carry out computations, whose outputs were transferred straightforwardly from the array processors to the camera by burning an imprint in the film. For the lack of computer capacity, the film was where the outputs of the computations were stored, and it is where they are after so many years. Figure 1, showing the fractal basin boundary and the basins of attraction of a forced damped pendulum equation, is the result of such computations in that computer array. The pendulum equation involved a trigonometric function, which was time consuming to evaluate numerically, the computation of such a figure can be done in seconds on a smart phone today, but it took an overnight in that specialised computer array. Our pictures were exhibited at the *National Academy of Sciences*, in a museum in New York, and were part of a travelling exhibition throughout the United States. They were also the covers of a dozen of mathematical, scientific, and technical publications, including *Science*,[23] *IEEE Spectrum* (April 1987), *Science News*, *International Journal of Bifurcations & Chaos*,[47,48] and *Nonlinearity*.[49]

Fig. 1. Fractal basin boundary and the basins of attraction of the forced damped pendulum equation (1).

Notes: Parameter values: $\nu = 0.1$, $\omega = 1$, and $f = 2$.

During the two decades that I was at the University of Maryland, we held an interdisciplinary pizza lunch seminar. The event took two hours, starting with pizza and Chinese food lunch, followed by two twenty-minute talks, but each talk was accompanied by an unlimited number of questions and long discussions. The participants were people working on interesting dynamical problems in the Washington-DC area, including the statistical mechanics group of Michael Fisher, the space physics of Roald Sagdeev, the Richard Prange and Shmuel Fishman group in quantum chaos, Tim Sauer from George Mason University, Louis Pecora and Thomas Carroll from the Naval Research Laboratory, Steven Schiff from the Children's National Medical Centre, and colleagues from the mathematics department. Because of the world-wide recognition or our Maryland group, we did not have to invite, or even to offer a discretionary fee to prestigious speakers; people from all corners of the globe would come to visit us and talk at our seminars. For the same reason, young researchers, mainly from Europe and Brazil, would come for year-long visits. The Brazilian researchers were the ones responsible for implanting the successful research in chaotic dynamics in their country.

4. The World Perspective

Chaotic dynamics was being developed throughout the world, though Europe became the hotbed of chaotic dynamics, especially in the eighties and nineties. During that time, I came in contact with outstanding people doing major research work. I should mention two major groups, part of my experience, which was having a more lasting influence in chaotic dynamics. One if the ETHZ group of Ruedi Stoop. The other is the Potsdam group of Jürgen Kurths. I met him and Ulrike Feudel at an EQUADIFF conference in Szeged in 1988; I also met the distinguished physicist Tamas Tél in Budapest in the same trip. Those encounters led to close collaboration with them that lasts until today. Stoop and Kurths' groups have been playing a major role in the diffusion of nonlinear dynamics, not only in Europe but worldwide. Of major significance, while staying one year in Kurths's group in 1996–1997 as a result of being awarded the Senior Humboldt Prize, I came across Otto Rössler's work on dynamical systems, including his most famous Rössler differential equations system. Afterwards, many of our works, including those on model-based data analysis and data-based modelling, have been using the Rössler and Lorenz systems as paradigms to validate data analysis developments. The Rössler system is currently being employed in nonlinear dynamics research worldwide, but particularly in China, where the mathematics and science developments have been astounding. Another major group, also part of my current experience, is the one led by Ying-Cheng Lai at Arizona State University, which is certainly the most influential one in chaotic dynamics and complex network in the United States nowadays. That group has consistently been producing outstanding work across many areas of research in nonlinear dynamics and complex networks. I was fortunate to be part of a recent major breakthrough with his group, by opening up a new interdisciplinary research field,[50] relativistic quantum chaos, bringing together the pillars of the twenty-century physics: relativity, quantum mechanics, and chaotic dynamics.

In conclusion and of major significance, I met Aziz Alaoui at Le Havre-Normandie Université for the first time in 2009, at one of his conferences on complex systems and applications. The last time I returned to Le Havre was on a momentous occasion in which I shared the stage with Otto Rössler, James Yorke, and Guanrong Chen, when we all four were awarded the Doctor *Honoris Causa* degree from that University in 2014 (see Fig. 2).

A scientific career does not have to be orthodox and unwavering. I occasionally suggest to my students, including to my young and only son Mateus, that one does not have to be reluctant to make changes upon

Fig. 2. Ceremony for the *Honoris causa* academic degree. From the left to the right: Celso Grebogi, James Yorke, Otto E. Rössler and Guanrong Chen. Le Havre, 24th June 2014.

alternative career choices. It is alright to allow oneself to embrace, of one's own volition, divergent possibilities to fully develop one's potential and expectations. One of my earlier papers on chaos, *Chaotic Attractors in Crisis*,[22] articulates precisely that, the interconnection between possible alternatives and apparently contradictory possibilities, between the creation and the destruction, the complementary yin and yang to lifelong fulfilment.

References

1. S. Weinberger, *Gravitation and cosmology: Principles and applications of the general theory of relativity*. John Wiley, London (1972).
2. C. W. Misner, K. S. Thorne, and J. A. Wheeler, *Gravitation*. W. H. Freeman, New York (1973).
3. B. R. Hunt and J. A. Yorke, Maxwell on chaos, *Nonlinear Science Today*. **3**, 1–4 (1993).
4. F. Diacu and P. Holmes, *Celestial Encounters: The Origins of Chaos and Stability*. Princeton University Press (1996).
5. J. Barrow-Green, *Poincaré and the Three Body Problem*. History of Mathematics, Vol. 11. American Mathematical Society (1997).
6. B. V. Chirikov, A universal instability of many-dimensional oscillator systems, *Physics Reports*. **52**(5), 263–379 (1979). doi: 10.1016/0370-1573(79)90023-1.
7. V. I. Arnol'd, *Mathematical Methods of Classical Mechanics*. Graduate Texts in Mathematics, Vol. 60. Springer-Verlag, New York (1989). Translated by A. Weinstein and K. Vogtmann.
8. A. Lichtenberg and M. Lieberman, *Regular and chaotic dynamics*. Applied Mathematical Sciences, Vol. 38. Springer-Verlag (1992). doi: 10.1007/978-1-4757-2184-3. First edition in 1983.
9. A. N. Kaufman, A half-century in plasma physics, *Journal of Physics: Conference Series*. **169**, 012002 (2009). doi: 10.1088/1742-6596/169/1/012002.
10. R. G. Littlejohn, A guiding center hamiltonian: A new approach, *Journal of Mathematical Physics*. **20**(12), 2445–2458 (1979). doi: 10.1063/1.524053.
11. L. D. Landau and E. M. Lifshitz, *Théorie des champs*. Physique Théorique, Vol. 2. MIR (1960).
12. J. M. Greene, A method for determining a stochastic transition, *Journal of Mathematical Physics*. **20**(6), 1183–1201 (1979). doi: 10.1063/1.524170.
13. J. R. Cary, Lie transforms and their use in hamiltonian in perturbation theory. Technical Report DOE/ET-0074, National Technical Information Service, Sprigngfield, Virginia (1978).
14. G. Darboux, *Leçons sur la Théorie Générale des Surfaces*. Gauthier-Villars, Paris (1912).
15. C. Grebogi and R. G. Littlejohn, Relativistic ponderomotive Hamiltonian, *Physics of Fluids*. **27**(8), 1996–2004 (1984). doi: 10.1063/1.864855.

16. C. Grebogi, Relativistic ponderomotive Hamiltonian of two interacting electromagnetic waves, *Physical Review A.* **31**(2), 914–920 (1985). doi: 10.1103/PhysRevA.31.914.
17. S. W. McDonald and A. N.Kaufman, Spectrum and eigenfunctions for a Hamiltonian with stochastic trajectories, *Physical Review Letters.* **42**(18), 1189–1191 (1979). doi: 10.1103/PhysRevLett.42.1189.
18. M. V. Berry, Forms of light, *The Sciences.* **19**(8), 18–20 (1979). doi: 10.1002/j.2326-1951.1979.tb01782.x.
19. R. Thom, Structural stability, catastrophe theory, and applied mathematics, *SIAM Review.* **19**(2), 189–201 (1977). doi: 10.1137/1019036.
20. E. J. Heller, Bound-state eigenfunctions of classically chaotic Hamiltonian systems: Scars of periodic orbits, *Physical Review Letters.* **53**(16), 1515–1518 (1984). doi: 10.1103/PhysRevLett.53.1515.
21. T.-Y. Li and J. A. Yorke, Period three implies chaos, *The American Mathematical Monthly.* **82**(10), 985–992 (1975).
22. C. Grebogi, E. Ott, and J. A. Yorke, Chaotic attractors in crisis, *Physical Review Letters.* **48**(22), 1507–1510 (1982). doi: 10.1103/PhysRevLett.48.1507.
23. C. Grebogi, E. Ott, and J. A. Yorke, Chaos, strange attractors, and fractal basin boundaries in nonlinear dynamics, *Science.* **238**(4827), 632–638 (1987). doi: 10.1126/science.238.4827.632.
24. C. Grebogi, S. W. McDonald, E. Ott, and J. A. Yorke, Final state sensitivity: An obstruction to predictability, *Physics Letters A.* **99**(9), 415–418 (1983). doi: 10.1016/0375-9601(83)90945-3.
25. C. Grebogi, E. Ott, S. Pelikan, and J. A. Yorke, Strange attractors that are not chaotic, *Physica D.* **13**(1-2), 261–268 (1984).
26. C. Grebogi, S. W. McDonald, E. Ott, and J. A. Yorke, Exterior dimension of fat fractals, *Physics Letters A.* **110**(1), 1–4 (1985). doi: 10.1016/0375-9601(85)90220-8.
27. C. Grebogi, E. Ott, and J. A. Yorke, Metamorphoses of basin boundaries in nonlinear dynamical systems, *Physical Review Letters.* **56**(10), 1011–1014 (1986). doi: 10.1103/PhysRevLett.56.1011.
28. C. Grebogi, E. Ott, and J. A. Yorke, Unstable periodic orbits and the dimensions of multifractal chaotic attractors, *Physical Review A.* **37**(5), 1711–1724 (1988). doi: 10.1103/PhysRevA.37.1711.
29. C. Grebogi, S. M. Hammel, J. A. Yorke, and T. Sauer, Shadowing of physical trajectories in chaotic dynamics: Containment and refinement, *Physical Review Letters.* **65**(13), 1527–1530 (1990). doi: 10.1103/PhysRevLett.65.1527.
30. M. Ding, C. Grebogi, E. Ott, and J. A. Yorke, Transition to chaotic scattering, *Physical Review A.* **42**(12), 7025–7040 (1990). doi: 10.1103/PhysRevA.42.7025.
31. E. Ott, C. Grebogi, and J. A. Yorke, Controlling chaos, *Physical Review Letters.* **64**(11), 1196–1199 (1990). doi: 10.1103/PhysRevLett.64.1196.

32. S. P. Dawson, C. Grebogi, J. A. Yorke, I. Kan, and Hüseyin Koçak, Antimonotonicity: Inevitable reversals of period-doubling cascades, *Physics Letters A*. **162**(3), 249–254 (1992). doi: 10.1016/0375-9601(92)90442-O.
33. S. Hayes, C. Grebogi, and E. Ott, Communicating with chaos, *Physical Review Letters*. **70**(20), 3031–3034 (1993). doi: 10.1103/PhysRevLett.70.3031.
34. L. Poon and C. Grebogi, Controlling complexity, *Physical Review Letters*. **75**(22), 4023–4026 (1995). doi: 10.1103/PhysRevLett.75.4023.
35. Y.-C. Lai, C. Grebogi, J. A. Yorke, and S. C. Venkataramani, Riddling bifurcation in chaotic dynamical systems, *Physical Review Letters*. **77**(1), 55–58 (1996). doi: 10.1103/PhysRevLett.77.55.
36. U. Feudel, C. Grebogi, B. R. Hunt, and J. A. Yorke, Map with more than 100 coexisting low-period periodic attractors, *Physical Review E*. **54**(1), 71–81 (1996). doi: 10.1103/PhysRevE.54.71.
37. Z. Toroczkai, G. Károlyi, A. Péntek, T. Tél, and C. Grebogi, Advection of active particles in open chaotic flows, *Physical Review Letters*. **80**(3), 500–503 (1998). doi: 10.1103/PhysRevLett.80.500.
38. S. Banerjee, J. A. Yorke, and C. Grebogi, Robust chaos, *Physical Review Letters*. **80**(14), 3049–3052 (1998). doi: 10.1103/PhysRevLett.80.3049.
39. Y.-C. Lai and C. Grebogi, Modeling of coupled chaotic oscillators, *Physical Review Letters*. **82**(24), 4803–4806 (1999). doi: 10.1103/PhysRevLett.82.4803.
40. D. V. Anosov, Geodesic flows on closed Riemannian manifolds with negative curvature, *Proceedings of the Steklov Institute of Mathematics*. **90**, 1–235 (1969).
41. R. Bowen, Markov partitions for axiom A diffeomorphisms, *American Journal of Mathematics*. **92**(3), 725–747 (1970).
42. R. Bowen, *On Axiom A Diffeomorphisms*. Regional Conference Series in Mathematics, Vol. 35. American Mathematical Society, Providence, R.I. (1978).
43. S. M. Hammel, J. A. Yorke, and C. Grebogi, Do numerical orbits of chaotic dynamical processes represent true orbits? *Journal of Complexity*. **3**, 136–145 (1987).
44. S. M. Hammel, J. A. Yorke, and C. Grebogi, Numerical orbits of chaotic processes represent true orbits, *Bulletin of the American Mathematical Society*. **19**(2), 465–469 (1988).
45. B. A. Cipra, Computer-drawn pictures stalk the wild trajectory, *Science*. **241**(4870), 1162–1163 (1988). doi: 10.1126/science.241.4870.1162.
46. T. Shinbrot, C. Grebogi, J. Wisdom, and J. A. Yorke, Chaos in a double pendulum, *American Journal of Physics*. **60**(6), 491–499 (1992). doi: 10.1119/1.16860.
47. Y.-C. Lai and C. Grebogi, Chaotic scattering in time-dependent Hamiltonian systems, *International Journal of Bifurcation & Chaos*. **01**(03), 667–679 (1991). doi: 10.1142/S0218127491000488.

48. J. C. Alexander, J. A. Yorke, Z. You, and I. Kan, Riddled basins, *International Journal of Bifurcation & Chaos.* **2**(4), 795–813 (1992). doi: 10.1142/s0218127492000446.
49. H. E. Nusse and J. A. Yorke, A numerical procedure for finding accessible trajectories on basin boundaries, *Nonlinearity.* **4**(4), 1183–1212 (1991). doi: 10.1088/0951-7715/4/4/008.
50. Y.-C. Lai, H.-Y. Xu, L. Huang, and C. Grebogi, Relativistic quantum chaos — An emergent interdisciplinary field, *Chaos.* **28**(5), 052101 (2018). doi: 10.1063/1.5026904.

https://doi.org/10.1142/9789819823130_0020

Chapter 20

FROM NOISE TO CHAOS TO BIOLOGY

Hanspeter Herzel

Humboldt-Universität Germany
h.herzel@hu-berlin.de

The chapter reflects the early days of chaos theory in East Germany. Discussions of most influential readings and of chaos teaching already in 1984 illustrate the situation at the Humboldt University Berlin at these days. Furthermore, the role of oscillator theory in biology is reviewed using examples from voice research, heart dynamics, and chronobiology.

1. My Own Chaotic Trajectory

In 1984, I graduated from Humboldt University Berlin in the field of statistical physics under the supervision of Werner Ebeling. His group was very interdisciplinary with many international connections, which was exceptional in East Germany.[1] Consequently, the ideas of Ilya Prigogine, Manfred Eigen, Peter Schuster, Yuri Klimontovich, and Hermann Haken on self-organization and spatio-temporal patterns shaped my education.[2] Inspired by Evgeni Selkov, the focus of my thesis was studying the role of noise and deterministic chaos in biochemical systems[3] (see also Fig. 1). We emphasized that the dynamics of biochemical networks is characterized by strong alterations of phase space expansion and contraction. This implies that the calculation of long-term averages such as Lyapunov exponents should be complemented by quantifications of the heterogeneity.

I asked myself how relevant the beautiful chaos theory is in the real world. For this reason, my research interests moved away from quite theoretical concepts (Fokker–Planck equation, Lyapunov exponents)

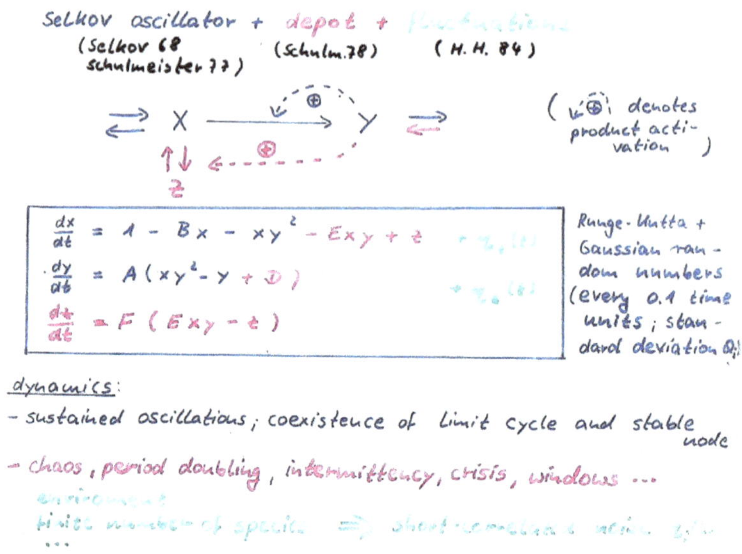

Fig. 1. Hand-drawn transparency on our research on biochemical oscillators (blue), an extension leading to period-doubling and chaos (red), and effects of fluctuations (green).

to time-series analysis, physiological rhythms, voice production and the circadian clock, fields of which all deal with the theory of coupled oscillators. This guided me through an exciting world of nonlinear phenomena in biomedical systems.

2. On the Shoulders of Giants — Some of My Most Influential Readings

Of course, I was fascinated by the widely-known early heroes of chaos theory: Huygens, Poincaré, van der Pol, Lorenz, Ruelle & Takens, and May (nicely discussed in Jackson 1989[4]). Below I wish to reminisce on some other visionary studies with direct effects on my own research.

As a native German speaker I could easily access a booklet by Georg Duffing[5] on nonlinear oscillators. Motivated by problematic resonances in mechanical engineering, he studied periodically driven oscillators and found experimentally that sometimes amplitudes were quite unstable. Linear stability analysis of a periodically forced oscillator predicted exponential instability. His findings can be interpreted as very early experimental observations of chaos and as an approximate calculation of positive Lyapunov exponents.

In the 30s, Erich von Holst studied locomotion in fish.[6] He observed complex patterns of frequency-locking in fins and sudden transitions by varying external parameters, such as water temperature. In order to understand the organization of frequency ratios, he constructed mechanical devices of coupled oscillators. The resulting diagrams of frequency locking patterns resembled Arnol'd tongue structures. He also coined the term "relative coordination" to describe slow toroidal modulations of rhythms.

Our research in biochemical oscillators was heavily influenced by papers of Otto E. Rössler from the 70s.[7] He predicted, with an impressive intuition, the chaos found in simple biochemical networks. Furthermore, he realized early on that the fractal dimension of attractors should be governed by the balance of manifold expansions and volume contractions. The exponential instability within chaotic attractors requires permanent folding leading to strange attractors. For example, in a 3-dimensional chaotic system 2-dimensional areas in phase space grow to infinity (the sum of the two largest Lyapunov exponents is positive) whereas volumes shrink to zero due to dissipation. Consequently, a fractal dimension between 2 and 3 can be expected. Thus, sums of Lyapunov exponents govern expansion and contraction of manifolds. I remember that Otto Rössler envisioned early on the formula of Kaplan & Yorke connecting Lyapunov exponents and fractal dimensions.

3. Teaching Deterministic Chaos

Already in 1984, my colleagues Thomas Nattermann and Bernd Koch taught a series of lectures on chaos theory in Berlin. From 1985 on, I continued and provided yearly lectures on dynamical systems at the Humboldt University (see Fig. 2).

My lectures were partly based on the great textbooks of Bergé *et al.*[8] and of Heinz-Georg Schuster.[9] The book by Schuster reproduced colorful spectrograms from acoustic cavitation pioneered by the Lauterborn group.[10] Four years later, our computer broke and I had to plot some figures at another institute in northern Berlin. Note, that reasonable computers such as HP 9845 were rare in East-Berlin. At the other computer I met by chance Kathleen Wermke, an expert in newborn cry analysis. She showed me her thesis with many spectrograms from newborn cries. I was deeply

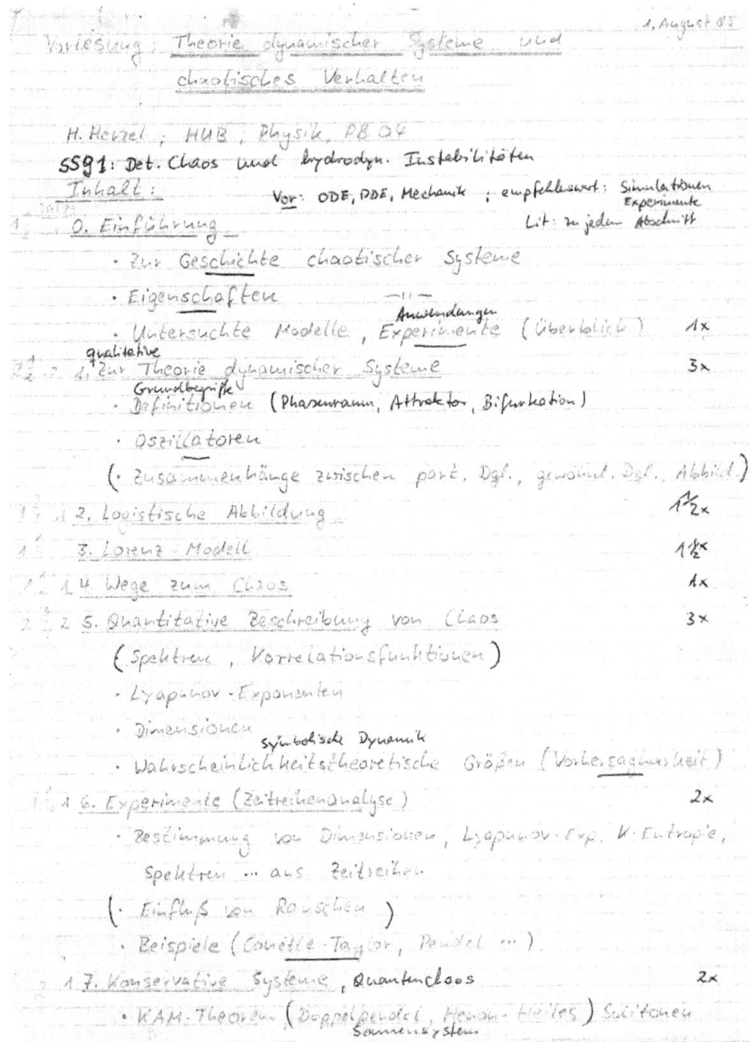
1. August 85

Vorlesung: Theorie dynamischer Systeme und chaotisches Verhalten

H. Herzel; HUB, Physik, PB 04

SS91: Det. Chaos und hydrodyn. Instabilitäten

Inhalt: Vor: ODE, PDE, Mechanik; empfehlenswert: Simulationen, Experimente

0. Einführung Lit: zu jedem Abschnitt
- Zur Geschichte chaotischer Systeme
- Eigenschaften —"—
- Untersuchte Modelle, Experimente, Anwendungen (Überblick) 1x

1. zur qualitative Theorie dynamischer Systeme 3x
- Definitionen, Grundbegriffe (Phasenraum, Attraktor, Bifurkation)
- Oszillatoren

(· Zusammenhänge zwischen part. Dgl., gewöhnl. Dgl., Abbild.)

2. Logistische Abbildung 1½x

3. Lorenz-Modell 1½x

4. Wege zum Chaos 1x

5. Quantitative Beschreibung von Chaos 3x

(Spektren, Korrelationsfunktionen)
- Lyapunov-Exponenten
- Dimensionen
- Wahrscheinlichkeitstheoretische Größen (Vorhersagbarkeit), symbolische Dynamik

6. Experimente (Zeitreihenanalyse) 2x
- Bestimmung von Dimensionen, Lyapunov-Exp., K-Entropie, Spektren ... aus Zeitreihen

(· Einfluß von Rauschen)
- Beispiele (Couette-Taylor, Pendel ...)

7. Konservative Systeme, Quantenchaos 2x
- KAM-Theorem (Doppelpendel, Henon-Heiles, Sonnensystem) Solitonen

Fig. 2. Scan of the outline of a lecture series on deterministic chaos at Humboldt University Berlin in 1985.

impressed seeing in many of these cries bifurcations including period-doubling, "biphonation" (independent frequencies) and deterministic chaos (see Fig. 3).[11]

The corresponding spectrograms were quite similar to data from acoustic cavitation I had seen in the Schuster-book. My excitement regarding

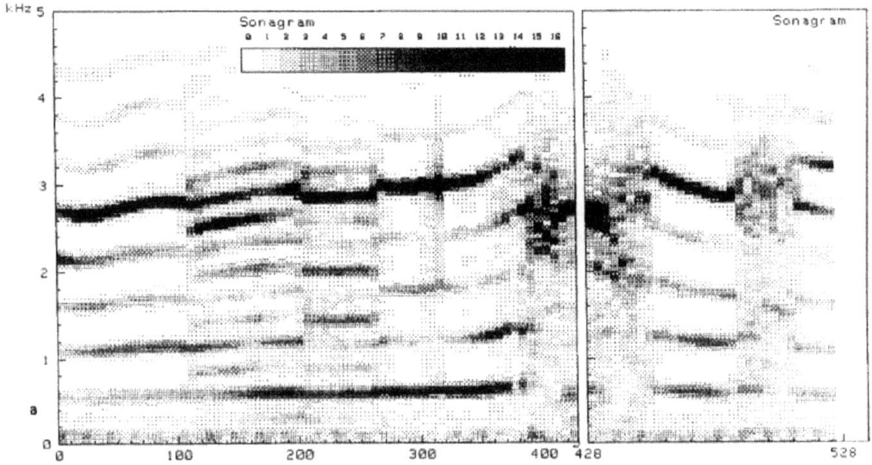

Fig. 3. Spectrogram from a newborn cry period-doubling and chaos.
Source: From Mende *et al.*[11]

these quite naturally occurring bifurcations in newborn cries initiated my research on the nonlinear dynamics of voice production.

While teaching dynamical systems in the 80s, I enjoyed the synergy of statistical physics and mathematical theory.[12] I had many exciting discussions with physicists such as Rainer Feistel, Peter Grassberger, or Vadim Anishchenko, and mathematicians including Hans-Guenther Bothe and Ludwig Arnold. Meeting colleagues from the Soviet Union was relatively easy. I spent, for example, a few months in Moscow and met there Sascha Michailov, Ruslan Stratonovich, and Yakov Sinai. In order to meet colleagues from the West (e.g. Alain Arnéodo and Ludwig Arnold) I had to travel to occasional international meetings in Warsaw, Budapest, and Eisenach.

In these years excellent reviews by Doyne Farmer and Michael Mackey brought together ergodic theory and empirical studies of deterministic chaos.[13,14] Several of these early papers from the 80s are still discussed in my group, where we study nonlinear phenomena in biological rhythms.

Getting the journals and books from the West was possible in East-Berlin but somewhat difficult. I had to visit several libraries to scan the relevant journals such as *Physical Reviews* and *Physica D*. Furthermore, copy machines were rarely available. Thus I made hand-written notes as illustrated in Fig. 4. The strange non-chaotic attractors discussed in

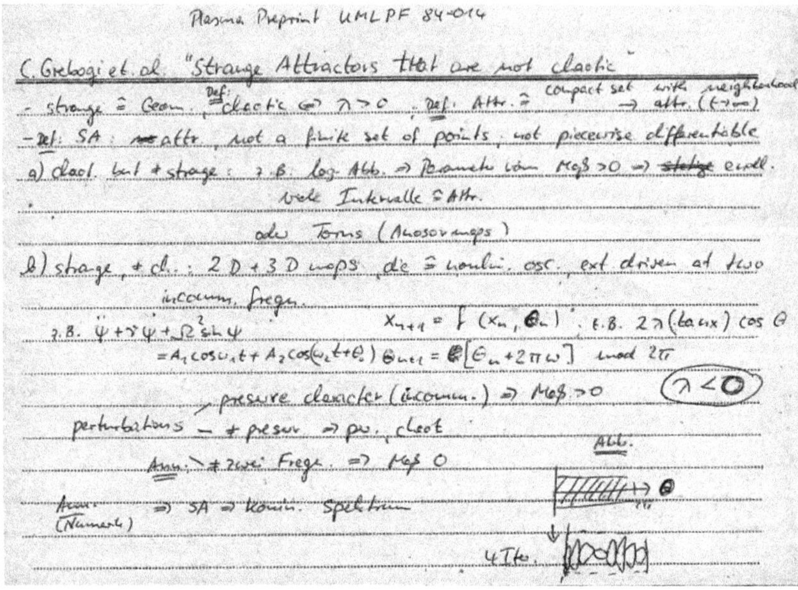

Fig. 4. 1984 — hand-written notes instead of copy machines.

that paper of Celso Grebogi,[15] interestingly enough, are again relevant in our current research. We study now nonlinear models of circadian clocks driven by multiple periodic inputs ("zeitgebers") and find quite complex attractors.

4. Chaos Theory Bridged East and West

The theory of pattern formation in dissipative systems was quite popular in the former Soviet Union and in the GDR. Some early experts were Leven, Koch & Pompe in Greifswald, who studied chaos in parametrically forced mechanical oscillators.[16] Moreover, the groups of Reinhart Heinrich and Tom Rapoport in Berlin pioneered the mathematical analysis of biochemical networks.[17] I remember intense debates about why in these large networks with many nonlinearities and feedbacks, where chaos might be expected, it nevertheless seems to be exceptional. Reinhart Heinrich speculated that the broad range of time-scales in cellular networks might suppress chaotic dynamics. Recent single cell studies indicate that complex temporal patterns are more frequent than anticipated in the 80s.[18]

Attending conferences in Western countries was quite difficult for us. Because of this, exchange of (p)reprints was very important. We systematically sent around postcards requesting "reprints and related papers". In many cases we got back thick envelopes with preprints and even floppy discs with software packages. I still have many yellow preprints from Peter Grassberger from the 80s with pioneering work on generalized dimensions and information theory.[19] I was particularly proud when I got myself a reprint request from Jack Hudson early in my career. He was interested in our simulations of the Belousov–Zhabotinsky map.[20] Later, as he started regularly visiting Berlin, we became friends and I remember a nice evening in the center of East-Berlin where he insisted on paying for the beer — not straightforward, since Western visitors had to first exchange currency to East-Mark. When the Wall came down, we met regularly together with Katharina Krischer from the group of Gerhard Ertl to discuss the role of chaos in chemical systems.[21]

Another opportunity for developing bridges between East and West was meeting at conferences in Eastern Europa. At the Banach Center in Warsaw I heard there the latest results on chaos in the logistic map by Michal Misiurewicz and we discussed Lyapunov exponents and dimensions with François Ledrappier. I remember also a great chaos-symposium in Budapest organized by Peter Szepfalusy and Tamas Tel. There was also a series of Wartburg-conferences organized by Manfred Peschel. In 1984 Peter Richer and Hans-Otto Peitgen[22] joined one of these Wartburg-meetings and we enjoyed their presentation of the beautiful world of fractals with Mandelbrot sets or "Apfelmaennchen" (which literally translated from German would be "apple manikin"). Following the tradition of the medieval "singer's war on the Wartburg", at the conference we sang songs from all over the world. Peter Richter was declared the winner of this new-age "singer's war" and his prize was an "Apfelmaennchen" constructed from real apples.

5. Chaos in Biology

Complex phenomena in biological systems stimulated early chaos research. For example, the studies of Van der Pol and Van der Mark were motivated by heart dynamics. Later Bob May analyzed iterated maps as models of population dynamics. Chronobiology, the study of biological rhythms at all scales, motivated Art Winfree to explore phase response curves and singularities.[23] However, the actual relevance of chaos theory in biology is still debated since biomedical systems are intrinsically complex and rarely generate the stationary signals that are required for attractor analysis. Nevertheless, combinations of experiments and theory provided exciting examples of nonlinear phenomena in biology. Many of these early findings are summarized in the book *From Clocks to Chaos* by Leon Glass and Michael Mackey.[24]

In particular, heart arrhythmias are intimately related to iterated maps and bifurcations of coupled oscillators.[25] These studies were successful since heart dynamics are relatively autonomous and can even be explored in isolated heart cells. Vibrations of vocal folds in the larynx are quite autonomous as well. Even in excised larynx preparations, bifurcations and chaos have been described.[26] Furthermore, the myoelastic-aerodynamic theory of flow-induced oscillations by Ingo Titze provides a quantitative framework.[27] The nonlinear dynamics of voice production became my favorite research topic after seeing spectrograms of newborn cries in the 80s. It turned out that chaos is common in pathological voices, in modern singing, and in animal vocalizations including bird songs.[28]

My current research field is theoretical chronobiology. In many model organisms, one can observe self-sustained rhythms with a period of about 24 h. These intrinsic oscillations, termed "circadian clocks," can be driven by light and temperature cycles.[29] Consequently, the theory of coupled oscillators becomes the appropriate framework in chronobiology as well.[30] Even though frequency locking beyond 1:1 entrainment and toroidal rhythms ("splitting") have been described,[31] the detection of deterministic chaos is difficult due to the requirement of rather long time-scales.

Retrospectively, the chaos-boom several decades ago established nonlinear dynamics as an integral part of theoretical biology and it continues to be part of our research interests. 40 years ago, I started to apply chaos theory to biological networks. In particular, the theory of coupled oscillators has guided our studies in gene-regulatory networks, heart dynamics, voice production, and circadian rhythms.

References

1. W. Ebeling, *Strukturbildung bei irreversiblen Prozessen.* Teubner, Leipzig (1976).
2. R. Feistel and W. Ebeling, *Physics of Self-Organization and Evolution.* Wiley (2011).
3. J. G. Reich and E. E. Sel'kov, *Energy Metabolism of the Cell.* Academic Press (1981).

4. E. A. Jackson, *Perspectives of Nonlinear Dynamics.* Cambridge University Press, Cambridge (1989). doi: 10.1017/CBO9780511623974.
5. G. Duffing, *Erzwungene Schwingungen bei veränderlicher Eigenfrequenz und ihre technische Bedeutung.* Vieweg, Braunschweig (1918).
6. E. von Holst, Die relative Koordination, *Ergebnisse der Physiologie und experimentellen Pharmakologie.* **42**, 228–306 (1939). doi: 10.1007/BF02322567.
7. O. E. Rössler, Chaotic behavior in simple reaction system, *Zeitschrift für Naturforschung A.* **31**, 259–264 (1976).
8. P. Bergé, Y. Pomeau, and C. Vidal, *L'ordre dans le chaos.* Hermann (1984).
9. H. G. Schuster, *Deterministic Chaos: An Introduction.* Wiley-VCH (1984).
10. W. Lauterborn and E. Cramer, Subharmonic route to chaos observed in acoustics, *Physical Review Letters.* **47**(20), 1445–1448 (1981). doi: 10.1103/PhysRevLett.47.1445.
11. W. Mende, H. Herzel, and K. Wermke, Bifurcations and chaos in newborn infant cries, *Physics Letters A.* **145**(8), 418–424 (1990). doi: 10.1016/0375-9601(90)90305-8.
12. J.-P. Eckmann and D. Ruelle, Ergodic theory of chaos and strange attractors, *Reviews of Modern Physics.* **57**, 617–656 (1985).
13. J. Farmer, E. Ott, and J. A. Yorke, The dimension of chaotic attractors, *Physica D.* **7**(1), 153–180 (1983). doi: 10.1016/0167-2789(83)90125-2.
14. M. C. Mackey, The dynamic origin of increasing entropy, *Reviews of Modern Physics.* **61**, 981–1015 (1989). doi: 10.1103/RevModPhys.61.981.
15. C. Grebogi, E. Ott, S. Pelikan, and J. A. Yorke, Strange attractors that are not chaotic, *Physica D.* **13**(1–2), 261–268 (1984).
16. R. W. Leven, B.-P. Koch, and B. Pompe, *Chaos in dissipativen Systemen.* Vieweg-Teubner Verlag (1989). doi: 10.1007/978-3-322-84175-9.
17. R. Heinrich, S. Rapoport, and T. Rapoport, Metabolic regulation and mathematical models, *Progress in Biophysics and Molecular Biology.* **32**, 1–82 (1978). doi: 10.1016/0079-6107(78)90017-2.
18. M. L. Heltberg, S. Krishna, L. P. Kadanoff, and M. H. Jensen, A tale of two rhythms: Locked clocks and chaos in biology, *Cell Systems.* **12**(4), 291–303 (2021). doi: 10.1016/j.cels.2021.03.003.
19. P. Grassberger and I. Procaccia, Characterization of strange attractors, *Physical Review Letters.* **50**(5), 346–349 (1983). doi: 10.1103/PhysRevLett.50.346.
20. H. Herzel, W. Ebeling, and T. Schulmeister, Nonuniform chaotic dynamics and effects of noise in biochemical systems, *Zeitschrift für Naturforschung A.* **42**(2), 136–142 (1987). doi: doi:10.1515/zna-1987-0205.
21. M. Eiswirth, K. Krischer, and G. Ertl, Transition to chaos in an oscillating surface reaction, *Surface Science.* **202**(3), 565–591 (1988). doi: 10.1016/0039-6028(88)90053-2.
22. H.-O. Peitgen and P. H. Richter, Fraktale und die Theorie der Phasenübergänge, *Physikalische Blätter.* **42**(1), 9–22 (1986). doi: 10.1002/phbl.19860420108.

23. A. T. Winfree, Integrated view of resetting a circadian clock, *Journal of Theoretical Biology.* **28**(3), 327–374 (1970). doi: 10.1016/0022-5193(70) 90075-5.
24. L. Glass and M. C. Mackey, *From Clocks to Chaos: The Rhythms of Life.* Princeton University Press (1988).
25. M. Guevara, L. Glass, and A. Shrier, Phase locking, period-doubling bifurcations, and irregular dynamics in periodically stimulated cardiac cells, *Science.* **214**(4527), 1350–1353 (1981). doi: 10.1126/science.7313693.
26. H. Herzel, D. Berry, I. Titze, and I. Steinecke, Nonlinear dynamics of the voice: Signal analysis and biomechanical modeling, *Chaos.* **5**(1), 30–34 (1995). doi: 10.1063/1.166078.
27. I. R. Titze, *Principles of Voice Production.* Prentice-Hall (1998).
28. W. T. Fitch, J. Neubauer, and H. Herzel, Calls out of chaos: The adaptive significance of nonlinear phenomena in mammalian vocal production, *Animal Behaviour.* **63**(3), 407–418 (2002). doi: 10.1006/anbe.2001.1912.
29. Circadian clocks. In A. Kramer and M. Merrow (eds.), *Handbook of Experimental Pharmacology.* Springer (2013).
30. U. Abraham, A. E. Granada, P. O. Westermark, M. Heine, A. Kramer, and H. Herzel, Coupling governs entrainment range of circadian clocks, *Molecular Systems Biology.* **6**, 438 (2010).
31. A. E. Granada, T. Cambras, A. Díez-Noguera, and H. Herzel, Circadian desynchronization, *Interface Focus.* **1**(1), 153–166 (2011). doi: 10.1098/rsfs.2010.0002.

https://doi.org/10.1142/9789819823130_0021

Chapter 21

A KNOTTED ROAD TO CHAOS

Robert Gilmore

Drexel University, USA
r2.gilmore@gmail.com

The Science of Chaos didn't even exist (except possibly in the mind of Poincaré) when the Ancients, whose memories are collected here, came of age. Therefore we all have different stories to tell about our strange and tortuous path to the present. This is mine.

1. Introduction

My trajectory as a physicist was strongly perturbed by Fortunato Tito Arecchi. He is a world-class laser physicist who visited M.I.T. for a year around 1970. He saw that most lasers were modeled as a set of N independent 2-level atoms in a cavity interacting with a single mode of the electromagnetic field. For such models the maximum output intensity behaved as $I \sim N^2$. He understood that 2-level atoms could be described by the Lie group SU(2). He wondered if a model with 3-level atoms could have an output behavior like $I \sim N^3$. Since SU(3) and the octet model were sweeping the particle physics community at the time, he tried to get answers from the nearby particle physicists. He described his failure this colorful way (T = Tito; PP = particle physicist):

$$\begin{bmatrix} \text{Knock on a door. Door opens a few inches.} \\ \quad\text{T: } \textit{Hi. My name is Tito.} \\ \text{PP: } \textit{?} \\ \quad\text{T: } \textit{I'm a laser physicist.} \\ \text{PP: } \textit{??} \\ \quad\text{T: } \textit{Can you teach me about SU(3)?} \\ \text{PP: } \textit{Why?} \\ \quad\text{T: } \textit{To solve laser problems.} \\ \text{Door closes} \end{bmatrix}^n$$

with $n \sim 10$. Finally a friend directed Tito to the guy teaching the group theory course in the Physics Department (me). We met; he explained the problem; we realized that for any model with $[\mathrm{SU}(p)]^{\otimes N}$, $I \sim |A|^2$ and $A \sim N$. So the answer was no and giant laser output patents were not in our future.

But this collaboration was too useful to quit there. He came up with another question. Roy Glauber had transformed the study of electromagnetism by introducing (field) coherent states in 1963.[1] Could we create analog coherent states for the atomic side of the laser interaction? The answer was yes and the answer was beautiful. We did this for SU(2)[2] and for SU(n)[3] and for any Lie group.[4,5]

2. Laser Physics

These new tools were soon used to great advantage to study laser problems. The proper hamiltonian consisted of quantum operators describing both the field and the atomic side of the interaction. Simpler semiclassical hamiltonians were obtained by taking expectation values of the hamiltonian using either the field coherent states or the atomic coherent states. By taking both, the hamiltonian was mapped to a complex function depending on the coherent state parameters. Determining the operating state was reduced to a variational problem. Learning how the operating state changed as the parameters changed was a bifurcation theory problem.[6] We did a lot of useful work in this field; besides it was fun.

At this time a strong connection between the laser physics community and the nonlinear dynamics community was established by Hermann Haken.

He had deep interests in laser physics, nonlinear dynamics, self-organization and Synergetics. He showed[7] that there was a deep connection between one of the standard laser physics models in a certain limit and the behavior of fluids as described by the Lorenz equations.[8] This connection provoked a number of experimental searches for Lorenz-like output behavior of various types of lasers.[9,10]

3. Catastrophe Theory

During this period I encountered a reference to 'catastrophe theory'. What is this?: I asked of anybody I encountered. The typical answer was: *?*. Eventually, somebody pointed me to Thom's book,[11] not yet translated into English.

After reading it I understood nothing, and put this down to my halting French. By perserverence and luck I was directed to Timothy Poston, then in the process of writing his book on the subject with Ian Stewart.[12] Tim gave me copies of several important draft chapters. They were so well-written and straightforward that the concepts were easily assimilable. (Tim was a model for the benevolent wizards in the later *Harry Potter* series.) He also directed me to a forthcoming work of Erik Christopher Zeeman[13] which put the subject to work through many imaginative examples — many too imaginative for the staid physics community. I later read the English translation of Thom's work[11] and realized that my French was up to the challenge.

In the spring of 1976 a friend at KU Leuven invited Thom to deliver a colloquium about catastrophe theory. He gave me a heads-up about the event. Thom arrived. I attended. Thom lectured. I understood nothing. In leaving I ran into my friend. He asked what I got out of the experience. "*Nada!*" His reply: "*Don't feel so bad. I know what Thom talked about. And I know that I knew. I taught this material to my graduate students in the last two weeks. And I didn't understand anything that he said.*" Vladimir I. Arnol'd had a similar experience which he wrote about[14] and also described to me in person when he visited the University of Pennsylvania to give the Rademacher Lectures in 1989.

The important take-aways from this diversion were:[15]

(1) The most visible singularities are the stable nodes but the most important are the unstable saddles because their eigendirections help define basin boundaries.

(2) Bifurcations on manifolds could have canonical forms.
(3) All the important ones were discretely classifiable
(4) The classification overlapped enormously with the classification of simple Lie groups.[4]

While trying to master this subject (1975–1980), I realized that catastrophes beautifully describe the dynamical behavior of gradient dynamical systems $dx_i/dt = F_i = -\partial V/\partial x^i$. This lead naturally to the question of what happens when $\nabla \wedge F = F_{i,j} - F_{j,i} \neq 0$, and this lead to dynamical systems that could exhibit chaotic behavior. This was my introduction to chaos, both of maps and flows, and the Lorenz attractor.[8]

During the summer of 1980 I made a social return visit to M.I.T. and became aware of the problems of the "siloization of science". This was at a time when new fads in mathematics "more important than anything since the development of calculus" (or maybe even before) were sweeping through the country at a rate of about 1 fad/6 months: Catastrophe Theory; NonStandard Analysis; Chaos Theory. I was asked to propose a candidate to present a colloquium about Chaos Theory to the Physics Department. Instead of answering, I pointed to a window high up in a building on the other side of a small parking lot. Here is the subsequent discussion:

- *Who's there?*
RG: *Lorenz.*
- *Who's he?*
RG: *He described 'sensitivity to initial conditions' and introduced the Lorenz equations and described the attractor.*
- *How do you spell his name?*
RG: *L-o-r-e-n-z.*

4. The Feigenbaum Shockwave

The study of chaos changed dramatically around this time. The enormously powerful tools of renormalization group theory[16,17] were applied to iterative maps in the late 1970s on both sides of the Atlantic.[18–21] These results rapidly lead to several new invariant quantities, such as the scaling ratio $\delta = 4.669\ldots$. These quantities are eigenvalues of the unstable eigenvectors at the fixed point of a functional iteration operator. Once 'universality' was claimed, a sea change occurred. In my days as a student (early 1960s) my colleagues were reluctant to embark on a prolonged study of nonlinear stuff.

The attitude was: I could spend a lifetime learning about XYZ system and learn nothing about XZY system. With the universality claim "*It was a very happy and shocking discovery that there were structures in nonlinear systems that were always the same if you looked at them the right way.*"[22]

The community of experimental scientists took this as a challenge, and the race was on. Experiments were carried out in fluid dynamics, chemical reactions, electrical engineering, and various branches of physics. Time scales for the experiments depended on the field, from very long (fluids) to very short (lasers). Some of the early experimental tests of the universality prediction are reprinted in the excellent collection by Cvitanović.[23]

One set of experiments was carried out by Arecchi and Jorge R. Tredicce and colleagues.[24] This experiment confirmed universality within experimental error. Tredicce moved to Drexel University to work with Lorenzo M. Narducci in 1985. I moved to Drexel somewhat earlier (1981). My reasons were in part: to continue working on laser problems with Narducci; to continue nuclear physics studies depending on group theory — SU(6) and its coherent states; and to continue work in classical thermodynamics — in particular, the symmetries of the Le Châtelier relations[25] and the relaxation back to a new equilibrium, dictated by gradient dynamical system equations of the form $(Kd/dt - L)\boldsymbol{x} = 0$, where K is a matrix of kinetic coefficients, L a matrix of linear response coefficients and $\mathbf{x}$ is a vector of displacements from the new equilibrium.[26]

In doing these experiments the Arecchi/Tredicce group had collected a great deal of data, only a small part of which went into the test of universality. Since data are expensive and it's never a good idea to discard any, Tredicce wondered how he could understand the remainder. Through the intermediary, Elia Eschenazi, we got together (about 1986) and he showed me his 'excess' data,[24] then asked: "*Bob, can you help me explain my data?*" I replied: "*I don't know. I'll try, but it might take a long time.*" He said he was prepared to wait. This was an exciting challenge that I eventually turned my complete attention to.

5. A Long Time

The data showed, among other things, multiple coexisting basins of attraction surrounding orbits of various low periods that sometimes came into existence or winked out of existence with a small change of parameter (Fig. 1).[24,27,28] Experience with catastrophes indicated the presence of saddle-node bifurcations. We also saw catastrophes of type A_3; in this

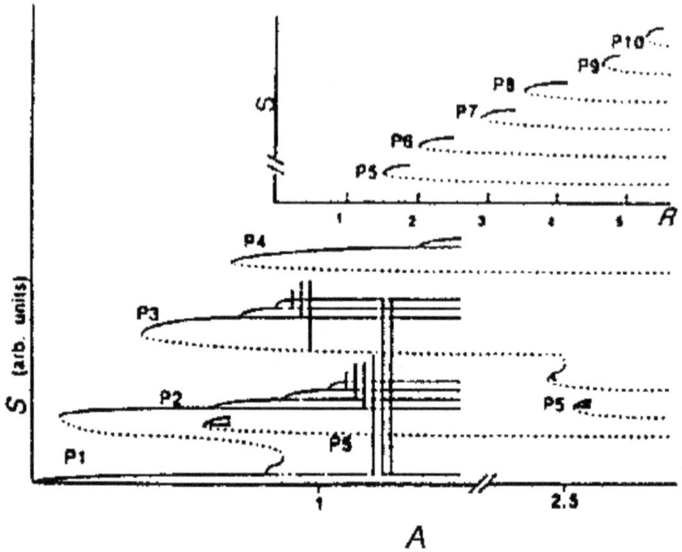

Fig. 1. A representation of some of the laser data, with S the output intensity and A a bifurcation parameter. Shown are saddle-node (A_2) bifurcations creating orbits of periods 2–10, period-2 and -5 snakes, period-doubling sequences, coexisting period-5 orbits.

context they were period-doubling bifurcations. To make sense of it all, we (Hernan Solari and myself) asked: *What would Einstein do? He would look for the observables and ask how they interacted with each other.*

The observables were the periodic orbits, and the most important ones were the unstable periodic orbits — again, a lesson from catastrophe theory. As for their interaction, here is what looked promising. Since the laser was a periodically driven 2-dimensional system, a period-p orbit would intersect a constant t plane (Poincaré section) in exactly p distinct points. To determine how two different orbits of periods p and q interacted with each other, in a plane t =const. we chose an intersection of the period p orbit and an intersection of the other periodic orbit, and connected these intersections by a vector. As t increased, the vector would rotate in the plane. After $p \times q$ periods the vector must return to its original orientation after R (an integer) full rotations. We defined the *relative rotation rate* as $\text{RRR} = \frac{R}{pq}$ for the initial conditions chosen. This is the average rotation per period. This index depends on the initial conditions, so for orbits of periods p and q there are pq relative rotation rates.[29] The ensemble of these fractions for any pair of orbits had very restrictive and informative properties. Furthermore, they could be extracted from experimental data and compared with models of the system. In this way we were able to show that the dynamics of the periodically driven laser were those of a suspension of the Smale horseshoe.[30] Not surprisingly, there was a simple relation between the linking numbers of two orbits and the sum of their relative rotation rates.

This method was applied to other periodically driven systems.[31] Then it was extended to autonomous three dimensional dynamical systems as well. The Rössler dynamical system[32,33] with the standard parameters exhibited this behavior. These methods could distinguish between 'left-hand bends' and 'right-hand bends'. However, they worked only for chaotic attractors with 'a hole in the middle'. Specifically, analysis using relative rotation rates could not be applied to the Lorenz attractor.[8]

At this point we became aware of the Birman–Williams theorem.[34,35] This became a key tool for us. We used it as follows. We could extract a set of low-period unstable orbits from a chaotic attractor and then pairwise compute their 'experimental' linking numbers. Then we could propose a branched manifold that might be the projected limit of the attractor (Fig. 2). The branched manifold immediately determined the nature and location of the components of the Poincaré section. We could identify periodic orbits of whatever period in the branched manifold. Since these existed in a branched manifold, the symbol name of the orbit was at hand. We could also compute a table of their pairwise linking numbers. At that point we would attempt to assign a symbol name to each of the experimental orbits. A following comparison of the experimental linking numbers with those derived from the branched manifold would show either that we 'nailed' the analysis or had to go back to the drawing board. The net result was that, with Gabriel Mindlin, we were able to classify the topological structure of chaotic attractors by integers.[36] We were then able to analyze chaotic time series and search for the integer representation of the dynamics that generated these data.[37–39]

In fact, it became even better. Linking numbers of a lot of orbits *overdetermine* a branched manifold. The linking numbers of a very small number of periodic orbits (e.g. 3 for a branched manifold with two branches) could be used to determine the appropriate branched manifold. Comparing the excess number of 'experimental' linking numbers with their counterparts from the branched manifold is then used either to reject or confirm (better: fail to reject) the proposed relation between the experimental attractor and its branched manifold model. This is a 'gold standard' for experimentalists.

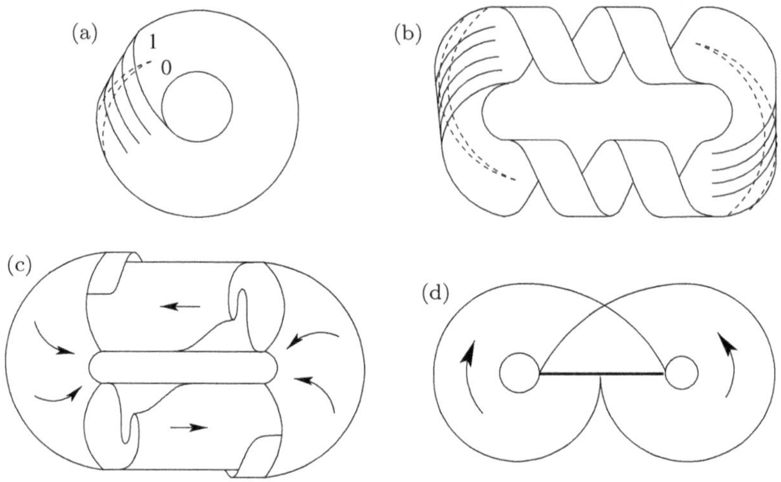

Fig. 2. Branched manifolds describing (a) Rössler, (b) periodically driven Duffing, (c) van der Pol attractors, and (d) Lorenz attractors and for particular parameter values. *Source*: Redrawn from Ref. 40.

One reason is that the measures are integers. There is no roundoff error, as always accompanies things like fractal dimension estimations (guesses), which at any rate tell you nothing about the topology of the attractor.[40]

We even believed that this theoretical technology applied to chaotic attractors in n (> 3) dimensions provided there was only one unstable direction. We proposed this to Robert Williams when we ran into him at a conference. After a short while he got back to us and confirmed it was the case for 'strongly contracting dynamical systems', that is, dynamical systems for which the Lyapunov dimension is less than 3. I like this result very much, but have not seen a proof, and have a lingering doubt. Lyapunov exponents are average values. That means there could be a part of phase space whose effective dimension is larger than 3, where transversality no longer leads to strict organization of periodic orbits.

6. Back to Group Theory

All my efforts to extend these topological ideas into dimensions greater than 3 have lead to frustration. Following the group theory playbook, we even created the beginnings of a representation theory for chaotic attractors.[41,42] In this theory the results depend strongly on embedding dimension.

The strongest results relating group theory to chaotic dynamics came about through Christophe Letellier showing me a large number of dynamical systems with various degrees of symmetry.[43] We embarked on a study of what happens when you 'mod out' the symmetry. And also what happens when you take a simple chaotic dynamical system of low genus (for example, the Rössler system) and create a 'covering dynamical system' with whatever symmetry you choose.[44]

One immediate benefit was that it allowed a systematic procedure to process data from lasers whose underlying dynamics was analogous to the Lorenz attractor. The physics depends on the amplitude A but the observables are related to the intensity $I \sim |A|^2$ — so cover-image relations are directly applicable to physical systems.[38,45]

This work was also a lot of fun, and took me back to my roots in a way I never could have imagined — point group symmetries of molecules and space group symmetries of solids.

7. Overview About Topology

Cartan's sublimely elegant discrete classification of simple Lie groups[4] and Arnol'd's beautiful discrete classification of zero-modal catastrophes[14,15] have inspired me to search for a discrete classification of some subset of dynamical systems exhibiting chaotic behavior. A discrete classification now exists for dissipative dynamical systems in $\mathbb{R}^3$. There are three levels in this classification scheme and each level involves knots.

7.1. *Bounding torus*

At the macro level are bounding tori.[40,46] For such systems a chaotic attractor is contained within an open set whose closure is a bounding torus. The torus is identified by its genus and also the linking organization of its set of longitudes and meridians.

7.2. *Branched manifold*

The Birman–Williams theorem can be used to project the chaotic attractor down to its representative branched manifold. The branched manifold can be treated as an assembly of building blocks (Legos©, see below) that are connected together in a way that can be described by a 'dressed' transition matrix. The 'dressing' consists of twisting and crossing information. All periodic orbits are identified by a symbol sequence, where the symbols identify the various branches of the branched manifold — the Lego© components that comprise the branched manifold. The branched manifold can be used to construct the linking information of all knots in the projected flow.

7.3. *Basis sets of orbits*

Usually some orbits allowed in the branched manifold are not in the original attractor, or they may be 'pruned away' by a change in some parameter. The ensemble of orbits in the attractor is described, up to some period p, by a finite list of orbits that 'force' the presence of all these orbits. Forcing relations[40,47–49] have been determined in simple cases by inspecting a matrix of linking numbers.

8. Discussion with Otto

Over the years Otto Rössler and I have discussed the 'topolization' of low-dimensional dynamical systems theory: the evolution from a geometric understanding to a topological understanding. He was the first to introduce a 'cardboard model' to describe what we now call the Rössler attractor. The 'cardboard model' idea was introduced by Williams[50] to provide understanding of Lorenz dynamics. We joyfully drew diagrams on paper napkins of a basic Lego© unit (Fig. 3) which can be used to construct the branched manifold for the Rössler attractor by folding it back on itself (Fig. 3(a)) or by joining two together as originally drawn by Williams[50] (Fig. 3(b)), or combining three or more to describe even more complicated dynamical systems.

We also marveled that it was not too difficult to identify a branched manifold once an attractor had been constructed by integrating dynamical

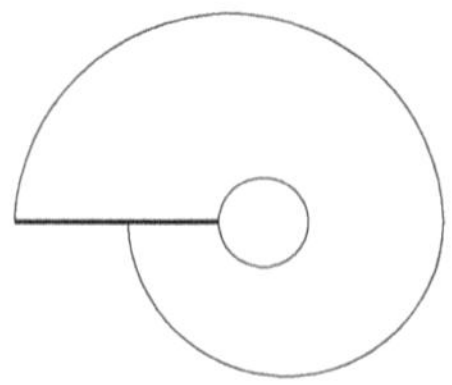

Fig. 3. A cardboard model of a basic building block (Lego©) for low-dimensional dynamical systems. An unstable focus sits inside the central hole. The projection from three down to two dimensions is along the stable direction of the focus. The expansion rate is almost 2 per cycle.

systems equations. But the reverse process — going from a branched manifold to a set of differential equations — seemed out of reach for the foreseeable future. There are some tentative moves in this direction. For example, attractors contained in a bounding torus of genus $2n+1$ with all fixed points in a straight line could be constructed by a method introduced by Aziz-Alaoui[51] that depended on a 'wiggley' driving function. And those with a symmetry could be constructed by a method introduced by Letellier and myself,[44] but the method was not intuitive and not pretty. We felt that the inverse process in (dynamical system equations) $\leftrightarrow$ (branched manifold) would be a long time in coming.

Acknowledgments

Robert Gilmore is endebted to the following friends and colleagues, listed in the order encountered on the road described: F. T. Arecchi, L. M. Narducci, J. R. Tredicce, H. G. Solari, E. Eschenazi, G. B. Mindlin, J. L. Birman, J. S. Birman, P. Glorieux, M. Lefranc, C. Letellier, V. Messager, O. E. Rössler, R. Williams.

References

1. R. J. Glauber, Coherent and incoherent states of the radiation field, *Physical Review.* **131**(6), 2766–2788 (1963). doi: 10.1103/PhysRev.131.2766.
2. F. T. Arecchi, E. Courtens, R. Gilmore, and H. Thomas, Atomic coherent states in quantum optics, *Physical Review A.* **6**(6), 2211–2237 (1972). doi: 10.1103/PhysRevA.6.2211.
3. R. Gilmore, Geometry of symmetrized states, *Annals of Physics.* **74**(2), 391–463 (1972). doi: 10.1016/0003-4916(72)90147-9.
4. R. Gilmore, *Lie Groups, Lie Algebras, and Some of Their Applications.* Wiley, New York (1972).
5. R. Gilmore, On the properties of coherent states, *Revista Mexicana de Física.* **23**, 143–184 (1974).
6. R. Gilmore, C. M. Bowden, and L. M. Narducci, Classical-quantum correspondence for multilevel systems, *Physical Review A.* **12**(3), 1019–1031 (1975). doi: 10.1103/PhysRevA.12.1019.
7. H. Haken, Analogy between higher instabilities in fluids and lasers, *Physics Letters A.* **53**(1), 77–78 (1975). doi: 10.1016/0375-9601(75)90353-9.
8. E. N. Lorenz, Deterministic nonperiodic flow, *Journal of the Atmospheric Sciences.* **20**, 130–141 (1963). doi: 10.1175/1520-0469(1963)020⟨0130:DNF⟩2.0.CO;2.
9. C. O. Weiss and W. Klische, On observability of Lorenz instabilities in lasers, *Optics Communications.* **51**(1), 47–48 (1984). doi: 10.1016/0030-4018(84)90281-5.

10. C. O. Weiss and J. Brock, Evidence for Lorenz-type chaos in a laser, *Physical Review Letters.* **57**(22), 2804–2806 (1986). doi: 10.1103/PhysRevLett.57.2804.
11. R. Thom, *Structural Stability and Morphogenesis: An Outline of a General Theory of Models.* W. A. Benjamin, New York (1973). translated from the French by David Fowler.
12. T. Poston and I. Stewart, *Catastrophe Theory and Its Applications.* Pitman, London (1978).
13. C. E. Zeeman, *Catastrophe Theory: Selected Papers, 1972-1977.* Addison-Wesley, Reading, Massachusetts (1977).
14. V. I. Arnol'd, *Catastrophe Theory.* Springer-Verlag, New York (1986).
15. R. Gilmore, *Catastrophe Theory for Scientists and Engineers.* Wiley, New York (1981).
16. K. G. Wilson, Renormalization group and critical phenomena. I. Renormalization group and the Kadanoff scaling picture, *Physical Review B.* **4**(9), 3174–3183 (1971). doi: 10.1103/PhysRevB.4.3174.
17. K. G. Wilson, Renormalization group and critical phenomena. II. Phase-space cell analysis of critical behavior, *Physical Review B.* **4**(9), 3184–3205 (1971). doi: 10.1103/PhysRevB.4.3184.
18. C. Tresser and P. Coullet, Itérations d'endomorphismes et groupe de renormalization, *Comptes-Rendus de l'Académie des Sciences (Paris).* **287A**, 577–580 (1978).
19. P. Coullet and C. Tresser, Itération d'endomorphismes et groupe de renormalization, *Journal de Physique Colloques.* **39**, C5–25–C5–28 (1978).
20. M. J. Feigenbaum, Quantitative universality for a class of nonlinear transformations, *Journal of Statistical Physics.* **19**(1), 25–52 (1978). doi: 10.1007/BF01020332.
21. M. J. Feigenbaum, The universal metric properties of nonlinear transformations, *Journal of Statistical Physics.* **21**(6), 669–706 (1979). doi: 10.1007/BF01107909.
22. J. Gleick, *Chaos: Making a new science.* Viking Press, New York (1987).
23. P. Cvitanović, *Universality in Chaos.* Adam Hilger, Boston (1984).
24. F. T. Arecchi, R. Meucci, G. Puccioni, and J. Tredicce, Experimental evidence of subharmonic bifurcations, multistability, and turbulence in a q-switched gas laser, *Physical Review Letters.* **49**(17), 1217–1220 (1982). doi: 10.1103/PhysRevLett.49.1217.
25. R. Gilmore, Le Châtelier reciprocal relations, *The Journal of Chemical Physics.* **76**(11), 5551–5553 (1982). doi: 10.1063/1.442859.
26. R. Gilmore, Le Châtelier dynamics, *Molecular Physics.* **106**(8), 991–997 (2008). doi: 10.1080/00268970802190257.
27. I. I. Matorin, A. S. Pikovskiĭ, and Y. I. Khanin, Multistability and autostochasticity in a laser with a delayed-response active medium subjected to periodic loss modulation, *Soviet Journal of Quantum Electronics.* **14**(10), 1401–1405 (1984). doi: 10.1070/qe1984v014n10abeh006417.

28. H. G. Solari, E. Eschenazi, R. Gilmore, and J. R. Tredicce, Influence of coexisting attractors on the dynamics of a laser system, *Optics Communications.* **64**(1), 49–53 (1987). doi: 10.1016/0030-4018(87)90367-1.
29. H. G. Solari and R. Gilmore, Relative rotation rates for driven dynamical systems, *Physical Review A.* **37**(8), 3096–3109 (1988). doi: 10.1103/PhysRevA.37.3096.
30. S. Smale, Differentiable dynamical systems. I diffeormorphisms, *Bulletin of American Mathematical Society.* **73**, 747–817 (1967).
31. H. G. Solari and R. Gilmore, Organization of periodic orbits in the driven Duffing oscillator, *Physical Review A.* **38**(3), 1566–1572 (1988). doi: 10.1103/PhysRevA.38.1566.
32. O. E. Rössler, Chaotic behavior in simple reaction system, *Zeitschrift für Naturforschung A.* **31**, 259–264 (1976).
33. O. E. Rössler, An equation for continuous chaos, *Physics Letters A.* **57**(5), 397–398 (1976). doi: 10.1016/0375-9601(76)90101-8.
34. J. Birman and R. F. Williams, Knotted periodic orbits in dynamical systems I. Lorenz's equations, *Topology.* **22**, 47–82 (1983). doi: 10.1016/0040-9383(83)90045-9.
35. J. Birman and R. F. Williams, Knotted periodic orbits in dynamical systems II. Knot holders for fibred knots, *Contemporary Mathematics.* **20**, 1–60 (1983).
36. G. B. Mindlin, X.-J. Hou, H. G. Solari, R. Gilmore, and N. B. Tufillaro, Classification of strange attractors by integers, *Physical Review Letters.* **64**(20), 2350–2353 (1990). doi: 10.1103/PhysRevLett.64.2350.
37. G. B. Mindlin, H. G. Solari, M. A. Natiello, R. Gilmore, and X.-J. Hou, Topological analysis of chaotic time series data from the Belousov-Zhabotinskii reaction, *Journal of Nonlinear Science.* **1**, 147–173 (1991).
38. R. Gilmore, R. Vilaseca, R. Corbalán, and E. Roldán, Topological analysis of chaos in the optically pumped laser, *Physical Review E.* **55**(3), 2479–2487 (1997). doi: 10.1103/PhysRevE.55.2479.
39. R. Gilmore, Topological analysis of chaotic dynamical systems, *Reviews of Modern Physics.* **70**(4), 1455–1529 (1998). doi: 10.1103/RevModPhys.70.1455.
40. R. Gilmore and M. Lefranc, *The Topology of Chaos.* Wiley (2003). doi: 10.1002/9783527617319.
41. D. J. Cross and R. Gilmore, Representation theory for strange attractors, *Physical Review E.* **80**(5), 056207 (2009). doi: 10.1103/PhysRevE.80.056207.
42. D. J. Cross and R. Gilmore, Complete set of representations for dissipative chaotic three-dimensional dynamical systems, *Physical Review E.* **82**(5), 056211 (2010). doi: 10.1103/PhysRevE.82.056211.
43. C. Letellier and R. Gilmore, Symmetry groups for 3D dynamical systems, *Journal of Physics A.* **40**(21), 5597–5620 (2007). doi: 10.1088/1751-8113/40/21/011.

44. R. Gilmore and C. Letellier, *The Symmetry of Chaos*. Oxford University Press, New York (2007).
45. E. Roldán, G. J. de Valcárcel, R. Vilaseca, R. Corbalán, V. J. Martínez, and R. Gilmore, The dynamics of optically pumped molecular lasers. on its relation with the Lorenz-Haken model, *Quantum and Semiclassical Optics: Journal of the European Optical Society B*. **9**(1), R1–R35 (1997). doi: 10. 1088/1355-5111/9/1/001.
46. T. D. Tsankov and R. Gilmore, Strange attractors are classified by bounding tori, *Physical Review Letters*. **91**(13), 134104 (2003).
47. G. B. Mindlin, R. Lopez-Ruiz, H. G. Solari, and R. Gilmore, Horseshoe implications, *Physical Review E*. **48**(6), 4297–4304 (1993).
48. T. Hall, Weak universality in two-dimensional transitions to chaos, *Physical Review Letters*. **71**(1), 58–61 (1993).
49. M. A. Natiello and H. G. Solari, Remarks on braid theory and the characterization of periodic orbits, *Journal of Knot Theory and Ramifications*. **3**, 511–529 (1994).
50. R. F. Williams, The Lorenz attractor, *Lecture Notes in Mathematics*. **615**, 94–112 (1977).
51. M. A. Aziz-Alaoui, Differential equations with multispiral attractors, *International Journal of Bifurcation & Chaos*. **9**(6), 1009–1039 (1999). doi: 10.1142/S0218127499000729.

https://doi.org/10.1142/9789819823130_0022

Chapter 22

CHAOS RESEARCH AT THE NAVAL RESEARCH LABORATORY

Louis M. Pecora* and Tom Carroll†

U.S. Naval Research Laboratory, USA
**louis.pecora@nrl.navy.mil*
†thomas.carroll@nrl.navy.mil

Lou Pecora and Tom Carroll both started out in condensed matter physics, but dove into nonlinear dynamics in the 1980's. After studying chaos in magnetic resonance in a driven magnetic material, they realized they need a way to harness chaos that would catch the interest of their employer, the US Navy. With that in mind, they found a way to synchronize chaotic systems, hoping they could apply this concept to communications. This synchronization work led to a broad range of subsequent projects with oscillator synchronization in networks, applications of chaotic signals in radar and communications, and chaotic vibrations as structural monitoring probes. Recently, Carroll and Pecora have applied concepts from nonlinear dynamics to reservoir computing to better understand and design the analog behavior of these signal analysis devices.

1. Background

Chaos research at the U.S. Naval Research Laboratory (NRL) was started in the 1980s by Lou Pecora and Tom Carroll, physicists in the former Metals Physics Branch of the Materials Science and Technology Division (MSTD) and by Ira Schwartz originally in Optical Sciences division of NRL. We concentrate here on Carroll and Pecora.

Pecora had been working on positron annihilation, which gave information on bonding in metal alloys.[1] He also had pursued a side interest in several mathematical topics and had published a few papers on some of that work.[2–4] In 1984, he took a course in nonlinear dynamics, which was taught by Prof. Jim Yorke of the University of Maryland and who named the dynamical phenomenon we now call chaos. The course immediately helped him to understand general dynamical systems including exotic motions such as chaos. Much of the understanding was geometric in nature and that appealed to him since he was pretty good at visualizing objects and relating the geometric concepts to vector formulations of the problem. So, Pecora became very interested in nonlinear dynamics. Beyond appealing to his abilities, he found the field to be very broad and have applications in many scientific areas like electronic circuits, fluid dynamics, biological systems, and many condensed matter dynamical systems. This opening into so many directions was a marked change from his current research in positron annihilation, which was in fact a very narrow and small field. Nonlinear dynamics had so much more to explore. This motivated a transition in his research from positron annihilation to the field of nonlinear dynamics, also occasionally referred to as chaos theory.

The following is a conversation between Lou Pecora and Tom Carroll on how the nonlinear dynamics research effort developed and led to areas of research related to nonlinear dynamics that neither saw coming in the beginning.

LOU: I wanted to study nonlinear phenomena, especially chaotic motion, in a real system, one that would have interest to the Navy. The material yttrium iron garnet (YIG), a ferrimagnetic compound used in many radiofrequency applications, appeared to be a good candidate material for experimentation. Work by Prof. Carson Jeffries at the University of California, Berkeley, showed that the magnetic spin waves in YIG were a nonlinear dynamical system and could display chaotic behavior.[5] I got Fred Rachford, also in our Metals Physics Branch, interested in the experiments. However, Fred was a highly sought-after radiofrequency experimenter and was often too busy to pursue the experimentation full-time. I realized I needed someone to help Rachford, so I began to look for a postdoc.

TOM: In January 1987, I was a newly minted Ph.D. from the University of Illinois. I came to NRL as an American Society for Engineering Education postdoctoral researcher to work on the ferromagnetic resonance experiment that Rachford had set up. My thesis was on a completely different topic "optical spectroscopy of compounds with metal-metal bonds as they were

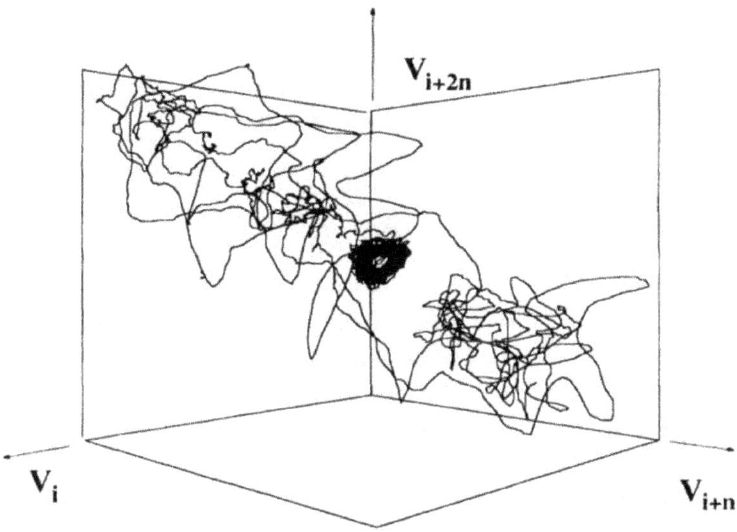

Fig. 1. Trajectory of a chaotic transient ending in a periodic attractor (from Ref. 8).

squeezed in diamond anvil cells"[6] but I had experience with magnetic resonance from an earlier project in graduate school.

Because Rachford had the experiment ready to go, I was able to jump right in and start taking data on chaos in spin wave interactions in YIG spheres in 1987. The results of the experiment did at first appear "chaotic," but not in the sense I was looking for. I did notice some strange transients, where the output from the experiment would at first appear chaotic, but then suddenly become periodic. Lou and I traveled to the University of Maryland to consult with Celso Grebogi and Ed Ott, who showed them that these transients were something that Grebogi, Ott, and Jim Yorke had actually predicted theoretically.[7] Ott explained the scaling of the transient times with applied power. Our data showed that the average length of these chaotic transients as micro-wave power was increased fit the predicted theoretical power law. By March 1987, Rachford and I had accumulated enough data to allow Lou to present the results at the American Physical Society that month, and, by December 1987, Carroll, Pecora, and Rachford had published in *Physical Review Letters* on these transients.[8]

LOU: Here was our first big publication in the nonlinear dynamics area and I was just a minor player. Tom and Fred Rachford did all the major work, which included some very clever experimental ideas along with

Tom's revealing reconstruction of chaotic transient trajectories. To this day I think it's the one of the best papers I've ever been privileged to be on. And it showed me that people with no background in nonlinear dynamics can begin to use its power to expose new physical behavior in real systems.

Early on in our efforts at NRL we were asked many times what research in nonlinear dynamics would help the Navy. After all, they were helping support the research itself. The mathematical complexities made it difficult to show how understanding the chaos in YIG could lead to results that would be useful to the Navy, so we needed to come up with some way to make chaos useful. Chaotic signals, while deterministic, were also unpredictable, so we reasoned that chaotic signals could be used to hide messages. An eavesdropper would see only a chaotic signal, which would look like noise. If the receiver could reproduce the original chaotic signal, signal and message could be separated, leading to a possibly secure form of communication. But it became clear that such a system would require transmitter and receiver to be exactly synchronized, something chaotic systems do not want to do: they naturally desynchronize quickly after each is started in the same state.

1.1. *The synchronization of chaotic circuits*

LOU: Tom and I talked about synchronizing chaotic systems, but we couldn't come up with any clear way to do that. I remember early January in the next year I attended a Dynamics Days conference. I came home tired from the trip and after dealing with my young daughter late at night I went to sleep thinking that somehow, we could drive a chaotic system with a signal from an identical system and maybe the two would synchronize. I managed to remember that idea the next day and in the next weeks some simple numerical experiments with iterated maps seemed to confirm that chaotic driving of identical nonlinear systems could cause them to synchronize. But we needed more than just numerical examples featuring simple maps.

TOM: Lou and I wanted an experimental example of synchronous chaos. I remembered an analog computer circuit that a professor had demonstrated when I was an undergraduate. The circuit used operational amplifiers, capacitors, and resistors to simulate the equations for a bouncing ball with damping, and the circuit output was displayed on an oscilloscope. I had thought ever since that I wanted to build an analog computer circuit, and chaotic synchronization gave me a reason. I found in the literature a report of a chaotic circuit developed by Prof. Robert Newcomb of the University of Maryland.[9] I built a pair of similar circuits: a drive circuit to hide a chaotic signal, and a response circuit which synchronized to the drive circuit in order to extract a message signal that had been added to the driving signal.[10] I even used this pair of circuits, along with a digital spectrum analyzer to demonstrate chaotic masking of an information signal in front of NRL's director of research, Dr. Timothy Coffey, as part of Lou's new funding presentation. In June 1992, I became a permanent employee of NRL, due in part to the success of this demonstration.

LOU: The work on the synchronization of chaotic circuits was still unpublished, but caught the interest of some people from the Space and Naval Warfare Systems Command (SPAWAR) through Dr. Mike Melich of the Naval Postgraduate School in Monterey, California. Tom and I were given some funding to pursue our idea of communicating with chaotic signals. Along the way, several communications and signal processing experts came in for demonstrations. One came to evaluate the work for use in secure communications. We were not allowed to ask his name or what agency he represented. We took to calling him Dr. X, although not in his presence. The report by Dr. X to the SPAWAR group showed some interesting viewpoints, but seemed inconclusive as to the use of chaotic circuits for secure communications. As other SPAWAR experts came through, I made a bet with one of them that they could not find a sine wave signal buried in a chaotic signal. The bet was a beer. I lost. That proved two points. One was that simply adding a communication signal to a chaotic signal was a weak way to 'encrypt' the communications signal (the latter could be discovered using standard linear analysis). The second point was that when you find something new and interesting, do not think it will do everything you want it to. I still have to buy that beer. I later briefed the NSA on the technique that we had invented.

In 1990, Dr. Michael Shlesinger of the US Office of Naval Research rounded up several workers in nonlinear dynamics in laboratories associated with the US Navy. His objective was to launch a new conference that would concentrate on experimental work in nonlinear dynamics. Much of the work and most of the conferences in the field to that time concentrated on theory and numerical calculations. This brought together Drs. William Ditto, Louis Pecora, Mark Spano, and Sandeep Vohra who organized the First *Experimental Chaos Conference* which took place in Crystal City in October 1991. The conference has been held biennially almost without interruption since then in various locations in the US, Europe, and South America always emphasizing experimental and observational work in nonlinear and complex systems.

TOM: The NRL director of research had decided to let us publish our work on synchronized chaotic systems. We wrote a paper and submitted it to *Physical Review Letters.*[11] Judged by both referees as immediately publishable without need of revision, the paper on chaotic synchronization attracted considerable attention, including from *Science News*[12] and the Washington Post. At the time, debate about the multi-billion-dollar cost of the (never completed) superconducting supercollider was in the news, so the Post reporter was impressed that the tabletop synchronization experiment, built for under US$ 100, could produce new science. He even referred to my lab as "slightly dingy," which drew good-natured complaints from other division members who claimed their labs were very dingy, too.

LOU: In 1992, I was invited to join Dr. Bill Ditto (then associated with the Naval Surface Warfare Center (NSWC) in White Oak, Maryland) to write a *Scientific American* paper[13] on applications of chaotic systems. In the same issue of *Scientific American*, in the Amateur Scientist column, Joe Neff (a student of Ditto's) and Tom had an article on building a chaotic synchronizing circuit for the Amateur Scientist column.

1.2. *Chaotic synchronizing circuits and systems and their phenomena*

TOM: Soon after, Lou and I became advisors to Dr. Jim Heagy, who became a postdoc at NRL. Jim proved capable and productive, producing several articles for *Physical Review Letters* on various complex and intriguing phenomena of chaotic systems which are synchronized. One of Jim's studies showed that the prediction of the final state of a system of four coupled, chaotic oscillators was exceedingly difficult because the space of starting points leading to synchronized chaos for the oscillators was riddled (literally) with an infinite number of points which led to a state in which the oscillators were not chaotic and only partially synchronized. Jim's study was one of the first publications[14] showing the existence of such a situation, although the phenomenon was theoretically predicted several years before.

We also got another postdoc who was in some ways the experimental mirror image of Jim Heagy, and that was Dr. Gregg Johnson, an excellent experimentalist in nonlinear circuits and systems. Gregg worked on a way to synchronize chaotic systems in a robust way, using ideas from control theory. The system worked so well that one circuit could drive another and they remained synchronized even when parameters in the driving system changed values by 50%. If the drive system changed its behavior, the driven system followed, despite having parameters for which the isolated system would have had vastly different behavior. This resulted in a contribution to *Physical Review Letters*,[15] but it also resulted in losing Gregg Johnson to another research group at NRL who recognized Johnson's skills and, after one year of his postdoc, hired him on.

LOU: From 1996 to 2002, Tom and I worked on various types of behavior in coupled, synchronized chaotic circuits and systems. In 1998, we made another breakthrough, one that stimulates research projects to this day. We developed a theory that describes how a large network of oscillators or circuits could synchronize, even when chaotic, an idea that could explain the synchronization phenomenon in any other network of those oscillators no matter how complicated. We published again in *Physical Review Letters*[16] on the work in which Tom developed a network of four chaotic circuits (based on the Rössler system) that followed the theory and showed that, indeed, the predictions of synchronization (or not) as coupling strengths were varied were accurate. We named the mathematical construct that underpinned the theory "the Master Stability Function" (Fig. 2), and the name or its abbreviation, MSF, is now commonly used in the literature. The theory was later applied by Mauricio Barahona (of Imperial College London) and me to networks, which Mauricio and I described as small world systems in a 2002 article in *Physical Review Letters*.[17] Our paper showed why these systems would synchronize better than most other arrangements of connected oscillators.

1.3. *Oscillators*

LOU: In the early 2000's I decided to get away from the topic of synchronization and investigate applications of nonlinear dynamics to other systems. I worked with a group at NRL on using chaotic vibrations to detect damage in large structures like bridges or buildings. I also worked with Ed Ott and Tom Antonsen at the University of Maryland to describe tunneling of waves (or particles in quantum mechanics) through barriers connecting confining cavities. This topic comes under the area of quantum or wave chaos.

However, in 2013, after seeing a talk by Francesco Sorrentino of the University of New Mexico I got inspired to return to the topic of synchronization in the form of complicated networks of connected oscillators.

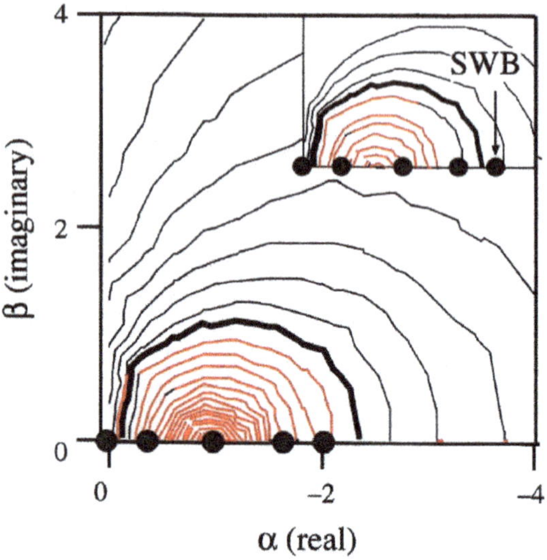

Fig. 2. Master stability function showing eigenvalue placement in the complex plane for x-coupled Rössler systems with diffusive coupling where the coupling strength is set to have all the transverse modes stable with eigenvalues in the stable (red contours) area. The remaining mode at $\alpha = \beta = 0$ is the eigenvalue for the synchronization manifold. The inset shows the emergence of a short wavelength bifurcation (SWB) as the coupling strength is increased and the largest eigenvalue, which is associated with the shortest spatial wavelength of the eigenvector moves into the unstable (black contours) region.

Here it turns out the network of oscillators can break up into clusters or subsets in which the oscillators in the same cluster are all synchronized, but not synchronized to oscillators in other clusters or subsets even though all oscillators are connected in some way in the network. I teamed up with a laser group at the University of Maryland under the direction of Prof. Raj Roy and Prof. Tom Murphy, along with graduate student Aaron Hagerstrom and Francesco Sorrentino, to return to the topic of synchronization of chaotic systems, but this time looking for more complex patterns that we called cluster synchronization in networks. This behavior has been studied for special cases for several years in the nonlinear dynamics world, but general approaches to the common phenomenon were lacking. We devised an approach using computational group theory, that is, group theory analysis done on a computer, which could analyze very large, complex networks for symmetries that exposed what synchronized clusters were possible. This resulted in a paper in *Nature Communications*[18] and another paper in *Science Advances* (a journal of the American Association for the Advancement of Science). The topic is rich and applicable to the world of electric power grids, neuronal networks, and any complex network of oscillators in which knowledge of synchronization patterns is important. Once again, the field of nonlinear dynamics proved a good source of new and unexpected phenomena.

TOM: After the master stability work, I looked into applying chaos to communications or radar. Encoding information onto chaotic signals was easy, thanks to the work of Scott Hayes on symbolic dynamics,[19] but decoding was hard. Chaotic radar also seemed like a promising area, but since radar is a very mature technology, it was difficult to explain these completely new concepts to radar engineers. I have been able to do some useful work in radar recently with Dr. Chandra Pappu, an engineer with a degree in radar.[20]

One of Pecora's acquaintances at NRL, Tom Reinecke, was involved with a project on reservoir computing, and immediately thought we should be involved. Reservoir computers arose from the field of machine learning, so they are usually described using the same terms that are used to describe neural networks, but reservoir computers are nonlinear dynamical systems, so the language of dynamical systems should be more useful. To date I have found the work to be challenging because reservoir computers are very high dimensional systems (potentially hundreds of dimensions), but I have made some progress in understanding how they work.

Recently, Carroll and Pecora won a Clarivate Award for Citation Laureates in the Physics[a] category for their 1990 paper, "*Synchronization in Chaotic Systems*,"[11] which had been awarded for its large number of citations (over 7,600). Clarivate runs the Web of Science and keeps track of citations on all scientific papers.

Acknowledgments

The work by L. Pecora and T. Carroll was supported directly by the Office of Naval Research (ONR) and by ONR through the Naval Research Laboratory's Basic Research Program.

References

1. L. Pecora and A. C. Ehrlich, Reconstruction of momentum distribution from point-geometry positron-annihilation data, *Physical Review B*. **19**, 719–729 (1979). doi: 10.1103/PhysRevB.19.719.

[a]https://clarivate.com/webofsciencegroup/citation-laureates/.

2. L. M. Pecora, A nonstandard infinite dimensional vector space approach to gaussian functional measures, *Journal of Mathematical Physics.* **23**(6), 969–982 (1982). doi: 10.1063/1.525465.
3. L. M. Pecora, Determination of the quantum density matrix from experiment: An application to positron annihilation, *Physical Review B.* **33**, 5987–5993 (1986). doi: 10.1103/PhysRevB.33.5987.
4. L. M. Pecora, 3D tomographic reconstruction from 2D data using spherical harmonics, *IEEE Transactions on Nuclear Science.* **34**(2), 642–650 (1987). doi: 10.1109/TNS.1987.4334688.
5. C. D. Jeffries, Chaotic dynamics of instabilities in solids, *Physica Scripta.* **T9**, 11–26 (1985). doi: 10.1088/0031-8949/1985/t9/002.
6. R. T. Roginski, T. L. Carroll, A. Moroz, B. R. Whittlesey, J. R. Shapley, and H. G. Drickamer, High-pressure spectroscopic studies of dihalotetrakis(pivalato)dirhenium(iii) metal-metal quadruply bonded complexes, *Inorganic Chemistry.* **27**(21), 3701–3706 (1988). doi: 10.1021/ic00294a006.
7. C. Grebogi, E. Ott, and J. A. Yorke, Super persistent chaotic transients, *Ergodic Theory and Dynamical Systems.* **5**(3), 341–372 (1985). doi: 10.1017/S014338570000300X.
8. T. L. Carroll, L. M. Pecora, and F. J. Rachford, Chaotic transients and multiple attractors in spin-wave experiments, *Physical Review Letters.* **59**(25), 2891–2894 (1987). doi: 10.1103/PhysRevLett.59.2891.
9. R. Newcomb and S. Sathyan, An RC Op Amp chaos generator, *IEEE Transactions on Circuits and Systems.* **30**(1), 54–56 (1983).
10. T. L. Carroll and L. M. Pecora, Synchronizing chaotic circuits, *IEEE Transactions on Circuits and Systems.* **38**(4), 453–456 (1991).
11. L. M. Pecora and T. L. Carroll, Synchronization in chaotic systems, *Physical Review Letters.* **64**(8), 821–824 (1990).
12. I. Peterson, Chaotic systems that stay in step, *Science News.* **137**(12) (1990).
13. W. L. Ditto and L. M. Pecora, Mastering chaos, *Scientific American.* **269**(2), 78–84 (1993).
14. J. F. Heagy, T. L. Carroll, and L. M. Pecora, Experimental and numerical evidence for riddled basins in coupled chaotic systems, *Physical Review Letters.* **73**(26), 3528–3531 (1994). doi: 10.1103/PhysRevLett.73.3528.
15. G. A. Johnson, D. J. Mar, T. L. Carroll, and L. M. Pecora, Synchronization and imposed bifurcations in the presence of large parameter mismatch, *Physical Review Letters.* **80**(18), 3956–3959 (1998). doi: 10.1103/PhysRevLett.80.3956.
16. L. M. Pecora and T. L. Carroll, Master stability functions for synchronized coupled systems, *Physical Review Letters.* **80**(10), 2109–2112 (1998).
17. M. Barahona and L. M. Pecora, Synchronization in small-world systems, *Physical Review Letters.* **89**(5), 054101 (2002). doi: 10.1103/PhysRevLett.89.054101.
18. L. M. Pecora, F. Sorrentino, A. M. Hagerstrom, T. E. Murphy, and R. Roy, Cluster synchronization and isolated desynchronization in complex networks with symmetries, *Nature Communications.* **5**(1), 4079 (2014). doi: 10.1038/ncomms5079.

19. S. Hayes, C. Grebogi, E. Ott, and A. Mark, Experimental control of chaos for communication, *Physical Review Letters.* **73**(13), 1781–1784 (1994). doi: 10.1103/PhysRevLett.73.1781.
20. C. S. Pappu, T. L. Carroll, and B. C. Flores, Simultaneous radar-communication systems using controlled chaos-based frequency modulated waveforms, *IEEE Access.* **8**, 48361–48375 (2020).

www.ingramcontent.com/pod-product-compliance
Lightning Source LLC
Chambersburg PA
CBHW060541050126
37699CB00004B/66
9789819823123